20th International Scientific and Research Conference "Metallurgy: Technologies, Innovation, Quality" (Metallurgy-2017)

IOP Conference Series: Materials Science and Engineering
Volume 411

Novokuznetsk, Russia
15 – 16 November 2017

ISBN: 978-1-5108-7367-4
ISSN: 1757-8981

Printed from e-media with permission by:

Curran Associates, Inc.
57 Morehouse Lane
Red Hook, NY 12571

Some format issues inherent in the e-media version may also appear in this print version.

Printed by Curran Associates, Inc. (2018)

Institute of Physics
Dirac House, Temple Back
Bristol BS1 6BE UK

Phone: 44 1 17 929 7481
Fax: 44 1 17 920 0979

techtracking@iop.org

Additional copies of this publication are available from:

Curran Associates, Inc.
57 Morehouse Lane
Red Hook, NY 12571 USA
Phone: 845-758-0400
Fax: 845-758-2633
Email: curran@proceedings.com
Web: www.proceedings.com

Table of contents

Volume 411

The 20th International Scientific and Research Conference "Metallurgy: Technologies, Innovation, Quality. Metallurgy-2017"

15–16 November 2017, Novokuznetsk, Russia

Accepted papers received: 8 August 2018
Published online: 19 October 2018

Preface

Papers

IOP Conf. Series: Materials Science and Engineering **411** (2018) 011001 doi:10.1088/1757-899X/411/1/011001

Preface

The 20th International Scientific and Research Conference "Metallurgy: Technologies, Innovation, Quality. Metallurgy – 2017", was held on 15-16 November 2017at Siberian State Industrial University, Novokuznetsk, Russia.

For the first time the conference was held in 1997 and over the last 20 years has gained popularity not only in Kuzbass and Russia, but also in the countries of near and far abroad. There is a constant interest in the results of its work on the part of foreign scientists, managers, specialized research centers and companies. Traditionally, for several days in autumn Siberian State Industrial University turns into a platform for lively professional discussion and acquaintance of the scientific and industrial community with the latest research results and technological solutions in the field of production, processing, welding of metals and composite materials, resource saving, recycling and environment protection, where the dominant modern trends are determined and forecasts for the future are provided.

The conference was conducted with the financial, organizational and information support of the Russian Foundation for Basic Research, the Administration of Kemerovo Region, JSC "EVRAZ ZSMK", JSC "RUSAL Novokuznetsk", JSC "Chermetinformatsiya", JSC "Kuzbass Technopark", Publishing House of the SB RAS, scientific and technical journals *Izvestiya. Ferrous Metallurgy*, *Bulletin of SibSIU*, *IOP Conference Series: Materials Science and Engineering*, West-Siberian Branch of the Russian Academy of Natural Sciences, Council of young scientists of Kuzbass.

Scientists in the field metallurgy and leading experts from industrial enterprises of Russia, China, Japan, Great Britain, Canada, Israel, Poland, Kazakhstan, Tajikistan, Ukraine, Belarus, Latvia, representing 87 educational and scientific organizations, industrial enterprises from 43 cities of Russia and foreign countries sent more than 200 reports to the Organizing Committee.

During the conference 4 sessions were held, discussions were organized at the departments of the university, excursions to the city enterprises took place. Chairman of the Program Committee, Academician of the Russian Academy of Sciences, the head of UrIMET L.A. Smirnov (Ekaterinburg) and Chairman of the Organizing Committee, Professor, Rector of SibSIU E.V. Protopopov (Novokuznetsk) took part in the opening of the conference and the work of its sessions.

The analysis of the reports allowed the following scientific and practical issues to be identified that require urgent solutions on the basis of concentration of state and business

IOP Conf. Series: Materials Science and Engineering **411** (2018) 011001 doi:10.1088/1757-899X/411/1/011001

resources, consolidation of efforts of university and branch science with the production sector taking into account the real state of domestic and world metallurgy:

- Priority improvement and development of metallurgical industries focused on import substitution (production of high-quality rail metal and long-length rails, expansion of the range of steels and alloys with special properties, rolling of pipes with a large diameter, etc.);
- Accelerated development of structural materials production with a set of properties in demand in the Far North and the Arctic (steels and alloys of low-temperature reliability, new flux and electrode materials for fusion welding, protection technologies for products and structures, etc.);
- Introduction of the principles of innovative creation and development of enterprises of the metallurgical complex (management based on the cluster approach, mini- and micro-factories, etc.);
- Mastering of new organizational and technological principles of design, creation and realization of production of high-quality products of responsible use with the automated production control systems and diagnostics;
- Development of surface hardening of structural materials to improve the operational properties of parts, products and mechanisms (wear and corrosion-resistant protective coatings, multi-layer materials, etc.);
- Effective use of nanotechnologies and nanomaterials in metallurgy (nanomodification, composite materials with a new level of service properties, etc.);
- Resource and energy saving based on optimization of thermal and technological operation of metallurgical units;
- Creation of a closed cycle in metallurgy, involving an increase in the environmental safety of metallurgical industries and waste recycling (processing of slags, sludges, dusts, lining, etc.).

The Organizing committee would like to thank all participants for the high activity and highly appreciates the contribution of all the authors and research teams into the development of technologies. We hope that the exchange of experience and thoughtful discussion of relevant scientific issues will be a powerful impetus to their effective solution, and the published proceedings of the conference will help to promote the achievements of the world industry.

Prof. Dr. Evgeniy Protopopov
Chair of Metallurgy2017
Novokuznetsk, Russia

Peer review statement

All papers published in this volume of *IOP Conference Series: Materials Science and Engineering* have been peer reviewed through processes administered by the proceedings Editors. Reviews were conducted by expert referees to the professional and scientific standards expected of a proceedings journal published by IOP Publishing.

Physical modeling of the processes of metal melt movement during continuous casting

E V Protopopov, V V Chislavlev and S V Feiler

Siberian State Industrial University, 42 Kirova street, Novokuznetsk, 654007, Russia

E-mail: chisl.vv@yandex.ru

Abstract. Studies of hydrodynamic processes in the intermediate ladle of a continuous casting machine by methods of physical low-temperature modeling was carried out. A qualitative and quantitative evaluation of efficiency of the use of various hydrodynamic elements is made using the method of studying the distribution of the liquid residence time in the unit.

1. Introduction

One of the main functions of intermediate ladle of a modern continuous casting machine is the refining of liquid metal from pre-crystallization non-metallic inclusions. The successful solution of this most important task depends to a large extent on the features of metal melt movement and mixing in the volume of intermediate ladle. Control of melt mixing processes can be carried out by installing special refractory elements (thresholds, jet strippers, partitions, etc.), the configuration and placement of which largely determines the distribution of flows in the intermediate ladle. Development of the configuration of the refractory elements in the intermediate ladle, providing favorable hydrodynamic conditions for the removal of nonmetallic inclusions, is based mainly on the results of physical and mathematical modeling [1, 2].

2. Methods of research

Investigation of hydrodynamic processes was carried out on the laboratory complex [3], including the physical model of 28-t intermediate ladle of the bloom caster at EVRAZ ZSMK JSC. Visualization of the flow of modeling fluid is done by introduction of a dye.

Estimation of the degree of homogenization and minimum residence time of the liquid portion in the volume of the intermediate ladle was carried out by conductometric analysis. Estimation of the configuration efficiency of the intermediate ladle for refining of the metallic melt from nonmetallic inclusions was carried out using the method of studying the distribution of the liquid residence time in the unit.

Physical models of refractory elements were made of organic glass and the separate components implemented by fused deposition modeling (FDM) using 3d-printing technology (figure 1), that helped to ensure the accuracy of the geometrical model parameters.

The study of the metal hydrodynamics was carried out for steel casting conditions at a speed of 0.7 m/min with the supply of metal to the intermediate ladle through a protective tube and the production of a blank with a cross section 300×360 mm.

The simulation was carried out with different designs of the intermediate ladle:

1) without refractory elements (basic option);

2) with the installation of thresholds (figures 1a, 1b), different height (h) and the cutoff angle of the top (α) at a distance of 144 mm from the axis of the protective tube: A) h=228 mm, $\alpha=0^o$; B) h=228 mm, $\alpha=30^o$; C) h=171 mm, $\alpha=0^o$; D) h=171 mm, $\alpha=30^o$; E) h=114 mm, $\alpha=0^o$; F) h=114 mm, $\alpha=30^o$; G) h=57 mm, $\alpha=0^o$; H) h=57 mm, $\alpha=30^o$;

3) with the installation of runner pots of various heights (figure 1, c): A) 64 mm; B) 96 mm; C) 128 mm;

4) with the installation of full-profile partition plate with overflow holes directing the flow of modeling liquid to the surface at angle 30°, various configurations (figure 1, d) at a distance of 144 mm from the axis of the protective tube: A) diameter of 5 holes in the upper row 20 mm and 4 holes in the lower – 20 mm; B) diameter of 5 holes of the upper row 20 mm and 4 holes of the lower row – 32 mm; C) diameter of 5 holes of upper row 32 mm and 4 holes of lower row – 40 mm; D) diameter of 4 holes of the upper row 40 mm and 3 holes of lower row – 32 mm; E) diameter of 4 holes of the upper row 40 mm and 3 holes of lower row – 20 mm; F) diameter of 4 holes of the upper row 32 mm and 3 holes of the lower row – 20 mm.

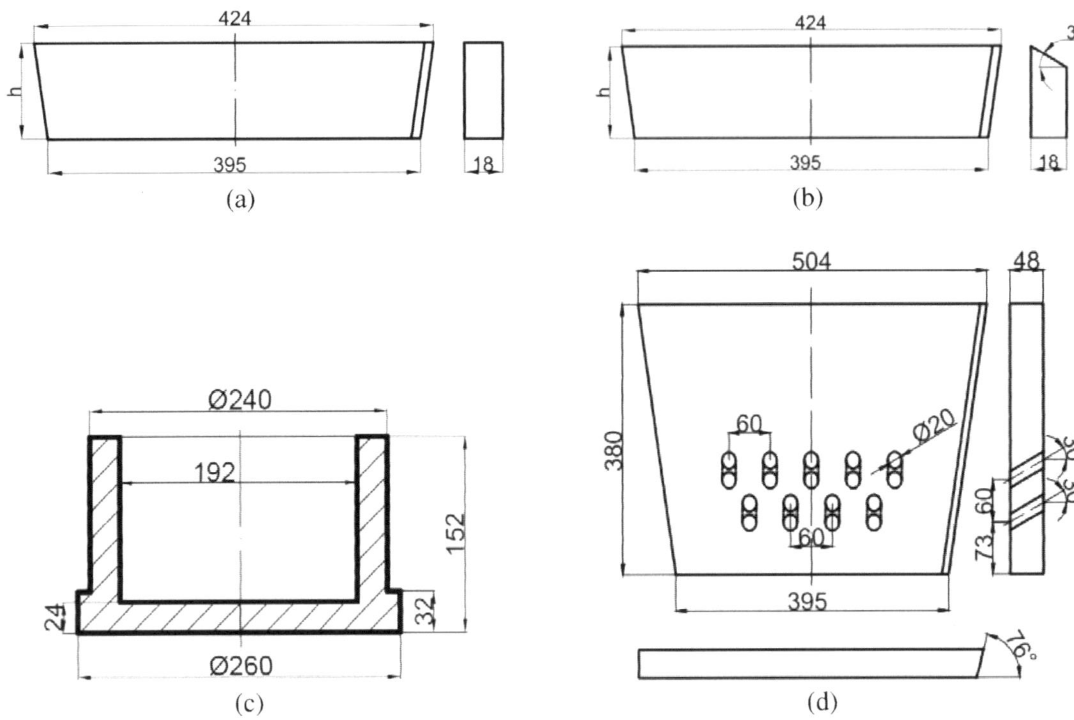

Figure 1. Models of refractory elements in the intermediate ladle: (a) a threshold with a straight top; (b) threshold with oblique top; (c) runner pot; (d) full-profile partitions.

3. Results and discussion

At the first stage, the hydrodynamic processes in the intermediate ladle of the basic design were investigated. It was found that with the basic design of the intermediate ladle, the time for reaching the central and peripheral casting cups is 4 (6) and 33 (52) seconds of the model time, respectively (figure 2, a). Here and further, the time without brackets indicates the simulation conditions, in brackets the time values correspond to the actual conditions of the metallurgical process. The flow of the central nozzles is achieved in a short time interval, forming "short" paths, which in industrial conditions results in the ingress of pre-crystallization non-metallic inclusions transported by the metal flow into the crystallizer, and further into the continuously cast billet.

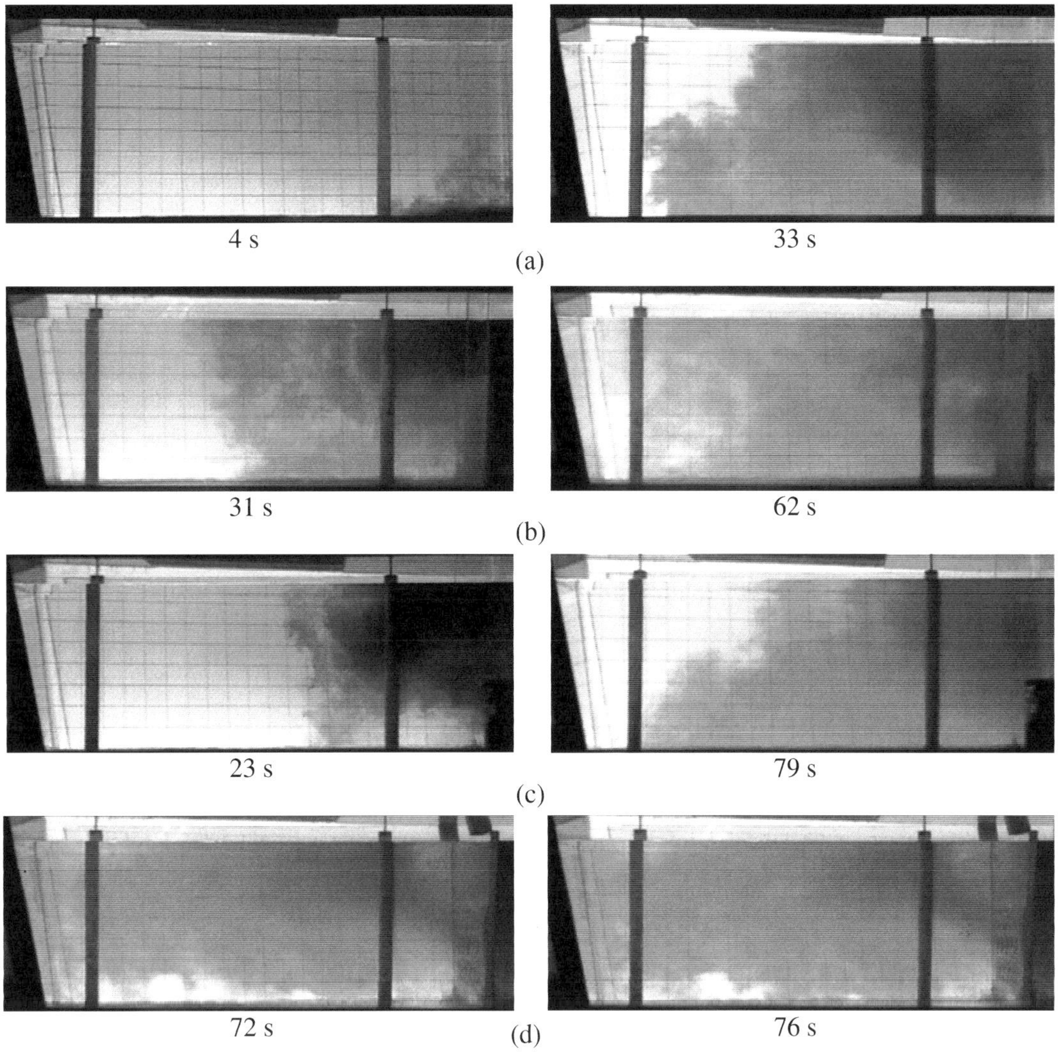

Figure 2. Hydrodynamic pattern in the intermediate ladle: (a) without hydrodynamic element; (b) using thresholds; (c) using a runner pot; (d) sing full-profile partitions.

The analysis of the data obtained by the conductometric method confirmed the minimum time for the liquid to reach the central and peripheral nozzles of the intermediate ladle – 4 (6) and 33 (52) seconds, respectively, and the uneven distribution of metal flows in the intermediate ladle of the basic configuration.

Further studies of metal hydrodynamics in the intermediate ladle were carried out using thresholds, runner pots and full-profile partitions plates of various configurations. The results of the studies at typical moments of time (reaching of nozzles) are given in figure 2.

The use of thresholds (figures 1a, b) made it possible to increase the time of reaching the portions of modeling liquid of central and peripheral nozzles to 31 and 62 seconds, respectively (figure 2 b). Using thresholds with an oblique top (figure 1 b) made it possible to increase the time for reaching a portion of the modeling liquid of the nozzles on average by 2 seconds. When colliding with a threshold, the main flow loses part of the kinetic energy and is directed along the threshold to the surface of the modeling fluid, reflected from which falls to the bottom and is directed to the central nozzles, then the flow in the bottom layers is directed to the peripheral nozzles.

When installing the runner pot (figure 1 c), the time for reaching the portion of modeling liquid of the central and peripheral nozzles reaches 23 and 79 seconds, respectively (figure 2 c). An increase in the minimum residence time of the portion of modeling fluid is associated with a decrease in the flow velocity at the outlet of the runner pot volume as a result of suppressing part of the jet energy during circulation in it. When leaving the runner pot, the main flow of the modeling liquid is directed to the surface and being reflected moves to the central nozzles and then to the peripheral ones. With this flow regime, the area with active vortices decreases, the turbulence decreases and, as a consequence, the minimum time for the metal portion to remain in the intermediate ladle increases.

When installing full-profile partitions (figure 1 d), the intermediate ladle is divided into 3 chambers (a receiving chamber and 2 casting chambers). In the receiving chamber there is an active circulation of the modeling fluid, through the overflow holes the flow enters into the casting chambers (figure 3 c). The jets passing through the holes in the partitions are directed to the surface and reach it in the zone of the central part of the casting chamber between the stoppers. When the surface is reached, the conditional separation of the main flow into two parts occurs. One part moves along the surface and, reaching the end walls, rushes to the peripheral nozzles and then moves at a lower velocity in the near-bottom layers near the rear walls to the central nozzles. The other part of the main flow, reflected from the surface, is directed to the central nozzle.

Figure 3 shows the time for reaching a portion of the modeling fluid of the central and peripheral nozzles for the options under study. From the data given, it can be seen that the most rational distribution of flows is achieved when using full-profile partitions with 4 holes in the bottom row with a diameter of 20 mm and 5 holes in the upper row with a diameter of 32 mm (option 4B), while the time for reaching the central and peripheral nozzles is 72 and 76 seconds respectively, differing by 4 s, which indicates active circulation of the modeling fluid, its homogenization and the absence of "short" paths.

Analysis of the results of physical modeling obtained by the conductometric method and the method for studying the distribution of the liquid residence time in the unit allows us to conclude that the volume of stagnant zones in the intermediate ladle of the basic design is ~ 28% (figure 4), and for industrial conditions the presence of a "short" way to the central nozzles does not allow the homogenization of the metal melt in chemical composition and temperature to be fully ensured, as well as the refining of the metallic alloy from nonmetallic inclusions, and the use of additional refractory elements is required to organize the movement of metal flows, to increase the minimum residence time of a portion of metal in the volume of the intermediate ladle and create zones of active melt circulation and reduce the volume of stagnant zones.

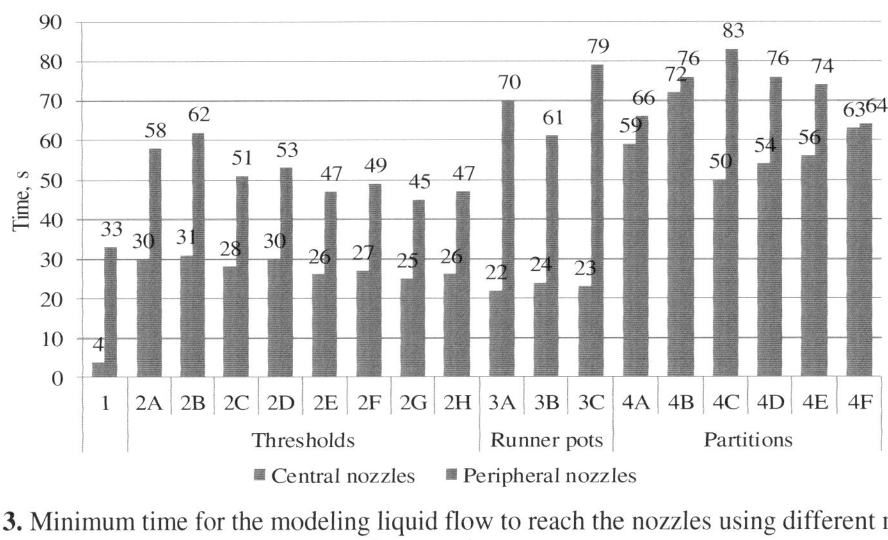

Figure 3. Minimum time for the modeling liquid flow to reach the nozzles using different models of refractory elements.

An increase in the height of threshold models to 228 mm contributed to an increase in the time for the portion of the modeling liquid to reach the nozzles (figure 3) and to reduce the volume of stagnant zones to 23% (figure 4).

The use of a runner pot allowed the time to reach a portion of the modeling liquid of the central and peripheral nozzles to be increased. In this case, an increase in the height of the runner pot model to 128 mm helps to reduce the volume of stagnant zones to 25% (figure 4). However, the effectiveness of the runner pot is of a short duration.

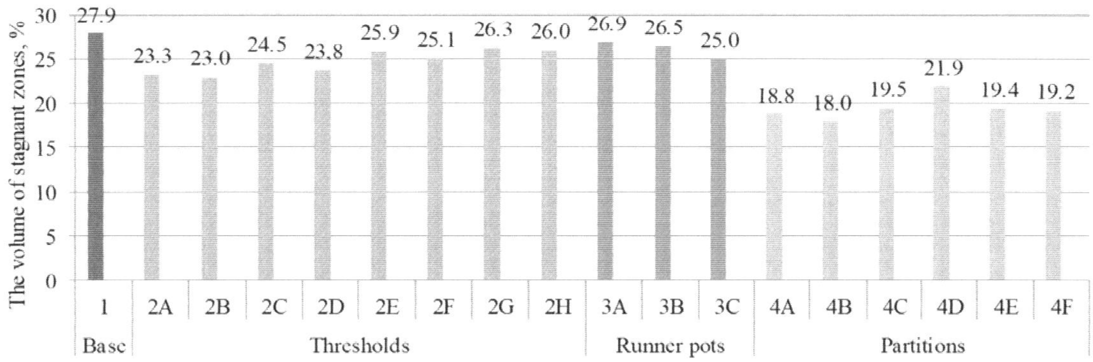

Figure 4. The volume of stagnant zones in the intermediate ladle using different models of refractory elements, %.

According to the results of physical modeling, it was established that the best results are achieved with the use of full-profile partitions with 2 rows of holes – 20 mm in diameter for the lower row (4 holes) and 32 mm for the upper row (5 holes), in this case the flow of modeling liquid goes to the surface, creating a closed loop of circulation, covering almost the entire volume of the intermediate ladle (figure 2, d). The volume of stagnant zones is 18% (figure 4).

4. Conclusion

As a result of the modeling it was established that the most favorable hydrodynamic conditions are achieved when using full-profile partitions. At the same time, the receiving chamber acts as a runner pot, metal circulates in a closed zone, actively mixing, which promotes the enlargement of nonmetallic inclusions. Getting through the overflow openings into the filling chambers, the metal flow goes to the metal-assimilating slag interface and moves near the surface, which facilitates the refining of the melt. The use of partitions with 2 rows of holes with diameter of (50) mm for the lower row and (80) mm for the upper row makes it possible to reach the central and peripheral streams almost simultaneously – (121) sec. and (126) sec. for central and peripheral flows, respectively.

References

[1] Eldarkhanov A S et al 2017 *Metallurgy of Machine Building* **4** 13–16
[2] Efimov V A and Eldarkhanov A S 2004 *Technologies of Modern Metallurgy* (Moscow: New Tecnologies) p 784
[3] Feiler S V and Chislavlev V V 2015 *Modern Problems of Theory and Practice of Education in the University* **18** 67–72

IOP Conf. Series: Materials Science and Engineering **411** (2018) 012002 doi:10.1088/1757-899X/411/1/012002

Converter steelmaking process: state, dominant trends, forecasts

E V Protopopov, S N Kuznetsov, S V Feiler, L A Ganzer and A N Kalinogorsky

Siberian State Industrial University, 42 Kirova street, Novokuznetsk, 654007, Russia

E-mail: protopopov@sibsiu.ru

Abstract. The analysis of the current state and prospects of development of converter industry in the world is performed. Technological features of steel smelting in Russia and the main directions of process improvement are considered.

1. Introduction

At present, world steel production is 1628 million tonnes, which is an order of magnitude higher than the production of aluminum, magnesium and the production of plastics [1]. At the same time, according to the data of the World Steel Association in 2016 there was a certain increase in the output of steel compared to 2015. It is characteristic that the last 50 years the steel smelting has grown mainly in developing countries [2], the main contribution to this process has been made by the PRC, which accounted for 49.6% of world production in 2016. Accordingly, the steelmaking in such countries and regions as Japan, NAFTA, EU, the CIS decreased by about half with relatively small changes in absolute figures.

2. Results and discussions

Thanks to the contribution of China, the share of smelting of converter steel has increased in the world, which reached 74.3% in 2016; the share of electric steel was 25.3%, the open-hearth furnace steel – 0.4% (table 1).

As can be seen from table 1, steelmaking in 28 EU countries amounted to 162 million tonnes, with a share of converter steel of 60.3%. The ratio of the smelting processes in the EU varies from country to country, which is explained by the historically developed range of production, i.e. requirements of consumers of rolled metal, and the trajectory of development of ferrous metallurgy. For example, in Germany converter steelmaking industry prevails with a share of 70.1% out of 42.1 million tonnes due to the focus on the production of sheet products for the automotive industry.

The share of converter steel in Russia is 66.9% with a total production of 70.8 million tonnes.

World experience of many years in the operation of oxygen converters has shown the merits of the latter: high productivity, sufficiently high lining stability, relative simplicity of construction and operation of equipment, technological flexibility in the composition of processed cast irons. At the same time, there is a number of significant drawbacks, which include the low efficiency of mixing the converter bath; the unevenness of the chemical composition and temperature of the metal; excessive oxidation of metal and slag during smelting of low-carbon steels; some problems in controlling the behavior of the bath, especially when foaming and generating emissions; limited opportunities in terms of improving the heat balance of melting.

Table 1. Structure of world steel production in 2016 [1].

Country	Steel production, mln. tonnes	Converter steel, %	Electrical steel, %	Open hearth steel, %
Austria	7.4	91.0	9.0	-
Belgium	7.7	69.3	30.7	-
Czech Republic	5.3	94.4	5.6	-
Finland	4.1	67.1	32.9	-
France	14.4	66.1	33.9	-
Germany	42.1	70.1	29.9	-
Italy	23.4	24.3	75.7	-
Netherlands	6.9	98.7	1.3	-
Poland	9.0	56.8	43.2	-
Slovakia	4.8	93.7	6.3	-
Spain	13.6	33.9	66.1	-
Sweden	4.6	67.3	32.7	-
United Kingdom	7.6	80.6	19.4	-
EU (28 countries)	**162.0**	**60.3**	**39.7**	**-**
Turkey	33.2	34.1	65.9	-
Other	4.5	44.1	55.9	-
Other European countries	**37.7**	**35.3**	**64.7**	**-**
Russia	70.8	66.9	30.8	2.4
Ukraine	24.2	71.7	6.8	21.4
Other CIS countries	7.3	56.9	43.1	-
CIS	**102.4**	**67.3**	**26.0**	**6.7**
Canada	12.6	55.4	44.6	-
Mexico	18.8	26.2	73.8	-
USA	78.5	33.0	67.0	-
NAFTA	**109.9**	**34.4**	**65.6**	**-**
Argentina	4.1	56.7	43.3	-
Brazil	31.3	77.3	21.1	-
Other	3.8	7.3	92.7	-
Central and South America	**40.9**	**67.5**	**31.3**	**-**
Egypt [a]	5.0	11.4	88.6	-
South Africa	6.1	62.2	37.8	-
Other African countries	1.9	33.8	65.9	-
Africa	**13.1**	**38.5**	**61.4**	**-**
Iran	17.9	12.2	87.8	-
Saudi Arabia	5.5	-	100.0	-
Other Middle Eastern countries[a]	8.1	-	100.0	-
Near East	**31.5**	**6.9**	**93.1**	**-**
PRC[a]	808.4	94.8	5.2	-
India	95.6	42.7	57.3	-
Japan	104.8	77.8	22.2	-
South Korea	68.6	69.3	30.7	-
Taiwan	21.8	64.2	35.8	-
Other Asian countries	25.9	6.9	93.1	-
Asia	**1 125.0**	**84.6**	**15.4**	**-**
Total	**1 628.3**	**74.3**	**25.3**	**0.4**

[a]assessment

In order to eliminate the shortcomings of the upper oxygen blowing in the late 60's of the last century, a new converter processes with a bottom supply of fuel-oxygen blowing spread. However, the forecasts of significant advantages of bottom oxygen-fuel blowing were not justified. At the same time, a number of positive results (the possibility of intensification of blowing, increased prime yield savings in ferroalloys, etc.) served as an incentive for the development of combined converter processes.

In the 80's of the last century, a converter process was developed with an upper blowing of oxygen and a bottom blowing of inert gas. According to this technology, up to 80% of the world's converters worked [2]. Bottom blowing intensified the bath mixing, increased the yield of iron and manganese, facilitated the control of slag formation and the process of dephosphorization. However, the appearance of the process of bath slagging to improve the stability of the converters lining led to the fact that the producers began to sacrifice bottom blowing in favor of the lining slagging process. Nevertheless, integral components of modern technology are: the combination of the upper oxygen blowing with bottom blowing of inert gases; use of special designs of blowing devices; cutoff of slag to reduce its entry from the converter into the ladle during discharge; automation of the process to ensure sustainable results; an emission prevention system to increase yield, improve process control and ensure the personnel safety; use of special models to predict various scenarios of the converter shop from the viewpoint of minimizing the costs of resources and time for the production and execution of technological operations [3].

The development of modern converter production aims at achieving traditional goals: improvement of products quality, reduction of production costs, increase in the productivity and improvement of environmental protection. In the world, the use of steel scrap is steadily increasing, reaching 590 million tonnes in 2016. But the ratio of the mass of liquid iron to the mass of the smelted steel increases and reached 74%. At the same time, in the crisis year 2009 it was 82%. With regard to converter steel production, it is expedient to provide process flexibility for the use of a metal charge in a wide ratio of different melting coolers and with a reduced proportion of cast iron. Replacing cast iron scrap, as it is known, dramatically reduces the energy intensity of steel and labour costs for its production.

In accordance with the developed provisions in the work of converter plants in Russia in 2017 for the first time in the last few years the tendency was observed to reduce the consumption of cast iron (table 2) [4]. The greatest decrease was achieved at EVRAZ NTMK (-10.4 kg/t).

Figure 1 shows the results of calculating the energy intensity of converter steel [4]. Despite the reduction in the consumption of pig iron, the energy intensity of the converter steel at EVRAZ ZSMK increased by 10 kg f.e./t as a result of the increase in the energy intensity of pig iron.

On the whole, the energy intensity of 1 tonne of converter steel according to the results of work in the first quarter of 2017 was 626.7 f.e./t, which is 7.1 f.e/t lower than the in same period in 2016 (figure 1).

Table 2. Change in the specific consumption of pig iron in the production of converter steel, kg / t.

Company	1st quarter of 2016	1st quarter of 2017	Change
As a whole on branch	924.6	923.1	-1.5
EVRAZ ZSMK	831.9	829.4	-2.5
MMK	894.8	886.1	-8.7
CherMK	906.5	899.7	-6.8
NLMK	942.2	938.9	-3.3
ChelMK	969.7	1003.9	+34.2
EVRAZ NTMK	1085.8	1075.4	-10.4

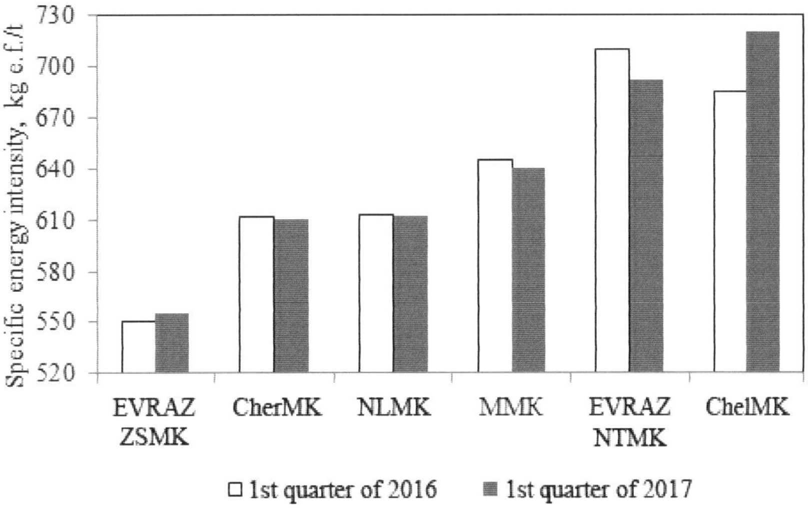

Figure 1. Specific energy consumption of converter steel.

The recent increase in the requirements for the quality of products in the metal market predetermines the corresponding requirements for the quality of steel scrap on the content of non-ferrous metal impurities, as well as physical properties. However, the amount of recycled scrap is reduced due to the improvement of the technology of continuous casting of steel, and in the amortization scrap due to repeated remelting the content of individual elements, contributing to the decrease in the performance properties of metal products, is increased. At the same time, the effectiveness of technologies for preparing steel scrap for converter smelting in domestic metallurgy does not always solve these problems.

Under the current conditions, it seems expedient to improve the traditional composition and technological regime for the formation of a solid metal charge with the use, in particular, of iron-containing products for the processing of dump converter slags, the processing technology of which remains technologically primitive and ecologically imperfect [5].

Despite the modern developments, cast iron remains the most important component of the metal charge when steel is smelted. At present, the converter process is considered as a complex process, including the preparation of cast iron, the actual smelting process and the out-of-furnace refining of steel, including its processing in the intermediate ladle of the continuous casting machine.

Extensive desulphurisation of cast iron, as well as its desiliconization and dephosphorization, was widespread. Reducing the silicon content in cast iron to 0.30-0.35% provides not only the possibility of subsequent secondary dephosphorization and effective low-slag technology in the converter, but also allows significant savings in the smelting of cast iron to be obtained. In general, this is becoming one of the main areas for improving technology in the blast furnace – basic-oxygen converter.

To obtain low concentrations of dissolved gases and nonmetallic inclusions in the converter steel, it is treated after being discharged from the unit in various types of vacuums. In the world, 27% of the steel produced is evacuated, in Japan – up to 50%.

The ladle furnace with evacuation is the most versatile, it is possible to carry out practically all the steel processing, especially if it is possible to load slag from the ladle. The improvement of facilities is mainly aimed at creating conditions conducive to the acceleration and deepening of metal refining. This is primarily an event that contributes to the intensification of metal mixing in the ladle. They are implemented by improving the equipment for inert gas injection, the location of blowing devices, the period and intensity of gas injection; for the units of batch and circulating evacuation – the improvement of the design of the chamber and inlet branch.

To improve the processability of casting of converter steel on continuous casting machine it is recommended to reduce the cross-section of billets with increase in casting speed in order to keep the productivity; it is advisable to use curvilinear machines for all types of blanks; to use intermediate buckets of increased capacity and height; crystallizers with a variable cone, with a more perfect swing mode; electromagnetic slowdown of steel flows in the mold and soft compression of the non-solidified ingot in the secondary cooling zone; thermal action on the workpiece in the end zone of solidification; use of technologies of "hot landing" of blanks in heating devices before rolling [3].

Much attention in the world metallurgy is paid to reduction of refractory materials costs in the production of converter steel. The requirements for the stability of the converters lining have predetermined the following world trend in the use of refractories: from the use of resin-bonded dolomite, then periclazodolomite and periclase, backed periclase impregnated resin in vacuum, to refractories of periclase-carbonaceous composition with antioxidant additives that are widely used at present.

Thanks to the use of periclase carbonaceous refractories, in combination with the "hot" repairs, which is an important component of the converter lining servicing, a record-breaking performance was achieved – 20 thousand (USA), 30 thousand (China), 7 thousand (Russia) smeltings [3]. A certain role in improving the lining stability is also play laser systems for measuring the profile and walls of local wear of refractories, which automatically scan and output data on their condition for the operative application of corrective measures.

It is characteristic that in the Russian metallurgy, high-magnesia fluxes are widely used in the smelting of steel in converters; this allows the MgO content to be increased in the slag, its aggressive effect to be reduced on the liner, and a refractory protective coating on the lining to be obtained when the slag is blown [6]. "Hot" repairs also include welding, semi-dry and/or flame gunning with the supply of gunite masses based on backed magnesite or dolomite using horizontal and vertical types.

To improve the efficiency of hot repairs of the converter lining, it is recommended to use special designs of gas-cooled tuyeres that provide for joint (separate) application of slag skull by blowing the final slag with nitrogen jets and flare gunning of lining with cheap gunning mix based on unfired magnesian slag-forming materials [7].

3. Conclusion

In accordance with the developed provisions, the scientists of the university scientific school together with the specialists of EVRAZ ZSMK have developed various options for steelmaking processes in oxygen converters using natural raw materials and man-made waste that ensure resource saving and rational use of the technological environment in the region. The inventions found application in the enterprise, which made it possible to increase the efficiency of converter production and the quality of metal products.

At present, research of the process of steel smelting in oxygen converters is carried out using iron-containing products for processing converter slags in solid metallurgy, and the development of new refractory coatings and materials, methods for hot repairs of the converter lining and devices for their implementation, which reduce the cost of refractories.

References

[1] *World Steel Association* / http://www.worldsteel.org
[2] Elanskiy D G and Elanskiy G N 2016 *Steel* **12** 21–29
[3] Velichko A G et al 2011 *Ferrous Metallurgy* **4** 50–55
[4] Ivanova I M 2017 *Ferrous Metallurgy* **7** 10–15
[5] Kuznetsov S N et al 2017 *Bulletin of the Russian Academy of Natural Sciences. West-Siberian Branch* **20** 95–101
[6] Vozchikov A P et al 2017 *Steel* **4** 16–20
[7] Protopopov E V et al 2014 *Problems of ferrous metallurgy and materials science* **4** 21–25

IOP Conf. Series: Materials Science and Engineering **411** (2018) 012003 doi:10.1088/1757-899X/411/1/012003

Improvement of the smelting efficiency based on the use of composite nozzles in the oxygen lances of high-capacity converters

V V Solonenko, E V Protopopov, S V Feiler and M V Temlyantsev

Siberian State Industrial University, 42 Kirova street, Novokuznetsk, 654007, Russia

E-mail: uchebn_otdel@sibsiu.ru

Abstract. The nozzle designs of oxygen lances are proposed, which ensure the improvement of technical and economic parameters of converter smelting and metallurgical quality of steel. The developed solutions make it possible to ensure the operation of composite lance nozzles without possible erosion wear of the supercritical part of the nozzle with an acceptable off-design degree. The cylindrical nozzle allows the supersonic flow of oxygen to be stabilized, the length of the initial and transitional sections of the jet to be increased, which is important for organizing a "tough" blowing of the converter bath at the final stage of the operation in order to reduce the oxidation of metal and slag.

1. Introduction

At present, the most widespread in oxygen-converter units with top blowing are multi-jet lances with de Laval nozzles. The lances of this design ensure the achievement of supersonic exhaust velocities, the maximum kinetic energy of the oxygen jet, and the deep penetration of the blowing into the melt. The number of nozzles in the multi-jet lances is of great importance here. In general the number of nozzles is selected taking into account the converter charge, the specific blowing rate and the distance from the level of a calm metal bath to the furnace mouth.

The increase in the number of nozzles (more than six) improves the degree of oxygen absorption and the hydrodynamic situation in the bath of the converter, accelerates the slagging, but leads to the reoxidation of the bath, emissions, a decrease in the yield of liquid steel. With a small number of nozzles (less than five), the "hardness" of the blowing, as a rule, increases, the slagging conditions deteriorate, which leads to an intensive removal of metal from the reaction zone, an increase in the metallization of technological equipment, and a decrease in the yield of liquid steel.

Among scientists and technologists there is no consensus on the optimal or rational number of nozzles and clear quantitative criteria for its determination. In [1, 2] a lance with 9-12 nozzles is recommended for converters with a capacity of 300-400 tonnes at a blowing rate of 5-7 $m^3/(t \cdot min)$. The authors of [3] believe that the number of nozzles in the lances of heavy-duty converters should be at least seven. Lances with 8 – 13 nozzles passed through the industrial testing in units of different capacity; however, having shown good technological results, they had low resistance of the tips because of the complexity of their design and the cooling system of the latter [1 - 5].

2. Theoretical analysis

When calculating and designing the optimal and rational designs of oxygen lances, it is necessary to take into account that in the real operating conditions, in most cases, it is not possible to provide a calculated mode for the gas jets outflow from nozzles of the top of lance [6, 7].

In the design mode of gas outflow from the nozzle, the static pressure at the nozzle exit section (P_{exit}) should be equal to the environmental pressure (P_{env}) at the level of the nozzle into which the outflow occurs. In this case, the off-design degree of the nozzle $Z = P_{exit}/R_{env} = 1$ [8]. Due to the fact that there is no pressure drop at the nozzle exit section ($P_{exit} - R_{env} = 0$), the entire static head pressure of the gas flow after the critical nozzle exit section passes into the high-speed, providing the given dynamic flow pressure to the melt. Thus, the design mode corresponds to a strictly defined value of the degree of nozzle expansion at a given oxygen pressure before the nozzle.

When the pressure of the oxygen jet is greater at the nozzle exit section or less than the ambient pressure on the nozzle, there are cases of off-design gas outflow and the operation of the blowing nozzle. The nozzle operates in the underexpanded mode if the gas pressure at the nozzle exit section exceeds the ambient pressure, i.e. $P_{exit} > P_{env}$ and $Z > 1$ [9]. In this case, the expansion of the gas to the ambient pressure continues beyond the nozzle, producing compression shocks and depression in the gas flow, and therefore additional energy losses of the jet occur.

At small values of the pressure difference ($P_{exit} - R_{env}$), the speed losses due to underexpansion are compensated for by the absence of frictional losses at the end of the nozzle, which allows the length of the supercritical portion of the nozzle to be significantly reduced. If the gas pressure at the nozzle exit section is less than the ambient pressure ($P_{exit} < P_{env}$ and $Z < 1$), then the nozzle works with overexpansion. At this outflow regime, the boundary layer of the gas jet breaks away from the walls and forms an oblique shock of compression and depression zone inside the nozzle. The latter circumstance is accompanied by entering into the nozzle of high-temperature gases with droplets of metal and slag, which are present in the working space of the unit, which leads to a rapid burn of the nozzle edges, distortion of the velocity diagram in the jet section and burnout of the lance head. Given the wear of the nozzle and the loss of kinetic energy of the flowing jet, the work of the blowing nozzle in the overexpansion mode must be considered inadmissible [6, 7].

In order to improve the stability of the gas flow, the characteristics of the oxygen jets parameters and to prevent changes in the geometry of the blowing nozzle due to the heat, it is recommended [6, 7, 9]:

- to design the blast nozzles in accordance with the design geometry that satisfies the actual state of the oxygen flow ahead of the nozzle and the prescribed oxygen flow rate;
- to prevent erosion wear of the exit section of the nozzle during operation, to design shortened blast nozzles operating in the underexpanded mode with an acceptable off-design degree (Z within 1.05 – 1.10).

For blowing conditions in 350-tonne unit of BOF No. 2 at JSC " Consolidated EVRAZ ZSMK", a stability improvement of the oxygen jets characteristics that flow from the 4 nozzle top of the lance can be achieved as a result of:

- calculation of a critical nozzle diameter with the upper limit of the oxygen flow rate for blowing the bath (1200 m^3/min);
- calculation of the exit diameter of the shortened de Laval nozzles with the lower limit of oxygen flow (700 m^3/min);
- supply of the shortened de Laval nozzle with a cylindrical orifice, with the condition of providing the total length of the composite nozzle in the supercritical part (expanding and cylindrical sections) equal to the length of the calculated de Laval nozzle at an oxygen flow rate of 1200 m^3/min through the 4 nozzle head (figure 1).

As a result, the operation of composite nozzles of the lance without possible erosion wear of the supercritical portion of the nozzle in the range of oxygen flow rates of 1200-700 m^3/min with an allowable degree of inaccuracy is ensured. In this case, the cylindrical nozzle allows the supersonic flow of oxygen to be stabilized, increasing the initial and transitional sections of the jet, which is important for organizing a "tough" blowing of the converter bath at the final stage of the operation in order to reduce the oxidation of the metal and slag.

3. Results and discussion

For the indicated blowing conditions, the nozzle calculations and the working design of the head of the 4-nozzle oxygen nozzle were performed using the procedure [6, 7] with a modified profile of the supercritical part of the nozzle. The total length of de Laval nozzle with a cylindrical orifice in the supercritical part was 142 mm, including a shortened de Laval nozzle, calculated for the oxygen flow range of 1200 – 700 m^3/min, and a cylindrical orifice with a length 58 mm (figure 2).

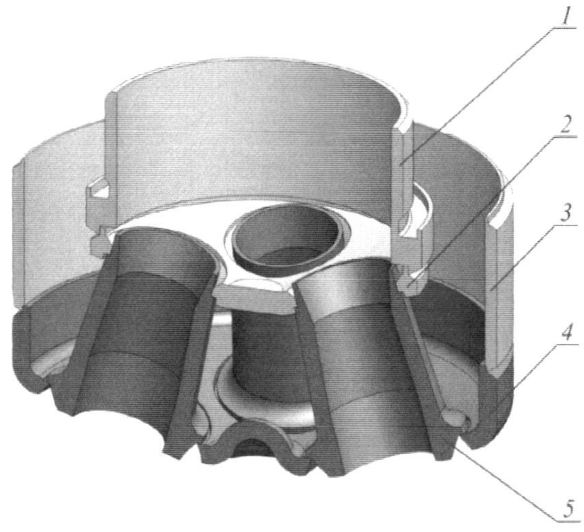

Figure 1. Design of a top of lance with 4 nozzles and a cylindrical orifice: 1 – connection; 2 – flange; 3 – the insert; 4 – the bowl; 5 – de Laval nozzle with a cylindrical orifice.

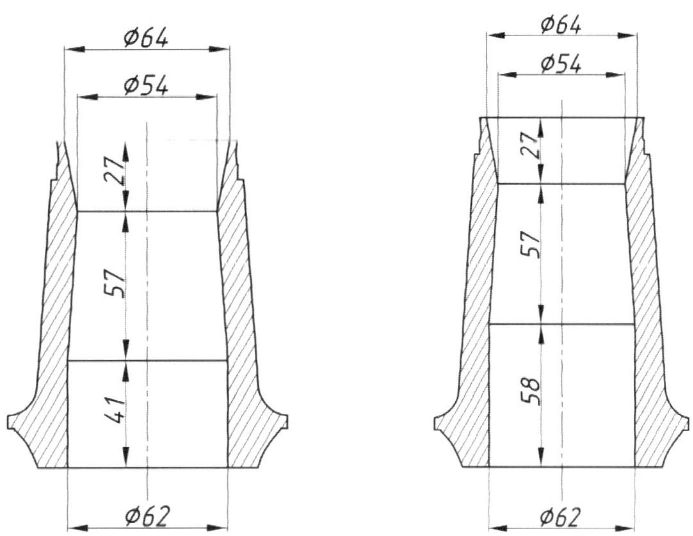

Figure 2. Design of de Laval nozzle with a shortened (a) and elongated (b) cylindrical orifice.

Using the developed design of the blowing device in industrial conditions, the tests were conducted. From the analysis of the results (figures 3, 4) obtained during the blowing of the metal melt using a lance with composite nozzles, it is evident that when using de Laval nozzle with a cylindrical orifice, the oxidation of the slag at an equal carbon concentration in the metal in the turndown (figure 3) is lower than when using a lance with traditional de Laval nozzles.

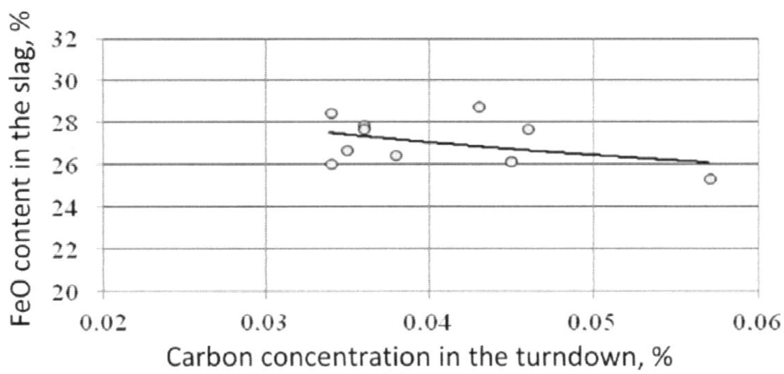

(a)

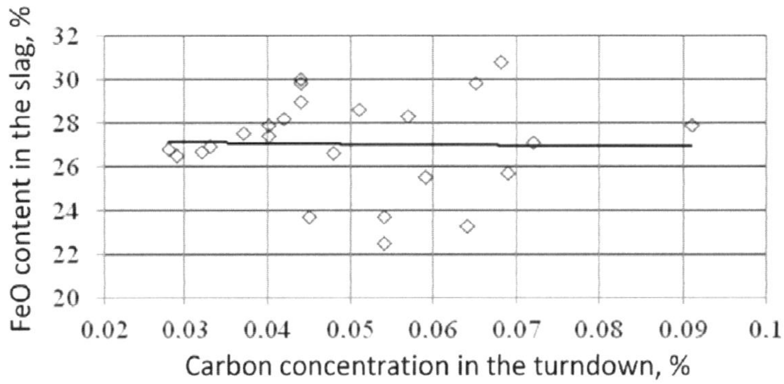

(b)

Figure 3. Dependence of slag oxidation on the concentration of carbon in the metal in the first turndown using traditional (a) de Laval nozzles and (b) with a cylindrical orifice.

Thus, with a carbon content in the metal in the range 0.04 – 0.07%, the minimum value of slag oxidation when using a nozzle with a cylindrical orifice was 22.5%, and with a traditional nozzle 25.4%.

When using nozzles with a cylindrical orifice, some increase in the manganese oxide content in the slag was found at a decrease in the carbon content in the metal (figure 4). At the same time, an increase in oxygen consumption for smelting does not practically affect the oxidation of the slag (figure 5).

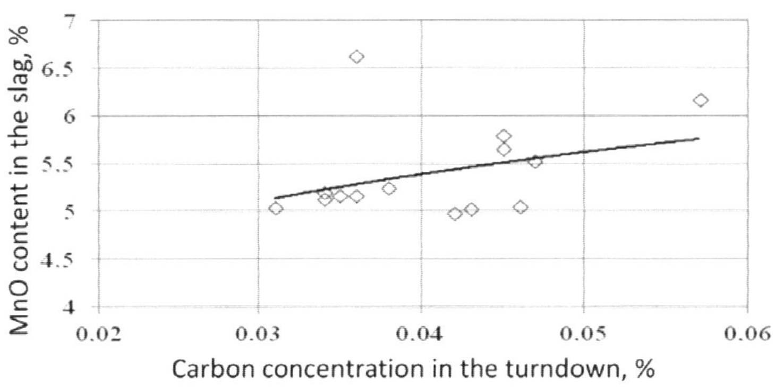

(a)

IOP Conf. Series: Materials Science and Engineering **411** (2018) 012003 doi:10.1088/1757-899X/411/1/012003

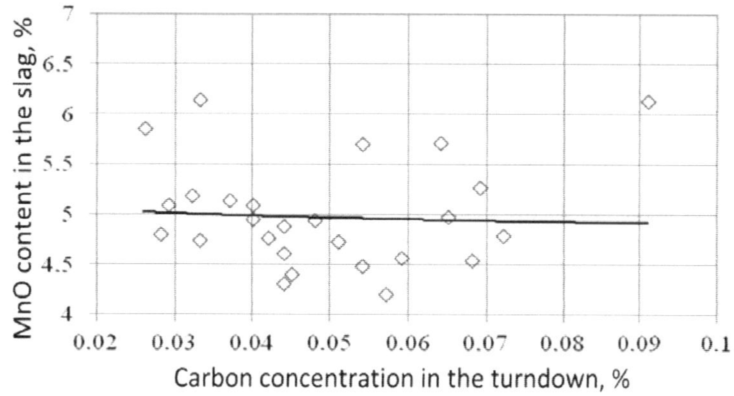

(b)

Figure 4. Dependence of MnO content in the slag on the concentration of carbon in the metal in the first turndown using traditional (a) de Laval nozzles and (b) with a cylindrical orifice.

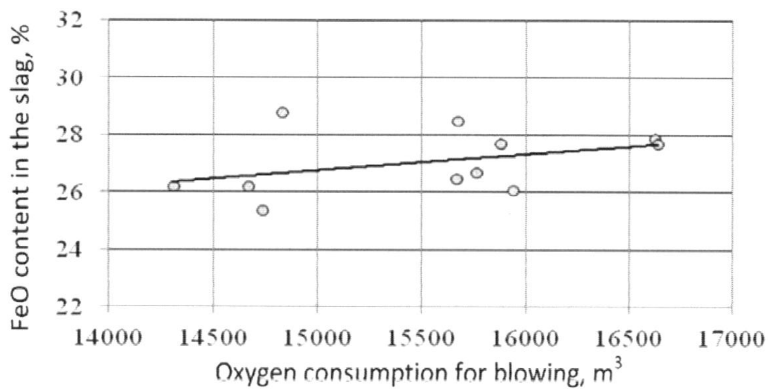

(a)

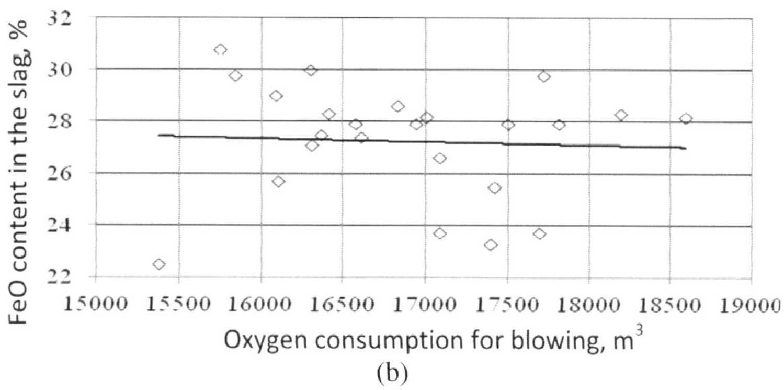

(b)

Figure 5. Dependence of slag oxidation on oxygen consumption for blowing using traditional de Laval nozzles (a) and with a cylindrical orifice (b).

4. Conclusion

The results analysis of the pilot industrial smeltings using a lance with de Laval nozzles equipped with a cylindrical orifice showed the effectiveness of the developed solution. At the final stage of blowing the use of a composite nozzle allows a "tough" mode of oxygen jet outflow to be organized in order to reduce the slag oxidation, which provides an improvement of technical and economic parameters of converter smelting and the metallurgical quality of steel.

References

[1] Baptizmanskiy V I and Shchedrin G A 1973 *Steel* **1** 20–23
[2] Friedl E and Schmidt G 1972 *Ferrous Metals* **15** 40–45
[3] Mikhailovsky V N 1973 *Bulletin "Chermetinformation". Series 6. Information* **4** pp 1–44
[4] Cottedzh D P, Ekols G L et al 1993 *Bulletin of NTI "Ferrous Metallurgy"* **1(1125)** 21–25
[5] Tabata Y, Marsh R C, Kelly P et al 1998 *Steelmaking Conference Proceedings* pp 451–457
[6] Protopopov E V V et al 2013 *Bulletin of Siberian State Industrial University* **4(6)** 3–7
[7] Feiler S V et al 2016 *Bulletin of the Russian Academy of Natural Sciences. West-Siberian Branch* **18** pp 111–117
[8] Yavoysky V I et al 1974 *Theory of Purging of a Steel-melting Bath* (Moscow: Metallugiya) p 495
[9] Gorbik A S et al 1971 *Collection of Sci. Works of the Institute VNIPIChERMETENERGOOCHISTKA"* (Moscow: Metallurgiya) pp 246–253

IOP Conf. Series: Materials Science and Engineering **411** (2018) 012004 doi:10.1088/1757-899X/411/1/012004

The study on manganese behavior in oxidizing conditions of steelmaking processes

V I Dmitrienko, E V Protopopov, A V Dmitrienko and Yu N Nosov

Siberian State Industrial University, 42 Kirova street, Novokuznetsk, 654007, Russia

E-mail: dmvliv@gmail.com

Abstract. The possible limits of increase in the residual manganese content in the oxidizing conditions of steelmaking processes were studied by thermodynamic calculations. In the conditions of the current production the manganese behavior during carbon oxidation in the converter (300 tonnes) and electric steelmaking furnace (100 tonnes) was investigated. The industrial data confirmed the thermodynamic calculations. The developed technological methods in case of their implementation will allow the consumption of manganese ferroalloys to be reduced.

1. Introduction

With the increase in the production of steels and improvement of their quality, there is an increase in the consumption of ferroalloys. For Russia a serious problem is the supply of manganese ferroalloys, because its own production is extremely insufficient, and its increase is hampered by the lack of manganese ore reserves required for ferroalloy production.

A partial solution to the problem can be achieved by developing new resource-saving technologies that reduce the consumption of manganese ferroalloys.

This can be achieved by increasing the residual manganese at the end of oxidation processes and, especially, by using the technology of direct alloying with manganese oxide materials.

Such materials can include both manganese ore and concentrates, as well as technogenic wastes and secondary resources of the mining and metallurgical industry.

A special advantage of this direction is that for such technologies resources of small deposits of regional importance can be used. For Kuzbass metallurgical industry these may be manganese ores from the Durnovsky and Selezensky deposits, as well as ores from the neighboring Altai Territory.

In this regard, the issue of the manganese behavior in the oxidizing conditions of steelmaking processes is very relevant.

It should be emphasized that various options for optimizing the technologies of oxidation process and direct alloying have been tested in relation to the steel production conditions in both converters and open-hearth and electric steelmaking furnaces. However, in recent years, serious innovations have been introduced in the production technology. Therefore, the need for further study of this issue in relation to modern metallurgical processes is relevant.

2. Results and discussion

The thermodynamic analysis applied to Mn was carried out on the basis of tendency towards equilibrium of system C – Mn – O.

Usually when considering the manganese oxidation a reaction is used:

$$[Mn] + (FeO) = (MnO) + [Fe]\ (lgK = (6440/T) - 2.95) \tag{1}$$

i.e. with the slag phase. However, in modern fast-flow steelmaking processes with a large consumption of gaseous oxygen the direct oxidation reactions can be used:

$$[Mn] + [O] = (MnO)\ и\ [C] + [O] = \{CO\} \tag{2}$$

The obtained results are shown in figure 1.

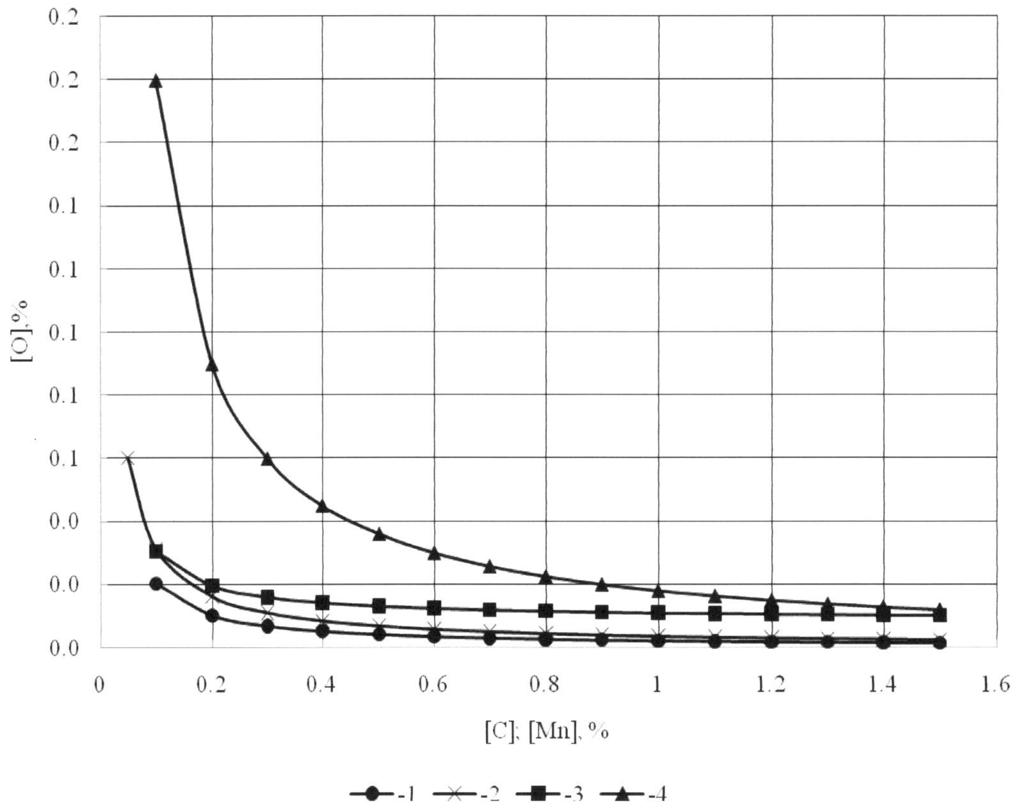

Figure 1. The ratio of [C] – [O] – [Mn].
1- [C] c [O] by the reaction [C] + [O] = {CO} at Pco = 1 atm [2];
2 - [C] – [O] according to practical data for 300 tonne – converter;
3 - [C] – [O] according to practical data for arc steel smelting furnaces
[O] = - (0.00216/[C]) + 0.00884 [1];
4 - [Mn] – [O] by the reaction of [Mn] + [O] = (MnO) [2].

The analysis of the results shows that practically in the entire range of carbon and manganese concentrations, characteristic for oxidizing conditions of steelmaking processes, it is carbon that determines the oxygen content in the metal. Manganese in steel can not be oxidized at a concentration up to ~ 1.0%. Only when the carbon content is lower than 0.1%, the content of manganese should decrease.

The equilibrium content of carbon and manganese is shown in figure 2.

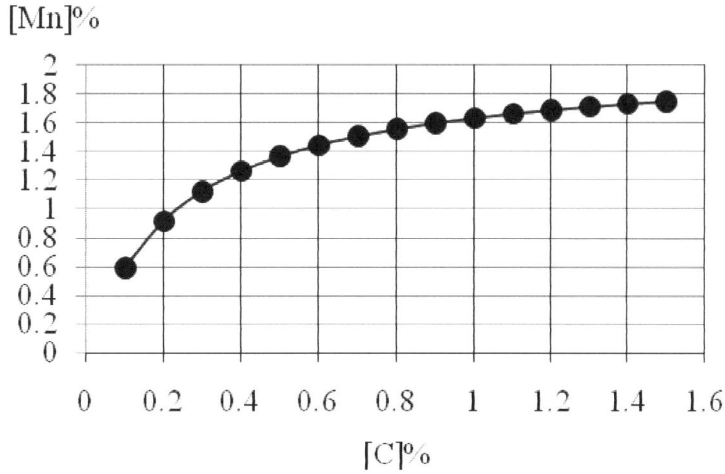

Figure 2. Equilibrium relationship [C] – [Mn] in the oxidizing conditions.

The joint behavior of carbon and manganese has been studied in the conditions of operating facilities – converter (converter with capacity 300 tonnes) and electric steelmaking (100 tonne arc furnace). The data on the manganese and carbon behavior in converter melting (300 tonnes) are shown in figure 3.

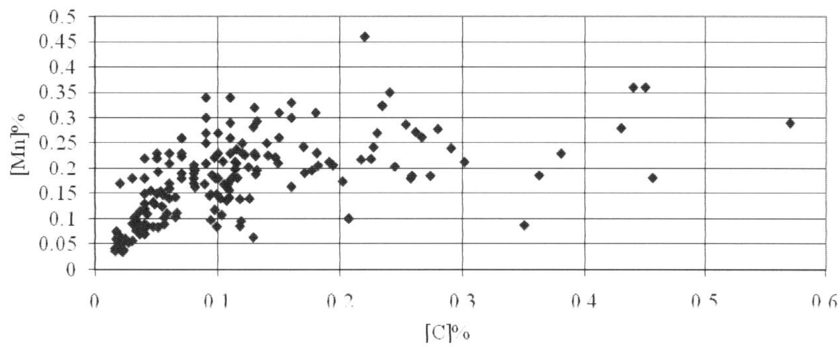

Figure 3. The ratio of [C] – [Mn] to the converter melting.

Data on the manganese and carbon behavior in the electric steelmaking furnace (100 tonnes) are given in figure 4.

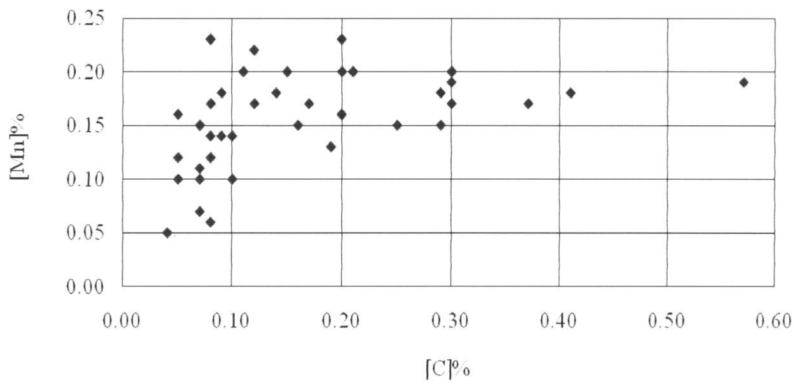

Figure 4. The ratio of [C] – [Mn] during melting in the electric arc furnace (100 tonne).

IOP Conf. Series: Materials Science and Engineering **411** (2018) 012004 doi:10.1088/1757-899X/411/1/012004

The analysis of the obtained industrial data shows that they correlate well with the presented above thermodynamic calculations.

It is clearly seen that the content of manganese practically does not change during the oxidation of carbon. Only after achieving a carbon content of 0.1%, a rapid manganese oxidation begins. The concentration of manganese with carbon of less than 0.1% becomes lower than the calculated values. It reaches 0.05%. This can be explained by the fact that at this carbon concentration the limiting stage of the oxidation process changes. The rate of the process begins to be limited by the rate of mass transfer of carbon in the metal. Excess oxygen begins to interact actively with other elements, such as manganese and iron.

3. Conclusion

The analysis showed that the main directions for optimization of the oxidation process technology in order to save manganese ferroalloys are:

- increase in the manganese content in the metal due to the use of a metal charge with the increased content of manganese;
- direct alloying with manganese from oxide manganese materials;
- control of carbon and, accordingly, stopping of the process at its maximum permissible concentrations at the final stages of metal blowing with oxygen.

References

[1] Povolotsky D Ya, Roshchin V E and Malkov N V 1995 *Electrometallurgy of Steel and Ferroalloys* (Moscow: Metallurgy) p 592

[2] Grigoryan V A, Belyanchikov L N, Stomakhin F Ya 1987 *Theoretical Foundations of Electric Steelmaking Processes* (Moscow: Metallurgy) p 271

Innovations in institution of higher education as a tool of economic and social development of the region (experience of project-based education technology implementation in SibSIU)

E V Protopopov, A V Feoktistov, G V Galevsky, O V Gordeeva, M B Vasilyeva and O Ya Gutak

Siberian State Industrial University, 42 Kirova street, Novokuznetsk, 654007, Russia

E-mail: rector@sibsiu.ru

Abstract. The article substantiates the need of a new model of education process in the university, providing its integration into the regional economy and social sphere. The main assumptions of project-based education introduction into institutional practice are referred to, the scope and special aspects of implementation of the named education technology as a tool of innovative transformations in the Siberian state industrial university are provided.

1. Introduction

Today in the debate on the prospects for higher education development the most acute topics are the ones disputing conditions of universities innovative transformation. Changing the role higher school requires complete rethink of parameters of education environment, including conformity of educational technologies to the challenges of innovation, technological and social development on regional and national level. It became obvious that efficiency of university education model is seen in improving the level of graduate's knowledge and skills, providing successful performance of professional tasks in the world of socioeconomic and political transformations [1, p. 38].

Siberian state industrial university (hereinafter referred as SibSIU) persistently develops and implements the model of an educational integrator for innovative development of the Siberian region, which claims a set of problems to be solved: the first is aimed at university scientific and technological development, the second targets a stable interaction of the university with regional authorities, clusters, industries, and the third demands for modern education technologies to be introduced into educational process. Taking into account the tasks listed, a decision was made by the University Methodical Board to introduce a new approach into curricular management. Project base education technology was chosen as a technological foundation for the model of the university transformation into a center of innovations genesis.

Making emphasis on innovative nature of project based education technology, it is necessary to clarify that the very idea of project base education is not neither of novelty nor of originality, since it has already been fully and comprehensively described in national literature, and the content and forms of it has repeatedly been tested and evaluated from the point of view of pedagogical conditions of its implementation [2, 3, 4]. At the same time, implementation of project based education technology as a mean of innovative transformations in education has not been subjected to specific pedagogic study, and its application toolkit, in regard of the requirements of federal state educational standards, is not currently described. Moreover, it was established that "education technology as a system can have different states, when values of its' parameters can vary significantly depending on purpose and conditions of its application" [5, p.38], while "under different conditions, any modeled education technology of all its possible states, can practically take only their limited number for successful performance, this is restricted by its main purpose [Ibid.].

Consequently, project based education model as a set of consequent stages of the competences design in regard to specific degree program or a group of degree programs in multidisciplinary university can be considered innovative.

2. Project-based approach institutional implementation. Discussion and assumptions

The process of implementation of project based education at Siberian state industrial university required preliminary stage at which the Methodical Boards met, and the concept of project based education was discussed. In discussions, peculiarities of existing university education environment were determined as factors of innovative transformations stimulation [5, p.49], i.e. the role of project technologies in professional education of the student was clarified, feasibility and efficiency of their application in Siberian state industrial university were determined, criteria for project based education results evaluation were identified, elements of degree programs aimed at the new technology implementation were designed.

Forum of employers of the South of Kuzbass was held in the SibSIU in February 2016, following which employers requirements for the quality of university graduates in regard to capacity of project based education technology were determined.

As a result of this stage, conditions for project based education introduction in SibSIU were determined, the choice was made taking into account the following ideas of innovation:

1. Project based education expands and transforms university responsibilities, which contributes to effective solution of the key tasks of regional university development [6]. This education technology is the mechanism that ensures integration of educational operation of the Siberian state industrial university into regional economy and social sphere.

2. Project based education provides an interactive education environment, the leading element of which is simulated reproduction of various situations, forms and patterns of professional operation. Such forms contribute to professional development and professional self-determination of the student, raising quality of his education [7].

3. Project based education technology allows us to develop a brand new structural and functional model for students' competences design in multidisciplinary university.

4. Development of new teaching and methodological aids as a result of introduction of innovative educational technology provides an opportunity to improve education process in the Siberian state industrial university. Practical experience of teachers in education innovations implementation provides prerequisites for initiation of other pedagogical innovations.

Let us consider specifics of each of the presented innovation positions.

As the practice of educational standards requirements implementation shows, today there is a need for transformation of the field of university responsibility. Introduction of project based education makes it possible to bring SibSIU to fundamentally new positions in the education market, such as a position of core discussion platform, with focus on discussion and generation of solutions to key problems of city and regional development. In other words, project based approach as an education technology serves as a tool for the SibSIU positioning as a regional multidisciplinary university, providing points of growth and conditions for evolutionary development of the Kuzbass economy, social and cultural environment. It is obvious, that application of project based education technology transforms the SibSIU into an innovative platform for profession-oriented projects accumulation, which ensures:

- academic staff involvement in solving acute problems of regional economy;
- popularization of university R&D and its positioning as a community of respected and highly demanded experts;

- university information transparency, strengthening ties between the university and employers.

The expected results in regard to the SibSIU transition into the regional university are as follows:

1. Generation of a spot of attraction for talented youth and nurturing future leaders of regional economy within a territory of advance development of youth creativity.
2. Developing system of continuous education based on implementation of project based education programs for all categories of regional customers.
3. University transformation into the center of integration network of knowledge-based and inter-sectoral innovations.
4. Building positive image of university in regional community, expanding geography and the domains of its influence.
5. Transformation of university management based on innovative tools of cluster and sectoral development.

Another equally important effect of project based education is generation of conditions for professional development and professional self-determination of university students. The task of professional development of a university graduate has been solved in the context of tasks aimed at building contextual professional education, focused not on transfer of knowledge, but on finding knowledge and its application in situations simulating real professional operation. The main goal of professional education rethinking is to find ways for future professionals to develop a pro-active position [8].

Project based education as a person-oriented technology promotes professional development and activates professional self-determination of a student. The very essence of professional development evidences this statement, as it has been recognised as an integration of cognitive and professional motives. This process involves certain stages, each of which is specified by certain psycho-physiological and socio-psychological characteristics that ensure successful implementation of professional operation.

In the context of university educational environment, project based education forms personal professional motivation, it stimulates the needs, attitudes, interests and motivates person to study and master future profession. Along with the research (cognitive) side of the project task solution, there is always an emotional (personal) and creative one. It is the emotional and the creative components of project based education that determines significance of the student's results and opportunities for their achievement.

Education management through the project based education makes it possible to transform education case into the situation of a student's personal development. At the same time, education project acts as a mean of professional and personal development, project base education facilitates acquisition of personal experience and relevant professional competencies as well. Project based education is a mechanism that directs student's professional orientation, as it contributes to development of professional self-awareness, to manifestation of sense of student's belonging to a particular profession, and desire to develop professional competences, and form a special system of professional values. Due to its novelty, feasibility, viability, and opportunity to evaluate goals set by professional community, project based education is assigned a pedagogical category that works to build professional competence of an engineer [9].

Project orientation of education is aimed at development of students' project culture, which is professionally significant quality of the individual, including a value-based attitude to project operation, acquisition of project knowledge and skills on individual creativity level, ambition to transform reality and develop personal project abilities [10, 11].

Consequently, implementation of educational projects fills university education with sense of professionalism, helps to master future professional sphere of operation in integrity and dynamics, thereby forming alertness for future professional.

The third innovative idea concerning implementation of project oriented education states that this technology allows institution to develop the brand new structural and functional model for students' competencies development in multidisciplinary university. It transforms education process into a management of competencies acquisition. In this aspect, advantages of project based education as a model of competence approach implementation are obvious:

1. It makes search of different ways of solving set of educational tasks the main component of education instead of knowledge acquisition.
2. It is aimed at propagation of self-motivated independent work of students, the main purpose of which is to solve the task (theoretical or practical) through a product generation. The expected final result of such transformation of educational process is high independence of graduates, acquiring ability to organize their own professional operation and seek new tools for professional functioning.
3. It is an individual and an operation-oriented technology, since it involves students into creative educational and cognitive activities as subjects of their personal and professional development.
4. It simulates individual or collective professional operation. Efficiency in competencies development is achieved through approximation of education process to practical professional activities. The main result that a student receives through the project educational activity is acquisition, consolidation or development of knowledge significant on practical level, skills required in chosen professional sphere and experience of self-management.
5. It facilitates restructuring of interaction between an academic and a student, it requires absolutely different alertness of an academic and a student to their new roles, determined by competence based approach.
6. It provides for the implementation of various forms, methods and techniques aimed at developing creative, research and analytical abilities of students.
7. Embedded in education operation, it radically changes all the components of education process by expanding students' self-dependent work, introduction of modern forms of its management and quality control.

Meanwhile, despite the advantages, listed above, project training has not yet found sufficient theoretical justification and introduction in practice of higher education within the competence approach on national level. Transformation of educational standards of higher education, change in education programs content did not directly affect the forms of pedagogic interaction. According to our observations, traditional education system still prevails in higher education, which, as is known, is reproductive or cognitive in nature and does not contribute to implementation of competence approach in higher education introduced by the federal state standard of higher education.

The reasons for not using innovative technologies are objective: in structure of education content, results are recorded in form of graduate student competences acquisition, but systematically organized student's activity necessary for that acquisition at a specific level, is not indicated. Thus, the same education results can be achieved in various ways. In addition, one competence may be acquired during study of several disciplines (modules), thus having an integrative and interdisciplinary nature.

In our opinion, system default and uncertainty of methodological tools in achieving education outcomes leads to the situation where curricula are designed in a formal way. Introduction of project based education technology allows the system approach competencies acquisition to be provided, since its model provides for new content and structural components of education process correlating with project based education logics, including step-by-step assessment of achievements and results.

Finally, we turn to the idea, which completes the list of innovations presented above, essential for implementation of project based education. As it has already been noted, development of new teaching and methodological products and other results of introduction of innovative education technology, has enabled improvement of education process in Siberian state industrial university.

Taking into account innovative ideas listed above, the SibSIU academics have developed the concept of project based education and have determined sequence of deliberate institutional changes in education process; definitions of new competencies, target levels of their development have been stated, "projects portfolio" has been generated.

Since 2016, the "Project Activities" module has been introduced into the curricula of all degree programs, the module belongs to compulsory disciplines of variable part of curriculum. This module forms a certain structural integrity of degree program, is comprehensive because it logically overlaps content of other disciplines and is aimed at development of "universal" competencies (indicative base of activity) associated with project implementation. The goals and objectives of disciplines of the "Project Activities" module correlate with a concrete practical result – implementation of various types of projects. Students' activities results were presented within the academic calendar, during weeks of projects demonstration.

This comprehensive concept corresponds to the activity paradigm of competence approach, contributes to overcoming fragmentation, consolidates education process and stimulates development of student' individual educational path. Since the competencies related to the project activity are formed during the entire education process, forms of control of level of its development have been established for each stage, and for its final evaluation within the framework of final state attestation.

Particular attention is paid to methodological parameters of the "Project Activities" module, such as list of competencies, discipline workload, forms of control. Since this discipline begins in the first year and is of propaedeutic function, lectures content has been defined, interactive forms of workshops, as well as forms of self-study, corresponding to freshmen level have been developed.

Thus, the model of project based education was stated in the following documents regulating the university education process:

- degree programs descriptions;
- curricula and programs of disciplines of the "Project Activities" module;
- means of evaluation;
- guidelines for academics and recommendations for students' self-study.

Teaching aids and methodological materials were tested through the pedagogical experiments, participants of which were academics themselves. Methodological challenges revealed throughout the experiment enabled development of methodological instructions for workshops. As a result of team work, a freshmen "Project Activity" course was introduced into the curricula of all the degree programs of the SibSIU and is now being successfully implemented.

Project based education has been introduced as a structural and educational component in all degree programs of the university, the entire academic staff of the SibSIU is involved in implementation of innovative education technology. For dissemination of experience of project based education implementation in the university, various ways of presenting the results of innovative education activities were used: they were reported in forums and conferences, presented at competitions. In addition, the SibSIU has launched "Education projects accelerator", as a form of corporate system of faculty professional development in the field of project education technologies design. This form of interaction facilitates full accelerated immersion of the academic staff in project activities as participants of real interdisciplinary project teams. Information on each project progress was published on the university website and through social networks.

Students' and academics project activities results were presented at specially arranged events. Thus, based on the results of the "Project Activities" course acquisition in the first semester, first year students had independently prepared video projects "My future profession". To select the best project, an open online vote was arranged on the university web page.

3. Conclusion

Evaluation of project based education implementation efficiency has not been conducted so far. The intermediate results are as follows:

1. Curricula content of degree programs has been optimized, continuity of project activity throughout the framework of degree programs has been revealed for different degree levels, such as bachelor's, master's, engineer's and Phd programs.
2. Mechanisms for interdisciplinary integration in education process have been generated.
3. Innovative education products have been developed (consistent study package for the "Project Activities" module, and "Projects Portfolio" has been generated).
4. Index of academic staff activity in application of innovative education technologies has been increased.

IOP Conf. Series: Materials Science and Engineering **411** (2018) 012005 doi:10.1088/1757-899X/411/1/012005

5. Opportunities for various forms of students' self-study have been arranged.
6. Level of education process participants' satisfaction has been raised.
7. Education process transparency is provided, new mechanisms of the university interaction with the employer have been formed.
8. A strategy of the SibSIU positioning as a university of innovations has been manifested.

References

[1] Loshchilova M A and Dobrycheva I V 2016 *Professional Education and Employment of Youth: XXI Century. Issues of Advanced Education of Personnel for Russian Economy (Regional Aspect). Proc. Int. Sci. Conf.* (Kemerovo) pp 38–41
[2] *Mieryn L A*, Bykova N N and Zarukina E V 2015 *Modern University Education Technologies: Manual* (St. Petersburg: St. Petersburg State University of Economics Press) p 169
[3] Blokhin A L 2005 *Project Method as a Person-oriented Pedagogic Technology: Dis. ... cand. sci. (Pedagogics)* (Rostov on Don) p 154
[4] Emelyanova N V 2011 *Bulletin of Higher School* **3** 82–84
[5] Bordovskaya N V 2016 *Modern Educational Technologies: Manual* (Moscow: KNORUS) p 432
[6] Stepanova S N 2012 *Transformation of the Idea of University in Transforming Education Space* (Tomsk: TML-Press) p 128
[7] Khitrintseva A V 2010 *Academic Journal of Western Siberia* (Tyumene: M-center) pp 6–7
[8] Bagdasaryan N G and Gavrilina EA 2010 *Higher Education in Russia* **6** 23–26
[9] Tsygulyova M V 2010 *Bulletin of Tomsk State Pedagogical University* **10** 56–62
[10] Markov A P 2003 *Bulletin of Kazan State University* **1** 56–60
[11] Stenina T L 2011 *Development of Students' Project Culture* (Ulyanovsk: Ulyanovsk State Technical University Press) p 243

IOP Conf. Series: Materials Science and Engineering **411** (2018) 012006 doi:10.1088/1757-899X/411/1/012006

Analysis of modern production of zirconium carbide

T I Alekseeva[1], G V Galevsky[1], V V Rudneva[1], A N Cherepanov[2], L Stafetsky[3] and S G Galevsky[4]

[1]Siberian State Industrial University, 42 Kirova street, Novokuznetsk, 654007, Russia
[2]ITAM SB RAS, 4/1 Institutskaya street, Novosibirsk, 630090, Russia
[3]"Neomat Co", Miera iela 34, Salaspils, LV-2169, Latvia
[4]St. Petersburg Mining University, V.O., 21 Line, 2, St. Petersburg, 199106, Russia

E-mail: kafcmet@sibsiu.ru

Abstract. The analysis and systematization of information on the methods of zirconium carbide production are analyzed. It is shown that the basic method is carbothermal synthesis, which is realized in various options (furnace synthesis from traditional and highly dispersed batches, melting-crystallization, plasma heating, in the electrothermal fluidized bed). The SHS process, mechanosynthesis, plasmosynthesis, a vapor-gas phase plating method are considered.

1. Introduction

One of the most important tasks of modern materials science is obtaining materials for usage in extreme conditions – at high temperatures and voltages, under the influence of aggressive media, etc. In the solution of these problems the leading role belongs to the compounds of refractory metals with boron, carbon, nitrogen, silicon – borides, carbides, nitrides and silicides, which have heat resistance and high temperature strength, specific physical and chemical properties alongside with the high hardness and refractoriness. Among carbides of transition metals, zirconium carbide by combination of special properties is included in the leading group, combining such properties of practical importance as hardness, refractoriness, corrosion resistance in liquid, gas and metal environments, durability.

The purpose of this paper is to analyze information on the methods of zirconium carbide production. For the analysis more than 20 scientific articles and monographs published in the period from 1968 to the present were selected, a patent search was carried out.

2. Research and analysis of technological solutions in the zirconium carbide production

The analysis of scientific and technical literature shows that the basic method for the production of zirconium carbide is furnace carbothermic synthesis, which is realized in the following technological options [1-3]:

1) from dioxide and carbon char at a temperature of 2173 K, in the vacuum for 1.5 hours (industrial level);

2) from a highly dispersed charge (ZrO_2+C) obtained by the sol-gel method at a temperature of 1373 K in the vacuum (pressure 0.5 kPa) (laboratory level);

3) from zirconium dioxide and carbonaceous matter when the compacted batch is heated on a water-cooled copper substrate by the argon plasma jet to a temperature of 2770 K;

4) from zircon sand ($ZrO2 + SiO2$) and coal or carbonaceous waste during melting in the electric arc furnace and subsequent crystallization (industrial level);

5) from zirconium dioxide and carbonaceous matter (petroleum coke, low-ash carbonated coal of reksil) in the electrothermal fluidized bed at a temperature of 1673-2073 K.

In the first option the carbizided batch pellets are subjected to grinding and classification to obtain carbide powders within the size range from 3 to 63 μm. The second option provides zirconium carbide in the form of a micropowder with a particle size of 2-3 μm, the third – in the form of a nanopowder with a particle size of less than 100 nm.

The method of plasma heating did not receive industrial implementation. Melting in the electric arc furnace followed by melt crystallization is currently the only way of obtaining zirconium carbide when used in the materials for abrasive processing, spraying and surfacing, protective coatings. Industrial application for the production of zirconium carbide reactor with an electrothermal fluidized bed, characterized by the presence of a highly reactive medium – the microplasma formed during electrical discharges between fluidizable carbonaceous particles are not clear. The first variant of carbothermic synthesis of zirconium carbide, which in fact has a 60-year technological history, forms the basis of modern industrial technology for the production of its micropowders [1]. At the same time, reductive synthesis requires sufficiently high temperatures and technological vacuum, as well as mechanical dispersion of carbidization products, leading for superhard carbides to their substantial contamination due to the milling yield of grinding bodies and surface oxidation. Specification of the method feature creates significant technological difficulties and, apparently, hinder the wider use of zirconium carbide in modern materials science.

The method for the production of zirconium carbide by the self-propagating high-temperature synthesis (SHS process), comprising mixing zirconium powder of up to 50 μm in size and a carbonaceous substance of a size up to 1 μm, loading the prepared charge into the reactor, synthesis in the gaseous atmosphere, cooling, mechanical dispersion and classification of zirconium carbide powders, has not been widely used and is in limited demand for obtaining mainly preparative volumes for research purposes [2]. At the same time, this method can be very useful for SHS compaction of special-purpose structural elements containing zirconium carbide in their composition.

The mechanical synthesis of zirconium carbide, which is usually realized in high-energy mills using zirconium-carbon batch, focused mainly on preparation of preparative amounts of carbide nanopowder of 5-200 nm size range for studying the features of its physicochemical properties in the nanostate and the justified search for new applications, during the last 20 years has not undergone any significant technological and hardware changes and upgrades and, according to its indicators, can still be attributed to laboratorial usage [4].

Plasma synthesis of zirconium carbide is the first attempt in its technological history to create an uninterrupted process. Technological fundamentals of plasma synthesis, laid down at the turn of the 70-80s of the last century, suggest the joint processing of $ZrCl_4$ and propane-butane technical mixture in the plasma flow of argon and hydrogen, quenching of products by plasma treatment and their trapping, non-reactor thermal treatment in the inert medium [5]. The method itself, the equipment and the hardware-technological scheme for its implementation were first mastered for the production of customized batches in the conditions of the Special Design and Technological Bureau of inorganic materials of the Academy of Sciences of Latvia, and then later and now for low-tonnage production in the conditions of the company “NEOMAT Co” (Latvia). The achieved indices allow the plasma synthesis to be considered as the most promising of the known methods for obtaining zirconium carbide in the nanostate state.

The method of obtaining zirconium carbide by precipitation from the vapor phase of $ZrCl_4+CH_4+H_2$ at a temperature of the order of 1573 K is focused mainly on the formation of carbide film coatings on substrate products and has no independent value in the total production of zirconium carbide [2, 3].

3. The research and analysis of the domestic and world market of zirconium carbide

Figure 1 presents information on the manufacturers of zirconium carbide, the technological options realized by them for its production and the main applications. Presented by the leading foreign manufacturers of nanopowders of metals and their compounds, scientific and technical information contains the following technological information about plasma processes for the production of zirconium carbide.

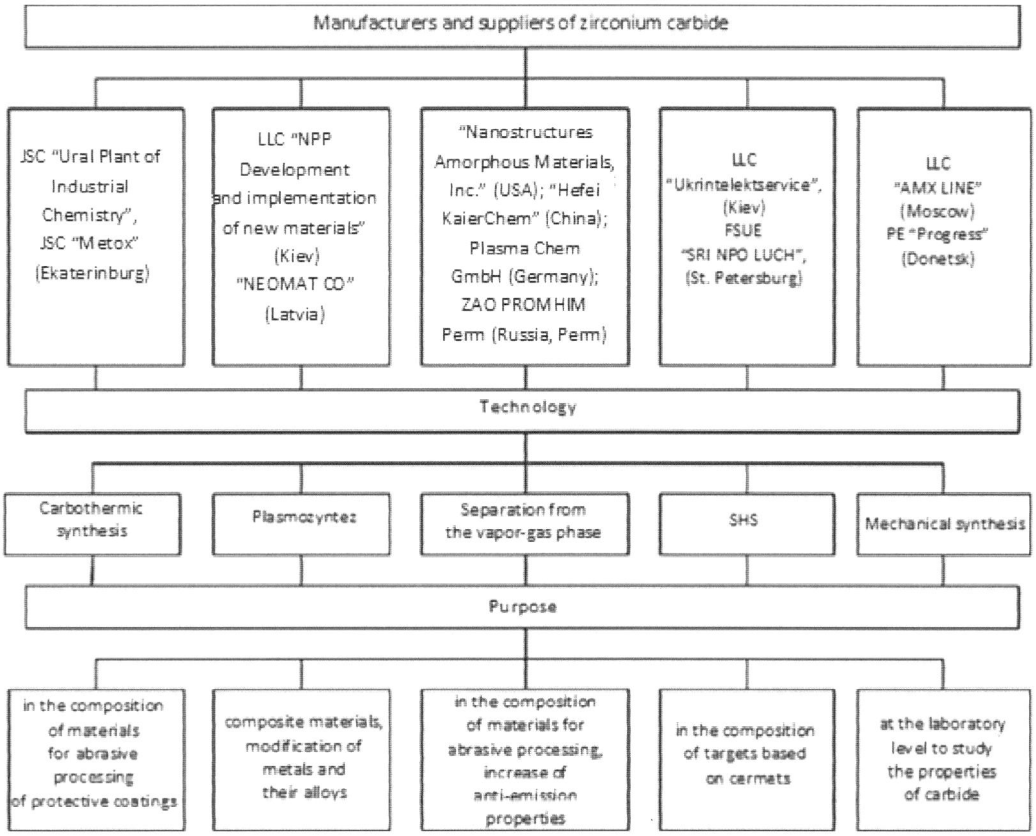

Figure 1. Information about the manufacturers of zirconium carbide, the technological options realized by them for its production and the main fields of application.

Among the most promising areas, the leading ones are occupied by the methods for obtaining zirconium carbide nanopowder in the plasma flow of argon and hydrogen (plasma synthesis), and the vapor-gas phase plating. Currently, the world's leading manufacturers of carbide are research and production companies – "Nanostructured & Amorphous Materials, Inc." (USA), "Hefei Kaier Nanotechnology & Development Ltd. Co" (China), "NEOMAT Co" (Latvia), "PlasmaChem GmbH" (Germany), which produce 97% of ZrC with a dimension of 10-60 nm in batches from 5 to 100g and several kilograms. To date, these manufacturers set the price range from $ 400 to $ 2 000/kg [6].

The market of CIS states is represented by several enterprises and laboratories of Ukraine: LLC "NPP Development and implementation of new materials" (Kiev), "Ukrintelektservice" LLC, (Kiev), "Progress" PE (Donetsk), which produce ZrC powder compliant with TU 6-09 -03-03-408-754. These specifications apply to the qualification of "pure" zirconium carbide, used in the composition of high-temperature alloys and in the composition of cathodes with a high work function of electrons.

The domestic market is mostly represented by small enterprises and laboratories that receive zirconium carbide by carbothermic method, plasma synthesis, mechanosynthesis, SHS, vapor-gas phase plating [7].

The company "IPK YUMEKS" (Ufa) sells high-refractory ZrC in the form of a gray powder at a contract price. The Ural Industrial Chemistry Plant (UZPKh) produces and supplies zirconium carbide of the "Pur." classification. "UZPKh" JSC is a modern chemical enterprise of a full cycle, which products are in demand not only in the markets of Russia and the countries of customs union, but also in high-tech far-abroad countries: Japan, Australia, Canada, USA. The sale of products to the domestic market is carried out through the official Trading House – "RIVIERA" LLC, Moscow. "Alterhim" LLC (Dzerzhinsk) produces and supplies zirconium carbide of qualification "Pur.", corresponding to TU 6-09-03-408-75, in lots from 1 kg at wholesale and retail prices. "Redmetur" LLC (Ekaterinburg) is engaged in the supply of quality products from rare earth metals throughout the Russian Federation and CIS countries. The main products are metals (molybdenum Mo, niobium Nb, tantalum Ta and magnesium), carbides (TaC, NbC, VC, TiC, ZrC, Cr_3C_2, Mo_2C), etc. Zirconium carbide is supplied as fractionated powders 3-5, 40-60, 40-100 μm at the agreed price.

4. Conclusion

Zirconium carbide ZrC, studied and put into usage by the scientific school of the famous Russian material scientist Samsonov G.V. more than 50 years ago, is still in demand in the technology of various materials: metal-ceramic instrument and structural, refractory and abrasive, for the modification of coatings. Analysis of modern scientific and technical information reflects the tendency of the transition from zirconium carbide to coarse-grained to micro- and nanocrystalline, which is determined by the desire of scientists and technologists to achieve a qualitatively new level of performance properties of materials and coatings on its basis.

References

[1] Kosolapova T Ya 1968 *Carbides* (Moscow: Metallurgy) p 300
[2] Panov V S 2006 *Refractory Metals of the IV-VI Group and their Compounds. Structure, Properties, Methods of Preparation: Textbook* (Moscow: MISiS) p 63
[3] Samsonov G V, Epik A P 1973 *Refractory Coatings* (Moscow: Metallurgy) p 400
[4] Avvakumov E G 1991 *Mechanical Methods of Activation of Chemical Processes* (Novosibirsk: Nauka) p 264
[5] Miller T N 1984 *Application of Low-Temperature Plasma in Technology of Inorganic Substances* (Moscow: Nauka) p 620
[6] *Neomat Nano Powders. Produts* http://www.neomat.lv
[7] *Official Website of Flagma* https://flagma.ru/karbid-cirkoniya-so233713-1.html

IOP Conf. Series: Materials Science and Engineering **411** (2018) 012007 doi:10.1088/1757-899X/411/1/012007

Application of zirconium carbide: assessment, determination of dominant trends and prospects

T I Alekseeva[1], G V Galevsky[1], V V Rudneva[1] and S G Galevsky[2]

[1]Siberian State Industrial University, 42 Kirova street, Novokuznetsk, 654007, Russia
[2]St. Petersburg Mining University, V.O., 21 Line, 2, St. Petersburg, 199106, Russia

E-mail: kafcmet@sibsiu.ru

Abstract. The analysis and systematization of information on the application of zirconium carbide in modern technology: in the production of hard alloys and products from them, in technologies of anti-emission coatings, surface modification of powder materials. Prospective directions are indicated: a component in composite materials technologies, a modifying phase of galvanic metal matrix composite coatings, a functional component of wettable coatings for cathodes of alumina reduction cells.

1. Introduction
Priority directions of modern multidisciplinary engineering require materials capable of stable operation in the conditions of high temperatures (more than 1000 K), high loads, aggressive media. In this regard, the role and importance of synthetic refractory superhard compounds, carbides, borides, nitrides, oxides and their compositions, is constantly increasing. Among carbides of transition metals, zirconium carbide, by combination of special properties, is included in the leading group, uniquely combining such practically significant properties as hardness, refractoriness, corrosion resistance in liquid, gaseous media and metallic melts, wear resistance.

The earlier conducted analysis of available scientific and technological literature containing the study results of the properties and methods for the production of zirconium carbide confirms that the main ideas were formed in the 1960s and 1970s mainly by the efforts of the scientific school headed by Samsonov G.V., Corresponding Member of the Academy of Sciences of Ukraine, contain characteristics of properties and conditions for obtaining coarse-grained powders and compacted samples. Th systemic studies of zirconium carbide in the nanocrystalline state were practically not carried out. Along with this in the scientific literature the information about the real applications of zirconium carbide in modern technology is extremely limited. In connection with this, the purpose of this paper is to analyze the most technologically and economically attractive areas of application of zirconium carbide and to predict new approaches to its use.

2. Application of zirconium carbide in modern technology: the actual state and forecast
In the technology of hard alloys, the choice of the carbide component, optimization of the structure of the metal bond, is described in detail in [1]. The increase in the level of performance properties is achieved by optimizing the structural state of the hard alloy. The main variable parameters are: the amount, dispersion, hardness, modulus of elasticity of the carbide phase, composition and mechanical properties of the metal binder, the thickness of the interlayer layer. The degree of dispersion of carbide powders obtained, as a rule, by carbothermic reduction or synthesis from elements, in most cases

amounts to several micrometers and rises to 0.3-1.0 microns during grinding-mixing operation of carbide charge. In the process of grinding the thin metallic structure of the carbide component undergoes significant changes, as evidenced by the observed usual increase in microconjugation and the degree of amorphization, the accumulation of linear and point defects, and defects such as grain boundaries. Such changes ensure the sintering of carbide mixtures in technologically acceptable conditions with the required set of properties.

According to the data of [1, 2] zirconium carbide is positively tested in the composition of the following hard alloys:

1) in the composition of hard alloys of the TTC group, usually containing up to 5-20% TaC, 3-15% TiC, 8-12% Co, the rest WC, instead of expensive TaC; alloys of this group are quite universal as they are used for processing both carbon steels and various cast irons;

2) in the composition of hard alloy WC-ZrC-Co of the TC group, which in its properties is practically equivalent to the WC-TiC-Co alloys;

3) in the composition of hard alloy of group THA SOT 30 (tungstenless hard alloys) on the carbonitride base – $(Ti_{0.95}Zr_{0.02}Nb_{0.10}Ta_{0.09})C_{0.53}N_{0.47}$; the alloy has δ1000 MPa, HRA 90 units, the average grain size of the carbonitride phase is 2-4 µm, the cutting resistance is 4 times higher than that of the alloy KNT 16;

4) in the composition of the hard alloy of group THA TiC-VC-ZrC-NbC-Ni-Mo (grade T and TP);

5) in the composition of carbidechrome hard alloys of the group KKhN10-40 Cr_3C_2 – Ni (Ni + Co + Fe + Mo + Cu) for partial replacement of Cr3C2; the alloys have high hardness, scale resistance, resist to wear and corrosion in aggressive melts and liquids, have a low propensity for setting and magnetization and are used mainly for the manufacture of drawing-blocks, dies and mandrels for drawing ferrous and non-ferrous metals, molds and other wear-resistant parts.

Along with this, a rather cautious attitude of specialists to zirconium carbide as a component of various hard alloys should be noted, which, apparently, is due to the non-manufacturability of its production and poor wettability by melts of the iron group, which makes it difficult to select a binding metal.

In coating technology the methods for the formation of effective high-reliability coatings of the following types are proposed:

- intermetallic anti-emission composition Pt_3Zr on the grids of powerful generator lamps [3]. The method for manufacturing an anti-ejection coating on mesh electrodes from molybdenum comprises forming a cataphoretic layer of carbide of zirconium with a thickness of the order of 10 µm; its baking in a vacuum at a temperature of 1773 K; forming a cataphoretic platinum layer; baking at a temperature of 1273-1573 K. In the composite coating obtained in this way, the anti-emission layer is platinum, and zirconium carbide acts as a diffuse barrier between the platinum and the core of the grid;
- abrasive composition of ZrC on particles of diamond powders [4]. On the particles of diamond abrasive produced by the company "Element Six", the coating from zirconium carbide is performed by the chemical vapor deposition. The fraction of zirconium carbide layer with a diamond particle size of 40 µm is 0.77%. The ZrC layer has a thickness of the order of 1 µm. During the X-ray structural analysis the diamond, ZrC, Zr were found in the composition of particles. Diamond powders with a carbide coating are widely used for the manufacture of cutting, grinding and drilling tools;
- corrosion-resistant composition Zr-ZrC on ceramic powders. Zirconium carbide is used when depositing metallic zirconium coating on ceramic powder materials [5]. The use of ZrC is conditioned by the need to increase the corrosion resistance of ceramic powder.

In the technology of superhard materials the application of nanocrystalline nitrides, carboronitrides, carbides, boronitrides of transition metals of IV-Va groups and their compositions with metallic powders as components and initiating additives in the sintering of materials based on boron nitride and diamond has been introduced on an industrial scale, which will allow a number of advantages to be obtained [6]:

- to reduce the processing time of the charge at high pressure and high temperature up to 30-60 s due to the high surface activity of the powders and thus increase the productivity and service life of high-pressure apparatuses;
- to increase the yield of suitable products to 75-80% due to the high activating ability of powders, reducing sensitivity of process to the accuracy of holding the pressure and temperature;
- to increase the hardness of compacts and their wear resistance when processing high-strength steels and alloys with HRC > 60 by reducing the powder content in the charge to 1-3% without reducing the efficiency of the process.
- to reduce the size of boron nitride crystals in compacts, which increases the strength and wear resistance of the cutting tool by changing the nature of its wear;
- to create new composite superhard materials containing up to 30% or more of nanodispersed components characterized by increased toughness, chemical inertness with respect to the treated surfaces, modified electrophysical properties.

The use of nanocrystalline borides and carbides CrB_2, Cr_3C_2, TiC, SiC as modifying additives ensuring the production of high-density coatings with a high complex of physical and mechanical properties through the formation of a fine-grained structure with low porosity has been introduced into the technology of galvanic composite metal-matrix coatings in the processes of nickel plating, galvanizing, chromium plating. The combination of a nanocrystalline state with a direct-formed complex of properties inherent in nanoborides and nanocarbides ensures the stable achievement of a number of positive effects when using them as a modifying electrodeposited metal phase matrix [7, 8]. The low concentration of nanopowder in the electrolyte simplifies the operation of galvanic baths and reduces the loss of nanopowder due to the removal of electrolyte with parts. The presence of nanopowder in the electrolyte improves its productivity due to the increase in the upper limit of cathode current density. Testing of titanium diboride nanopowder instead of nanodiamonds in the process of obtaining protective galvanic coatings of different tools shows that with comparable characteristics and service life of a hardened tool, the following technological and economic advantages are achieved when replacing nanodiamonds: the technology of preparation of electrolyte suspension is simplified and accelerated, the deposition rate of the coating increases by 2.5 times, the corrosion resistance of coatings increases by 1.5 times, the cost of 1 m^3 electrolyte-suspension decreases by 4 times. When replacing 1 kg of nanodiamonds with nanopowder titanium diboride, the economic efficiency is 105 thousand rubles. Economic efficiency is determined for the cost of nanopowders of titanium diboride 500, diamonds $ 2 000.

A considerable amount of research devoted to the technology of aluminum production by electrolysis of fluoride cryolite-alumina melt has been carried out and many technical suggestions have been made to increase the service life of the cathode of aluminum cell by protecting it with a coating wetted by aluminum. Such coating should be chemically inert to aluminum and electrolyte melts, hard, high-electroconductive, wear-resistant, well wetted with aluminum. As a part of the functional basis of such coating, borides and carbides of titanium and zirconium can be used. The results and technical solutions described in the literature propose so far only the use of titanium diboride. Taking into consideration the mentioned above facts, it can be assumed that further improvement of the cells design and technical and economic parameters of electrolytic aluminum production process can be associated with the expansion of the range of materials used and their availability due to development of efficient production technologies that reduce their real value to less than $ 100/ kg [9].

3. Conclusion

The conducted analysis of the scientific and technical literature demonstrates the positive technological testing of zirconium carbide with standard granulometry in the composition of hard alloys, anti-emission, abrasive and corrosion-resistant coatings, which makes it possible to predict such areas of its effective application in the nanostate as the production of polycrystalline superhard materials and products thereof, the electrodeposition of composite metal coatings in electroplating, the formation of protective wettable coatings of cell cathodes in the metallurgy of aluminum.

IOP Conf. Series: Materials Science and Engineering **411** (2018) 012007 doi:10.1088/1757-899X/411/1/012007

References

[1] Kosolapova T Ya 1968 *Carbides* (Moscow: Metallurgy) p 30
[2] Panov V S and Chuvilin A M 2001 *Technology and Properties of Sintered Solid Alloys and Products Made of Them* (Moscow: MISiS) p 428
[3] Shiryaeva L S, Nozdrin I V, Gallevsky G V and Rudneva V V 2011 *Bulletin of the Mining and Metallurgical Section of the Russian Academy of Natural Sciences. Metallurgy: Proceedings* (Moscow – Novokuznetsk: SibSIU) **28** 79–91
[4] Bystrov Yu A and Winds N Z 2005 *Method for Obtaining an Intermetallic Anti-emission Coating* patent of the Russian Federation No 2261940 IPC C23C28 / 00
[5] Egan D P et al 2010 *Method for Producing Super-hard Coated Abrasive* pat. of Ireland No 2005825 IPC C23C14 / 24
[6] Gavrilin S S and Deniskin V P 2008 *Method of Precipitation of Metal Coatings on Ceramic Powder Materials* pat. the Russian Federation No 2342349 Russia IPC C04B41 / 88
[7] Rezchikova T V and Sokolovsky B M 1984 *Plasma-Chemical Processes in Nitride Technology: Proceedings* (Institute of the Academy of Sciences of the USSR: Chernogolovka) p 142–164
[8] Galevsky G V, Rudneva V V and Garbuzov A K 2015 *Izv. vuzov. Ferrous Metallurgy* **58(5)** 335–40
[9] Efimova K A, Galevskii G V, Rudneva V V and Kozyrev N A 2015 *Applied Mechanics and Materials* **778** 75–81
[10] Ivanov S Yu et al 2013 *A Method for Producing a Titanium Diboride Powder for a Wettable Cathode Material of Aluminum Cell* patent of the Russian Federation No 2498880 IPC C04B35 / 58

Thermodynamic modeling of plasma synthesis of zirconium carbide

T I Alekseeva, G V Galevsky and V V Rudneva

Siberian State Industrial University, 42 Kirova street, Novokuznetsk, 654007, Russia

E-mail: kafcmet@sibsiu.ru

Abstract. Thermodynamic analysis of equilibrium and quasi-equilibrium compositions of multicomponent systems for various technological options of hydrocarbons pyrolysis and zirconium carbide synthesis is carried out. The objects of thermodynamic modeling were the systems C-H-N, Zr-C-H-N and Zr-O-C-H-N. The formation of ZrC is possible when the vapors of zirconium and cyanogen interact at a temperature below 4000K.

1. Introduction

To date, in the production of refractory materials two basic options for technological application of high-temperature heating have been formed: the long-term heat treatment of the compacted solid batch of the given chemical and granulometric composition by the directed plasma jet and rapid synthesis in the conditions of turbulent chemically active plasma flow (so-called plasma synthesis) [1-3]. The second option along with the intensification and the possibility of realizing the process in a continuous mode ensures the production of the target products in the nanodisperse state.

To assess the technological feasibility of plasma synthesis of zirconium carbide, its thermodynamic modeling was carried out. The plasma synthesis of high-temperature compounds has the following specific features that require compulsory methodological accounting when performing its thermodynamic modeling:

- extremely limited residence time of dispersed feedstock in the evaporation zone and reaction mixture in the zone of formation of the target product in the flow of gas-coolant with a temperature of 5500-2000K, which is several micro- or milliseconds, respectively, makes it possible to assume the dominant role of the temperature factor over time and, hence, the possibility of achieving equilibrium;
- high probability of formation of the target products in the analyzed conditions during interactions in the gas phase or with its participation;
- the need to implement plasma synthesis in the systems consisting of several chemical elements introduced with the processed raw materials and plasma forming gas and generating multiphase and multicomponent systems, predetermines the unconditional application of computer technologies for carrying out thermodynamic modeling;
- the special importance of the results of thermodynamic modeling of plasma synthesis processes in the absence of a real possibility to describe their kinetic regularities and mechanism with sufficient reliability;
- the predicted nature of thermodynamic modeling results of plasma synthesis processes, which makes it possible to consider them as technological guidelines requiring experimental confirmation.

2. The modeling technique

Thermodynamic modeling of synthesis processes was carried out to predict the optimal parameters for the production of zirconium carbide (the ratio of components and temperature), determine the equilibrium parameters of the process (the degree of conversion of the raw material into carbide, the composition of gaseous and condensed products), estimate the contribution to the carbide formation of gas-phase reactions providing an effective processing of dispersed raw materials in the conditions of plasma technologies.

The objects of thermodynamic modeling were the systems C-H-N, Zr-C-H-N and Zr-O-C-H-N due to the possibility of using hydrocarbon raw materials – methane, plasma forming gas – nitrogen as zirconium-containing raw materials of zirconium and its dioxide. The thermodynamic estimation of carbide formation processes in the Zr-C-N and Zr-O-C-H-N systems is carried out for the first time.

The compositions of gaseous and condensed products necessary for analysis were calculated by the "constant method". The "constant" method is based on the joint solution of the following equations.

- law of mass action

$$n_i = \left(\frac{P}{N}\right)^{\sum_{i=1}^{q} a_{ij}-1} \prod_{\gamma=1}^{q} n_j^{a_{ij}} / K_i, \quad 1 \le i \le m\text{-}q; \tag{1}$$

- material balance

$$\sum_{i-1}^{m} a_{ij} n_i + \sum_{i-1}^{f} n_{is} = C_j, \quad 1 \le j \le q; \tag{2}$$

- the total number of moles of the gas mixture

$$\sum_{i-1}^{m} n_i = N, \tag{3}$$

- existence of a condensed phase

$$K_i^S = \frac{P}{N} n_i, \tag{4}$$

- Dalton law

$$\sum_{j=1}^{m} P_j = P, \tag{5}$$

where n_i – the number of moles of the corresponding component in the gas phase; n_{is} – the number of moles of the corresponding component in the condensed phase; N – the total number of moles of all components in the gas phase; m – the number of components in the gas phase; K_i – equilibrium constants of the dissociation reactions of complex gaseous components into atoms; a_{ij} – matrix of stoichiometric coefficients of dissociation reactions of complex substances; j – the number of components that are present in the condensed phase; C_j – the number of atoms of the element j in the system, divided by the Avogadro number; K_i^S – partial saturation pressure over the corresponding condensed phase.

It should be noted that for plasma synthesis processes changes in the conditions for the existence of condensed phases can be expected due to their formation in the form of nano-sized particles, i.e. with a great curvature of the surface for which the partial pressure over the condensed phase exceeds the saturation pressure above the flat surface. However, at the present time there is no uniform opinion on the size threshold (~ 1 or 2-10 nm) and the corresponding reference data. Earlier, the "constant" method was repeatedly successfully used by the scientific school of Professor G.V. Galevsky for thermodynamic modeling of high-temperature interactions in complex carbide and boride-forming systems Si-O-C-H-N, Si-C-H-N, Cr-O-C-B-H-N, Cr-B-H-N, etc., and therefore it was chosen as the main for thermodynamic study of carbide-forming systems in the production of zirconium carbide in the conditions of plasma flow.

The temperature dependence of interaction products compositions for the systems under study was calculated using the PLASMA computer simulation softawer, supplemented by the development of an embedded database required for analyzing the processes of obtaining oxygen-boron-nitrogen-carbon-containing compounds of refractory metals. The temperature range of 2000-6000 K was considered at a constant total pressure in the system equal to 0.1 MPa. The equilibrium constants of the reactions of elements compounds formation, borrowed from publications and reference books [4-8], were used as initial ones, in particular for ZrC – from [5], ZrN – from [6].

The components of the gas and condensed phases taken into account in the calculations are presented in table 1, and their ratios in table 2. The ratios of the initial components were determined in accordance with the stoichiometry of reactions of the target products formation and the parameters of the technological equipment (flow of the plasma-forming gas).

Table 1. Compositions of the gas and condensed phases of the systems under study.

System	Phase composition	
	Gas	Condensed
C–H–N	H, H^+, H_2, N, N_2, NH, NH_2, NH_3, C, C^+, C_2, C_3, C_4, C_5, CH, CH_2, CH_3, CH_4, C_2H, C_3H, C_4H, C_2H_2, C_2H_3, C_2H_4, CN, CN^{-1}, NCN, NCC, C_2N_2, C_4N_2, HCN, HCCN, C_3HN	C
Zr–C–H–N	H, H^+, H_2, N, N_2, NH, NH_2, NH_3, C, C^+, C_2, C_3, C_4, C_5, CH, CH_2, CH_3, CH_4, C_2H, C_3H, C_4H, C_2H_2, C_2H_3, C_2H_4, CN, CN^{-1}, NCN, NCC, C_2N_2, C_4N_2, HCN, HCCN, C_3HN, Zr, ZrH	C, Zr, ZrC, ZrN
Zr–O–C–H–N	H, H^+, H_2, N, N_2, NH, NH_2, NH_3, C, C^+, C_2, C_3, C_4, C_5, CH, CH_2, CH_3, CH_4, C_2H, C_3H, C_4H, C_2H_2, C_2H_3, C_2H_4, CN, CN^{1}, NCN, NCC, C_2N_2, C_4N_2, HCN, HCCN, C_3HN, O, O_2, NO, NO_2, H_2O, CO, CO_2, Zr, ZrO, ZrO_2	C, Zr, ZrC, ZrN, ZrO, ZrO_2

Table 2. Specified ratio of components in the systems under study.

Ratio of components, mole	Note
C:H:N=0.75:3:20	By stoichiometry for the reaction CH4 = C + 2H2 (6)
C:H:N=0.75:6:20	A 2-fold excess of hydrogen compared with the stoichiometry for reaction (6)
Zr:C:H:N = 0.25:0.25:1:20	By stoichiometry according to the reaction $Zr + CH_4 = ZrC + 2H_2$ (7)
Zr:C:H:N = 0.25:0.375:1.5:20	A 150% carbon amount compared to the stoichiometry for reaction (7)
ZrO:C:H:N = 0.25:0.50:0.75:3:20	According to the stoichiometry for the reaction $ZrO_2 + 3CH_4 = ZrC + 2CO + 6H_2$ (8)
Zr:O:C:H:N= 0.25:0.50:0.20:0.75:20	75% shortage of carbon compared with the stoichiometry for reaction (8)
Zr:O:C:H:N= 0.25:0.50:0.375:1.5:20	50% shortage of carbon compared with the stoichiometry for reaction (8)

Zr:O:C:H:N = 0.25:0.50:0.47:1.88:20	37.5% lack of carbon compared with the stoichiometry for reaction (8)
Zr:O:C:H:N = 0.25:0.50:0.56:2.25:20	25% shortage of carbon compared with the stoichiometry for reaction (8)
Zr:O:C:H:N = 0.25:0.50:0.66:2.63:20	12.5% shortage of carbon compared with the stoichiometry for reaction (8)

3. The results of modeling and their analysis

The thermodynamic analysis of the C-H-N system was performed to obtain predictive information about the possible composition of the gas phase formed as a result of methane pyrolysis and its interaction with the heat-carrier gas. The temperature dependence of equilibrium and quasi-equilibrium compositions is shown in figure 1.

The results of calculating the equilibrium compositions of the system (figure 1 a, b) revealed the presence of HCN, CN, C_3H, C^2H in the gas phase at a temperature of 2800-4200 K, with about 98% of the carbon being bound to the hydrogen cyanide. The decomposition of the hydrogen cyanide occurs at temperatures below 2800 K and leads to the formation of free carbon in the condensed state.

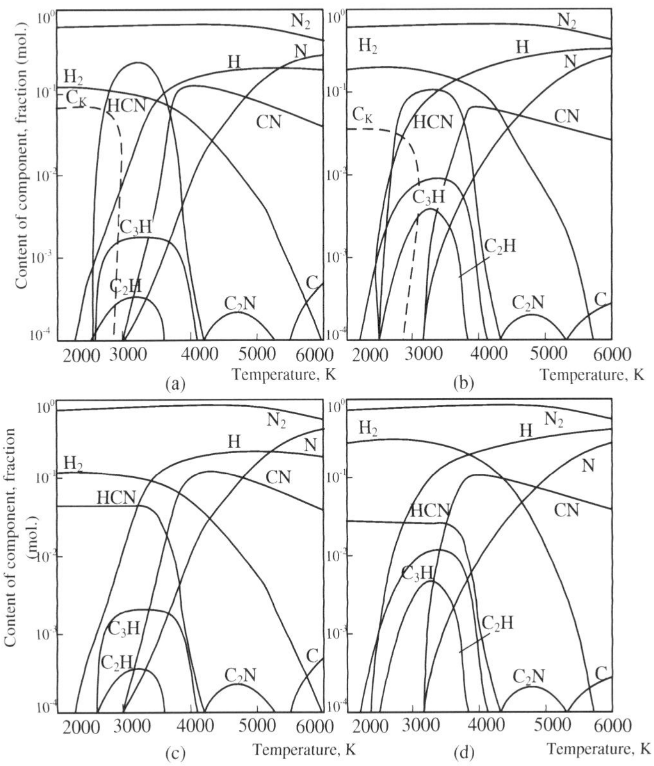

a – C:H:N=0.75:3:20; b – C:H:N=0.75:6:20
c – C:H:N=0.75:3:20; d– C:H:N=0.75:6:20

Figure 1. Equilibrium (a, b) and quasi-equilibrium (c, d) composition of the system C-H-N, depending on the ratio of components and temperature.

The calculation results of quasi-equilibrium compositions (figure 1 c, d), which differ significantly in the temperature range of the hydrogen cyanide stability (2000-2800 K), make it possible to obtain the composition of the final products and the main process indices corresponding to the experimental data [8].

The results of the thermodynamic calculations of the Zr-C-H-N system are shown in figure 2. In this system the formation of ZrC is possible by the reaction

$$Zrg+CNg=ZrCk+0.5N2t \quad (9)$$

A 100% ZrC yield is achieved with the stoichiometric ratio of components Zr:C in the temperature range 2300-4000 K. With an excess of carbidizer (ratio Zr:C=0.25:0.375) in the conditions of equilibrium, the product of the synthesis is a carbide-carbon composition containing 2.8% of free carbon. In the conditions of quasiequilibrium with an excess of carbide, cyanide is stable in the temperature range 2300-3800 K, which makes it possible to obtain carbide not contaminated with pyrolytic carbon. At a temperature below 2300 K, the interaction of carbide with nitrogen is thermodynamically possible according to reaction

$$ZrCk + N2 = ZrNk + 0.5N2g \tag{10}$$

The results of the thermodynamic calculations of the Zr-C-O-H-N system are shown in figure 3. As the ratio of Zr:O:C changes from 0.25:0.50:0.19 to 0.25:0.50:0.75 the degree of conversion of zirconium to carbide varies from 0.25 to 1. Zirconium carbide is stable in the temperature range 2300-4000 K. The formation of ZrC is possible by reactions (9) – up to 98.5% and (11) – up to 1.5%

$$ZrOg + 2CNg = ZrCk + COg + 0.5N2g \tag{11}$$

At a temperature below 2300 K, the reaction (10) becomes thermodynamically possible leading to the formation of a ZrN nitride.

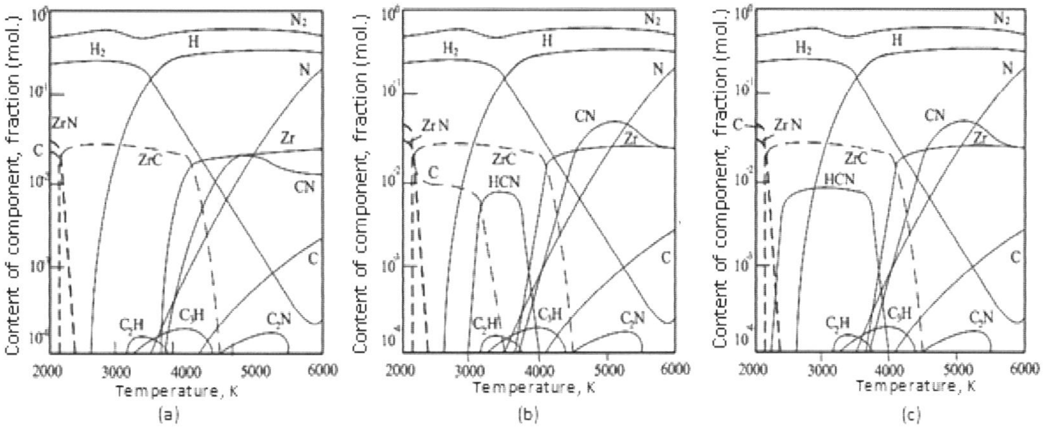

Figure 2. Results of thermodynamic calculations of the Zr-C-H-N system: a) the equilibrium compositions of the gas and condensed phases depending on the temperature at a ratio of Zr:C:H:N=0.25:0.25:1:20; b) the equilibrium compositions of the gas and condensed phases depending on the temperature at the ratio Zr:C:H:N=0.25:0.375:1.5:20; c) quasi-equilibrium compositions of the gas and condensed phases depending on the temperature at a ratio of Zr:C:H:N=0.25:0.375:1:20.

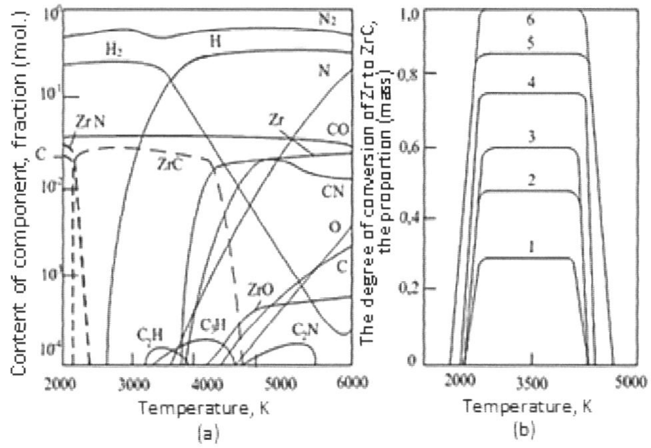

Figure 3. Results of thermodynamic calculations of the Zr-C-O-H-N system: a) equilibrium compositions of the gas and condensed phases depending on the temperature at a ratio of Zr:C:O:H:N=0.25:0.75:0.50:3:20; b) the dependence of the degree of conversion of Zr to ZrC on the ratio Zr:C=0.25:0.20 (1); 0.25:0.375 (2): 0.25:0.47 (3): 0.25:0.56 (4); 0.25:0.66 (5); 0.25:0.75 (6).

4. Conclusion

Thermodynamic study of equilibrium and quasi-equilibrium compositions of multicomponent systems for the pyrolysis of hydrocarbons and various technological options for the production of zirconium carbide has been carried out.

It is established that:

- in the C-H-N system under equilibrium conditions, 100% gasification of carbon is achieved in the temperature range of 2800-3800 K due to the formation of hydrogen cyanide and hydrocarbon radicals. Up to 98% of carbon is present in the gas phase as HCN. Quasi-equilibrium conditions that exclude the formation of condensed carbon make it possible to expand the temperature range of the thermodynamic stability of hydrogen cyanide to 2300 K;
- in the Zr-C-H-N system, the formation of zirconium carbide is possible at temperatures below 4000 K in equilibrium and quasi-equilibrium conditions with stoichiometric and excessive amounts of carbon; with an excess of carbon at a ratio Zr:C=0.25÷0.375 in the conditions of equilibrium a carbide-carbon composition containing 2.8% carbon is formed, in the conditions of quasi-equilibrium – zirconium carbide;
- in the system Zr-C-O-H-N is also stable in the temperature range 2300-4000 K; when the ratio of Zr:O:C is changed from 0.25:0.50:0.19 to 0.25:0.50:0.75, the degree of conversion of zirconium to carbide will change from 0.25 to 1;
- in both systems, the formation of zirconium carbide is thermodynamically possible through gas-phase reactions involving zirconium and cyanide vapors, i. e. according to the "steam-crystal" scheme, which allows the achievement of its high yield in the real conditions of plasmosynthesis to be predicted.

References

[1] Andrievsky R A and Ragulya A V 2005 *Nanostructural Materials* (Moscow: Akademiya) p 192
[2] Hartmann U 2008 *Charm of Nanotechnology* ed L N Patrikeeva (Moscow: BINOM, Laboratory of Knowledge) p 173
[3] Rudskoy A I 2007 *Nanotechnologies in Metallurgy* (St. Petersburg: Science) p 186
[4] Avvakumov E G 1991 *Mechanical Methods of Activation of Chemical Processes* (Novosibirsk: Science) p 264
[5] Wiks K E and Blok F E 1965 *Thermodynamic Properties of 65 Elements and Their Oxides, Halogenides, Carbides and Nnitrides* (Moscow: Metallurgy) p 126
[6] Barin J and Knacbe O *Thermochemical properties of Inorganic Substances* 1973 (New-York: Academic press) p 647
[7] JANAF *Thermochemical tables. – Wash.Gov.print. off, 1966 – 1975* https://flagma.ru/karbid-cirkoniya-so233713-1.html
[8] Glushko V P 1971 *Thermodynamic and Thermophysical Properties of Combustion Products: Reference Book: in 3 volumes* (Moscow: VINITI)
[9] Suris A L 1985 *Thermodynamics of High-temperature Processes* (Moscow: Metallurgy) p 568

IOP Conf. Series: Materials Science and Engineering **411** (2018) 012009 doi:10.1088/1757-899X/411/1/012009

Reduction processing of technogenic microsilica with the use of a brown coal semi-coke

A E Anikin[1], G V Galevsky[1], V V Rudneva[1] and S G Galevsky[2]

[1]Siberian State Industrial University, 42 Kirova street, Novokuznetsk, 654007, Russia
[2]St. Petersburg Mining University, V.O., 21 Line, 2, St. Petersburg, 199106, Russia

E-mail: kafcmet@sibsiu.ru

Abstract. Thermodynamics is investigated and optimal temperature-time regimes of carbidization of briquetted charge "microsilica – brown coal semi-coke" are determined. In Si-O-C and Si-O-C-H systems the carbide formation process is dominant: with the stoichiometric composition of the charge, the maximum content of silicon carbide reduction products is achieved at 1700 K. The complete carbidization of the charge in the conditions of heat treatment is achieved at a temperature of 1923 - 1973 K with a duration of 20 - 15 minutes. The technological expediency, conditions and chemical enrichment indices of carbidization products are established. The phase, chemical, granulometric compositions and morphology of silicon carbide particles are studied. It was found that carbidization forms a carbide of the cubic structure (β-SiC). The content of carbide after enrichment is more than 90%. Silicon carbide is obtained in the form of a micropowder with a specific surface area of 8000 - 9000 m^2/kg from irregularly shaped particles with a size range of 0.2-1.0 μm.

1. Introduction

The production of silicon carbide – one of the most important in the structure of modern metallurgy is about 800 thousand tonnes per year. The largest areas of silicon carbide application are metallurgy (about 45% of global demand), production of abrasives (up to 30%) and refractories (up to 25%). Specificity of thermomechanical, electrophysical, physical and chemical properties of silicon carbide predetermines the following directions of its traditional and perspective application: materials on bundles; structural and functional ceramics; composite materials and coatings; surface and volumetric modification of materials.

The current production of silicon carbide is almost entirely based on the carbon-thermal method of Acheson. According to this method the production of silicon carbide is carried out by a periodic block process in electric resistance furnaces with a capacity of 4.5 - 5.5 MVA at a temperature of 2500 - 2900 K. The obtained silicon carbide is crystals that have grown into so-called druses (pieces) which, after disassembling the electric furnace are subject to fragmentation, enrichment and sifting by fractions. The products of abrasive plants are carbide grains of various sizes: abrasive grit (size 2500 - 160 μm), grinding powders (size 160 - 28 μm), micropowders (size 63 - 10 μm), fine micropowders (size 10 - 1 μm). Currently there are three types of silicon carbide: black, green and electrotechnical.

For a long time the demands of metallurgy, refractory and ceramic production in silicon carbide were satified by abrasive grinding materials, which increased the cost of refractories and ceramics, and in metallurgy in some cases made the use of silicon carbide technologically and economically impractical. The situation was aggravated by the constant increase in electricity prices and requirement strengthening of national environmental legislation. In this regard the main producers of silicon carbide – companies "Saint – Gobian" (France), "Exolon – ESK" (USA – Germany), "Carborundum Co" (USA), "H. Starck Co" (Germany), etc., in the 80's developed and mastered the technological processes of the production of so-called non-grinded silicon carbide, obtained, as a rule, from highly dispersed batches by furnace synthesis in the form of micropowders with subsequent chemical enrichment. Research in this direction was carried out in Russia as well and ended with positive scientific and applied results: decrease in the synthesis temperature from 2500 - 2900 to 1900 K, i.e. by 600 – 1000°, duration of synthesis from 26 - 29 to 1 hour, electricity consumption in 2 - 4 times; achievement of silicon carbide content in synthesis products to 90.0% by weight; production of silicon carbide in the form of micropowders of fineness 1 - 5 μm; possibility of application of non-commercial and technogenic raw materials for the production of silicon carbide, for example, low-quality sands, shungite rocks, microsilica, coke dusts and fines, semi-coke; possibility of realization of non-stop production of silicon carbide, for example, using tunnel, tubular rotating, shaft furnaces. In this regard, it was decided to perform implementation activities at a number of enterprises of the Ministry of Industry and Trade. However, the socio-economic changes that took place in the country at the turn of the 1980s-1990s did not allow the plans to be realized.

Consequently, the scientific and technological justification and implementation of the production of domestic non-grinded silicon carbide remains an urgent task of great practical importance. In connection with this, the purpose of this work was to study the processes of carbidization of technogenic microsilica by a semi-coke from brown coals of the Berezovsky deposit of the Kansk-Achinsk basin, the technological properties of which were investigated earlier by the authors and described in detail in [1, 2]. The technological and economic feasibility of using semi-coke as a carbon reducing agent was confirmed in [3 - 6].

2. Thermodynamic modeling of carbide formation in Si-O-C, Si-O-C-H systems

Thermodynamic modeling was carried out to predict the optimal parameters for the production of silicon carbide (the ratio of components, temperature and pressure in the system), to determine the equilibrium parameters of the process (the degree of conversion of the raw material into carbide, the compositions of gaseous and condensed products), to estimate the contribution to the carbide formation of individual reactions providing conditions under investigation, efficient processing of raw materials. In connection with the use of a brown coal semi-coke with a volatile content up to 10% as a reducing agent and carbidizer in the process, the objects of investigation were Si-O-C, Si-O-C-H systems.

The equilibrium compositions of the Si-O-C and Si-O-C-H systems were calculated by the "constant" method. Calculations were carried out using the computer simulation software for high-temperature complex chemical equilibria "PLASMA" (IHTTiM SB RAS), which has a built-in database of interaction products for oxide, boride, carbide, and nitride-forming systems. In the calculations, the temperature range 800-2000 K was considered at different pressures in the system: from 0.1 to 0.0001 MPa. The calculations were carried out at intervals of 100°. The initial data were the reaction equilibrium constants of compounds formation from the elements. The calculations took into account the possibility of formation in the gas phase in the Si-O- C system of Si, Si^+, Si^-, Si_2, Si_3, O, O_2, C, C^+, C_2, C_3, C_4, C_5, SiO, SiC, SiC_2, Si_3C, CO, CO_2, in the system Si-O-C-H of Si, Si^+, Si^-, Si_2, Si_3, O, O_2, C, C^+, C_2, C_3, C_4, C_5, H, H^+, H_2, CH_2, CH_3, CH_4, C_2H, C_3H, C_4H, C_2H_2, C_2H_4, SiO, SiC, SiC_2, Si_3C, CO, CO_2, in the condensed phase in both systems – Si, SiO_2, C, SiC systems. The choice in the investigated systems of the ratios of the initial components was determined by the stoichiometry of the formation reaction of silicon carbide and the composition of the charge materials. The fruitfulness of this approach in studying the processes of carbide formation was confirmed earlier in [7, 8].

The results of calculations of the equilibrium compositions of the Si-O-C and Si-O-C-H systems for the total pressure 0.1 MPa are given in figures 1 and 2, and for pressure 0.01 and 0.0001 MPa – in tables 1 and 2. It can be seen that in both systems the carbide formation process is dominant

$$SiO_{2(s)} + 3C_{(s)} = SiC_{(s)} + 2CO_{(g)}. \quad (1)$$

With the stoichiometric composition of the charge, the maximum content of silicon carbide in products is achieved at 1700 K, and with 10% shortage of carbon at 1900 K. Introduction of hydrogen into the system does not actually affect the carbide formation process, which is caused by an extremely low content (less than 0.001 mole) of hydrocarbons and hydrocarbon radicals in the temperature range of the carbide formation. In the Si-O-C system the equilibrium degree of silicon conversion to carbide does not exceed 0.97%, which corresponds to a content of silicon monoxide in the gas phase of 0.02 mol. As a consequence, it is impossible to obtain a single-phase silicon carbide with no free carbon from the charge of the stoichiometric composition (SiO_2 + 3C) (table 2).

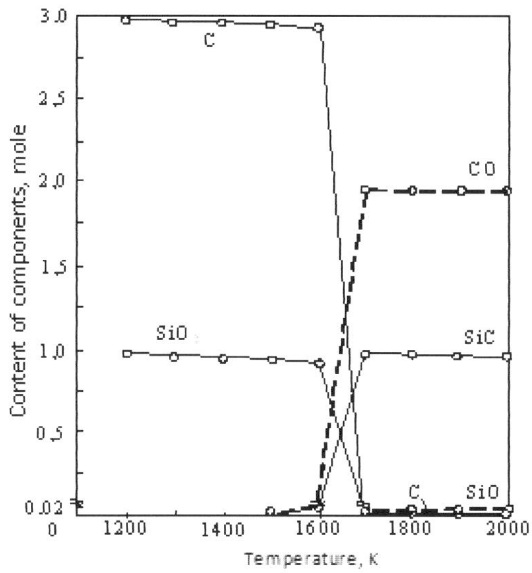

Figure 1. Dependence of the equilibrium composition of the Si-O-C system on the temperature with ratio Si:O:C = 1:2:3 and pressure P = 0.1 MPa (— condensed, – – – gaseous products).

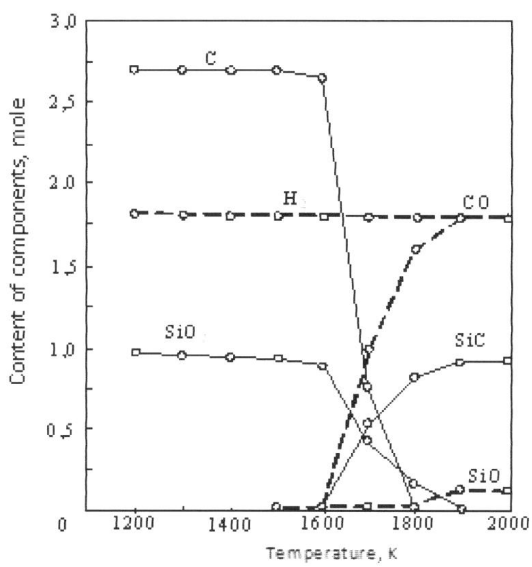

Figure 2. Dependence of the equilibrium compositions of the Si-O-C-H system on the temperature with ratio Si:O:C:H = 1:2:2.7:3.6 and pressure P = 0.1 MPa (— condensed, – – – gaseous products).

This can be avoided by using the charge with some (~ 10%) shortage of carbon – a reducing agent (table 2). In this case, the area of coexistence of silicon carbide and excessive silica appears in the system. And the temperature range of coexistence of these phases is wider the higher the pressure in the system. Above these temperatures, silicon carbide itself is a reducing agent for silica, i.e. the following reaction becomes possible

$$2SiO_{2(s)} + SiC_{(s)} = 2SiO_{(g)} + CO_{(g)}. \quad (2)$$

As a result of this reaction, silicon carbide is purified from excessive silica. It can be particularly effective to conduct it in the vacuum.

Table 1. Equilibrium concentrations (in %) of condensed reaction products in the Si-O-C system with component ratio Si:O:C = 1:2:3.

T, K	P = 0.01 MPa			P = 0.0001 MPa		
	SiO_2	C	SiC	SiO_2	C	SiC
800	62.5	37.5	0	62.5	37.5	0
900	62.5	37.5	0	62.5	37.5	0.02
1000	62.5	37.5	0.01	62.2	37.3	0.52
1100	62.4	37.4	0.2	0	0.3	99.71
1200	61.0	36.7	2.3	0	0.4	99.60
1300	0	0.2	99.8	0	0.6	99.40
1400	0	0.2	99.8	0	1.0	99.0
1500	0	0.3	99.7	0	1.3	98.7
1600	0	0.4	99.6	0	2.0	98.0
1700	0	0.6	99.4	0	2.8	97.2
1800	0	0.9	99.1	0	3.7	96.3

Table 2. Equilibrium concentrations (in %) of the condensed reaction products in the Si-O-C-H system with the components ratio Si:O:C:H 1:2:2.7:3.6.

T, K	P = 0.01 MPa			P = 0.0001 MPa		
	SiO_2	C	SiC	SiO_2	C	SiC
900	64.94	35.06	0	64.94	35.06	0
1000	64.94	35.06	0	64.94	35.06	0.001
1100	64.94	35.06	0.002	64.84	35.00	0.16
1200	64.74	36.04	0.04	62.54	33.86	3.60
1300	64.54	35.94	0.30	12.47	0	87.53
1400	62.84	34.53	2.63	8.95	0	91.05
1500	51.27	26.61	22.12	0	0	100.00
1600	11.35	0	88.65	0	0	100.00
1700	0	0	100.00	0	0	100.00
1800	0	0	100.00	0	0	100.00
1900	0	0	100.00	0	0	100.00

3. Method for studying microsilica carbidization

The study of microsilica carbidization was carried out by the discrete method, including the thermal treatment of the briquetted charge at the given temperatures and durations and hardening of the carbidization products followed by their chemical analysis.

For studies microsilica formed in the production of silicon (MC-Cr) and high-silica ferrosilicon (MK-FS) were used. They contain 93.92 and 93.00 wt% of silicon dioxide, respectively, with a specific surface area of 25,000 m^2/kg, as well as a semi-coke of brown coal (BSC) from the

Berezovsky deposit of the Kansk-Achinsk basin, containing %: carbon – 81.9; volatile – 9.5; ash – 8.6, with a specific surface area of 264000 m^2/kg. In all cases, the stoichiometric charge was used to produce the carbide composition.

Examination of the temperature and time conditions of carbidization of briquettes with microsilica of MC-Cr, MC-FS-BSC was carried out at temperatures 1883, 1923, 1973 K with a heat treatment time of 5, 10, 15, 20, 25, 30 min.

Heat treatment of the charge was carried out in the electric resistance furnace with a cylindrical graphite heater with an internal diameter of 0.075 m (Tamman furnace). Temperature control in the working space of the furnace was carried out using a tungsten-rhenium thermocouple BP 5/20. Cylindrical briquettes weighing approximately $2.5 \cdot 10^{-3}$ kg and an outer diameter of 0.015 m were placed in graphite crucibles. Three crucibles were simultaneously installed in the working space of the heated electric furnace, were heated in it for 3 – 4 minutes to the required temperature, and then kept in it for a specified time period – from 5 to 90 minutes. At the end of the specified time, the graphite crucibles were removed from the electric furnace, installed in corundum glasses, which were tightly closed with stoppers and placed for 10 minutes in the cold water for rapid cooling of the carbidization products. This made it possible to prevent the oxidation of carbidisation products by oxygen and atmospheric moisture. After cooling, the products of carbidization were extracted without losses from crucibles, weighed, placed in platinum tanks, poured with a solution of a mixture of nitric acid with a concentration of 7.5 – 12.5 M and hydrofluoric concentration 2.0 – 3 M, taken in a volume ratio of 1: (3 ÷ 5), and boiled for 2.5 - 3.0 hours.

The treatment of silicon carbide with a mixture of nitric and hydrofluoric acids ensures the removal of impurities of aluminum, magnesium, calcium, iron, manganese, nickel, chromium, copper, sodium, potassium, silicon and their oxides and is recommended for silicon carbide powders of fineness from 5 to 200 μm. In the presence of nitric acid, metal impurities and their oxides form soluble nitrate complexes, while when treated with hydrofluoric acid, insoluble fluorides can be formed. After cooling, the insoluble residue, which is silicon carbide with an admixture of free carbon, was filtered out on a paper filter, washed on a filter with distilled water and acetone, dried to constant weight and weighed. After that, the undissolved residue was subjected to oxidative firing for refining from free carbon at a temperature of 1073 - 1123 K for 1.5-2.0 h. After the loss of mass, the content of free carbon and silicon carbide was calculated in the carbidization products. According to the average content of silicon carbide for three crucibles, its yield was calculated as the ratio of the mass of the practically obtained carbide to the theoretically possible ratio.

4. Experimental studies of carbidization of MC – BSC charges and discussion of the results

The results of the studies are presented in figures 3 and 4. In both cases the carbide formation process starts from the first minutes and by the 5th minute the carbide yield at the carbidization temperatures of 1873, 1923, 1973 K is 5.23 – 7.21, 17.06 – 22 , 30, 20.20 – 23.51%. The process of carbidization is completed at a temperature of 1973 K by the 15th minute, 1923 K – by the 20th minute. The content of carbide in the products of thermal treatment of charge is 92.50 for MC-FS and 94.90% for MC-Cr. Such temperature-time conditions for carbide formation are confirmed by the results of a study of the phase composition of heat treatment products. Thus, it was found that in both cases by the 10th minute the dominant phase is silicon carbide of cubic structure (β-SiC), partially changing to hexagonal one (α-SiCII) at a temperature of 1923 – 1973 K and a heat treatment time of 50 – 90 min. During the carbidization of the charge microsilica MC-FS-BSC, α-iron accompanies it. In all the samples studied, there is also a vitreous phase, formed, apparently, by silicates of calcium, magnesium and iron.

As a result of chemical enrichment, content of SiC in carbide exceeds 90%, i.e. level, regulated for abrasive micropowders with a grain size of 1 – 2 μm [9]. The efficiency of enrichment from impurities of oxides and iron is rather high and amounts to 87 – 95%. Silicon carbide is characterized by an increased silica content of more than 7%, which makes it possible to consider it as a promising material for the production of silicon carbide refractories based on silica, containing usually, %: SiC – 84 – 89, SiO_2 – 6 – 12 [10].

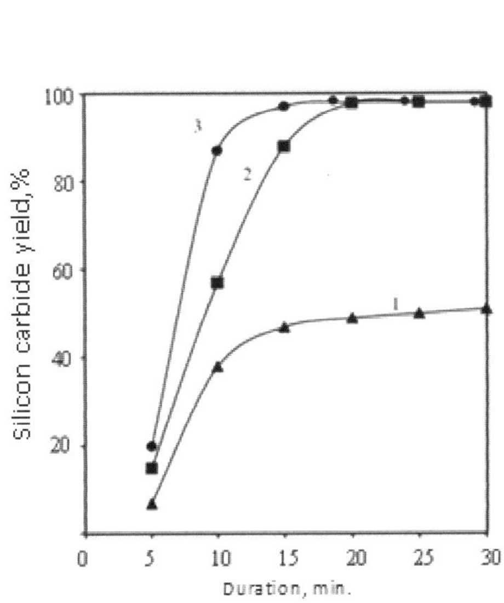

Figure 3. Dependence of the yield of silicon carbide on the temperature and duration of heat treatment for the charge of microsilica MK-FS-BOD (1 – 1873 K (Δ) = (7.21 ÷ 53.13) ± (0.20 ÷ 1.13), 2 - 1923 K (□) = (17.06 ÷ 96.82) ± (0.55 ÷ 1.35), 3 – 1973 K (○) = (20.20 ÷ 97.01) ± (0.61 ÷ 1.45)).

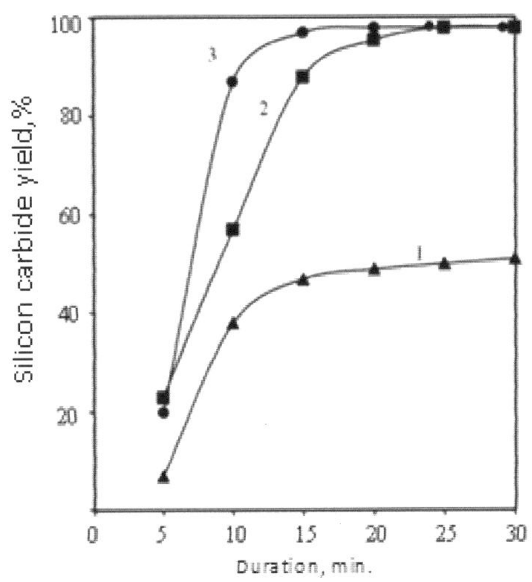

Figure 4. Dependence of the yield of silicon carbide on the temperature and duration of heat treatment for the charge of microsilica MC-Cr-BSC (1 – 1873 K: (Δ) = (5.23 ÷ 46.26) ± (0.19 ÷ 1.11); 2 – 1923 K: (□) = (22.30 ÷ 96.78) ± 0.59 ÷ 1.33); 3 – 1973 K: (○) = (23.51 ÷ 97.62) ± (0.67 ÷ 1.38)).

Optimal conditions for obtaining and enriching silicon carbide and the range of variation of its main characteristics for a series of five parallel experiments are given in Table 3. It can be seen that the investigated technological options provide silicon carbide with reproducible phase, chemical and granulometric compositions and can be considered as a technological basis for the design of industrial production of silicon-free silicon carbide from the briquetted highly dispersed charge "microsilica – brown coal semi-coke".

Table 3. Optimal conditions for obtaining and enriching silicon carbide and its characteristics.

Conditions for obtaining and enriching silicon carbide and its main characteristics	SiC (MC–FS+BSC)	SiC (MC–Cr + BSC)
Charge composition,%:		
Microsilica (MC)	58.80	58.04
Brown coal semi-coke (BSC)	42.20	41.96
Specific surface area of charge components, m^2/kg		
microsilica	20000 – 22000	
brown coal semi-coke	255000 – 265000	
The temperature of charge carbidization, K	1923 – 1973	
Duration of charge carbidization, min	25 – 20	
Chemical enrichment of carbidization products		
concentration of hydrochloric acid solution,%	35	
the ratio S:F	1:2	
temperature, K	383	
duration, hour	3	
Phase composition	β - SiC	

Continuation of Table 3.

Conditions for obtaining and enriching silicon carbide and its main characteristics	SiC (MC–FS+BSC)	SiC (MC–Cr + BSC)
Chemical composition, %:		
SiC	90.42 - 90.86	90,94 - 91,18
Al_2O_3	0.10 -0.25	0,10 -0,15
CaO	0.32 - 0.53	0,37 - 0,49
MgO	0.12 - 0.18	0,03 - 0,10
Fe	0.08 - 0.12	0,05 - 0,11
SiO_2	7.32 - 7.94	6,97 - 7,13
C free	0.41 - 0.54	0,63 - 0,81
Si free	0.13 - 0.17	0,23 - 0,27
The yield of silicon carbide, %	96.84 - 97.12	97,31 - 97,63
The size and morphology of silicon carbide particles:		
specific surface area, m^2/kg	8000 - 9000	
size range, μm	0.2 - 1.0	
form	irregular	

5. Conclusion

Thermodynamic and experimental studies of the processes of reducing processing of technogenic microsilica on silicon carbide with the use of brown coal semi-coal were carried out.

It was established that in both Si-O-C and Si-O-C-H systems chosen for thermodynamic analysis, the carbide formation process is dominant. With the stoichiometric composition of the charge the maximum content in the reduction products of silicon carbide is reached at 1700 K, and with 10% shortage of carbon – 1900 K. The second composition of the charge is technologically more preferable, since it ensures the production of carbide that does not contain impurities.

The carbidization of the briquetted charges with microsilica (MC-FS and MC-Cr) – brown coal semi-coke was carried out at temperatures 1883, 1923, 1973 K and duration 5 – 30 min. Optimum temperature and time conditions of carbidization are determined: temperature 1923 – 1973 K at a duration of 20 – 15 minutes.

The expediency and the optimal conditions and indices of chemical enrichment of carbidization products were determined: exposure to hydrochloric acid with the concentration at least 35% at the temperature 353 K, ratio of S:F = 1:2, duration of 3 hours, carbide content in enrichment products of more than 90%, removal of oxides and iron impurities by 87 – 95%.

The phase, chemical, granulometric compositions and morphology of silicon carbide particles of both technological production variants were studied. It was found that during the carbidisation the carbide of the cubic structure (β-SiC) is formed, partially resrtuctured at a temperature of 1923 – 1973 K and the heat treatment time 50 – 90 min into a hexagonal one (α-SiC_{II}). The content of carbide after enrichment is 90.42 – 90.86% with the use of microsilica MC-FS and 90.94 – 91.18% using microsilica MC-Cr. In both cases, silicon carbide is obtained in the form of a micropowder with a specific surface of 8000 – 9000 m^2/kg, particle size range 0.2 – 1.0 μm, having an irregular shape.

References

[1] Anikin A E and Galevskii G V 2014 *Bulletin of the Siberian State Industrial University* **3** 52–59
[2] Anikin A E and Galievsky G V 2014 XVIII *All-Russian Scientific and Practical Conference in Metallurgy: Technologies, Management, Innovations, Quality* (Novokuznetsk: SibSIU) pp 115-120
[3] Rudyka V I and Malina V P 2014 *Coke and Chemistry* **7** 15–25
[4] Romanov S M 2009 *Coal* **1** 15–17
[5] Strakhov V M 2008 *Coke and Chemistry* **9** 44–49
[6] Strakhov V M 2010 *Coke and Chemistry* **8** 29–33
[7] Nozdrin I V, Galevskiy G V, Shiryaeva L S and Terentyeva M A 2011 *Izv. vuzov. Ferrous Metallurgy* **10** 12–17
[8] Shiryaeva L S, Nozdrin I V, Galievsky G V and Rudneva V V 2012 *Bulletin of the Mining and Metallurgical Section of the Russian Academy of Natural Sciences. Metallurgy* **29** 94–101
[9] Garshin A P, Shumyacher V M and Pushkarev O I 2008 *Abrasives and Structural Materials Based on Silicon Carbide* (Volgograd: VolgGASU) p 189
[10] Sorlye M and Oya Kh A 2013 *Cathodes of an Aluminum Eelectrolysis Cell* ed P V Polyakov (Krasnoyarsk: Verso) p 720

IOP Conf. Series: Materials Science and Engineering **411** (2018) 012010 doi:10.1088/1757-899X/411/1/012010

Features of thermal expansion of special-purpose aluminum alloys after treatment of melt and heat treatment

V K Afanasyev, M V Popova, M A Malyukh and A N Prudnikov

Siberian State Industrial University, 42 Kirova street, Novokuznetsk, 654007, Russia

E-mail: starostina_ma1976@ mail.ru

Abstract. A combination of corrosion resistance, low specific gravity, and high dimensional stability over a wide temperature range is necessary for alloys for special purpose. The dimensional stability is determined, first of all, by the low temperature coefficient of linear expansion (TCLE). The paper presents the results of studies on the effect of melt treatment on TCLE of aluminum alloys for special purpose on the basis of the system Al-(11÷40)%Si. It is shown that melt treatment with an aqueous solution of $CuSO_4$ and modification of melt with a mixture of ($CaCO_3$ ·$MgCO_3$) leads to the decrease in TCLE for all alloys under consideration in the low-temperature test interval t_{test} = 50÷150 °C. Treatment of the melt by $CO(NH_2)_2$ vapours leads to the decrease of TCLE of alloys Al-40%Si in the whole temperature range from 50 to 450 °C. For high silicon alloys it is established that the decrease in TCLE occurs in the interval t_{test} = 350÷450°C after additional heat treatment

1. Introduction

The special purpose of aluminum alloys is determined by the requirement for a specific set of mechanical, physical, physical and chemical and technological properties necessary for the operation of products in the specified conditions, for example, at low or elevated temperatures, in devices and units in the aerospace industry.

For special purpose alloys a combination of corrosion resistance, low specific gravity and high dimensional stability over a wide temperature range is necessary, since the lightening of structures and devices put into orbit facilitates significant fuel savings [1]. The latter requirement is ensured by the minimum temperature coefficient of linear expansion (hereinafter – TCLE) and minimization of structural transformations in time. We have established that the effect on the melt with the help of various physical, mechanical and chemical methods alters the ratio of hydrogen, nitrogen and oxygen in the melt, which during crystallization will control the formation of the solid metal properties [2]. Thermal and mechanical characteristics of aluminum and its alloys are the subject of many research papers. In the opinion of the authors [3], alloys of Al-Mg-Si (aluminum alloys of the 6xxx series) possess properties attractive for their use in the aviation industry. These properties include medium and high strength, good corrosion resistance, improved weldability, good strength, and reduced residual stresses in large plates and sheet products. The mechanical properties of these alloys, fatigue strength, damage resistance and corrosion resistance were investigated.

When developing new materials with improved physical properties, special attention is paid to the hypereutectic alloys of the Al-Si system. However, with conventional methods of casting their structure is characterized by the presence of a coarse eutectic and large crystals of the siliceous phase. For the purpose of grinding primary silicon and eutectic crystals and providing the necessary complex of mechanical and technological properties, such alloys are subjected to modification, refining or progressive methods of heat treatment. A large number of methods for modifying silumin have been developed: the use of sodium-containing mixtures, the introduction of modifiers in the form of salts and ligatures, phosphorus-containing compounds, oxygen-containing reagents, additives and additives based on highly disperse components [4, 5]. It should be noted that the listed technologies, along with the advantages (convenience of modifiers introduction), have a significant drawback – the presence of harmful fluoride and chloride emissions into the atmosphere. Based on systematically conducted studies, the authors developed a method for treating a melt of hypereutectic silumin with a mixture of calcium and magnesium carbonates, which significantly improves their technological properties [6]. It is known that combined modifying agents, consisting of 2 or more elements, are superior in efficiency to each element separately. Calcium and magnesium are eutectic modifiers in hypereutectic silumins, and their introduction in the form of salts facilitates the crushing of primary silicon crystals.

In the development of special-purpose alloys, along with other requirements, they must have stable performance properties in the range of operating temperatures and pressures [7]. The use of castings is limited due to high natural gas saturation, which is undesirable in the conditions of vacuum. The most effective ways of changing the gas content are melt processing, selection of crystallization conditions and heat treatment [8]. It should be noted that the composition of the external medium has a significant influence on the formation of the structure and physicomechanical properties of aluminum alloys during heat treatment [9, 10]. In [10] we showed that heat treatment in media with an increased content of hydrogen and nitrogen accelerates the course of phase transformations in aluminum alloys, since it activates the diffusion of hydrogen in them. There is a large number of scientific publications confirming the active participation of hydrogen in the formation of the structure and properties of aluminum alloys [11-13]. In this regard, the purpose of this paper is to investigate the possibility of obtaining stable values of aluminum alloys TCLE *Al-(11÷40)%Si* in the working temperature range by treating the melt with compounds with an increased content of hydrogen, oxygen and nitrogen.

2. Materials and methods of research

Double alloys Al-(11÷40)%Si were chosen as the object of investigation. Charge aluminum was melted, silicon was introduced into it in an amount of 11, 20, 30, 40%, after its complete dissolution, the melt was treated with a mixture of calcium carbonate and magnesium carbonate, taken in the equal proportion in an amount of 1-7% of the mass of the melt. The treatment was carried out for 3-15 minutes at a temperature of 50-250 °C exceeding the melting point of aluminum (710-910 °C). To show the modification efficiency, the melt was simultaneously blown with vapours of aqueous solution of copper sulfate $CuSO_4$ in an amount of 0.1-0.2 wt% of the melt for 5-15 minutes at 800-1100 °C [14]. For comparison, the melt was blown with $CO(NH_2)_2$ carbamide vapours at 700...900 °C for 3...6 min [15]. Crystallization of alloys was carried out in the aluminum block mould at casting temperatures of 730...750 °C. From the obtained samples, samples were prepared for the dilatometric study. TCLE was determined using a differential optical photoregulatory dilatometer of Shevenar system, the error of determination was $\pm 0.1 \cdot 10^{-6}$ deg^{-1}.

To stabilize the properties of the investigated alloys Al-(11÷40)%Si, heat treatment was carried out, which consisted in heating for 10 hours at temperatures of 100, 150 and 200 °C followed by air cooling. The choice of treatment temperatures is due to the intervals of the most active diffusion redistribution of hydrogen in the metal, as well as its interaction with nitrogen, both intrinsic and introduced during melt processing [8].

3. Results and discussion

In the *low-temperature range* of the tests, a decrease in TCLE of alloys Al-20÷40%Si pretreated with an aqueous solution of copper sulphate in the smelting process was revealed, that can be seen in figure 1. Thus, the average TCLE of the ordinary Al-20% Si alloy in the range 50-150 °C has a value of $\overline{\alpha}_{50\text{-}150}$ = $18 \cdot 10^{-6}$ deg^{-1}, while the alloy obtained after treatment by copper sulfate has the values $\overline{\alpha}_{50\text{-}150}$ = $16.6 \cdot 10^{-}$

^{6} deg^{-1} . For the alloy Al-30%Si, there is a slight decrease in TCLE in the low-temperature interval. The greatest decrease in TCLE values is observed in alloy Al-40%Si. In this case, after the melt treatment, the average TCLE is $\bar{\alpha}_{50\text{-}150} = 11.3 \cdot 10^{-6}$ deg^{-1}, while the values of $\bar{\alpha}_{50\text{-}150}$ alloy of conventional preparation are $13.2 \cdot 10^{-6}$ deg^{-1}.

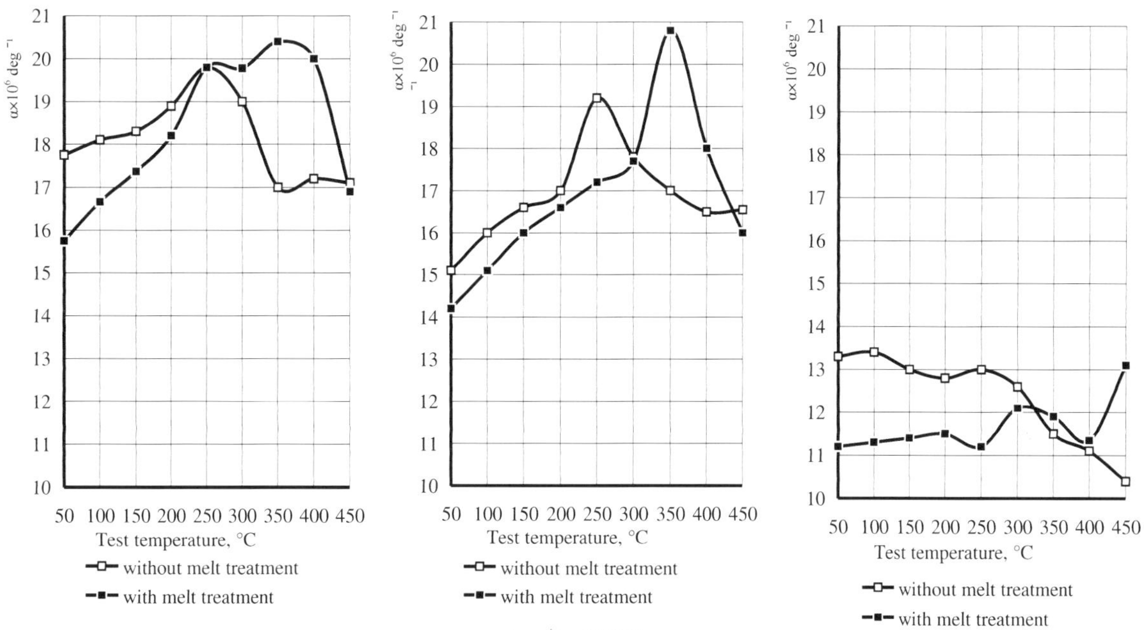

Figure 1. Influence of melt processing by vapours of aqueous solution $CuSO_4$ on linear expansion of alloys Al-Si.

This decrease may be due to the increase in the copper content in the alloys composition and the occurrence of aging processes in the range 200-300 °C upon cooling from the crystallization temperatures.

In the high-temperature range of the test, the values of TCLE are increased due to the treatment of the melt, which is not determinative, since the devices are not operated at such temperatures. The chosen mode of blowing the melt is optimal, because with a shorter time of blowing, the TCLE decreases insignificantly, while with a longer period of time the amount of slag increases sharply, which reduces the yield of the usable metal.

It has also been established that blowing of the melt slightly increases the microhardness of the substrate, the amount of siliceous phase and gas saturation [14].

The data on the results of the effect of melt treatment by carbamide on linear expansion of alloys Al-11÷40%Si are shown in figure 2. It can be seen that the chosen preparation method significantly reduces the TCLE of alloys Al-11% Si in the entire temperature range of the test. The greatest decrease is observed in the low-temperature interval 50÷150 °C from the values $19.0 \cdot 10^{-6}$ deg^{-1} to $15.7 \cdot 10^{-6}$ deg^{-1}. Processing of the melt of high-silicon silumins (20, 30, 40 % Si) leads to the appearance of an anomaly of linear expansion consisting in a sharp increase in TCLE at t_{vap} = 300 °C, which is characteristic of this group of alloys and hypothetically associated with the decomposition of the siliceous phase and the release of the interstitial elements in the basis metal. However, if we consider the low-temperature interval which is of interest to us, i.e. the operating temperature of the instrument equipment, then there is a uniform decrease in the TCLE for all processed alloys. It is known that silicon, which reduces TCLE aluminum to the greatest extent, with the introduction up to

12%, does not give the required values of thermal expansion, and the increase in its content beyond the eutectic composition leads to embrittlement of the alloy and, consequently, loss of producibility and serviceability. One of the reserves for fragility reduction is heat treatment.

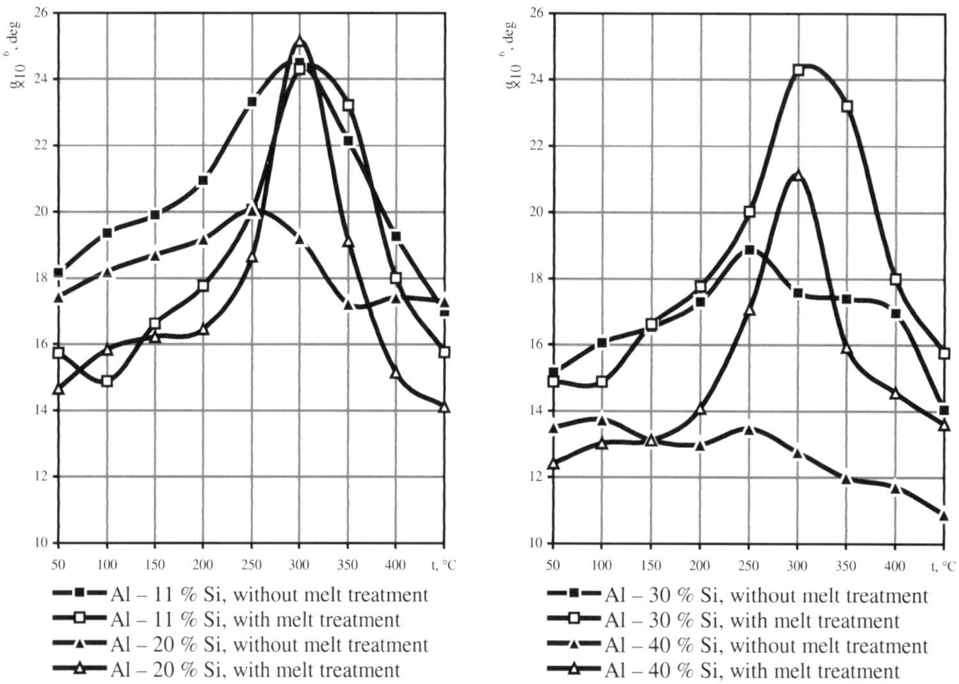

Figure 2. Effect of melt processing by vapors $CO(NH_2)_2$ on linear expansion of Al-Si alloys.

The effect of heat treatment, which consisted in heating from 100 to 200 °C for 10 h and cooling in the air, is manifested in the decrease in TCLE of the investigated alloys (table 1). Thus, for alloys Al-11%Si, the decrease in TCLE is observed in the interval t = 300...450 °C after heating at any of the selected temperatures, but the heat treatment at 150 °C is most effective. However, it should be noted that re-heating with the same parameters does not lead to a further decrease in TCLE. Heating of high-silicon silumin reduces the anomaly of linear expansion, which manifests itself at t = 300 °C, the more effective the higher the heating temperature.

After heating at 200 °C, this anomaly is completely eliminated, which is probably due to the formation of interstitial elements for the intrinsic compounds with a small TCLE. An increase in the heating time at 200 °C to 20 h does not give any additional decrease in TCLE, which indicates the sufficiency of the chosen treatment time for the diffusion processes to be completed.

Table 1. Influence of heating (10 hours, air) on the linear expansion of high-silicon alloys Al – Si (melt treatment by vapours $CO(NH_2)_2$).

Heating mode	$\alpha \times 10^6$ deg^{-1}, at temperatures of the test, °C								
	50	100	150	200	250	300	350	400	450
					Al-11%Si				
-	15.7	14.9	16.6	17.8	20.1	24.3	23.2	18.0	15.8
100 °C	15.2	17.7	18.7	19.4	21.5	25.6	18.1	15.4	13.4
150 °C	16.1	16.9	17.2	18.3	20.5	20.4	14.7	12.4	11.7

200 °C	15.7	17.1	17.3	17.8	19.0	20.0	19.5	16.7	13.1
Al-20%Si									
-	14.6	15.8	16.2	16.5	18.7	25.2	19.1	15.2	
100 °C	15.4	16.2	16.8	17.5	20.8	26.1	15.7	13.1	12.1
150 °C	17.1	17.4	17.3	17.9	20.8	23.9	14.7	14.1	13.4
200 °C	14.7	16.2	16.8	17.2	17.8	18.0	15.9	15.0	12.5
Al-30%Si									
-	14.9	14.9	16.6	17.8	20.0	24.3	23.2	18	15.8
100 °C	15.1	15.3	15.3	16.2	19.4	21.7	14.9	14.3	12.4
150 °C	15.2	15.7	16.0	16.8	21.0	20.0	15.2	14.0	13.2
200 °C	15.0	15.5	15.7	16.2	16.0	15.6	13.1	13.0	11.2
Al-40%Si									
-	12.4	13.0	13.1	14.1	17.1	21.1	15.9	14.6	
100 °C	12.4	13.7	13.9	14.9	21.3	17.4	14.3	13.6	12.6
150 °C	13.0	12.9	13.0	14.1	18.2	17.0	14.2	14.0	13.0
200 °C	12.1	12.8	13.1	13.6	14.5	14.9	15.0	15.0	14.1

Treatment of the melt by the mixture ($CaCO_3 \cdot MgCO_3$) is an additional factor in the reduction of TCLE in the operating temperature range. After processing by the proposed method, the values of α are 12 - 17% lower than for alloys of conventional preparation (figure 3). Increase of physicomechanical properties of Al-Si alloys ($CaCO_3 \cdot MgCO_3$) subjected to treatment by the mixture ($CaCO_3 \cdot MgCO_3$) in the process of smelting is connected apparently with better assimilation of hydrogen and oxygen introduced into the melt. Besides, particles of refractory oxides of Mg and Ca contained in the mixture serve as additional numerous centers of crystallization. The effectiveness of this method of melt processing for aluminum and Al-Si system alloys is indicated by the increase in the limiting degree of plastic deformation prior to failure by 7-25% on average and a decrease in TCLE by 7-14% [6].

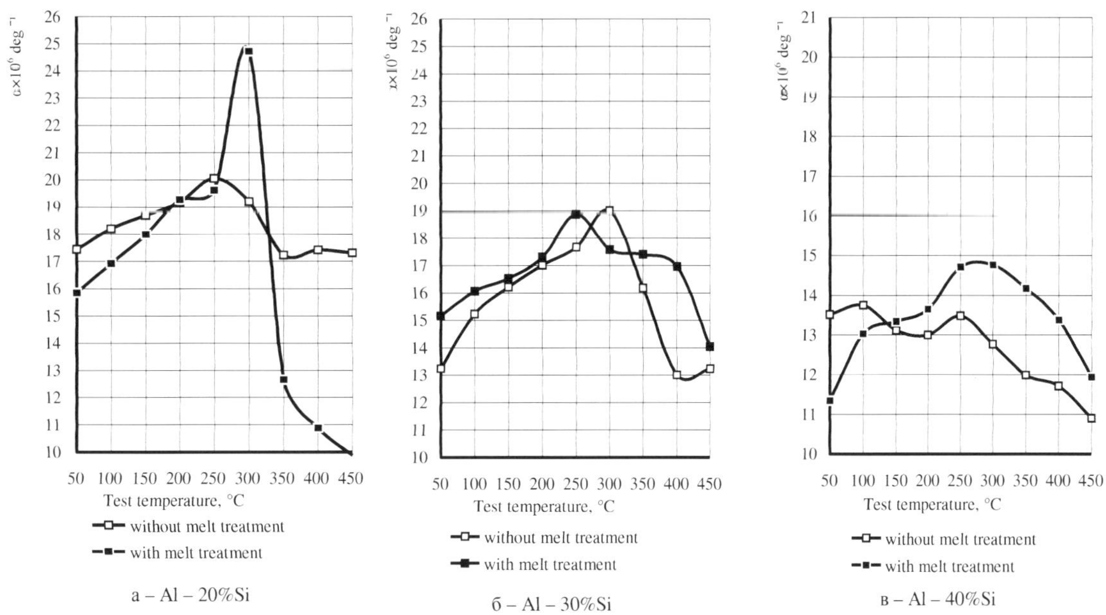

Figure 3. Effect of melt processing by the mixture ($CaCO_3 \cdot MgCO_3$) on the linear expansion of Al-Si alloys.

4. Conclusions

1. Comparing the obtained data on the influence of the considered methods of pretreatment of the melt, we can conclude that blowing with vapours of *aqueous solution of copper sulfate* for 5-15 minutes at a temperature 800-1100 °C leads to a decrease in TCLE of alloys Al-20÷40%Si in the low-temperature interval of testing. Thus, for the alloy Al-20%Si TCLE decreases from the values $\alpha_{50\div150}$ = $18\cdot10^{-6}$ deg^{-1} to $16.6\cdot10^{-6}$ deg^{-1} and for the alloy Al-40% Si from the values $\alpha_{50\div150}$ = $13.2\cdot10^{-6}$ deg^{-1} to $11.3\cdot10^{-6}$ deg^{-1}.

2. Processing of the melt with *carbamide vapours* leads to the decrease in TCLE of aluminum alloys Al-Si with the pre-eutectic composition throughout the temperature range. The greatest decrease is observed in the interval 50÷150 °C with the values $19.0\cdot10^{-6}$ deg^{-1} to $15.7\cdot10^{-6}$ deg^{-1}. Thermal treatment at 100÷200 °C promotes further reduction of TCLE: for low-silica alloys treatment at 150 °C is considered to be the most effective, for highsilicon – at 200 °C.

3. Processing of the melt by *the mixture of calcium and magnesium carbonates* is not only an environmentally friendly technological option of modifying alloys Al-20 ÷ 40% Si but also is an additional factor for reducing TCLE in the operating temperature range. After processing by the proposed method the values of α are by 12 - 17% lower than for alloys of usual preparation. Thus, in the interval 50÷150 °C the average TCLE of the alloy Al-20% Si decreases from $18\cdot10^{-6}$ deg^{-1} до $16.9\cdot10^{-6}$ deg^{-1}, for the alloy Al-30% Si from $16\cdot10^{-6}$ deg^{-1} to $14.2\cdot10^{-6}$ deg^{-1}, for the alloy Al-40% Si from $13.2\cdot10^{-6}$ deg^{-1} to $12.6\cdot10^{-6}$ deg^{-1}. The additional advantage of using the mixture of carbonates is not only in complex modification but also in the reduction of superheating temperature of the melt from 1100-1200 °C to 900 °C and in the reduction of the melting time from 5- 6 to 1-1.5 hours.

References

[1] Afanas'ev V K and Popova M V 2012 *Metallurgy of Machine Building* **6** 8–13
[2] Afanas'ev V K and Popova M V 2001 *Steel in Translation* **31(2)** 50–53
[3] Ehrstrom J and Warner T 2000 *Metallurgical Design of Alloys for Aerospace Structures ICAA* 7 vol **1** 5–16
[4] Popova M V and Kibko H B 2014 *Processing of Metals* **2(63)** 107–16
[5] Zu Fangqiu and Li Xiaoyun July 2014 *China Foundry* **11(4)** 287–95
[6] Popova M V, Gertsen V V, Doronchenko A V and Afanas'ev V K 1999 Patent of the Russian Federation No 98104521 appl. 05.03.1998, publ. 10.09.99
[7] Afanas'ev V K, Popova M V and Samon V A 2014 *Metallurgy of Machine Building* **5** 21–28
[8] Afanasyev V K, Malyukh M A, Popova M V, Dolgova S V and Lavrova N B 2017 *Metallurgy of Machine Building* **1** 33–40
[9] Starink M J, Sinclair I and Gregson P J 2000 *Aluminum Alloys: Their Physical and Mechanical Properties* vol 331-337 pp 97–110
[10] Afanas'ev V K, Popova M V, Gertsen V V, Dolgova S V and Leys V A 1983 *Processing of Metals* **4(61)** 28–34
[11] Hess P D and Tumbull G K 1974 *Paper from Hydrogen in Metals (American Society for Metals)* pp 277–87
[12] Afanas'ev V K, Popova M V, Starostina M A (Malyukh M A) and Krivicheva N V 2011 *Metallurgy of Machine Building* **3** 30–3
[13] Borisov G P 2005 *Metallurgy of Machine Building* **5** 11–20
[14] Ushakova V V, Popova M V and Luzyanina Z A 1995 *Izv. Vuzov. Ferrous Metallurgy* **4** 69
[15] Afanas'ev V K, Dolgova S V et al 2014 *Proc. on Actual Problems in Machine Building* (Novosibirsk: NSTU) pp 381–87

Selective recovery of valuable metals from crushed electronic waste

Z M Akhmetvaliyeva[1], N A Kulenova[1], Ya Takasaki[2], S V Mamyachenkov[3], O S Anisimova[3], L K Mudashiru[4], E L Fokina[5] and J Bast[6]

[1]East Kazakhstan State Technical University, 19 Serikbayev street, Ust-Kamenogorsk, 070010, Kazakhstan
[2]Akita University, 1-1 Tegata Gakuen machi, Akita, 010-8502, Japan
[3]Ural Federal University, 17 Mira street, Ekaterinburg, 620002, Russia
[4]Durham University, 3LE Stockton Road, the Palatine Centre, Durham, DH1, United Kingdom
[5]Saint Petersburg State University, 7-9 University embankment, Saint Petersburg, 199034, Russia
[6]Freiberg University of Mining and Technology, 6 Academy street, Freiberg, 09599, Germany

E-mail: zakhmetyali@gmail.com

Abstract: This research was carried out to recover valuable metals from the electronic waste (e-waste). E-waste samples were crushed and thermally activated under 450 °C within 1 hour. The optimal temperature of the thermal treatment, at which all organic phases are removed and metal's forms turned into oxides was determined. Leaching of thermally activated material using EDTA at room temperature, pH 7 for 1 hour resulted in greater than 95 % extraction of lead. Non-ferrous metals were extracted at 85 °C with a leaching solution of 2 M H_2SO_4 which resulted in recovery of more than 98 % of copper within 3 hours.

1. Introduction

The electronic waste (e-waste) is an attractive source of raw materials for the extraction of non-ferrous, precious and rare metals. According to [1, 2], there is about 20-50 % by weight of copper and other valuable components, including Pb, Sn, Zn, Al, Ag, Au, Pd, etc. are contained in the e-waste depending on the type of device.

There are various methods for the e-waste recycling well known today, which in the majority are based on the combination of pyrometallurgical, hydrometallurgical and electrometallurgical technologies [2, 4, 5].

Hydrometallurgical routes for the e-waste processing are described by a high degree of recovery of the target components (Cu, Zn, Pb and precious metals), by a flexible mode of operation and by the possibility of application on a small scale. These factors make hydrometallurgical processes more economically stable for processing a wide range of the e-waste, including low-grade ones. Nevertheless, there is no existing technology for the e-waste recycling, completely based on hydrometallurgical processes on an industrial scale.

Different acids could be used as the leaching agents for metals extraction from the e-waste - sulfuric acid, nitric acid, hydrochloric acid and their mixtures [6, 7]. Basically, during the leaching process several metals are simultaneously dissolute in the solution (Cu, Pb, Zn, Sn etc.), which leads to the further difficulties on the step of the re-extraction of metals from productive solutions. Considering this point, the authors of this article recommend to conduct the e-waste leaching in stages using different leaching agents for the selective extraction of metals.

Methods for the selective extraction of oxidized forms of lead from industrial products using various organic solvents (alkylamines) are given in several published works [8, 9]. One of the main advantages of leaching with organic solvents is the possibility of separation of Pb and Ag. Authors [9] also note that the use of ethylenediaminetetraacetic acid (EDTA) for lead leaching from industrial products of zinc production allows the selective separation of Pb, Cu, Zn and Fe^{n+}.

The main purpose of this research was to experimentally confirm the possibility of the recovery of valuable metals from the e-waste via physical pre-treatment, thermal treatment and chemical treatment (leaching with different solutions).

2. Experimental procedures

2.1. Materials and methods

The samples used in the experiment were printed circuit boards (PCBs) and automobile microcircuits weighed to a specified mass, sourced from various waste computers and automobiles regardless of brand. Component analysis of the PCBs used in this experiment determined the metal component to be 53 % by weight, and this consisted of base metals (Cu 32.87 %, Fe 5.49 %, Ni 0.69 %, Sn 3.55 %, Pb 2.05 %, Al 5.50 %, and Zn 2.44 %) and precious metals (Au 5.1 g/ton, Ag 1346 g/ton, and Pd 182.7 g/ton). Plastic materials were 20 % by weight and consisted of more than 16 % C-H polymers, which included polypropylenes and α-polypropylenes; less than 4 % halogenated polymers (mainly polyvinyl chloride (PVC), traces of polytetrafluoroethylene (PTFE), and polybromo compounds); and less than 1 % N-containing polymers, including nylon and polyurethanes. Also, metal oxides were 27 % by weight and included 11 % silica, 4.5 % alkaline and alkaline earth oxides, 4 % alumina, and 7.5 % other oxides.

To extract valuable metals, Pb was dissolved in EDTA solution (crystalline powder dissolute in water), Cu, Fe, Zn, Ni, and Sn were dissolved in H_2SO_4 (98 %, extra pure (EP) grade). Di-(2-ethylhexyl) phosphoric acid (DEHPA) was used to selectively extract Zn^{2+} and LIX extraction agent and kerosene were used in combination to selectively extract Cu^{2+} from SO_4^{2-} solution.

2.2. Experimental setup

Figure 1 shows a flowchart for the recovery of valuable metals from the e-waste samples, developed in the framework of this research. The e-waste samples were firstly shredded by plastic cutter "Scutter SA-22 (Stolz Co., Ltd)" and then crushed by a disc mill. Crushed e-waste samples were sorted by size using a sieve that allowed only pieces within 0.08-0.25 mm to pass through. Pieces larger than 0.25 mm were crushed again to a suitably smaller size and re-sieved.

Crushed materials were thermally treated in a laboratory muffle furnace SNOL 7,2/1100 at temperature range 400-450 °C. The time of the thermal treatment of the sample (1 hr) was determined according to the differential scanning calorimetry (DSC) results, which correspond to the rate of oxidation of the organic components. The thermally treated materials were used as the source from which valuable metals were extracted.

The leaching experiments were carried out in heat-resistant fluoroplastic beakers with capacity of 500 ml equipped with a pH meter, a top-drive electrical laboratory stirring device "IKA RW 16 basic" set at 500 rpm, and a condenser in a thermostatically controlled water bath. Analyses of components of the starting material were carried out using an atomic absorption spectrometer (AAS) and inductively coupled plasma-mass spectrometer (ICP-MS).

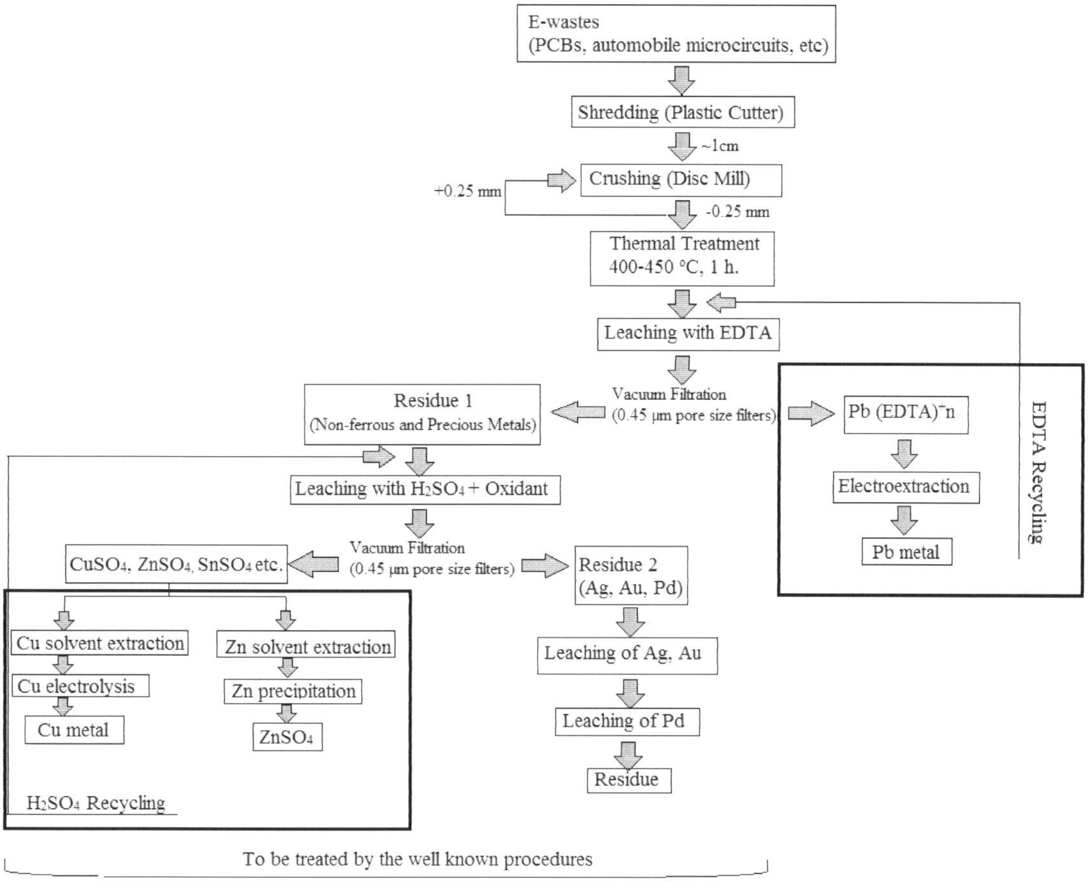

Figure 1. Flowchart for the recovery of valuable metals from the e-waste.

3. Results and discussion

3.1. Thermal treatment effect

The crushed materials (average particle size of 0.08-0.25 mm) were thermally treated in the laboratory muffle furnace SNOL 7,2/1100. The efficiency of the thermal treatment was investigated using the synchronous thermal analyzer Mettler Toledo for the thermo-gravimetric analysis (TGA) and DSC.

Laboratory tests of the thermal treatment of the starting material were conducted in the temperature range 25-700 °C to determine the temperature zone of the complete removal of organic phases from the e-waste materials. The process time in the furnace was determined by the rate of oxidation of organic components, based on the results of DSC, and was 1 hour. It was found that the maximum percentage of the removal of organic phases is observed in the temperature range 400-450 °C.

Phase analysis of the material after the thermal treatment at 450 °C showed the presence of oxide forms of Pb, Cu and Zn, etc. At the same time, the content of all valuable metals after the thermal treatment increased by 25 % by mass of the e-waste, including: Cu (41 %); Pb (2.56 %); Zn (3 %); Ag (1750 g/ton); Au (12.5 g/ton). TGA-DSC analysis results of the thermally activated material are presented in Figure 2.

The formation of the metals' oxide forms as a result of the thermal treatment makes possible to conduct at the next stage the leaching process with EDTA for the selective Pb extraction.

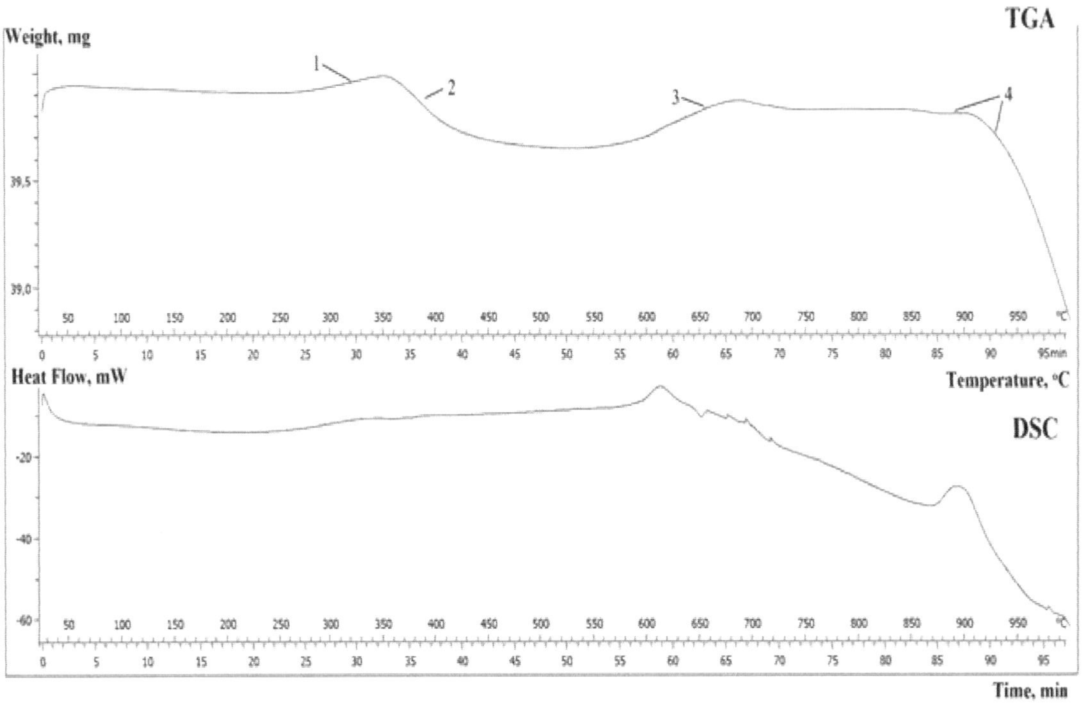

Figure 2. TGA and DSC curves of the thermally activated e-waste in air atmosphere, heating rate 10 °C, min^{-1}. TGA curve: 1 - weight increase by 0.19 %; 2 - weight loss by 0.84 %; 3 - weight increase by 0.55 %; 4 - weight loss by 2.47 %.

3.2. Selective Leaching of Pb with EDTA

Authors [10] investigated that EDTA selectively forms complex compounds with cations of non-ferrous metals in different pH ranges and the strength and solubility of the formed metal complexes depend on the pH and density of the solution.

Complexing reagent EDTA (concentration of 100 g/l) was used as the leaching solvent to selectively extract Pb oxide forms from the crushed thermally activated e-wastes. Leaching experiment was carried out at 25 °C and 60 min. An aqueous solution of caustic soda (concentration of 200 g/l) was supplied in order to achieve and maintain a predetermined pH value. pH was maintained constant during the experiment process, dosing the required volume of alkali (NaOH).

In order to investigate effects of pH and solution's density on the Pb (EDTA)+n complexes formation, experimental tests for different pH (5; 7; 10) and density values (liquid-solid ratio: 3:1; 7:1; 10:1) were carried out. Productive solutions after leaching with EDTA was filtered through 0.45 micron filter using a system of vacuum filtration. Residues were dried overnight and analyzed by X-Ray fluorescent analysis to determine content of Pb, Cu and Zn. Figure 3 shows the extraction rates of these three metals in different pH, while Figure 4 shows the effect of density (liquid-solid ratio) on the leaching of the metals with EDTA.

The optimal conditions for EDTA leaching were found to be a solution 100 g/l EDTA, pH – 7, L:S ratio 10:1 and a reaction temperature of 25 °C, which achieved 92-95 % selective extraction of Pb within 1 hr, while Cu and Zn remained in the residue.

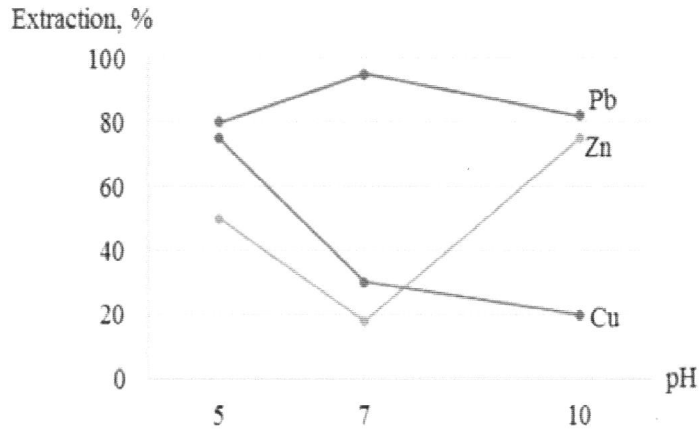

Figure 3. Dependence of metals extraction in the solution on the pH value at 25 °C and liquid-solid ratio (L: S) - 10:1.

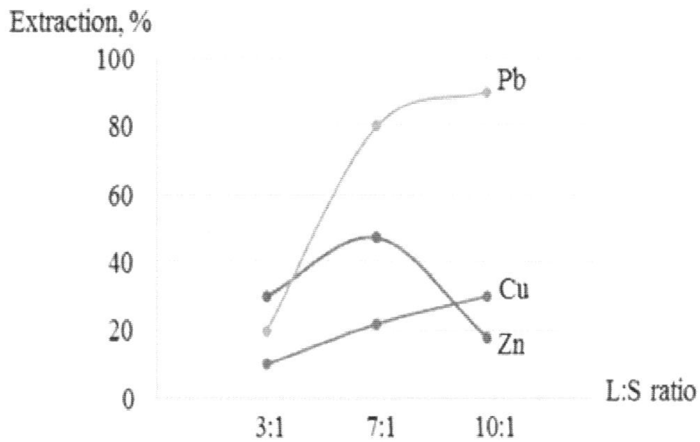

Figure 4. Dependence of metals extraction in the solution on the L:S ratio at 25 °C and pH 7.

3.3. Leaching of Metals with H_2SO_4

Sulfuric acid was used as the leaching solvent to extract the base metals, with air oxygen as an oxidant. The aeration of the system was achieved using a glass tube and an aquarium pump. The remaining solid material was filtered through 0.45 micron filter using a system of vacuum filtration, washed and leached in the next stage. Leaching with the H_2SO_4 solution was carried out at 40, 65, and 85 °C over a period of time. The H_2SO_4 concentration ranged from 0.5 to 2 M, 10 g of sample per L of leaching solution were added.

The main target component at the H_2SO_4 leaching stage is copper, because of its high content in the sample (32.87 %). In order to optimize the leaching parameters, 3D surface characterizing the dependence of copper extraction in the solution on the temperature and leaching time (see Figure 5a) and 3D surface characterizing the dependence of copper extraction on the temperature and H_2SO_4 concentration (Figure 5b) were constructed applying the program STATISTICS 10.0.

The optimal conditions for H_2SO_4 oxidative leaching were found to be a solution 2 M of H_2SO_4 and a reaction temperature of 85 °C, which achieved ~ 98 % recovery of Cu within 3 hr.

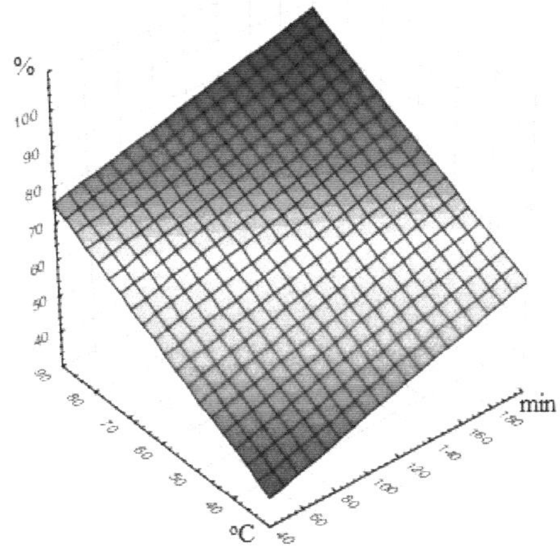

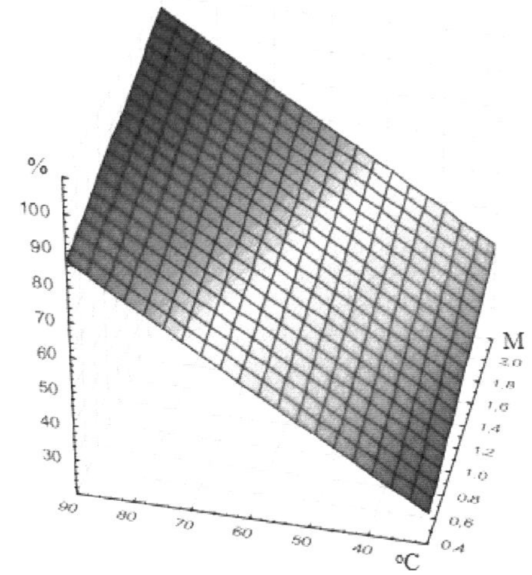

Figure 5. (a) Dependence of copper extraction to the solution (%) on the temperature (^{o}C) and leaching time (min); (b) Dependence of copper extraction to the solution (%) on the temperature (^{o}C) and H_2SO_4 concentration (M).

4 Conclusions

The present research developed a leaching procedure for the recovery of valuable metals from e-wastes via physical pre-treatment, thermal treatment and chemical treatment, and the results at each stage are as follows.

(1) The results of the thermal treatment showed that the maximum percentage of the removal of organic phases and formation of the metals' oxide forms are observed in the temperature range 400-450 °C. Presence of metals' oxide forms makes possible to conduct further step of the leaching with EDTA for Pb selective extraction.

(2) In the leaching reaction using 100 g/l EDTA at 25 °C, pH 7, L:S - 10:1, 95 % of Pb was selectively leached out within 1 hr.

(3) In the case of leaching with the leaching solution of 2M of H_2SO_4 solution at a reaction temperature of 85 °C, 98 % of Cu was leached out within 3 hr.

References

[1] Cui J and Zhang L 2008 *J. Hazard. Mater.* **158** 228–256
[2] Veit H M, Bernardes A M, Ferreira J Z, Tenorio J A and de Fraga Malfatti C 2006 *J. Hazard. Mater.* **137** 1704–09
[3] Wong M H, Wu S C, Deng W J, Yu X Z, Luo Q, Leung A O W, Wong C S C, Luksemburg A S and Wong A S 2007 *J. Environ Pollut.* **149** 131–140
[4] Dhawan N, Kumar V and Kumar M 2009 *Extraction and Processing Division (EPD) Congress: The Minerals, Metals and Materials Society* (Warrendale, PA, USA) pp 1107–09
[5] Hagelüken C 2006 *World Metall. Erzmetall* **59** 152–161
[6] Loleit S I 2010 *Development of Environmentally Friendly Technologies of Complex Extraction of Precious and Non-ferrous Metals from the Electronic Scrap* abstract of the dissertation of Dr. Sci. Eng. (Moscow: Moscow Institute of Steel and Alloys)
[7] Kolmakova A A 2004 *Physical and Chemical Regularities of the Electronic Scrap Processing Containing Noble Metals by a Mixture of Nitric and Sulfuric Acids* abstract of the dissertation of Cand. Sci. Eng. (Krasnoyarsk: State University of Non-ferrous metals and gold)

[8] Sherrit Gordon Mines Ltd. 1965 *A Method of Lead Production* pat. **243311** (Australia)
[9] Karelov S V, Anisimova O S, Mamyachenkov S V and Sergeev V A 2008 *J. Non-ferrous Metallurgy* **2** 20–23
[10] Dyatlova N M, Temkina V Ya, Kolpakova I D 1970 *Complexes* (Moscow: Chemistry)

IOP Conf. Series: Materials Science and Engineering **411** (2018) 012012 doi:10.1088/1757-899X/411/1/012012

Comparative metallographic analysis of the structure of St3 steel after being exposed to different ways of work-hardening

A E Balanovsky[1], M G Shtayger[2], M V Grechneva[1], V V Kondrat'ev[1] and A I Karlina[1]

[1]Irkutsk National Research Technical University, 83 Lermontov street, Irkutsk, 664074, Russia
[2]MC Mechel Steel, 1 Krasnoarmeysky street, 125167 Moscow, Russia

E-mail: kvv@istu.edu

Abstract. Results of the study of the change in the microstructure of St3 steel using different methods of heat treatment are presented. Comparison of hardness during bulk quenching and quenching by a highly concentrated heat source (plasma arc) is given. Description of the obtained microstructures is presented.

1. Introduction

Quenching can significantly increase the hardness and wear resistance of machine parts and tools. Moreover, the degree of increasing of hardness after quenching depends mainly on the carbon content in it. This is because the hardness of martensite depends on the degree of distortion of the crystal lattice. The smaller the content of carbon in the martensite, the less distorted its crystal lattice is and, consequently, the lower the hardness of steel is [1].

In this paper, the possibility of increasing the microhardness of low-carbon St3 steel by means of various technological methods of hardening is considered: bulk hardening, ultrasonic hardening [4], use of highly concentrated energy sources – plasma arc [5,6].

2. Methods and materials of research.

To study structural transformations in St3 steel, bulk hardening and surface hardening by plasma were carried out. Bulk hardening was performed in the furnace at various temperatures from 800 to 1000°C, holding time is 15 minutes, cooling in water. Plasma hardening was carried out on a plate with dimensions of 80x20x10 mm of St3 steel. Plasma surface quenching was carried out with the following parameters: plasma arc current 90A, argon flow rate 8 l/min; gap between the cut of the nozzle of the plasmatron and the detail is 5 mm; speed of movement of samples is 5 30 mm/s. With the given parameters, quenching process proceeds without metal surface flash [7,8]. Chemical composition of the material St3пс in percent GOST 380 - 2005 (table 1).

To carry out ultrasonic processing, a device consisting of an ultrasonic generator and a tool in the magnetostrictive transducer and indentor was used. Ultrasound processing of materials is based on the energy of mechanical oscillations of the instrument-indenter. The oscillations are made with an ultrasonic frequency (18-24 kHz) and an amplitude of oscillations (0.5-50) of 10-6 m. Energy is introduced in the cutting zone by means of the static pressing force of the tool to the surface of the workpiece.

Table 1. Chemical composition of the material St3пс in percent.

C,%	Si,%	Mn,%	Ni,%	S,%	P,%	Cr,%	N,%	Cu,%	As,%
0.14 –0.22	0.05 –0.15	0.4 – 0.65	to 0.3	to 0.05	to 0.04	to 0.3	to 0.008	to 0.3	to 0.08

3. Results of research and discussion.

Microstructure of St3 steel before and after quenching is shown in figure 1-5, structure of the initial metal consists of ferritic and pearlitic grains with an average hardness of 1900 - 2100 MPa.

Analysis of the microstructures shown in Figure 1, St3 steel is heated to a temperature of 800 °C. It is impossible to obtain a pure ferrite-martensitic structure even using water quenching of sample with a cross section of 15x15 mm. During water quenching, a small amount of excess ferrite is released from the austenite, and the remaining austenite decomposes into a martensitic-bainitic mixture. Investigations carried out showed that when heating in the intercritical temperature range of Ac1-Ac3, the formation of the first portions of the austenite in samples with the initial ferrite-pearlite structure takes place in pearlitic grains. Further increase of the amount of austenite is due to the migration of its boundaries into ferrite grains.

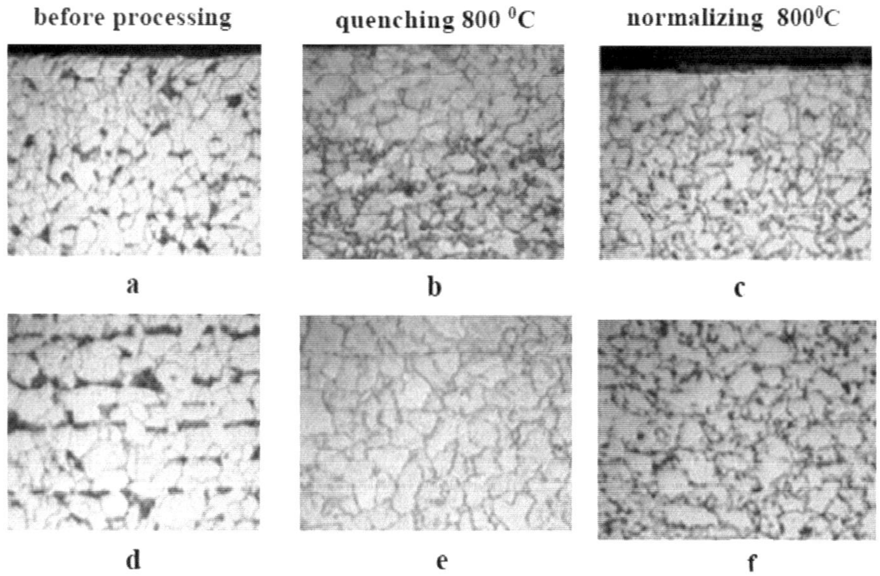

a, b, c – edge; d, e, f – middle

Figure 1. Microstructure of St3 steel after being heated to a temperature of 800 °C.

After cooling, low-temperature decomposition products of austenite (martensite-bainite), surrounded by new ferrite precipitates, are located along the boundaries of the old ferrite, which did not undergo transformation upon heating (figure 2). At the same time, figure 1 shows that the microstructure of the surface and the core differ from each other. On the surface, decarburization of the surface layer is seen both for the quenching regimes and for the normalization regimes. An increase in the heating temperature in the intercritical interval to 850 °C is accompanied by a decrease in the stability of the austenite, and after the excess ferrite is isolated, supercooled austenite is almost fully converted to bainite. With this heat treatment, quenching with the temperatures of the two-phase field ensures the formation of a ferrite-bainite structure (figure 2). During the cooling process, in air, from the lower half of the intercritical interval, mainly a ferrite-bainite structure with a small perlite content is formed. The amount of perlite increases with increasing of the heating temperature in the intercritical interval.

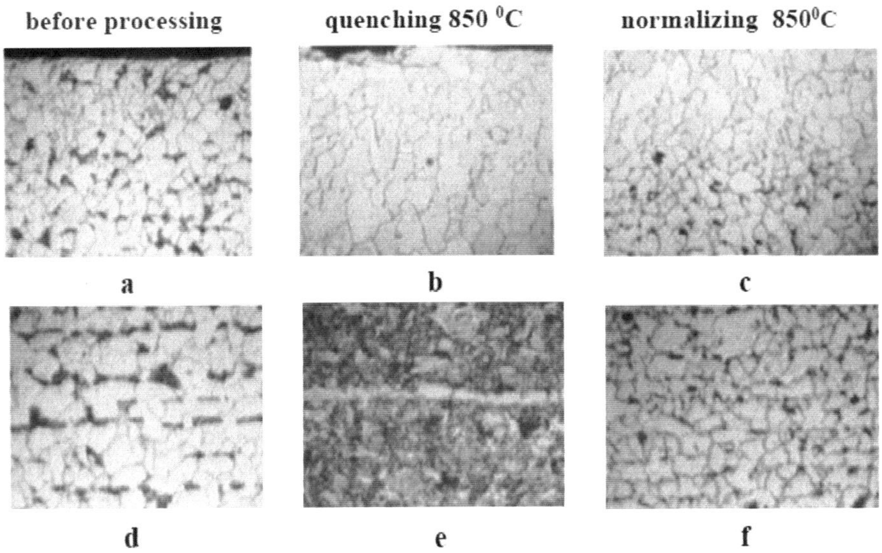

a, b, c - edge; d, e, f - middle

Figure 2. Microstructure of St3 steel after being heated to a temperature of 850 °C

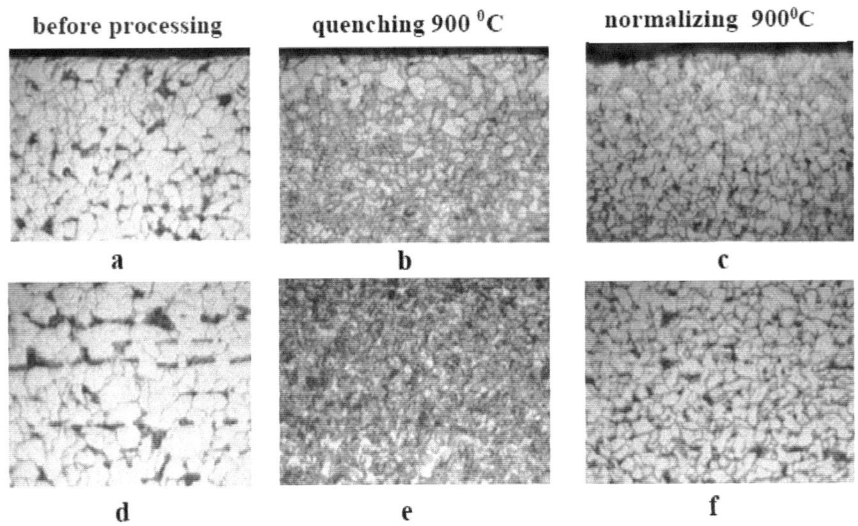

a, b, c – edge; d, e, f – middle

Figure 3. Microstructure of St3 steel after being heated to a temperature of 900 °C.

With increasing temperature of heating in the intercritical interval figures 3, 4. stability of the over-cooled austenite decreases, Mn rises, which is due to a decrease in the carbon content in the forming austenite, appearance of heterogeneity in chemical composition between the microvolumes of the sample and inside the grains.

IOP Conf. Series: Materials Science and Engineering **411** (2018) 012012 doi:10.1088/1757-899X/411/1/012012

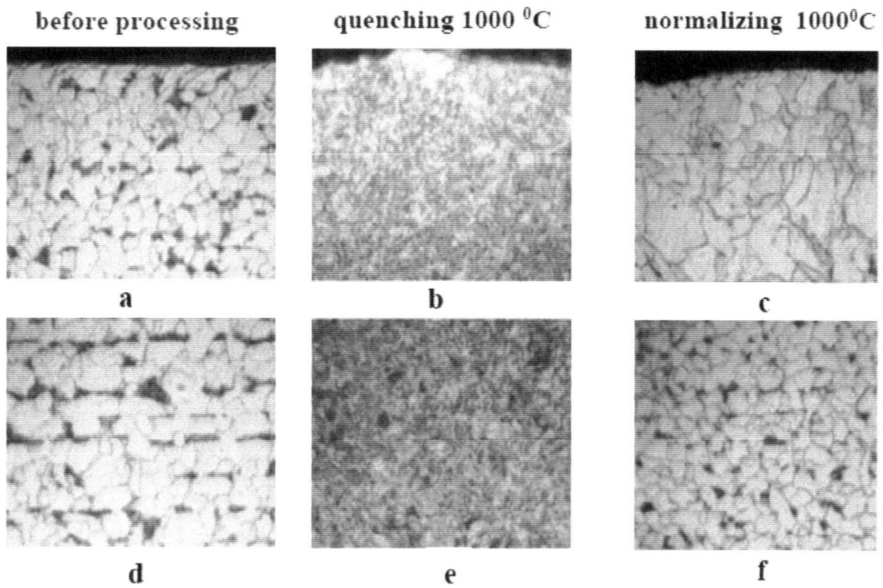

a, b, c – edge; d, e, f – middle

Figure 4. Microstructure of St3 steel after being heated to a temperature of 1000 °C.

During quenching the samples after its heating at 30 °C above Ac1 (figures 3, 4), lath martensite is formed with an "indirectly" located midrib with a high density of microtwins and dislocations, which binds to a low Mn because of the highest concentration of carbon in the first portions of the austenite formed from perlite. Decrease of the carbon content in austenite, along with increase of its amount and temperature of formation, leads, during the water quenching, to the occurrence of high-temperature martensite, in which individual laths with a high density of dislocations are assembled into piles. During the cooling process from the upper region of the intercritical interval, austenite decomposes according to the bainite mechanism. Figure 5 shows the values of the macrohardness of the samples after being heated to different temperatures. For comparison, plasma surface quenching of St3 steel at a processing speed of 15-30 mm/s was carried out. Figure 6 shows a macro image of a hardened zone and a microstructure along the depth of a hardened layer with hardness values.

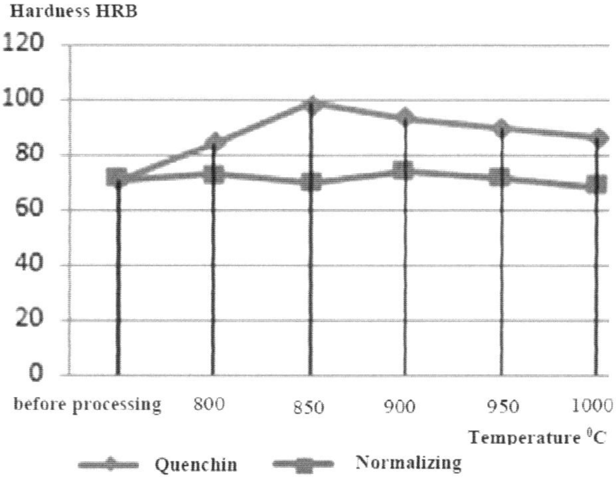

Figure 5. Quantitative values of macrohardness on the surface of samples of St3 steel after heat treatment.

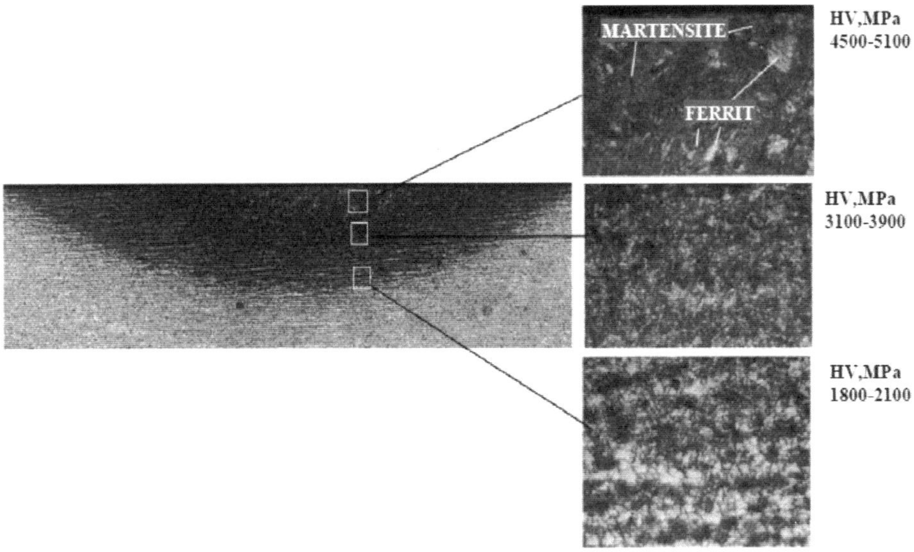

Figure 6. Microstructure of St3 steel after plasma quenching in arc mode.

This is due to the fact that in the process of plasma surface heating, there are different stages of austenization [3-7]. During the plasma quenching of St3 steel, previously subjected to quenching and tempering, tempering zone is identified at the boundary with the initial structure. The entire quenching zone is a region of homogeneous martensite with a microhardness of 500650 HV, which is higher than the hardness of martensite obtained by quenching. A transition zone is observed at the boundary with the original structure. A significant increase in the hardness of the surface layer in comparison with the results presented above is possible with the help of plasma surface carburizing [8,9]. Figure 7 shows the results of the plasma surface carburizing of St3 steel. The proposed treatment allows the formation of various gradient-phase states in the surface layer of low-carbon steels.

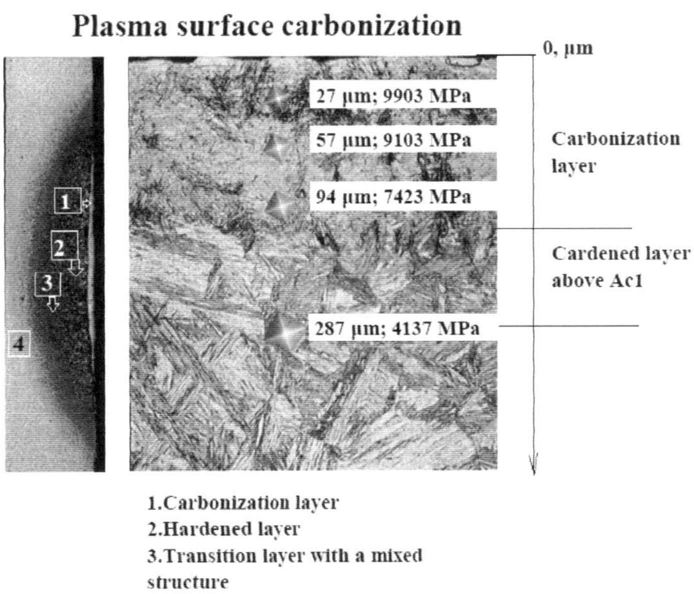

Figure 7. Plasma surface carburizing of St3 steel.

With the combined treatment of St3 steel in the regimes of: bulk quenching and subsequent ultrasonic treatment, microhardness of a thin surface layer is increased to 500 HV. The depth of the plastic deformation zone is (0.05 0.1 mmV), microhardness of ferrite grains is 380 ... 450. The increase in microhardness is associated with phase hardening.

4. Conclusion

Thus, comparative studies have shown the prospectivity of new methods for treating low-carbon steels with the use of a plasma arc and in the quenching and surface cementation regime and, also, for combined hardening by the example of St3 steel. Further studies in this direction will be continued with an aim to clarify the shape and dimensions of the martensitic and bainitic crystals during surface quenching, mechanism of saturation of the surface layer in the process of plasma carburizing.

References

[1] Pout D 1987 *Modification and Doping of a Surface by Laser, Ion and Electron Beams* (M.: Mashinostroeniye) p 424
[2] Chudina O V 2003 *Combined Methods of Surface Hardening of Steels Using Laser Heating. Theory and Technology* (M.: MADI (GTU) p 248
[3] Balanovsky A E 2006 *Plasma Surface Hardening of Metals* (Irkutsk) p 180
[4] Balanovsky A E 2015 *Strengthening Technologies and Coatings* **12(132)** 18–30
[5] Markov A I 1980 *Ultrasonic Processing of Materials* (M.: Mashinostroeniye) p 237
[6] Balanovsky A E 2016 *Strengthening Technologies and Coverings* **1(133)** 25–34
[7] Balanovsky A E 2016 *Strengthening Technologies and Coverings* **2(134)** 20–30
[8] Balanovsky A E and Gyu Vu 2017 *Letters on Materials* vol 7 **2** 175–179 DOI: 10.22226 / 2410-3535-2017-2-175-179
[9] Fedin V M and Borts A I 2009 *Metal Science and Heat Treatment* vol 51 **11-12** 544–552
[10] Vorob'ev V L et al 2012 *Physics and Chemistry of Material Processing* **6** 5–9

Plasma-arc surface modification of metals in a liquid medium

A E Balanovsky[1], M G Shtayger[2], V V Kondrat'ev[1], Vu Van Huy[1] and A I Karlina[1]

[1]Irkutsk National Research Technical University, 83 Lermontov street, Irkutsk, 664074, Russia
[2]MC Mechel Steel, 1 Krasnoarmeysky street, 125167, Moscow, Russia

E-mail: kvv@istu.edu

Abstract. The method of saturation of the steel surface with carbon under the action of a plasma electric arc in a liquid is considered. The microstructure and microhardness of the hardened layer after treatment using this method were studied. It is established that during the time of plasma action of 0.1-1c the surface layer is saturated with carbon and nitrogen to the level of concentration of high-carbon steel. Microstructures and characteristics of the metal surface after plasma modification in a liquid medium were studied. The main parameters of the modified layer were determined: depth of the layer is 15-1200 μm, microhardness reaches 7.000-14.000 MPa.

1. Introduction

The process of operation of machines and equipment is always accompanied by wear, which gradually leads to a decrease in the service life of machines. Therefore, increasing the wear resistance of wearing parts for various purposes is the most important task of modern mechanical engineering and other branches of engineering [1-10].

In this paper, we present the results of an experimental investigation of the surface layer structures formed in the process of plasma heating in liquid media. To saturate the surface layer with carbon or nitrogen, the workpiece is immersed in a liquid which contains carbon (toluene, mineral oil, etc.) or nitrogen (an aqueous solution of ammonium chloride, etc.).

2. Materials and methods of research

Plasma surface modification in liquid media was carried out on steels 1018 and 1020 according to the ASTM standard. Chemical composition of the material in percent according to ASTM A29 standard is shown in the table. To test the regimes of plasma modification, samples of 80x20x10 mm were used.

Table 1. Chemical composition of the material.

	1018	1020
Iron, Fe	98.81-99.26%	99.08-99.53
Carbon, C	0.18%	0.20%
Manganese, Mn	0.6-0.9%	0.30-0.60%
Phosphorus, P (max)	0.04%	0.04%
Sulfur, S (max)	0.05%	0.05%

Figure 1 shows the scheme of the process of plasma arc modification in liquid media. The workpiece is immersed in a layer of a flowing liquid (position 2, 3). Plasma arc is created by a plasmatron (position 1). Under the action of the gas-dynamic pressure of the plasma arc, a funnel is created in the liquid and a chemical-thermal influence on the surface layer of the metal occurs. Processing takes place in a semi-enclosed volume of fluid that is limited from all sides. When the plasma jet moves relatively to the surface of samples, heated to the temperatures of structural transformations, surface is immediately closed by a liquid, cooling them.

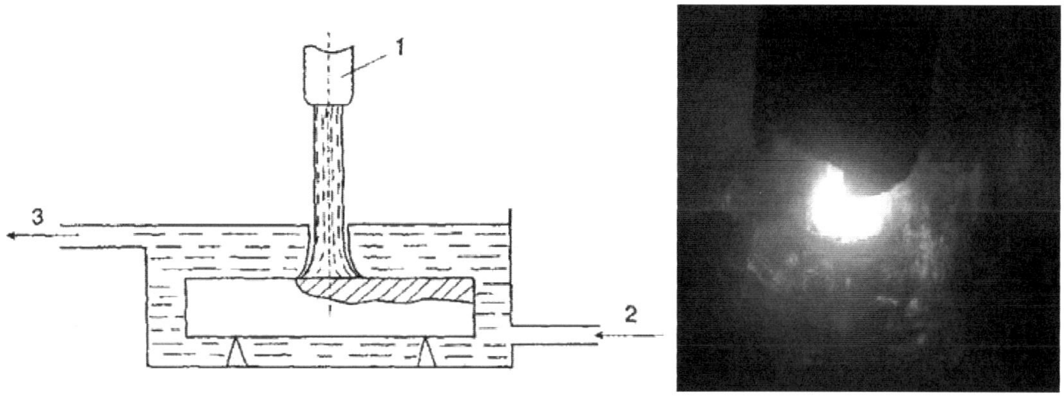

Figure 1. Scheme of the process of plasma-arc surface modification of a metal in liquid media.

The microstructure was examined on OLYMPUS GX51 microscopes with magnification of 100 and 500 times. Electron microscopic studies of the surface of metal samples were carried out using a scanning electron microscope JIB-4501. Microhardness was measured on a microhardness tester HMV-2T (Shimadzu) at a load of 0.01-120 g. Distance from the surface of the sample to the point with the maximum hardness values in comparison with the hardness of the base metal was considered as the depth of the modified layer.

3. Investigation of the structure

The conducted studies on plasma-arc modification in a liquid medium showed changes in the microrelief on the metal surface. After etching the sections (Figure 2), a white layer with a high microhardness of 4000-12000 MPa (under a load of 2H) is observed under the microscope in the cross section. Figure 2 shows the microhardness measurements for the depth of the modified layer (first digit is the depth of measurement in μm, second is the microhardness value). Depending on the carbonitride and nitride zones. The trans he regimes of plasma-arc modification, it is possible to obtain layers with different structures in the surface layer.

Directly on the surface, a non-etching ε-phase saturated with nitrogen is formed, followed by a supercooled γ-phase, under which nitrogenous martensite is. The photographs show an ultrafine-grained structure of ition from the nitrided layer to the underlying layers is smooth, which is one of the basic requirements for the microstructure of nitrided steel. The thickness of the diffusion zone varies from 50 to 200 μm. Structure of the quenched zone is martensite, residual austenite and carbides.

4. Nitriding

As a liquid medium, an aqueous solution of an ammonium salt of different concentrations was used.

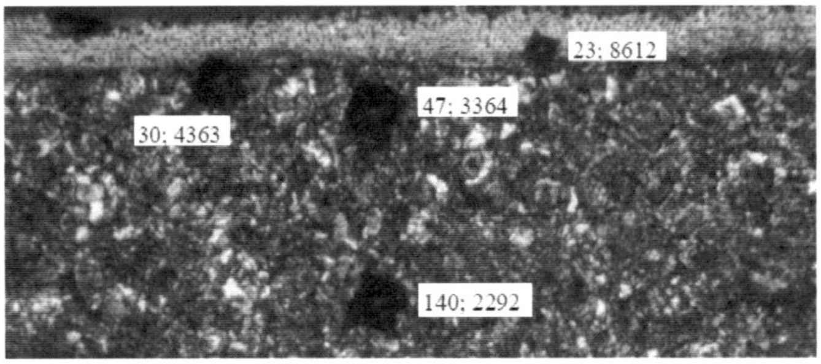

(a)

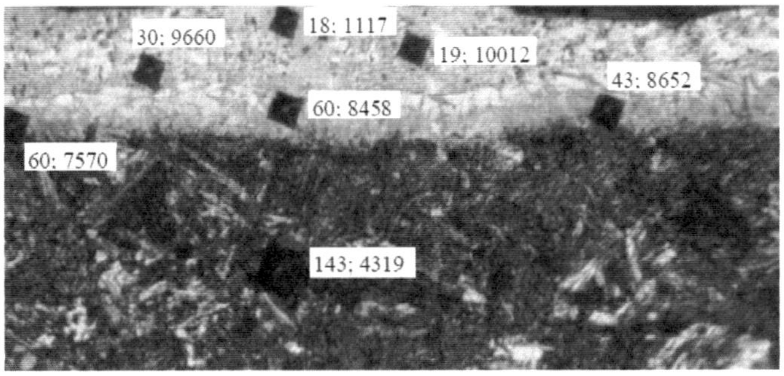

(b)

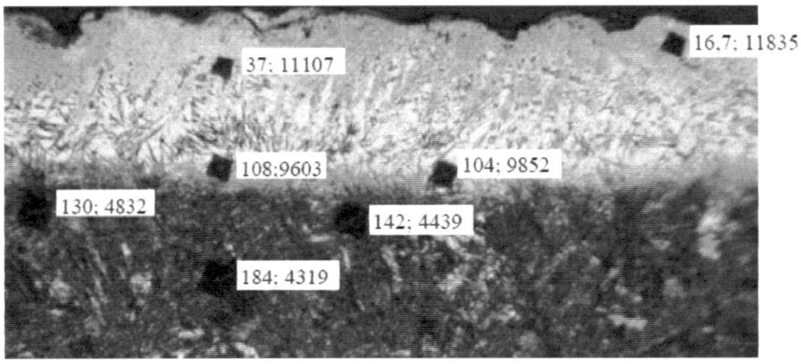

(c)

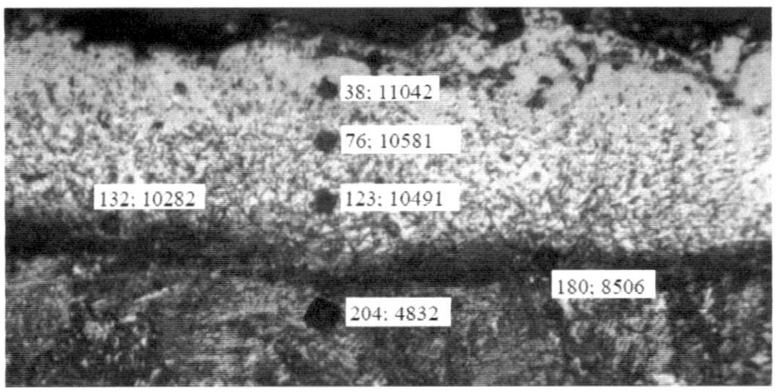

(d)

Figure 2. Microstructure of the modifying layer in various regimes of plasma-arc modification in liquid media.

Chemical-thermal treatment consists of three processes: dissociation - production of a saturating element in the active atomic state: 2NH3 ↔ 2N + $3H_2$, CH_4 ↔ C + $2H_2$, etc.; absorption - absorption of active atoms of the saturating element by the metal surface; diffusion - movement of atoms of the saturating element from the surface into the interior of the metal. The conducted studies have shown that an increase in the concentration of nitrogen in the treatment zone leads to an increase in the nitrogen content in the surface layers, which results in an increase in the depth of the layer and microhardness. Microstructure of the layer after the modification in liquid media is the same as after simple nitriding from the gas and solid phases [5], but there are some differences.

A nitrogen-modified reinforced layer which consists of two zones: bond zones (a white layer consisting of submicroscopic nitrides indistinguishable during the metallographic analysis) and a diffusion zone. After modification at a minimum processing speed (1-3 mm/s), white layer is absent, and the diffusion zone is characterized by greater homogeneity. According to X-ray diffraction analysis, a γ'-phase of the (Fe, Me) 4N type is present on the surface nitrided at a minimum speed of processing of 1020 steel, and at a processing speed of 10 mm s, in addition to the γ'-phase, an ε-phase of the (Fe, Me) 2-3N is presented too. In the first case, the intensity of the x-ray lines which correspond to the α-solid solution is several times higher than in the second one, and exceeds the intensity of the lines which correspond to the γ'-phase. Consequently, due to the change in technological saturation factors, it is possible to regulate the resulting layers in terms of structure and phase composition and optimize the properties of the nitride workpieces and tools.

5. Nitrocarburizing

A unique feature of nitrocarburizing during plasma-arc modification in a liquid is an increased concentration of nitrogen and carbon. The depth of the diffusion layer on 1020 steel was 0.6-1.1 mm, microhardness was 11000-12500 MPa. Microhardness increases with increase of heating rate. Heating at a lower rate increases the time, during which the nitrogen-carbon-containing liquid evaporates intensively, which increases the concentration of active carbon and nitrogen atoms at the boundary line between the saturated medium and the metal surface. However, an increase in the concentration of nitrogen and carbon in the region of the anode spot of the plasma arc [6] leads to an increase in the residual austenite (from 2.5 to 40% on 1020 steel), which reduces the microhardness. The depth of the diffusion layer on 1018 steel was 0.05-0.1 mm, and the microhardness was 10200-11100 MPa. The depth of the diffusion layer on 1020 steel was 0.06-0.12 mm, and the microhardness was 11200-13000 MPa.

The presented results show the possibility of surface plasma-arc modification in liquid carbon and nitrogen containing media. Processing this way causes appearance of a boundary condition of the workpieces surface, where thin surface layers can have traces of micro-melting without forming a pool of liquid metal in the form of a thin film. In this state, process of diffusion of carbon and nitrogen into the metal is greatly accelerated. Important parameters are the composition of the liquid medium, heating and cooling rate of the surface layer of the metal.

6. Conclusion

The conducted studies on the estimation of plasma-arc modification of the surface of metals in liquid media have shown the promising outlook of this method for the production of complicated in terms of microstructure of modified layers. Modified by carbon and nitrogen layers of thickness up to 200-250 μm with a microhardness of 8000-12400 MPa were obtained. By regulating the ratios of the components, which are part of the liquid medium, it is possible to obtain different surface layers of the metal with respect to structure and properties.

References

[1] Santhanakrishnan S and Narendra B 2013 *ASM International Handbook Committee* pp 476–503
[2] Lu S, Wang Z, Lu K and Mater J 2010 *Sci. Technol* **26** pp 258–263
[3] Bataev I, Golkovskii M, Bataev A, Losinskaya A, Dostovalov R, Popelyukh A and Drobyaz E

2014 *Surface & Coatings Technology* **242** 164–9 DOI: 10.1016/j.surfcoat.2014.01.038
[4] Ismail M I S and Taha Z 2014 *Int. J. Technol* **5** pp 79–87
[5] Balanovskii A E 2016 *Plasma Surface Hardening of Metals* (Irkutsk: IrSTU) p 180
[6] Balanovskii A E 2016 *High Temperature* vol 54 **5** pp 627–631
[7] Medvedev S I, Nezhivlyak A E, Grechneva M V, Balanovsky A E and Ivakin V L 2015 *Welding International* vol 29 **8** pp 643–9
[8] Kondrat'ev V V, Balanovskii A E, Ivanov N A, Ershov V A and Kornyakov M V 2014 *Metallurgist* vol 58 **5-6** pp 377–387
[9] Balanovskii A E and Huy Vu V 2017 *Letters on materials* **7(2)** 175–9
[10] Grechneva M V and Tokmakov V P 1992 *Welding production* **7** pp 8–12

Complex metallographic researches of 110G13L steel after heat treatment

A E Balanovsky[1], M G Shtayger[2], V V Kondrat'ev[1], S A Nebogin[1] and A I Karlina[1]

[1]Irkutsk National Research Technical University, 83 Lermontov street, Irkutsk, 664074, Russia
[2]MC Mechel Steel, 1 Krasnoarmeysky street, 125167, Moscow, Russia

E-mail: kvv@istu.edu

Abstract. Complex studies with the use of modern methods of metallographic analysis with manganese steel were carried out. The mechanism of phase and structural transformations in the temperature range of heating of 110G13L steel samples for quenching from 920 to 1100 °C is considered. The optimal heating temperature for saving energy was determined.

1. Introduction

At the present time, as before, the main place in the production of cutting tools for earthmoving machinery is occupied by manganese steels, most of which are 110G13L steel or Gatfield steel [1]. The usual mode of heat treatment of such steel is quenching in water with a temperature of 1100-1150 °C, while the heating rate to quenching temperature is limited to 70-150 °C/h [1,2]. This is because the carbides of cement type formed during crystallization of the casting have different thermodynamic activity of carbon [1]. Such low heating rates, coupled with a sufficiently high quenching temperature, as well as with often required long exposures, make the process of heat treatment of 110G13L steel sufficiently long and energy-intensive.

One of the ways to reduce the time of heat treatment, in our opinion, is to reduce the quenching temperatures to some reasonable theoretically justified values. Reasonability of quenching of all castings with one temperature of 1100 °C is not entirely justified, because according to the Fe-Mn-C diagram, for example, for a steel with a carbon content of 0.90%, structure of pure austenite can be provided by heating to 860 °C [4]. For the experimental verification of the proposed assumption, castings from 5 different meltings were selected in the foundry of the heavy engineering plant (Irkutsk) and investigations were carried out. In castings of different meltings, the carbon content varies from 0.90% to 1.25%.

2. Optical microscopy

Figure 1 shows the results of optical microscopy of samples heated to different temperatures. Figure 1a shows the microstructure of austenite grains of samples heated to a temperature of 920 °C, with scattered carbide particles, both in the grain volume and along the grain boundaries. Figure 1b – to a temperature of 980 °C, it can be found less carbide particles, i.e. the process of dissolution of carbides proceeds.

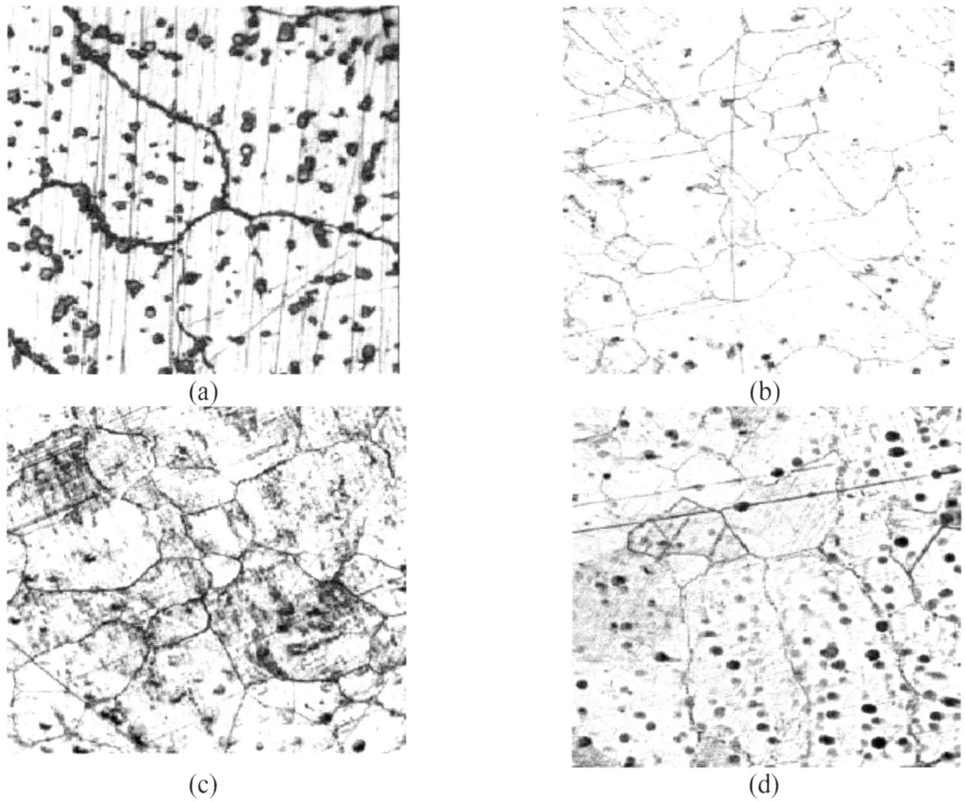

a – microstructure of austenite grains; b – dissolution of carbide particles; c, d – dark spots

Figure 1. Distribution of carbides in 110G13L steel samples heated to different temperatures.

In figure 1, the samples were heated to a temperature of 1050 °C, dark spots appear, the austenite grain increases, the carbide particles dissolve. After being heated up to a temperature of 1100 °C, multiple depressions appear in the area of the grain under investigation.

3. Electron microscopy
Figure 2 shows the photomicrographs of the surface of the steel after quenching. Structural studies have shown that against the background of the austenite structure, small carbide inclusions located along the grain boundaries are observed. This structure leads to a decrease in the strength properties of the metal. Such a structure is associated with incorrectly selected heat treatment regimes, i.e. the temperature differences, appeared during carrying the basket with castings from the furnace to the quench tank were not taken into account, temperature of the charge was lowered, which caused the carbides to fall out again. In austenite grains, traces of plastic deformation by sliding and twinning are visible. There are no obvious defects in the structure. It should be noted that the etched surface of the sample in figure 2a has traces of nonmetallic inclusions, as well as traces of darkening of the individual places in figure 2b and the depressions of the spherical shape in figure 2, c. The austenite grains are large and have the classical boundaries, figure 2, d (triple junction). The body of the grain is literally strewn with spherical depressions – figure 2, c, d, e. It's not possible to identify these spherical depressions as carbides, - with a larger zoom there is nothing in these places figure 2, d.

A large number of shear deformation bands were observed in the structure of the samples under study (figure 2, f). It is known that in 110G13L steel, toughness is approximately four times greater than in samples in whose structure these bands are absent. At the same time, a detailed analysis of the grain boundaries showed that the "loose ones" are eaten by spherical depressions. Can we consider them (spherical depressions) as the location of carbides that were artificially removed during mechanical polishing of the sample? The issue is debatable, i.e. the analysis of the literature shows that despite the preparation (mechanical polishing) of the samples, the carbide phases in 110G13L steel are retained on the surface, and they are clearly visible in the microscope (optical and electronic). In our case, it is impossible to speak unambiguously about the presence of a carbide phase in the austenite with the result of electron microscopy, as the spherical depressions can only theoretically be the locations of the carbides and which may have been removed as a result of polishing. A characteristic of the microstructure of all microsections was the presence of polygonal austenite grains, within which there were lines of shear (twinning). These lines were approximately parallel within each grain and had a definite orientation. Traces of plastic deformation lines of various degrees were observed in an overwhelming number of grains, regardless of place of study of the microstructure.

4. EBSD analysis

After the electron microscopy, the sample was repolished and underwent an additional ion polishing to conduct the EBSD analysis figure 3, a. As a criterion for small- and large-angle boundaries (SAB and LAB, respectively), disorientation of 15 degrees was used. Grains were understood as crystallites from all directions surrounded by LAB, and for the size of the grain, the diameter of the circle whose area is equivalent to the grain area on the EBSD map was assumed, figure 3b.

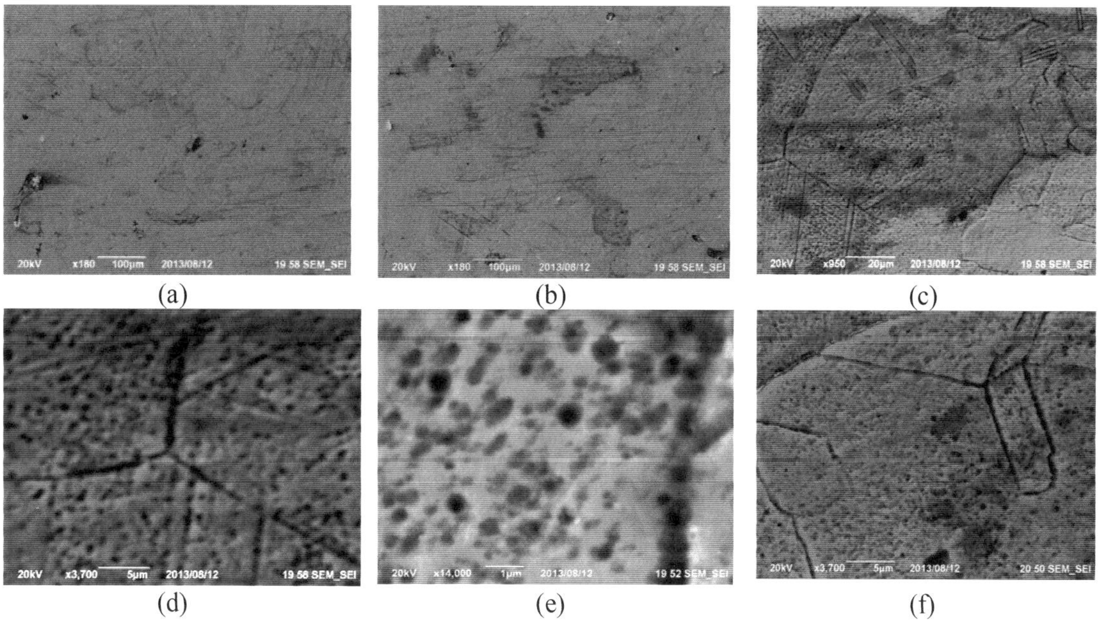

a – traces of nonmetallic inclusions; b – depressions of spherical form; c, d, e – body of the grain is literally covered with spherical depressions; f – bands of shear deformation

Figure 2. Electronic photos of microstructure of 110G13l steel.

The histogram of the distribution of the grain boundaries along the angles of misorientation in the structure of the steel sample is characterized by the presence of an acute peak near the misorientation angle of about 2° and indicates the unconditional predominance of small-angle boundaries whose content in the structure is ~ 90%.

The grain boundaries were superimposed with twin grain figure 3, c. In figure 3a, twin boundaries are represented by a common [111] direction and rotation angle of 600. Analysis of the results, presented in figure 3b, shows that after quenching, in the structure of the 110G13L steel sample, special twin boundaries Σ3 60° <111> with a concentration of 0.813% were formed. The stresses of the 2nd kind, figure 3c, were measured (in austenitic grains the maximum stresses are indicated in red). The visual image, figure 3c, shows that the sample is in a stressed state. Local stresses in the places of spherical depressions are shown in figure 3e. Figure 3e shows the general picture of grain distribution in the sample and their diameter. In the course of the research it has been established that, on the one hand, large grains occupy the entire volume, but this is only at first glance, there are more small grains and, besides having twins within the same grain, the program considers them as a single grain.

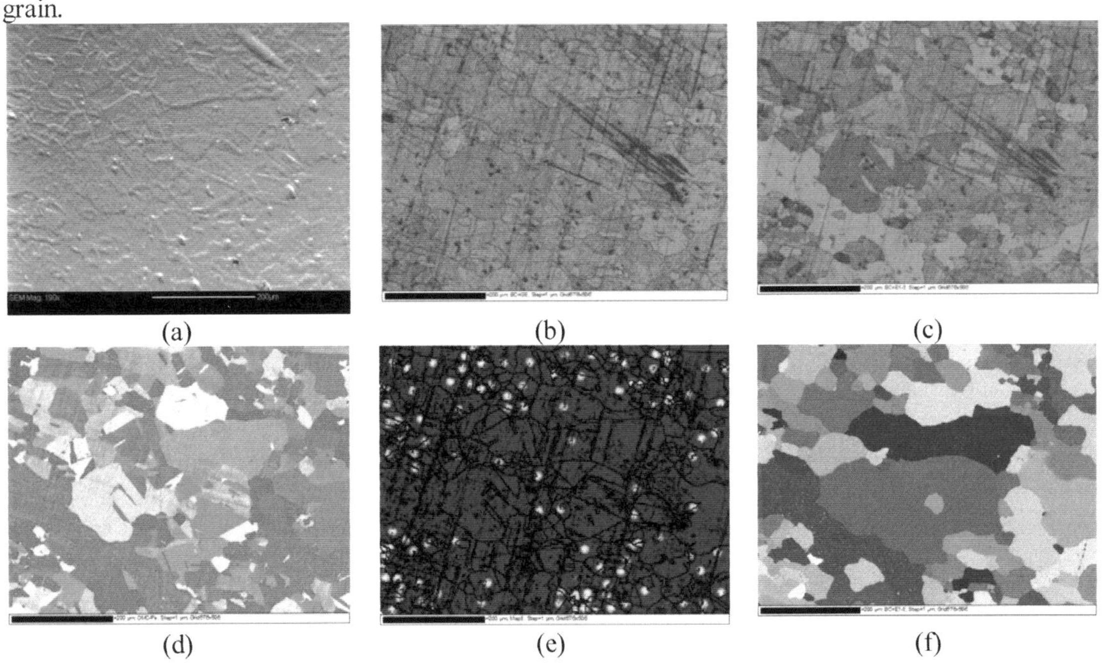

a – microrelief of the polished surface, b – grain structure with clear boundaries; c – grains with twin boundaries (rotation angle – 600), special twin boundaries Σ3 60° <111>; d – strain of the 2nd kind in austenite grains (red color); e – local stresses in the places of spherical depressions; f – general image of the grain distribution

Figure 3. EBSD analysis of samples from 110G13L steel.

5. Conclusion

Thus, the complex metallographic studies, carried out by us, made it possible to clarify the mechanism of phase and structural transformations in the temperature range of heating of 110G13L steel samples for quenching from 920 to 1100 °C. This makes it possible to correct the heat treatment regimes of 110G13L steel castings in real production conditions. Quenching temperature can be reduced by at least 50-90 °C, which will save energy.

References

[1] Chernyak S S and Broido V L 2001 *Irkutsk* p 352
[2] Mulyavko N M 2001 *Proceedings of the Chelyabinsk Scientific Center* **4(13)** 28–30
[3] Balanovsky A E 2015 *Strengthening Technologies and Coatings* **12(132)** 18–30
[4] Balanovsky A E 2015 *Strengthening technologies and coatings* **6(126)** 27–32

[5] Balanovsky A E 2016 *Strengthening Technologies and Coverings* **1(133)** 25–34
[6] Balanovsky A E 2016 *Strengthening Technologies and Coverings* **2(134)** 20–30
[7] Gyu V V, Balanovsky A E and Kondratiev V V 2017 *Metallurgy of Machine Building* **1** 9–15
[8] Balanovsky A E and Gyu Vu 2017 *Letters on Materials* vol **2** 175-9 DOI: 10.22226 / 2410-3535-2017-2-175-179
[9] Fedin V M and Borts A I 2009 *Metal Science and Heat Treatment* vol 51 **11-12** 544–52
[10] Kargapoltsev S K, Shastin V I, Gozbenko V E, Livshits A V and Filippenko N G 2017 *International Journal of Applied Engineering Research (IJAER)* vol 12 **17** 6499–503

Investigation of mechanical properties of cold-rolled, annealed and welded semi-finished products from the test alloys of Al-Mg system, economically alloyed with scandium

V N Baranov[1], S B Sidelnikov[1], V F Frolov[3], Y Y Zenkin[2], T A Orelkina[1], I L Konstantinov[1], D S Voroshilov[1], O V Yakivyuk[1] and I N Belokonova[1]

[1]Siberian Federal University, 79 Svobodny ave., Krasnoyarsk, 660025, Russia
[2]JSC "RUSAL Bratsk aluminum smelter", Bratsk, 665710, Russia
[3]LLC "RUSAL ETC", 37/1 Pogranichnikov street, Krasnoyarsk, 660111, Russia

E-mail: d.s.voroshilov@gmail.com

Abstract. The results of experimental investigation of mechanical properties of sheet metal from an experimental aluminum alloys Al–Mg system alloyed with scandium have been represented. It was established that the deformed and annealed semi-finished products of these alloys have higher strength properties. Testing of welded samples made from these semi-finished products showed high corrosion resistance and strength of welded joints.

1. Introduction

The problem of searching for new welded aluminum alloys having higher strength characteristics than existing industrial alloys of the Al-Mg system is most directly connected both with the creation of new, more sophisticated samples of new technology and with the solution of a number of national economic problems [1].

One of the most urgent problems of scientific and technical progress is the problem of increasing the weight perfection of missile and aircraft equipment, shipbuilding and the automotive industry. The main way to solve this problem is to increase the specific strength characteristics of structural materials by using new aluminum alloys with increased mechanical properties and corrosion resistance [2].

Studies in the field of creating alloys of the Al-Mg system doped with scandium and studying the properties of products made from them have been devoted many scientific publications, including foreign ones [3-12]. With the purpose of deeper studying the patterns of formation of mechanical and operational properties of products made of aluminum alloys economically alloyed with scandium, the employees of "RUSAL" together with scientists of the School of Non-Ferrous Metals and Materials Science of the Siberian Federal University have conducted a number of experimental studies in recent years with varying the chemical composition of alloys and regimes their processing, as well as produced batches of cast and deformed semi-finished products from them [13-16].

2. Experimental procedures

The ingots for rolling were obtained by casting alloys of the Al-Mg system with a scandium content of 0.1 to 0.24%. They were subjected to homogenization annealing according to the following regime: heating in furnace at a speed of 1,16 °C per minute to 350 °C; holding at this temperature for 11 hr;

reheating to temperature 425 °C with speed 1,25 °C per minute; holding at this temperature for 8 hr; air cooling.

Each ingot was rolled according to a deformation scheme adopted, including hot rolling a billet 40 mm thick heated to a temperature of 450 °C, up to a thickness of 10 mm, and then cold rolling of these strips to a thickness of 1 mm. Next, the strip was annealed at a temperature of 300 °C and holding time of 3 hr. As equipment for rolling, the DUO 330 mill was used (table 1).

Table 1. Technical characteristics of the sheet rolling mill DUO 330

Parameter name	Value
Electric motor power, kW	90
The mains supply voltage is three-phase, V	380
Current frequency, Hz	50
Width of barrel rolls, mm	520
Diameter of rolls, mm	330
Maximum roll separation, mm	70
Rotational speed of rolls, rpm	10
Maximum rolling force, MN	1.55
The maximum rolling moment, MNm	0.82

In the course of rolling, samples were taken for testing the mechanical properties and the temporary tensile strength R_m, yield strength R_p and elongation A_{10} for each sample studied of alloys.

Mechanical properties after cold rolling at different degrees of deformation ε are given in table 2 (HR – after hot rolling, CR – after cold rolling at different degrees of deformation).

Table 2. Mechanical properties of samples from experimental alloys in a deformed state

Alloy	Sc, %	R_m, MPa				R_p, MPa				A_{10}, %			
		HR	CR			HR	CR			HR	CR		
			40%	60%	80%		40%	60%	80%		40%	60%	80%
1.	0.07	344	426	459	454	172	244	254	267	12.5	9.3	8.7	5.0
2.	0.09	350	438	470	494	225	237	256	274	12.2	8.2	7.4	4.8
3.	0.12	356	447	477	501	232	232	251	267	12.0	8.2	7.0	4.6
4.	0.13	373	460	478	505	235	232	257	372	11.7	8.3	7.0	4.2
5.	0.14	374	464	479	510	233	246	264	270	11.2	8.2	7.0	4.5
6.	0.14	360	465	480	513	245	295	314	278	10.6	8.1	7.0	4.4
7.	0.22	376	467	485	520	254	392	412	472	7.6	7.0	5.0	4.3
8.	0.24	380	476	490	554	250	395	414	488	7.0	6.3	4.6	4.2

The results of studies of the properties of the final products of rolling (annealed strips with a thickness of 1 mm) are given in table 3. Their analysis shows that the level of properties for experimental alloys containing scandium reaches elongation of values of about 21%. The yield strength of these alloys varies from 248 to 318 MPa.

Table 3. Results of studies of properties annealed strips of 1 mm thick

Alloy number	Tensile strength R_m, MPa	Yield strength R_p, MPa	Elongation A_{10}, %
1.	407	261	21.2
2.	412	248	14.8
3.	425	275	14.3
4.	435	276	12.8
5.	462	277	12.1
6.	463	278	12.0
7.	474	307	10.3
8.	487	318	9.2

Data on the strength properties of semi-finished products from experimental alloy 3 containing 0.12% scandium, in comparison with the analog alloy 5083, which does not contain scandium, are given in table 4 and showed that the strength characteristics of economically alloyed alloy increased to a considerable extent, both in the deformed and annealed state.

Table 4. Strength properties of samples of rolled products from experimental alloys in a different state.

Sample thickness, mm	Condition	Strength properties, MPa			
		alloy 5083		experimental alloy	
		R_m	R_p	R_m	R_p
10	hot deformed	297	196	356	232
8	hot deformed	305	217	372	280
6	hot deformed	318	213	387	312
3	cold deformed	394	373	453	429
3	annealed	277	139	390	277
1	cold deformed	497	359	554	467
1	annealed	342	187	425	275

To study the properties of the welded seam, the resulting strips were welded along (method 1) and across (method 2) direction of deformation by means of argon-arc welding with filler wire from alloy 01570 (Figure 1), obtained by the method of combined casting and rolling-extruding with subsequent drawing. The mechanical properties of rods and wire are shown in table 5. Welding was performed using TIG-AC technology using a welding machine TIG200P.

Figure 1. View of welded samples.

Table 5. Mechanical properties of rods and wires from alloy 01570.

Alloy	Mechanical properties of semi-finished products				
	Rod Ø 9 mm		Wire Ø 3 mm		
	Tensile strength R_m, MPa	Elongation A_{10}, %	Wire condition	Tensile strength R_m, MPa	Elongation A_{10}, %
01570	320	9.9	Hardened	432	2.2
			Annealed	340	17.5

To assess the nature and degree of corrosion destruction of the investigated alloys, including in the weld zone, in accordance with State Standard 9.021-74 accelerated tests on the intergranular corrosion were performed (IGC). To do this, strips 3 mm thick were used, as well as samples welded from them with a welded joint along and across the direction of deformation. From the strips, tensile samples were prepared and placed in a solution containing sodium chloride solution and 0.3% hydrogen peroxide (58 g/l NaCl + 10 ml/l 33% solution H_2O_2). The temperature of the solution was 18-25°C, and the test duration is 24 hr. On all sides samples were washed with a solution, as shown in Figure 2. The samples were degreased with isopropyl alcohol before the IGC test and dried with filter paper. Then, etching was carried out for 2 minutes in a solution containing 50 ml of nitric acid (70%), 5 ml of hydrofluoric acid (48%) and 945 ml of water, at a solution temperature of 95±3°C; then the samples were washed in cold water and clarified in 25-30% nitric acid solution, then washed again and dried with filter paper. After the end of the test, the samples were washed under running water and dried in furnace at a temperature of 100±5°C within 5 minutes.

After that, tensile tests were carried out and the mechanical properties of the samples in the weld zone.

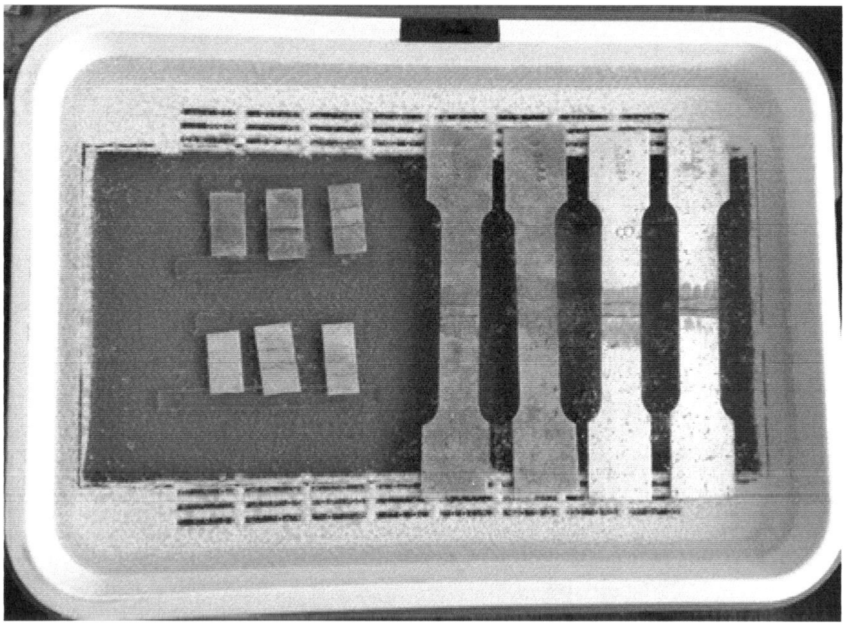

Figure 2. Scheme of laying samples for tests on the IGC.

The results of investigations for some alloys are given in table 6. Tests for intergranular corrosion of samples after welding showed high corrosion resistance of welded joints, while the weld strength is 0.75-0.85 of the strength of the base metal.

Table 6. Mechanical properties of welded samples from experimental alloys

Number of alloy/sample	Method for obtaining samples	Mechanical properties of samples after welding and annealing		
		R_m, MPa	R_p, MPa	A_{10}, %
2/1	1	362	246	6.0
2/2	2	353	293	3.3
3/1	1	361	250	6.0
3/2	2	334	245	2.6
5/1	1	351	242	5.7

5/2	2	333	240	3.3
6/1	1	344	250	5.8
6/2	2	339	246	3.1

3. Conclusion

Thus, the conducted studies made it possible to obtain data on the level and patterns of changes in mechanical properties from the total degree of deformation of rolled products from new alloys of the Al-Mg system doped with scandium in the range 0.10-0.24%, in a different state (hot deformed, cold deformed, annealed and welded). These data will be used in mastering the technology of casting and processing of these alloys in industrial conditions.

Acknowledgements
The manuscript was prepared using the results of the work carried out during the project 03.G25.31.0265 "Development of economically alloyed high-strength Al-Sc alloys for use in road transport and navigation" in the framework of the Program for the implementation of complex projects for the creation of high-tech production, approved by the Government of the Russian Federation on April 9, 2010 №218.

References

[1] Gorbunov Y A 2015 *Journal of Siberian Federal University. Engineering & Technologies* **5(8)** 636–45
[2] Belov N A 2010 *Phase Composition of Industrial and Advanced Aluminum Alloys* (Moscow: MISiS) p 511
[3] Chunchang S, Liang Z, Guohua W, Xiaolong Z, Antao C and Jiashen T 2017 *Materials Science & Engineering* **A680** 232–38
[4] Pedro H R P, Ying C W, Yi H and Terence G L 2017 *Materials Science & Engineering* **A685** 367–76
[5] Mondol S, Alamb T, Banerjee R, Kumar S and Chattopadhyay K 2017 *Materials Science & Engineering* **A687** 221–31
[6] Mengjia L, Qinglin P, Yunjia S, Xue S and Hao X 2017 *Materials Science & Engineering* **A687** 298–305
[7] Buranova Y, Kulitskiy V, Peterlechner M, Mogucheva A, Kaibyshev R, Divinski S V and Wilde G 2017 *Acta Materialia* **124** 210–24
[8] Fuller C B 2005 *Acta Materialia* **53(20)** 5415–28
[9] Ryset J and Ryum N 2005 *International Materials Reviews* **50(1)** 19–44
[10] Toropova L S, Eskin D S, Kharakterova M L and Dobatkina T V 1998 *Advanced Aluminum Alloys Containing Scandium: Structure and Properties* (Amsterdam: Gordon and Breach Science Publicshers) p 175
[11] Ibrokhimov S Zh, Eshov B B, Ganiev I N and Ibrokhimov N F 2014 *Izvestia of Samara Scientific Center of the Russian Academy of Sciences* **16(4)** 256–60
[12] Ibrokhimov S Zh, Eshov B B and Ganiev I N 2013 *Reports of the Academy of Sciences of the Republic of Tajikistan* **56(6)** 472–75
[13] Yakivyuk O V, Sidelnikov S B, Dovhzenko N N, Frolov V F, Krohin A Yu, Bezrukikh A I and Frolov V A 2016 *Non-Ferrous Metals and Minerals 2016, Collection of Abstracts* 262–63
[14] Frolov V A, Yakivyuk O V, Frolov V F and Voroshilov D S 2016 *Collection of Materials XVII International Scientific and Technical Ural School - a Seminar of Metal Scientists - Young Scientists. Part 2* 113–17
[15] Frolov V A 2016 *Innovative Processes of Metal Working With Pressure: Materials 2 of the Int. Scientific and Practical Conf.*, ed M V Chukin (Magnitogorsk: Magnitogorsk State Technical University named after G I Nosov) pp 23–24
[16] Baranov V N, Sidelnikov S B, Bezrukikh A I and Zenkin E Y 2017 *Tsvetnye Metally* **9** 83–88

Accelerated reclamation of iron ore enrichment waste

A S Vodoleev, O V Berdova and N A Yumasheva

Siberian State Industrial University, 42 Kirova street, Novokuznetsk, 654007, Russia

E-mail: botanik-egf@yandex.ru

Abstract. Reclamation includes a set of measures aimed at forming on the surface of industrial dumps a favorable root layer by various methods with subsequent use of this area. Conservation is the fixation of the dumps surface by mechanical means or by gardening the dumps surface with the preliminary application of a minimum layer of soil, peat, mineral fertilizers, growth substances, unconventional soil improvers – sewage sludge (SS). As a result of the work carried out on the surface of iron ore enrichment waste of Abagur Processing Agglomeration Factory, technosols were formed – artificial soils with a root layer of sown leguminous grasses, consisting of a mixture of SS and tailing material. The reclamated areas have been created on iron ore enrichment waste, which ensure erosion stability of the tailing dump surface and conservation of enrichment wastes. The chemical and agrochemical parameters of the technosols over the two years of their development have changed. Some alkalinization of the medium and a sharp decrease in the amount of organic matter and all forms of nitrogen in the substrate of the root layer of plants are observed.

1. Introduction

Any disturbed territory of technogenic origin passes in its development two phases – technogenic formation and a post-technogenic phase of development [1,2,3,4]. The leading mechanisms of transformation of man-made landscapes into natural ones are the biological processes leading to the restoration and development of biogeocenosis. The formation of cenoses of any level is determined by the soil-ecological effectiveness of reclamation, which depends on the level of utilization of reclamation resources. In connection with various purposes of reclamation, these or other technologies are used to create conditions for the development of soil-biological processes in disturbed areas.

Biomonitoring of reclaimed experimental sites makes it possible to optimize and accelerate the process of formation of crop phytocenoses on industrial waste, to monitor the condition and development of the grass stand, to monitor the restoration of fertility of disturbed lands, to predict and guide the development of succession processes in the reclaimed areas in the most expedient direction, to calculate the ecological and economic effect of the conducted reclamation works.

Now in Kuzbass and other industrially developed regions, huge production areas are occupied by industrial waste dumps, many of which may be near populated areas. At the same time, the stored waste causes great environmental damage to the environment and negatively affects the living conditions of the population. Their harmful effects can be neutralized in several ways: by recycling, reclamation or conservation.

Utilization is the secondary use of industrial waste in human economic activities. Reclamation includes a set of measures aimed at forming a favorable root layer on the surface of industrial dumps in various ways, for example by applying potentially fertile and fertile soil layers up to 1 m thick, followed by the use of this area for agriculture or other purposes [5]. Conservation is the fixation of

IOP Conf. Series: Materials Science and Engineering **411** (2018) 012016 doi:10.1088/1757-899X/411/1/012016

the dumps surface by mechanical means (film deposition, asphalting, gravel covering, etc.) or by greening the surface of the dumps [6] with the preliminary application of a minimum layer of soil, peat, mineral fertilizers, growth substances, unconventional soil improvers – sewage sludge, etc. [7].

2. Object of the study

The investigations were carried out at the tailing dump No. 1 of Abagur Processing Agglomeration Factory, which is located in Novokuznetsk. Abagur Processing Agglomeration Factory is a large industrial facility, which activity significantly complicates and worsens the environmental conditions of the environment and negatively affects the city environment and adjacent landscapes. Its vast tailings are filled with sandy loamy substrata and are constantly exposed to water and wind erosion, polluting the atmosphere, soils and water.

At the same time, the stored waste is a substratum containing a lot of valuable elements, which in the future will be processed. Therefore, these tailings can be considered as technogenic deposits. For this reason, radical reclamation works on these technogenic objects are inexpedient, it is necessary and sufficient to fix their surface by biological methods by forming sanitary-protective plantations.

According to the technological scheme at Abagur Processing Agglomeration Factory the production waste – tailings of magnetic separation in the amount up to 2.2 mln tonnes per year of iron ore through the slurry pipeline are sent to the tailing dumps. At present, three tailing dumps occupy an area of more than 350 hectares: tailing dump No. 3 is active, dumps No. 1 and No. 2 are isolated plateaus up to 20 meters in height with a total area of about 190 hectares. The reserve of tailings stored in them is estimated at about 100 million tons. By their chemical composition and properties, tailings can be classified as industrial raw materials with a wide range of possible applications. At present, the technology of extracting useful elements from the stored waste by the methods of gravitational enrichment, plasma treatment, etc., is being explored. Large-scale processing for 30 years of existence of tailing dump No. 1 has not been organized.

For the organization of reclamation works, it is possible to use the classification of disturbed territories, which takes into account the specifics of technogenic objects, and determines the direction of practical measures to restore this type of disturbances. According to the classification of industrial dumps by V.V. Tarchevski [8], this industrial object by its origin refers to dumps of the processing industry of a bulk type; by age – middle-aged (over 25 years); by shape – cup-shaped; by height – medium (up to 25 meters); by mechanical composition of the surface substrate – coarse-grained and sands (particles up to 0.1 mm); by acidity (pH) – acidic; by utilization – unused.

The tailing dump is filled with the help hydrotransport, which led to the differentiation of material according to the granulometric composition. In the soil-ecological relation, the material of the processed tailing dumps is characterized by a very high heterogeneity of practically all chemical, physicochemical, agrophysical and agrochemical parameters. This heterogeneity is determined by the specifics of the technology of formation of hydrodumps, which differentiates the material both in the area of the hydrodump and in its thickness. High density – 1.7 g/cm^3 and higher, makes this substrate almost root-impermeable, sharply reduces the volume of pore space and water permeability. For this reason, reclamation of tailing dumps requires the introduction of a special technological element that reduces this density, for example, mixing with other less dense substrates.

The tailing dumps substrate is classified as highly saline, phytotoxic. At the same time, the degree of phytotoxicity reaches very high values. Factors of phytotoxicity are chlorides and sulfates. The share of these salts is approximately the same. Their waste is classified as a class 4 toxicity.

The main reasons preventing the natural overgrowth of the surface of the Abagur tailing dump are:

- phytotoxicity due to salinization of the surface;
- high density of substrate;
- low content of plant nutrients;
- unfavorable microclimatic conditions, such as high temperatures in summer, insignificant accumulation of snow cover in winter, wind erosion, which not only carries out tailing material, but damages young shoots and leaves of plants with sand particles.

Phytotoxicity of rocks in combination with their high density are the main reasons for the long existence of technogenic desert of tailings.

For the formation of soil-vegetation layer, a technology of reclamation is necessary, which allows the conditions on the surface of the tailing dump to be improved and the long-term functioning of the phytocenosis in these microclimatic conditions to be ensured. Successful reclamation is possible by creation of a favorable root layer introducing organic-mineral mixtures during the technical stage and subsequent biological reclamation, i.e. creation of sustainable crop plants to prevent erosion and pollution of surrounding areas [9,10]. The creation of a grass cover on the surface of the tailing dump will radically improve the situation with snow accumulation in winter, which, together with the presence of organic substrate, will improve the efficiency of soil-plant layer restoration on its surface.

3. Technique of field experiment

It is known that the most rational way to restore the organic component of the dumps is to place on their surface a fertile layer of soil (FLS), removed during the construction of new dumps. This method was used for reclamation of slopes of the tailing dump No. 2. At the same time, a FLS removed from the tailing dump No. 3 was used. But this material is not enough for all areas that require reclamation.

Therefore, to create a root layer of seeded leguminous plants on the surface of the tailing dump, sewage sludge of city treatment facilities was used. As a result of the work carried out on its surface, technosols were formed – artificial soils with a root layer of sown grasses, consisting of a mixture of SS and tailing material. Below are the results of the study of the properties and modes of technosols of experimental sites in Abagur tailing dump No. 1.

The goal of creating techno-systems was to study two soil-ecological effects. First, it was necessary to find out how the sharp differentiation of the profile of the technosols affects two layers with different granulometric composition for the restoration of soil functions and, accordingly, for biological productivity. Secondly, the effect of different deposition rates of the SS layer on the same parameters.

The variants of technosols included three series (A-1, A-2, A-3), which differed from each other in the amount of introduced SS (with thickness 30, 20 and 10 cm, respectively).

Earlier it was shown [7] that the first factor limiting the development of the cultivated phytocenosis in all variants of experiments is the phytotoxicity of the tailing dumps and SS because of chlorides and sulphates. In the case of technozomes belonging to the A series and having a profile differentiated by the rocks, the phytotoxicity of the rocks in the root zone is due to the salts contained in the SS, and in the underlying one – due to the tailing dump rocks. It was predicted that a pre-plowing of the tailing dump surface and loose build-up of freshly dumped SS will create conditions for substrate self-melioration in the root layer. The investigations carried out by us for the 2 years of vegetation of leguminous grasses and, consequently, the development of soil regimes and soil-ecological functions, confirmed this prediction.

4. Results and discussion

The results of the analysis of the salt composition of aqueous extract obtained from the substrate of the root zone of technosols of the considered experiment show that the degree of phytotoxicity sharply decreased (table 1). This was facilitated by a decrease in the proportion of dense residue and a decrease in the concentration in solutions of chlorides and sulfates, primarily magnesium and sodium.

It is necessary to pay attention to the fact that in different technosols of this option the degree of decrease in the concentration of phytotoxic salts is different. The largest it was in the series A-1, the smallest – in A-3. In other words, the lower the thickness of the SS layer, the higher the leaching rate of salts.

However, this does not mean that the 30-cm thickness of the SS layer is superfluous. First, at such thickness the phytotoxicity of the substrate decreased substantially to the level that does not interfere with the satisfactory development of cultivated phytocenosis. Secondly, there is reason to believe that the process of leaching of salts will further continue, and if this leaching mode is maintained, the processes of optimizing the saline mode of the technozem will cover the entire thickness of the root layer. Thirdly, one of the most important goals of the experiment is to solve the problem of SS placement. Therefore, in the long term, in the soil-ecological plan, it is much more important to prevent the development of processes of secondary salinization, which can manifest itself quite realistically, provided soil drought and secondary compaction of the substrate in the root layer at any SS thickness.

Table 1. Comparative analysis of water extract of technozems in the root layer

The year of experiment	Option	Dense residue, %	Content, mg·equiv/100g						Toxicity	
			HCO_{3-}	Cl^-	SO_4^{2-}	Ca^{2+}	Mg^{2+}	Na^++K^+	Cl^-	SO_4^{2-}
The 1st year	A-1	1.731	0.72	1.00	16.12	14.65	1.57	1.62	3.3	0.2
	A-2	1.646	0.81	1.20	15.62	14.21	1.86	1.56	4.0	0.2
	A-3	1.753	0.76	0.98	16.52	14.75	1.52	1.99	3.2	0.3
The 2nd year	A-1	1.006	0.96	0.52	7.59	7.23	1.23	0.61	1.7	0
	A-2	1.070	1.00	0.65	8.16	9.50	1.36	1.05	2.2	0
	A-3	1.125	1.28	0.70	9.45	10.73	1.45	0.75	2.3	0.1

After 2 years of self-development of soil-ecological functions, the parameters characterizing the physical state of substrate in the root layer changed, the density of soil increased noticeably, and the porosity decreased (table 2).

Table 2. Comparative analysis of the basic physical properties of technosols.

The year of experiment	Option	Density of the solid phase, g/cm^3	Density of compaction, g/cm^3	Porosity, %
The 1st year	A-1	2.28	0.68	70.2
	A-2	2.20	0.73	66.8
	A-3	2.25	0.80	64.4
The 2nd year	A-1	2.33	1.03	55.8
	A-2	2.58	1.18	54.3
	A-3	2.70	1.05	61.1

However, despite this, the named physical parameters remain within the limits close to optimal, and compaction itself should be considered a consequence of the natural subsidence. It is important, in terms of further monitoring studies, to trace at what level of density the subsidence processes will stop and whether they will go beyond the boundaries of the optimum zone. As a prediction it can be assumed that the subsidence will continue until the active humus accumulation processes begin to develop in the root layer of the technosols. At the current stage of development of technosols the thickness of the SS layer has practically no effect on the intensity of subsidence phenomena.

The chemical and agrochemical parameters of technosols changed after 2 years of their development (table 3).

Table 3. The main chemical and agrochemical parameters of Abogur tailings after 2 years of vegetation of perennial grasses.

Option	pH	C, %	The content of mobile forms, mg/100g of substrate			
			NH_4^+	NO_3^-	K_2O	P_2O_5
A-1	7.61	3.5	3.5	7.1	32.8	40.9
A-2	7.66	4.1	3.2	6.4	33.0	38.5
A-3	7.28	5.8	5.3	3.0	32.7	41.2

Some alkalinization of the medium (pH values increase) and a sharp decrease in the amount of organic matter and all forms of nitrogen in the root layer are found.

At the same time, it is necessary to note the unusually high intensity of processes of organic substances mineralization contained in SS, and losses of various nitrogen forms. However, later on with the development of a stable crop plant, a certain balance of processes of humification and mineralization of organic substances will settle in the recultivated areas, which will also contribute to the stabilization of the nitrogen content in the root layer of technosols.

5. Conclusions

1. The carried out researches show that without reclamation measures the creation of a stable phytocenosis and the formation of a root layer of sown grasses in the tailing dump, which prevents the development of water and wind erosion, is impossible. The prospectivity of using SS as a soil amendments significantly improving the chemical and physical properties of iron ore enrichment waste was established.

2. Almost on all experimental sites there is a noticeable decrease in the intensity of the modes of biological activity and the functioning of the formed cultural phytocenoses.

3. As a result of SS introduction, the physical and nutritional modes on the surface of the tailing dump improve. In all options of SS placement methods, a positive effect has been achieved – the growth of terrestrial biomass of leguminous plants with an increase in the rate of SS introduction.

Thus, when depositing SS on the surface of the tailing dumps with sowing of grasses, the set goal of reclamation is achieved – the creation of a sustainable crop plant and the cessation of the transfer of pollutants to the adjacent areas.

References

[1] Androkhanov V A, Dvurechensky V G, Klekovkin S Yu, Vodoleev A S, Kudashkina S A and Stepnov A A 2006 *Problems of Regional Ecology* **5** 33–38
[2] Galanina T V 2007 *Use and Protection of Natural Resources in Russia* **1** 22–24
[3] Kurachev V M, Androkhanov V A and Dvurechensky V G 2003 *Biological Reclamation of Disturbed Lands* (Ekaterinburg) pp 239–247
[4] Trofimov S S 1975 *Ecology and Soil Resources of Kemerovo Region* (Novosibirsk: Nauka, SO) 300
[5] Vodoleev A S, Stepnov A A and Kudashkin S A 2004 *Problems of Ecology and Health of Industrial Cities and Ways to Solve Them* (Novokuznetsk) pp 28–31
[6] Merzlaya G E, Zyabkina G A, Nesterovich I A and Fomkina T P 1995 *Agrochemistry* **5** 102–108
[7] Vodoleev A S, Stepnov A A and Kudashkin S A 2003 *Biological Reclamation of Disturbed Lands* (Ekaterinburg) pp 41–51
[8] Tarchevsky V V 1970 *Vegetation and Industrial Pollution: Protection of Nature in the Urals* **7** 84–89
[9] Beaver S H 1964 *Transactions and Papers* (Institute of British Geographers) **34** 1–31
[10] Charter R A, Tabatabai M A and Hemming S J 1993 *Amer. Soc. Agron. Annu. Meet.* (Cincinnati) p 269
[11] Weaver S H 1964 *Transactions and Papers* (Institute of British Geographers) **34** 1–31
[12] Charter R A, Tabatabai M A and Hemming S J 1993 *Amer. Soc. Agron. Annu. Meet.* (Cincinnati) p 269

IOP Conf. Series: Materials Science and Engineering **411** (2018) 012017 doi:10.1088/1757-899X/411/1/012017

Current state of the world and domestic aluminium production and consumption

G V Galevsky[1], V V Rudneva[1] and V S Aleksandrov[2]

[1]Siberian State Industrial University, 42 Kirova street, Novokuznetsk, 654007, Russia
[2]RUSAL Novokuznetsk JSC, 7 Ferrosplavny proyezd, Novokuznetsk, 654000, Russia

E-mail: kafcmet@sibsiu.ru

Abstract. The analysis of the current state of world and domestic aluminium production and consumption is carried out, the dominant tendencies are determined and the forecast for in the immediate future is given.

1. Introduction

Due to its lowest cost in comparison with other non-ferrous metals and its physicochemical properties, aluminium is widely used in aircraft and automotive, transport industries, in manufacturing of home appliances, in electrical engineering etc. Aluminium often successfully replaces other metals – lead, zinc, copper, and also competes with steel products. Therefore, aluminium consumption and production throughout the world is increasing rapidly.

2. Analysis of aluminium production

In 2016, world aluminium production increased by 3.6% compared to 2015 and amounted to 59 million tonnes [1], 37 million tonnes of which are primary aluminium produced by the electrolytic method, and 22 million tonnes are secondary aluminium, i.e. aluminium alloys, obtained by processing scrap and waste. The production of secondary aluminium is a rapidly developing branch of non-ferrous metallurgy in advanced foreign countries. Japan completely abandoned the production of primary aluminium and moved to the production of secondary aluminium, and the United States currently has the largest volumes of secondary aluminium production. Russia and the factories of the United Company RUSAL located on its territory are one of the largest aluminium producers, occupying in the group of leaders – Chalco, Rio Tinto Alcan, Alcoa the second place in the world (3.724 million tonnes in 2016) with a share of 6.8% after the Chinese companies (share 45.0%) [2]. Russia and UC RUSAL are also the largest aluminium exporter, providing 15.3% of the world exports. At the same time, 82.7% of the output is exported, which makes the financial condition of UC RUSAL directly dependent on the global market environment. These indicators also indicate a low capacity of the Russian market of aluminium-containing products.

The world aluminium production in 2016 by region is presented in figure 1 [1] and can be conditionally divided into 2 parts: China and the rest of the world.

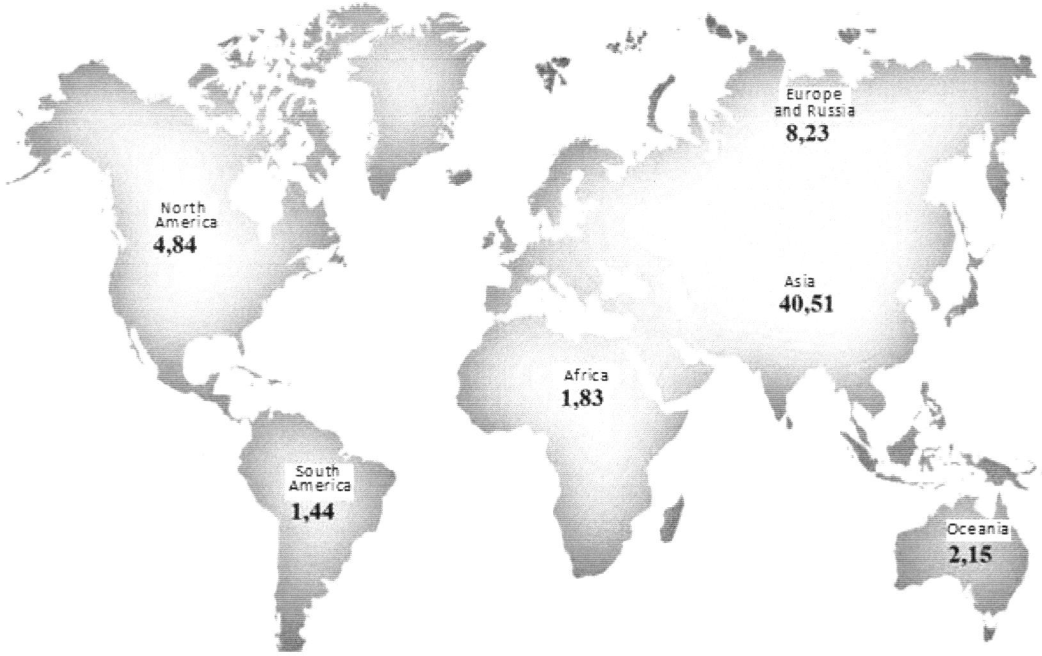

Figure 1. Aluminium production in the world in 2016, million tonnes.

Over the past 10 years, China has demonstrated incredible economic growth rates and has become the largest aluminium producer. According to the International Aluminum Institute and the analytical agency CRU, the world primary aluminium production excluding China in 2016 increased by 2.2% – 26.7 million tonnes [1]. According to Aladdiny agency, the aluminium production in China in 2016 was 32.3 million tonnes, increased by 5.5% as a result of the launch of new facilities. However, aluminium plants in China operate on thermal power, which leads to severe environmental pollution and the need for a reasonable increase in production [3]. The main aluminium producers after China are the USA and European countries, where demand is historically high due to the high level of technical and industrial development of these economies. Also Japan is a major consumer of aluminium, it imports all the primary metal it needs, liquidating its own capacity for its production. The reason for this is the lack of powerful and affordable sources of electricity on its territory and a significant tightening of the national environmental legislation. A good growth in aluminium production is shown by actively developing countries in South-East Asia. Companies from the Middle East have an important advantage: the ability to use relatively cheap electricity for production, produced by burning associated gas from oil fields.

Plans for the future development of many aluminium-producing corporations envisage an increase in aluminium production [3]. Thus, India is expected to cease exporting alumina and its processing in full and increase the aluminium production from 460 thousand tonnes to 2 million tonnes. Norse Hydro Corporation (Norway) at the Grevenbroich and Neuss plants in Germany increases the output of automotive aluminium rolled products from 50 to 200 thousand tonnes and introduces recycling of used aluminium cans with 50 thousand tonnes per year. UC RUSAL continues the commissioning of the production capacities of Boguchansky aluminium plant with the projected production of 750 thousand tonnes per year.

The price of aluminium on the London Metal Exchange in February 2017 reached $ 1905 per tonne [1]. This was due to the deficit of metal in the market: in 2016 it amounted to approximately 0.7 million tons. Figure 2 shows the dynamics of prices for aluminium in 2010-2017.

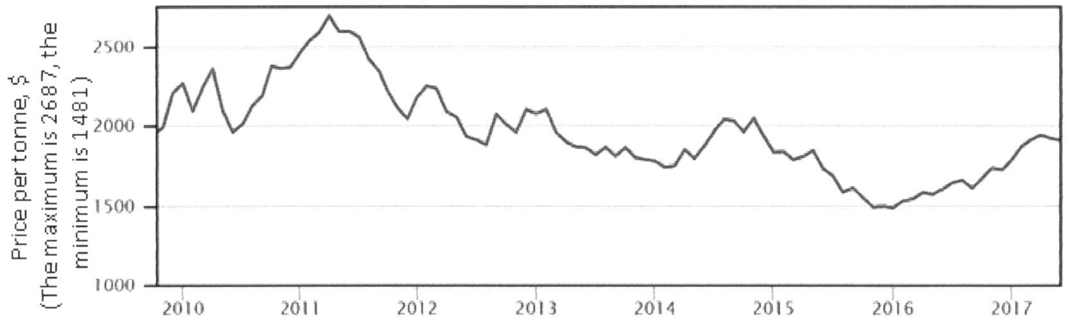

Figure 2. Performance of aluminium.

3. Analysis of aluminium consumption

Data on world aluminium consumption and production in the period 2010 – 2016 are shown in figure 3 [5].

According to the estimates [5] in 2017 the capacity of the global aluminium market could increase to almost 64 million tonnes, in 2018 – up to 66 million tonnes, in 2023 – up to 78 million tonnes. Demand for aluminium is provided by construction (+4 million tonnes), transport (+5 million tonnes), electric power (+2 million tonnes), packaging and machine building (+1 million tonnes) industries. The existing structure of the world aluminium consumption is presented in figure 4 [4, 5].

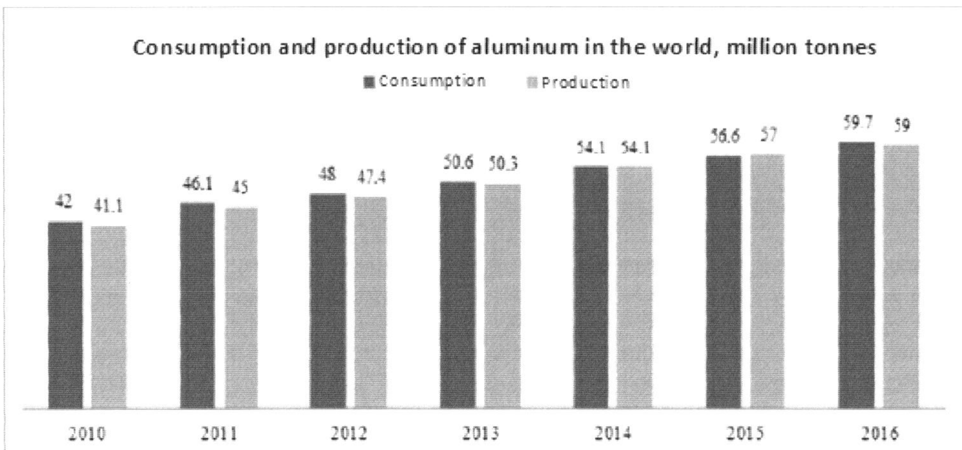

Figure 3. Diagram of world aluminium consumption and production.

The average per capita consumption of aluminium in the world at the end of 2015 was estimated at 7.7 kg. It is expected that by 2020 this figure will grow to almost 9 kg. The highest level of aluminium consumption is currently observed in countries with highly developed automotive industry – Germany and South Korea, where the per capita consumption level exceeds 26 kg per year. Next come the USA, Sweden, Japan, the United Arab Emirates, China – 15-16 kg per year [4]. Data on world average per capita consumption of aluminium are shown in figure 5.

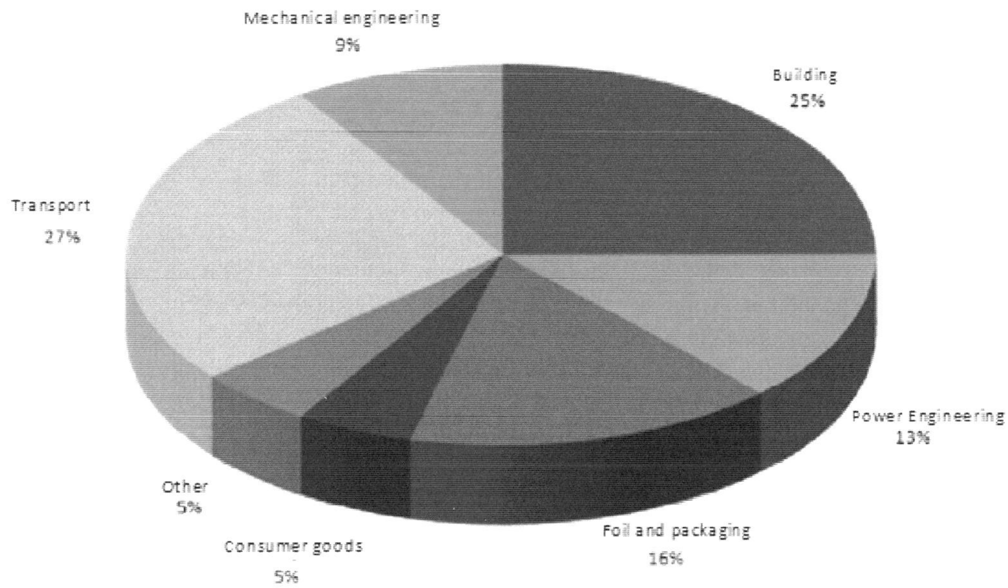

Figure 4. Diagram of world aluminium consumption by economic sectors.

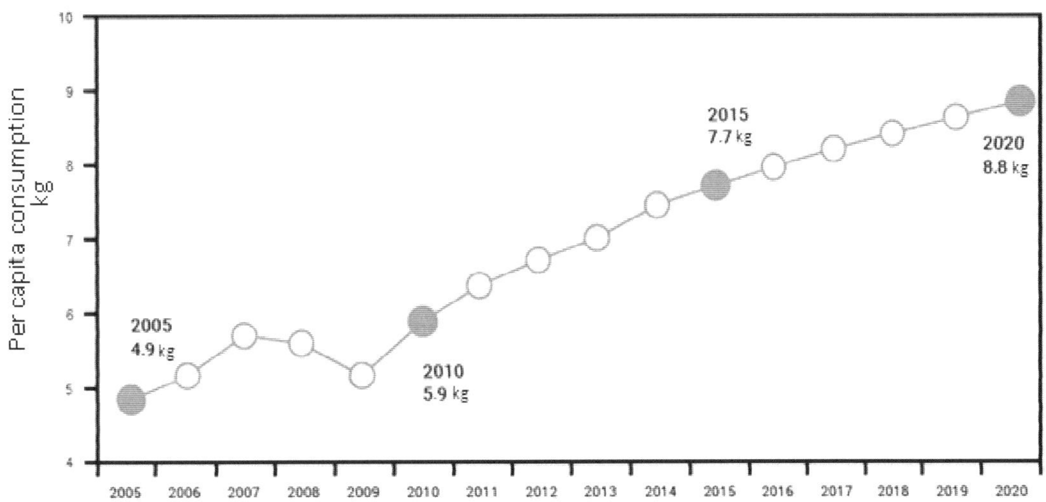

Figure 5. The graph of the change in the world average per capita consumption of aluminium.

The use of aluminium in the transport and engineering industries allows vehicles to save fuel and reduces emissions of harmful substances. Car and trailer hulls, wheel disks, hulls, bumpers, fuselage parts of aircraft, fuel systems, motor parts, marine vessels, space shuttles and rockets are made of aluminium alloys. According to calculations by the Automobile Manufacturers Association, every kilogram of aluminium used in car production leads to a reduction in its weight by kilogram, and each percent of car weight saving results in a 1.0% reduction in fuel consumption [7, 8]. This means that 100 kg of aluminium in the car save more than 1000 liters of gasoline for every 200 thousand kilometers and reduce carbon dioxide emissions by 2500 kg. Today a car produced in North America contains on average 116 kg of aluminium, Japanese and South Korean - 93, Western European - 90, and domestic - 30-40 kg.

In modern aircrafts aluminium accounts for an average 80% of the total weight, which, given the size of the airliners, is an impressive figure. So, for example, 75 tonnes of aluminium are used in Boeing 747, and in American shuttles the share of aluminium reaches 90% [7, 8].

Aluminium has also taken a very strong position in the construction. It is used everywhere: in building frameworks, window and door panels, roofing, facade and load-bearing structures, exterior decor elements, siding, stairs, air conditioning and heating systems. With the same load-bearing capacity, the weight of the aluminium structure is two to three times less than the weight of steel and up to seven times less than the weight of the reinforced concrete structure. The minimum design life of aluminium structures is estimated at 80 years [9]. They are resistant to climatic influences and "work" in a wide temperature range from –80°C to + 300°C. A potential drawback of the use of aluminium in construction is its high thermal conductivity. Therefore, in aluminium constructions intermediate thermal interlayers or thermal breaks can be used from materials with low thermal conductivity.

Packaging and energy are the following in terms of the volume of aluminium used by the economy. Aluminium has low density, attractive appearance, compatibility with food and beverages, high thermal conductivity, which makes it a demanded material for various types of packaging. In this area, the main positions are occupied by food foil and aluminium can for drinks: annually more than 230 billion cans are produced [9].

The electrical industry is one of the most important consumers of aluminium, which uses this metal for the production of conductor products, power lines, telephone wires, capacitors, radars, small motor cases [2, 9].

In aluminium and magnesium reduction cells, as well as in powerful arc furnaces for the production of silicon and some ferroalloys, tires from various aluminium alloys or from aluminium grade A5E are used to transfer large currents. Now the cross-section of such tires reaches significant sizes, and the cost is only one-third of the cost of copper tires equivalent in conductivity. Therefore, the use of aluminium tires restricts only the geometric dimensions of the units [2].

The scale of aluminium consumption for the production of consumer goods over the past decade has increased dramatically. Aluminium alloys are used for the production of household goods, furniture, components, parts and housings of household goods and equipment. Aluminium alloys have good thermal conductivity and high strength, so they are a good choice for applications in heating, ventilation and refrigeration systems.

Aluminium frying pans and cooking pots, baking trays and kettles, braziers and cooks are practical, cheap, functional, easy and convenient to use. In addition, aluminium has a high thermal conductivity – it is 2.4 times higher than that of steel. The aluminium pan absorbs only 7% of the heat received (four times less than steel) [9].

Aluminium powders and granules of a wide variety of sizes and shapes are of great interest [9]. They are used in metallurgy as alloying additives, for manufacturing semi-finished products and parts by pressing and sintering them, for obtaining a number of chemical compounds of aluminium, for synthesizing organometallic compounds and for catalysis, for increasing the corrosion resistance of steel and cast iron products, especially in atmospheric-exposed structures, as a pigment in coatings and inks used for painting equipment, as well as in printing.

Powder metallurgy makes it possible to reduce significantly the cost of parts and increase the coefficient of metal use, since the part obtained by pressing the powder in the die almost does not require subsequent machining. Aluminium powders have better compaction ability than iron-based powders. This allows the strength of the product to be increased and the possibilities of obtaining various semi-finished products with the same power of the press equipment to be expanded. Sintered aluminium alloys have unique characteristics, so that their area of application is constantly expanding, replacing in some cases titanium and high-strength steel grades [9].

Russia, being one of the world leaders in the production of primary aluminium, is considerably behind the countries of the European Union, North America, and China in its use in finished products.

IOP Conf. Series: Materials Science and Engineering **411** (2018) 012017 doi:10.1088/1757-899X/411/1/012017

At the same time, Russian companies export not only primary aluminium and alloys, but also high-tech aluminium semi-finished products, sheets and bars from special alloys for machine building. Along with this, the import of ready-made consumer aluminium products – foil, building structures, furniture profiles and fittings, radiators is quite large [10].

4. Conclusion

In the structure of world production the primary aluminium is 63%, secondary aluminium – 37%. The price for aluminium was $ 1905 per tonne (2017). The largest producer of aluminium is China. Russia and UC RUSAL provide 15.3% of world aluminium exports. The average per capita consumption of aluminium in the world is estimated at 7.7 kg. The world consumption of aluminium by sectors of the economy is, %: transport – 27, construction – 25, foil and packaging – 16, energy – 13, engineering – 9, consumer and other goods – 10.

References

[1] *World Aluminium – Primary Aluminium Production* http://www.world-aluminium.org/statistics/
[2] *RUSAL* http://www.rusal.ru
[3] Sizyakov V M, Bazhin V Yu and Vlasov A A 2010 *Metallurgist* **7** 42–47
[4] *How the Aluminium Market is Arranged* http://aluminiumleader.ru/economics/how_aluminium_market_works/
[5] *Review of the aluminium market. Aluminium Association* http://www.aluminas.ru/upload/iblock/4ab/2016-aluminium-market-review.pdf
[6] Ruyga I R and Hirevich E Yu 2015 *Concept* **8** 35–42
[7] Galevsky G V, Kulagin N M, Mintsis M Ya and Sirazutdinov G A 2008 *Metallurgy of Aluminium. Technology, Power Supply, Automation* (Moscow: Flint: Science) p 529
[8] *Aluminium. Thirteenth element. Encyclopedia* 2007 (Moscow: RUSAL Library) p 240
[9] *Finished Aluminium Products* http://aluminium-guide.ru/primenenie-alyuminiya-v-promyshlennosti-stroitelstve-i-bytu/
[10] *Aluminium Association of Russia: a Year of Fruitful Work* http://aluminium-guide.ru/alyuminievaya-associaciya-rossii-god-plodotvornoj-raboty/

IOP Conf. Series: Materials Science and Engineering **411** (2018) 012018 doi:10.1088/1757-899X/411/1/012018

Microsilica in the production of silicon carbide: the results of testing and evaluation of technological challenges

G V Galevsky[1], V V Rudneva[1] and S G Galevsky[2]

[1]Siberian State Industrial University, 42 Kirova street, Novokuznetsk, 654007, Russia
[2]St. Petersburg Mining University, V.O., 21 Line, 2, St. Petersburg, 199106, Russia

E-mail: kafcmet@sibsiu.ru

Abstract. The world production and consumption of silicon carbide is estimated. The expediency of using technogenic microsilica for the production of silicon carbide by the methods of furnace synthesis and plasma-metallurgical technology is shown. Schemes of mechanisms of carbon-thermal synthesis and plasma-metallurgical production of silicon carbide are proposed.

1. Introduction

The world consumption of silicon carbide is 600 – 700 thousand tonnes per year and is estimated at 670 million US dollars. The largest areas of silicon carbide applications are metallurgy (about 45% of the global demand), the production of abrasives (up to 30%) and refractories (up to 25%). The sales markets for silicon carbide powder materials are small-scale in terms of actual volume (less than 1%) but intensively developing and having a high cost estimate: powder with a particle size of less than 1 μm (so-called micronized carbide) for ceramics and nanopowder with particle size less than 100 nm (so-called "nanocarbide") for high-quality structural ceramics and electroplating. Silicon carbide in the form of nanocarbide opens up new directions for its application.

World production facilities for the production of silicon carbide of all types are estimated at 1.1 million tonnes per year. Geographical structure of their distribution is characterized by the following data (thousand tons per year): China – 400-500, Norway – 85, Russia – 70, Japan – 59, USA – 42, Venezuela – 41, Canada – 40, Ukraine – 32.5, Brazil – 30, Spain – 20, Poland – 20, Mexico – 20.

The leader in the world market for the production of silicon carbide is the French giant company Saint-Gobian. On the second place are the merged companies Exolon-ESK (USA) and Electroschmelzwerk Kempten GmbH (ESK) (Germany). In Russia the main producer of silicon carbide is the Volzhsky Abrazivny Zavod OJSC (Volzhsky city in Volgograd Region), which produces black, green and electrotechnical silicon carbide, abrasive grit, microgrits and abrasive ceramic and resinoid bond tools. The contribution of Russia to the world production of silicon carbide is about 70 thousand tonnes per year.

The prices for grain and silicon carbide powders in the last few years remain stable and amount to (US $ per tonne):

Black, with purity approx. 99 %	
Grade 1	1400 – 1500
Grade 2	1150 – 1300
Green, with purity approx. 99.5 %	1650 – 1850
Refractory	

Grade 1, with purity approx.98 %	1300 – 1400
Grade 2, with purity approx. 95 %	1050 – 1250
Micronized with purity approx 98 %	13000 – 15000
Nanocarbide with purity approx. 98 %	200000 – 250000

In the near future the increase in prices for silicon nanocarbide is predicted.

2. Purpose and statement of the problem

A new direction in the technology of silicon carbide, which allows the necessary characteristics of the product to be achieved and the requirements of resource-saving to be met, is the use of fine silica-containing materials of anthropogenic and natural origin as raw materials that have no value and are accumulated in great amounts. For example, over the course of twenty-five years the attention of researchers and technologists was focused on microsilica of various origins (technogenic, natural, artificially synthesized), its production conditions and properties were described in [1-5]. A common property of microsilica is its extremely high dispersity, which is 4-10 nm in artificially synthesized silica-carbon compositions, 100-200 nm in microsilica of technogenic origin, formed during silicon electrofusion and high-silicon ferroalloys, 1000-5000 nm in silica-containing natural schungites. The use of microsilica promotes the development of nanotechnologies in the production of silicon carbide and the creation of nanomaterials based on it.

To date, three directions of using microsilica in technological processes have been formed: the synthesis of organic-silicon compounds; secondary use in production cycles (for silicon smelting and ferrosilicon); synthesis of high-dispersed silicon carbide (figure 1).

3. Study of microsilica properties

Due to the inconsistency of the existing data on the characteristics of microsilica, its complex physical and chemical study was carried out, which included the determination of phase and chemical compositions, specific surface area, particle size and morphology. The microsilica obtained in the production of crystalline silicon Kr1 at RUSAL-Irkutsk JSC (MS-Si) and ferrosilicon of grade FS75 at Kuznetsk Ferroalloys JSC (MS-FS) were studied.

The technogenic microsilica of both species is represented by the following phases: β-cristobalite, α-quartz, iron oxides, silicates. In microsilica MS-FS there is magnetite, and in microsilica MS-Si – spinel $FeO{\cdot}Al_2O_3$. Both kinds of calcium oxides, oxides of phosphorus, magnesium, manganese, titanium, free carbon and silicon are present in microsilica. Quantitative differences in the chemical composition of microsilica refer mainly to the content of silica, iron compounds and free carbon (table 1). The specific surface of microsilica of both types is 20000-22000 m^2/kg. The study of photomicrographs obtained by scanning electron microscopy (SEM) shows that both types of microsilica in the state of delivery are represented by aggregates of spherical shape of various sizes, varying in a wide range – from 100 to 600 nm. Individual aggregates are formed by spherical particles predominantly 50-80 nm in size, the number of which in the aggregate depends on its size (figures 2, 3).

4. Carbidization of microsilica

Two technologies have been developed for the production of high-dispersed silicon carbide based on microsilica: plasma-metallurgical and furnace synthesis, the results of which are presented in table 2, and micrographs of micro- and nano-powder of silicon carbide – in figures 4, 5.

According to the existing thermodynamic concepts, the process of carbothermal reduction of silica occurs with the active participation of gaseous silicon oxides. The size effect arising in the one-component system "gas-dispersed crystalline phase" consists in changing the saturated vapor pressure above the surface of crystalline particles depending on the degree of solid phase dispersion. It is supposed that the speed of interaction can be increased by increasing the evaporation surface of SiO_2 and using a carbonaceous reductant with a high adsorption capacity and a developed surface.

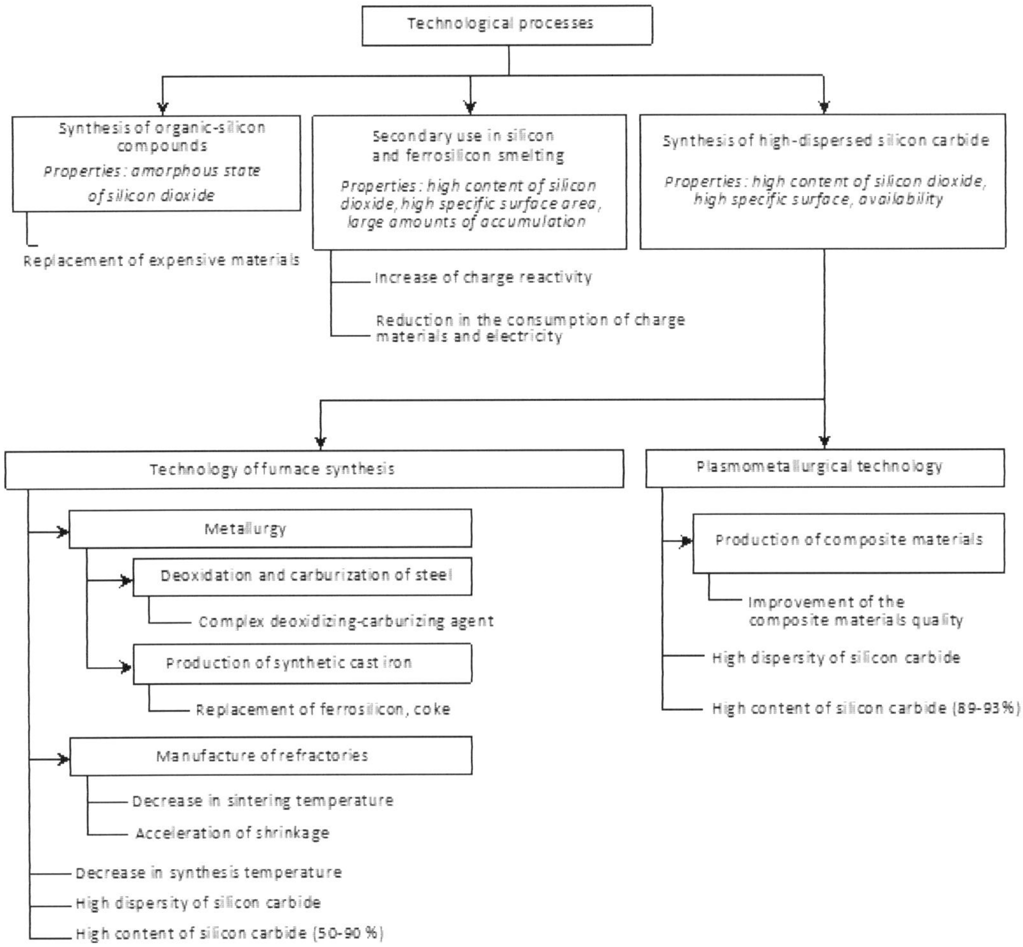

Figure 1. The use of microsilica in technological processes.

Table 1. Chemical composition of microsilica.

Element (compound)	Content,% wt.	
	Microsilica MS-Si	Микрокремнезём Microsilica MS-FS
SiO_2	93.41-95.33	91.72-93.63
C_{free}	1.96-3.28	0.56-1.18
Si_{free}	0.30-0.34	0.18-0.20
Ca(CaO)	0.30	0.52
$Al(Al_2O_3)$	0.40	0.68
$Fe(Fe_2O_3)$	0.36	0.6-1.4
$P(P_2O_5)$	0.18	0.20
Mn(MnO)	0.05	0.30
Mg(MgO)	traces	1.08
$Ti(TiO_2)$	0.01	traces

IOP Conf. Series: Materials Science and Engineering **411** (2018) 012018 doi:10.1088/1757-899X/411/1/012018

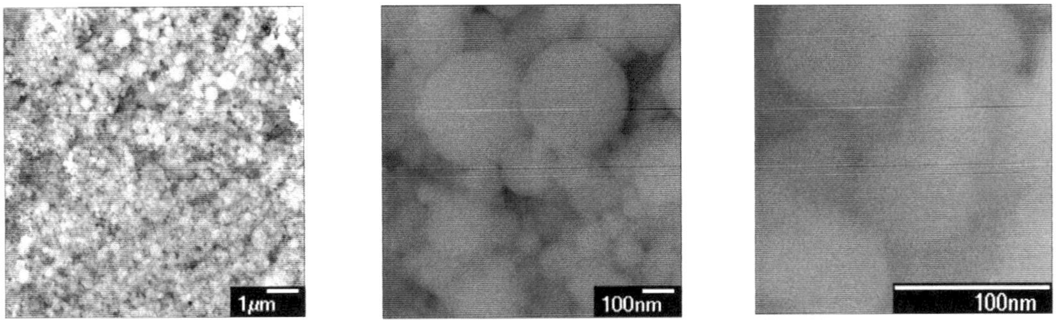

(a) (b) (c)

Figure 2. Microphotographs (SEM) of microsilica MS-Si: a – in the state of delivery; b – ensemble of particles and aggregates; c – single particles.

(a) (b) (c)

Figure 3. Microphotographs (SEM) of microsilica MK-FS: a – in the state of delivery; b – ensemble of particles and aggregates; c – single particles.

Table 2. Conditions for silicon carbide production and its basic characteristics.

Conditions for silicon carbide production and its characteristics	Plasma-metallurgical technology	Technology of furnace synthesis
Equipment	plasma-metallurgical reactor	electric resistance furnace
Raw materials containing		
silica:	microsilica of silicon and ferrosilicon production	microsilica of silicon and ferrosilicon production
carbon:	natural gas (94.5% wt. CH_4)	coke dust (82.69 % wt. C); carbon black
Process temperature, K	начальная температура плазменного потока 5400; температура закалки 2800-3000 initial temperature of the plasma flow is 5400; hardening temperature is 2800-3000	1843 – 1943
Duration of the process	$25 \cdot 10^{-3}$ sec	40-20 min
Phase composition	β-SiC, amorphous silica	α-SiC, mullite, silicate
Chemical composition, % wt.	carbide 90.87-93.23	carbide 89.77-92.02
Associated impurities, % mass	- silicon 0.61-1.04	- silicon 1.06-1.32

IOP Conf. Series: Materials Science and Engineering **411** (2018) 012018 doi:10.1088/1757-899X/411/1/012018

Conditions for silicon carbide production and its characteristics	Plasma-metallurgical technology	Technology of furnace synthesis
	- oxide 4.93-7.59 - carbon 0.61-0.95	- oxide 1.61-3.06 - carbon footprints
Specific surface area, m^2/kg	36000-38000	3000-4000
Average particle size, nm	61-65	200-900
Shape of particles	faceted	irregular, fragmented

(a) (b) (c)

Figure 4. Microphotographs of silicon carbide micropowder – SEM: a – ensemble of particles; b – single particles; c – surface of a single particle.

(a) (b) (c)

Figure 5. Microphotographs of silicon carbide nanopowder – SEM: a – morphological pattern of the aggregate; b – ensemble of nanoparticles; c – nanoparticles of a cubic shape.

According to the results of the research, a scheme of the mechanism of physical and chemical interactions in the carbon-thermal reduction of silicon oxide to carbide is proposed, which makes it possible to interpret the role of the size factor in the synthesis of silicon carbide (figure 6).

Using sufficiently small and well-mixed materials, the primary interaction is the solid-phase contact interaction of silicon dioxide with carbon $SiO_{2(s)} + C_{(s)}$ "1", which already produces gaseous monoxides of silicon and carbon at a temperature ≈1500 K, and at temperatures above 1800 K – silicon carbide and oxygen-deficient silicon-oxygen melt. The formation of silica with carbon films of silicon-oxygen melt in the contact zone increases the contact surface and intensifies the reduction process. Silicon carbide can also be formed due to the interaction of gaseous silicon monoxide with solid carbon "2". This interaction is also leading in the reduction process, and the degree of its development determines the completeness of the silicon extraction into carbide.

IOP Conf. Series: Materials Science and Engineering **411** (2018) 012018 doi:10.1088/1757-899X/411/1/012018

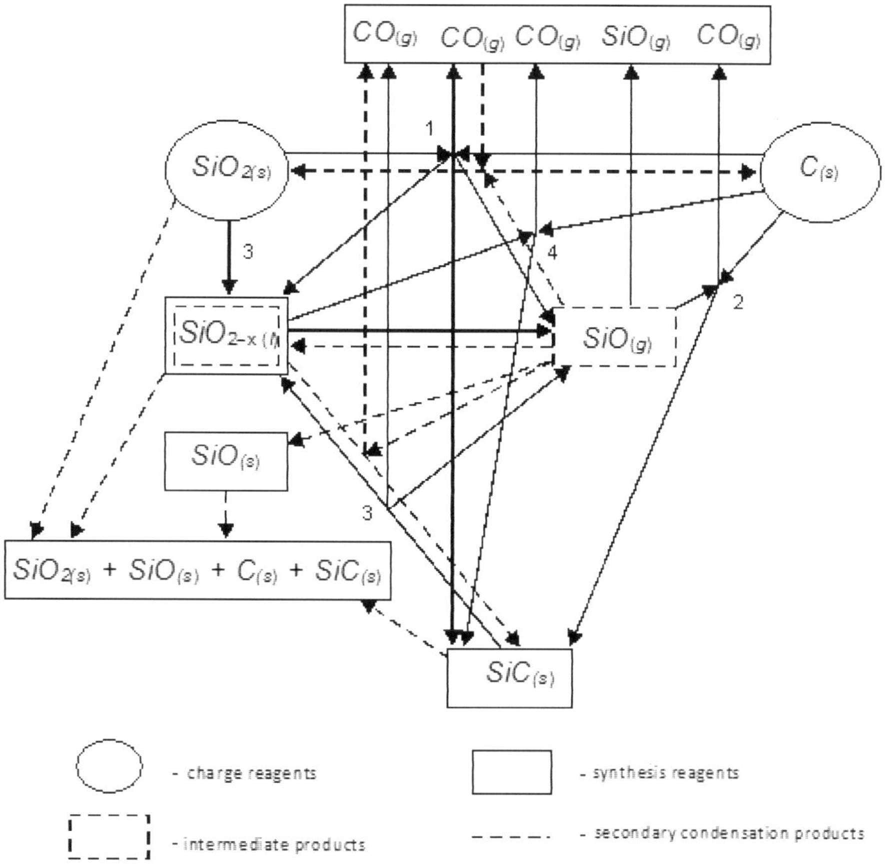

Figure 6. Scheme of interactions in the process of carbon-thermal synthesis of silicon carbide.

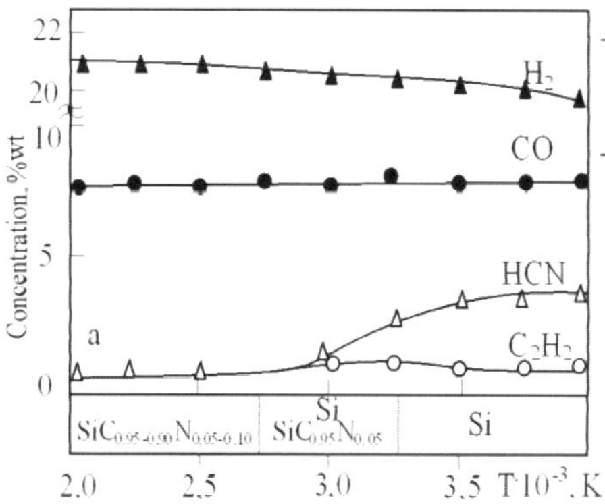

$\bar{a} \pm \Delta a$	Temperature, K			
	2000	3000	3250	4000
H_2	2.21±1.11 15.82±0.91	20.72±1.10 14.20±0.54	20.43±1.10 13.00±0.70	20.00±1.09 11.11±0.52
CO	7.54±0.40 1.11±0.06	7.31±0.41 1.32±0.06	7.53±0.39 1.19±0.06	7.67±0.36 1.40±0.06
HCN	0.51±0.03 0.62±0.03	1.33±0.07 2.83±0.14	2.74±0.14 7.11±0.32	3.71±0.19 8.93±0.41
C_2H_2	0.61±0.03 0.61±0.03	1.10±0.06 0.80±0.03	1.22±0.06 0.91±0.04	0.82±0.03 0.73±0.03

Figure 7. Temperature dependence of the composition of gaseous and phase composition of condensed products of microsilica reaction with methane in the nitrogen plasma flow ($\bar{a}$, ±Δa – average arithmetic values and confidence intervals of concentrations).

Thus, the reduction of silicon carbide to carbide is largely accomplished through the formation of a melt "4", which at high rates of chemical reactions and sufficiently long exposures can be processed without significant accumulation. The resulting silicon carbide at a later stage interacts with the silicon-oxygen melt "3" until it disappears completely.

Based on the analysis of morphological features of a nanoscale silicon carbide powder deposited in the form of faceted particles, that indicates their formation by the "vapour – crystal" mechanism and the temperature dependences of the gaseous and phase compositions of the condensed synthesis products in the temperature range 2000-4000 K (figure 7), a generalized hypothetical scheme is presented in figure 8. It seems expedient to distinguish in the plasma flow a number of spatially separated zones different in temperature conditions and, therefore, in the type of processes with a dominant development in them.

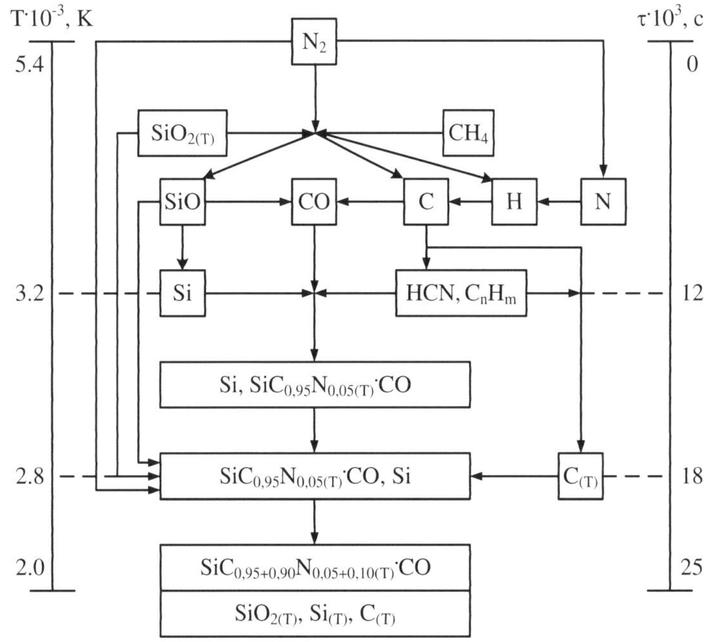

Figure 8. Scheme of interactions in the plasma-metallurgical production of silicon carbide

In the first, high-temperature zone, limited above by the initial temperature of the plasma flow, the processes occur that ensure the formation of reaction mixtures of the necessary compositions: evaporation, dissociation and reduction of silicon dioxide, disproportionation of silicon carbide, pyrolysis of methane, carbon "gasification". A characteristic feature of the gasification of silicon dioxide is the formation of the volatile monoxide SiO as an intermediate product. In the second zone, which is limited by the temperature interval 3200-2800 K, within which a significant decrease in the concentration of the main carbidizer, hydrogen cyanide, occurs, it is possible to assume with a high probability that the carbide formation proceeds during the interaction in the gas phase. In addition, sorption processes actively occur on the surface of the formed super-small carbide particles, that lead to their saturation with process gases and gaseous synthesis products. In the third zone, limited by the interval 2800-2000 K, nitriding of silicon carbide particles occurs, the nitrogen content of which is a function of the quenching temperature and, consequently, the duration of nitriding of the particles formed in the stream.

Thus, for the conditions of nitrogen plasma flow of an industrial reactor the features of carbide formation have been studied, and a single-channel version of the mechanism for the formation of silicon nanocarbide with the participation of silicon and cyanide vapours by the "vapour – crystal" scheme has been revealed.

5. Conclusion

Plasma-metallurgical technology allows silicon carbide to be obtained in the form of a nanopowder with a minimum of surface defects, and the control of the synthesis process (regulation of the gas phase composition, introduction of certain additives) provides the required composition. The silicon carbide powder obtained in this way meets the requirements for components of electrodeposited composite coatings and can be used in electroplating. When using the technology of furnace synthesis silicon carbide is obtained in the form of micropowder, the synthesis temperature can be reduced by 800-900° (compared to traditional furnace synthesis), and the duration – up to 20-40 minutes, which makes it possible to perform a continuous process, to organize environmentally friendly production, to reduce the power consumption by almost 2 times, to reduce significantly the cost of production and, accordingly, expand the possibilities of its use.

References

[1] *Bratsk Ferroalloy Plant. Production Statistics* http://www.infogeo.ru/metalls/proizv/?act=show&firm=2374
[2] *Chelyabinsk Electrometallurgical Combine. Production Statistics* http://www.infogeo.ru/metalls/proizv/?act=show&firm=2094
[3] *Microsilica* http://mk-78.ru/catalog/Mikrokremnezem
[4] *TU 5743-048-02495332-96 "Condensed Microsilica. Technical Conditions"* 1996 (Moscow: Scientific Research, Design and Technological Institute of Concrete and Ferroconcrete NIIZhB) p 16
[5] Anikin A E, Polyakh O A, Galevsky G V and Rudneva V V 2014 *Bulletin of the Mining and Metallurgical Section of the Russian Academy of Natural Sciences. Branch of Metallurgy* (Novokuznetsk: Publishing house of SibSIU) **33** 104–113

IOP Conf. Series: Materials Science and Engineering **411** (2018) 012019 doi:10.1088/1757-899X/411/1/012019

About the mechanism of titanium diboride formation in the conditions of plasma flow

G V Galievsky, V V Rudnev and K A Efimova

Siberian State Industrial University, 42 Kirova street, Novokuznetsk, 654007, Russia

E-mail: kafcmet@sibsiu.ru

Abstract. The analysis of technological results of titanium diboride production is provided, that indicates the uniqueness of the mechanism of boride formation for all methods of production depending on the composition and size of the charge materials, reaction conditions, etc. To formulate the ideas about the mechanism of boride formation processes, occurring during plasma treatment of titanium-boron-containing raw materials, the temperature dependence on the composition of gaseous and solid products, the composition of gaseous compounds thermally desorbed from the titanium diboride surface and the nanoparticles morphology were studied. Analysis of the results suggests a possible mechanism for the titanium diboride formation according to "vapour – melt – crystal" scheme, which includes the condensation of titanium vapour as aerosol, boronizing the nano-drops of the melt with borohydrides and crystallization of the titanium-boron melt. A generalized hypothetical scheme of boride formation containing 2 zones is developed: a high temperature zone (5400-3500 K) with formation of the reaction mixture in which processes of evaporation of a powdered titanium-boron-containing charge and "gasification" of boron into borohydrides occur, and a lower-temperature zone (3500-2000 K), in which the condensation of titanium vapour, a significant decrease in the concentration of borohydrides and the formation of titanium diboride take place.

1. Introduction

Modern metallurgy provides the world economic system with a variety of metal products for both mass and special purposes. According to the World Steel Association, in the structure of consumption of metal products 94% – ferroalloys, 5% – non-ferrous alloys, 1% is a diverse group of metal-containing materials with a specific set of properties.

This group includes titanium diboride TiB_2 studied and commercialized by the scientific school of the famous Russian material scientist Samsonov G.V. more than 50 years ago, and which is still in demand in the technology of various materials: ceramic-metal material tool and structural, refractory and abrasive, sprayed and fused materials, for modification and wetted by metal coatings [1-6]. At the same time, interest to application of titanium diboride is constantly growing: over the last 10 years, more than 115 papers have been published in the journals indexed in Scopus and Web of Science databases. They contain technological information on its production and use reflecting the transition trend from the application of coarse-grained titanium diboride to micro- and nanocrystalline one, which is due to the desire of scientists and technologists-practitioners to achieve a qualitatively new level of operational properties of materials and coatings on its basis.

The basis of modern production of titanium diboride is carbothermic, magnesium-thermal and gas-phase methods. Russian producers offer titanium diborides by the magnesium-thermal method of production. The strategically important nanocrystalline segment of the market is completely covered by foreign suppliers, including such companies as American Elements, Nanostructured & Amorphous Materials, Inc., PlasmaChem GmbH, NEOMAT Co. This necessitates the development of Russian nanotechnology of titanium diboride.

In connection with this, research and technological realization of boride formation processes in plasmetallurgical processing of titanium-boron-containing raw materials is an important scientific and practical task, which is of great importance for the development of domestic metallurgy of titanium and its multifunctional alloys, as well as the effective solution of innovative problems in applied materials science.

2. Purpose, hypotheses, methodology

The purpose of this work is to study the processes of boride formation in the treatment of titanium-boron-containing raw materials in the conditions of plasma flow, to describe the probable mechanism of their flow and to determine the possibility of technological control over the impurities content accompanying titanium diboride, including free boron.

A three-jet plasma-metallurgical reactor was used to experimentally study the processes of boride formation. Thermotechnical, resource and technological characteristics of the reactor are described in [7-9]. The plasma flow is generated using three EDP-104AM DC electric arc heaters (plasmatrons) with a power up to 50 kW each, installed in the mixing chamber at angle of 30° to the reactor axis. To supply the reactor with highly disperse raw materials and gaseous hydrocarbon, a water-cooled tuyere is used. To reduce the radial gradient temperature in the near-wall zone, the reactor channel is lined from the inside by a high-temperature thermal insulation material – cylindrical inserts of zirconium dioxide with an internal diameter of 0.054 m.

Titanium powders of PTN-8 grade (option 1), titanium dioxide of R-1 (option 2), boron of grade B-99 were used as titanium and boron-containing components of the charge. The powders have a size range and are characterized by the following distribution by fractions: titanium 0.5 – 4 μm and + 0.5 – 2 μm - 48.2%, +2 – 4 μm - 51.8%, dioxide 0.2 – 1.0 μm and + 0.2 – 0.5 μm, – 48.7%, + 0.5 – 1.0 μm, - 51.3%, boron 0.05 – 0.25 μm and + 0.05 – 0.125 μm – 53.2%, + 0.125 – 0.25 μm - 46.8%. Technical nitrogen was used as a plasma-forming and transporting gas (GOST 9293-84, rev.), natural gas with a methane content of up to 93.6% vol. – as a reducing agent, hydrogen – as a technological additive (GOST 3022-80, rev.).

During the production of TiB_2 in the conditions of plasma flows, the possibilities of studying the process of boride formation are even more limited which is due to the following reasons: 1) extremely high temperatures: the average mass temperature is 2800÷5400 K; 2) high movement speeds of turbulent reaction chemically active mixture – (40÷60 m/s); 3) short-term process, estimated at 15 ms; 4) extremely limited volume of the reaction zone, 0.0001 m^3; 5) the practical inaccessibility of the reaction zone for probe diagnostics due to the significant disturbing effect of a water-cooled probe technically designed with an external diameter not less than $0.6 \cdot 10^{-2}$ m; 6) complex composition of the gas phase, to be determined only after quenching, and possibly changing at the same time; 7) insuperable task with determining the content of unstable borohydrides in the gas phase. At the same time, the study of the mechanism of boride formation, alongside with the theoretical problem, is of great practical importance, since it makes possible to establish the possibility for controlling the impurities content, including the content of free boron.

The results of thermodynamic modeling of the processes of boride formation, described in [10-12], make it possible to put forward two scientific hypotheses concerning the TiB_2 formation in the conditions of charge plasma treatment:

- during interaction in the gas phase according to the "vapour – crystal" scheme

$$Ti_{vapour} + BH_{gas} \rightarrow TiB_{2\,cond.} + H_{2\,gas} \qquad (1)$$

- during interaction of titanium-containing metallic aerosol with gaseous borohydrides according to the "vapour – melt – crystal" scheme

$$Ti_{vapour} \rightarrow Ti_{melt} \quad (2)$$

$$Ti_{melt} + BH_2 \rightarrow [Ti - 2B]_{melt} + H_{2\ gas} \quad (3)$$

$$[Ti - 2B]_{melt} \rightarrow TiB_{2\ solid} \quad (4)$$

To confirm hypothetical representations, accumulate and analyse the necessary scientific information, the following studies were carried out:

1) The temperature dependence of the composition of gaseous and solid products of charge treatment was studied. For this purpose the process was carried out in the reactor of various lengths – from 6 to 24 calibers, which made it possible to change the quenching temperature of products from 4000 to 2000 K and to carry it out during the gas outflow from the reactor to the precipitation chamber at a rate 104-105 degrees/s. The nanoproducts deposited on the water-cooled surface of the precipitation chamber were analyzed to determine the phase and chemical compositions. At each tempering temperature, samples of the gas-powder mixture were also sampled using a probe. For these purposes a steel intensively cooled probe was used (water flow 0.050 kg/s), made in the form of a tube with an internal diameter 0.003 m and a length 0.5 m, sequentially connected to a filter, a glass container with an absorber (ethyl alcohol), a pipette, a flowmeter and vacuum pump KVN-8, creating vacuum up to 20 kPa. The probe was installed at a distance 1 caliber from the gas flow inlet into the precipitation chamber orthogonally to the flow in such a way that its gas inlet hole coincided with the axis of the reactor. The filter was made of two cones with their bases in contact, separated by filtering baffle from twill-woven gauze made of steel grade C-120 and ensured the separation of powders from the sample, protecting them from contact with the air atmosphere. The container with an absorber made it possible to capture boron-containing compounds. The sample of the dust-free gas stream was collected into a pipette. The composition of the gas phase was determined chromatographically (see figure 1). The curves H_2* and BH_2 * correspond to a 1.5-fold excess of boron.

2) The composition of thermal desorption products of samples sampled during probing (see table 1) was studied by mass spectrometric analysis.

3) The morphology of TiB_2 nanoparticles was studied by scanning electron microscopy (see figure 2).

3. Results and discussion

The analysis of the obtained results confirms the presence of gaseous boron-containing compounds in the nitrogen-hydrogen flow. In the mass spectra of products of vacuum thermal desorption, there are components with molecular weights of 12,13, presumably corresponding to BH and BH_2. Comparison of the temperature dependence of the concentrations of BH and BH_2* calculated from the experimental content in the gaseous boron phase and the change in the phase composition of the trapped powder products confirm the unconditional participation of gaseous boron compounds in boride formation. At the same time, the study of the shape and size of TiB_2 particles, obtained in the form of a nanopowder with spherical particles, unambiguously indicates their formation by the "vapor – melt – crystal" mechanism.

Zone 1 corresponds to the temperature range of 5400-3500 K. Processes of heat- and mass transfer and the formation of a reaction mixture of the given composition take place there. Zone 2 corresponds to the temperature interval of 3500-2000 K. A characteristic feature of the zone is a significant decrease in the concentration of boron hydrogen BH_2 in the gas flow, corresponding to the appearance of TiB_2 in products. Thus, we can conclude that the formation of TiB_2 is possible during the crystallization of the melt titanium – boron – product of the boronizing by borohydrides of a metallic aerosol formed in the flow with volumetric condensation of titanium vapour. The nature of the change in the concentration of BH_2 alongside with the temperature at 1.5 times excess boron in the charge confirms its stability within the temperature range and indicates a real possibility of limiting the free boron content in titanium diboride. The boride nanoparticles formed in the zone under consideration actively adsorb process gases and gaseous products of boride formation.

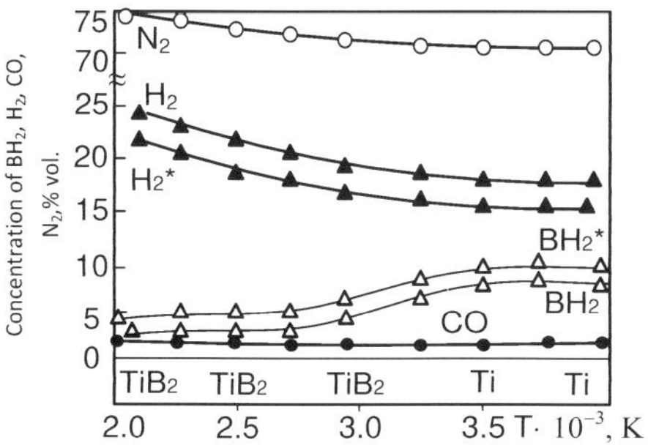

$\bar{a} \pm \Delta a$	Temperature, K			
	2000	3000	3500	4000
H_2	23.12±0.94	19.31±0.82	18.72±0.71	18.11±0.73
BH_2	0.82±0.06	3.62±0.10	7.00±0.39	7.31±0.42
H_2^*	21.05±	17.62±0.53	16.05±0.50	18.25±0.44
BH_2^*	3.37±0.10	6.54±0.28	9.12±0.49	10.21±0.63
N_2	74.35±1.23	74.00±1.36	73.65±1.21	73.51±1.44
CO	0.98±0.04	1.00±0.05	1.14±0.04	1.11±0.05

Figure 1. Temperature dependence of the composition of gaseous products and phase composition of solid products of titanium-boron interaction in the nitrogen plasma flow ($\bar{a} \pm \Delta a$ – average arithmetic values and confidence intervals of concentrations).

Table 1. Composition and quantities of vacuum thermal desorption products of TiB2 samples, non-contacting with the atmosphere.

Characteristic	TiB_2(1)	TiB_2(2)
Chemical composition, %		
TiB_2	96.26	93.27
B_{free}	1.05	0.45
Ti_{free}	1.13	-
C	-	1.22
O	0.44	-
N	1.12	1.24
TiO_2	-	3.82
Specific surface, m^2/kg	45000	37000
Particle[a] size, nm	37	45
Molecular masses of desorption products	2.12-13, 26-28, 32-36	2.12-12, 14- 16, 26-28, 32-36
The expected composition of the desorption products	H_2, BH, BH_2, N_2, CO, O_2	H_2, BH, BH_2, CH_3, CH_2, N_2, CO, O_2
Number of desorbed gaseous products	1.24	1.36

[a]The size was calculated from the specific surface area

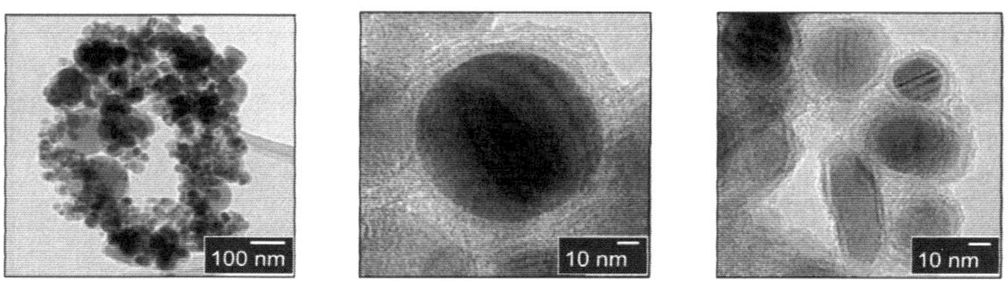

(a) (b) (c)

Figure 2. Micrographs of nanocrystalline titanium diboride TiB_2 (1) (a – particles and aggregates; b – morphological pattern of the aggregate, c – single particles).

The obtained results and their discussion make it possible to describe the process of boride formation in the form of the following scheme, which includes the two main temperature-space zones, presented in figure 3.

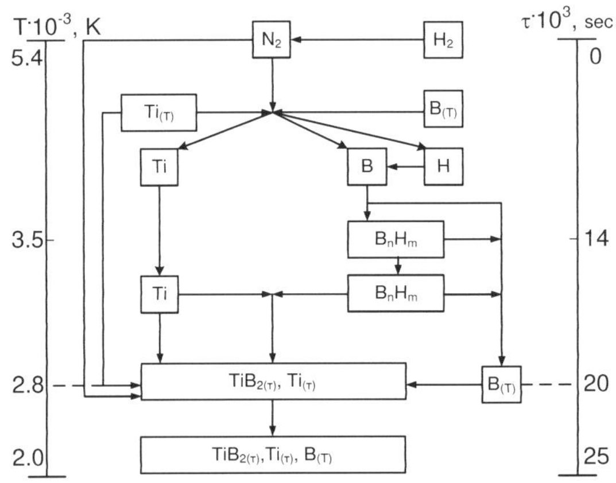

Figure 3. The proposed scheme for the interaction of titanium with boron in a flow of nitrogen-hydrogen plasma.

4. Conclusion

The process of boride formation with TiB_2 production in the conditions of plasma treatment of charge is characterized by features that substantially distinguish it from traditional technological options. The formation of TiB_2 is possible during crystallization of the melt titanium – boron – product of boronizing by the borohydrides of a metallic aerosol formed in the flow during volumetric condensation of titanium vapour. Alongside with the mechanism under discussion other reaction systems, that have not been identified by available means of diagnostics yet, are possible but typical for high-temperature non-ideal chemical-metallurgical reactors.

References

[1] Panov V S 2006 *Refractory Metals of IV – VI Groups and their Compounds. Structure, Properties, Methods of Production* (Moscow: MISiS) p 63

[2] Efimova K A, Galevsky G V and Rudneva V V 2016 *Bulletin of the Mining and Metallurgical Section of the Russian Academy of Natural Sciences. Section of Metallurgy* **B37** 90–102

[3] Efimova K A, Galevsky G V and Rudneva V V 2017 *St. Petersburg Polytechnic University Journal of Engineering Science and Technology* **2** 144–158

[4] Efimova K A, Galevsky G V, Rudneva V V and Alekseyeva T I 2015 *Bulletin of the Mining and Metallurgical Section of the Russian Academy of Natural Sciences. Section of Metallurgy* **B35** 112–9

[5] Krutskiy Yu L, Antonova E V, Bannov A G and Frolova I N 2014 *Proc. of the 1st Intern. Scientific-practical. Conf. on Actual Problems in Mechanical Engineering* (Novosibirsk: Izd-vo NSTU) pp 453–8

[6] Nozdrin I V, Rudneva V V and Galevsky G V 2014 *Chromium Boride – Nanotechnology, Properties, Application: Monograph* (Saarbrucken (Germany): LAP LAMBERT Academic Publishing) p 233

[7] Efimova K A, Galevsky G V, Rudneva V V and Alekseeva T I 2016 *Minsk International Heat and Mass Transfer Forum MIF-XV* pp 72–76

[8] Galevsky G V, Rudneva V V, Nozdrin I V, Galilevsky S G and Efimova K A 2016 *Proc. of XIV All-Ukr. Scientific-practical. Conf. (Kiev) on Special Metallurgy: Yesterday, Today,*

Tomorrow pp 248–259

[9] Glevsky G V, Efimova K A and Rudneva V V 2015 *Scientific and Technical Bulletins of SPbSPU. Metallurgy and Material Science* **2 (219)** 141–150

[10] Galevsky G V, Rudneva V V, Garbuzova A K and Efimova K A 2016 *St. Petersburg Polytechnic University Journal of Engineering Science and Technology. Metallurgy and Material Science* **1(238)** 90–99

[11] Efimova K A, Galevsky G V and Rudneva V V 2016 *St. Petersburg Polytechnic University Journal of Engineering Science and Technology. Metallurgy and Material Science* **4(254)** 233–244

[12] Efimova K A, Galevsky G V and Alekseeva T I 2015 *Proc. of V All-Russian Scientific-practical. Conf. with Intern. Participation on Prospects for Development of Technologies for Processing Hydrocarbon and Mineral Resources* (Irkutsk: IRNITU Publishing House) pp 18–20

Air oxidation of titanium nanodiboride

G V Galievsky, V V Rudnev and K A Efimova

Siberian State Industrial University, 42 Kirova street, Novokuznetsk, 654007, Russia

E-mail: kafcmet@sibsiu.ru

Abstract. The oxidation and thermal-oxidative stability of titanium diboride are studied during its storage and heating in the air. It was established that during the first 24 hours titanium diboride actively adsorbs oxygen and moisture: the oxidation of TiB_2 (1) reaches $11.06 \cdot 10^{-7}$, TiB_2 (2) $-8.15 \cdot 10^{-7}$, TiB_2 (3) $- 31.5 \cdot 10^{-7}$ kg O_2/m^2. Analytical dependencies were obtained for the calculation of oxidation. When heated in the air, titanium diboride nanocrystals are oxidized in the temperature range (623 – 673) ± 15 K, microcrystals at a temperature of 688 ± 5 K and higher.

1. Introduction

Nanomaterials, as a rule, have high chemical activity due to the following features of their crystal structure and energy state [1-4]:

1) The small size of the grains determines the greater development and the length of grain boundaries: at a grain size from 100 to 10 nm, the boundaries contain from 10 to 50% of atoms of nanocrystalline solid.

2) Grains in the nanostate have various atomic defects: vacancies, their complexes, dislocations, the quantity and distribution of which are qualitatively different than in the large grains of 5-10 μm and of greater sizes.

3) Exceptionally high diffusion mobility of atoms along the grain boundaries of a nanocrystalline solid, 105 to 106 times higher than in the ordinary polycrystals.

4) The excited state in the nanograins of atomic layers near their surfaces, which makes the crystals shrink and the atoms in them deviate from the position corresponding to the equilibrium state.

5) The constant propensity of nanosystems to the excess energy relaxation.

Taken together, this leads to the manifestation of nanoscale effects in such properties as propensity to aggregate, active interaction with atmospheric and process gases, increased solubility in liquid media, diffusion mobility in matrices of composite materials, ability to solid-state coalescence and sintering, etc. The specific physical and chemical properties of nanomaterials make it necessary to carry out their comprehensive study with mandatory determination of characteristics that are practically significant for their further application, determination of competitive advantages and real cost. During the last 15 years a significant number of works were published, containing the results of study on the properties of nanopowders of metals, carbides, borides, and their compositions, in particular [5-8]. Their analysis in general confirms the specifics of the properties of nanocrystalline substances and the need to take them into account in the subsequent technological processes with their participation. However, for each substance in the nano-state its temperature-time conditions for the appearance of size effects and specific level values are characteristic, which allows the results in each case to be viewed as new information of scientific and applied importance.

IOP Conf. Series: Materials Science and Engineering **411** (2018) 012020 doi:10.1088/1757-899X/411/1/012020

Taking into account the properties of nanocrystalline materials offered to consumers, since 2000 the range of certification characteristics of powdered nanoproducts has been expanded. Alongside with the traditional characteristics (the phase and chemical compositions, the specific surface) the oxidation and the temperature of the beginning of oxidation are added.

2. Purpose and research methodology

The purpose of this work is to study the oxidation and thermal-oxidative stability of titanium diboride of nano- and microscale level of dispersity in the air environment and comparative analysis of the results.

Oxidation is determined by the amount of oxygen (kg) per 1 m^2 of powder surface. Determination of the specific surface area of titanium diboride was carried out on the TRISTAR 3020 analyzer and is based on the method of low-temperature nitrogen adsorption.

The temperature of the beginning of oxidation corresponds to the change in the mass of nanopowder by 1% when heated in the air. As the characteristic of thermal-oxidative stability, the most probable temperature interval in which the oxidation of the nanopowder begins when heated in the air can also be taken into account. Oxidation of titanium diboride powders in the air was studied by thermogravimetric analysis and high-temperature X-ray diffraction using the Setaram LabSys Evo thermal gravimetric analyzer and DRON-3 X-ray diffraction meter with the GPVT-1500 add-on unit.

Titanium diboride nanotubes not contacted with the air and obtained by plasmasynthesis from the charge Ti+B (TiB_2 (1)) TiO_2+B (TiB_2 (2)) as well as its micropowder produced by magnesium thermal method (TiB_2 (3)), were used as objects of research. Phase and chemical compositions, specific surface area, initial oxidation of the samples are given in table 1.

Table 1. Characteristics of test samples of titanium diboride

Characteristics	TiB_2 (1)	TiB_2 (2)	TiB_2 (3)
Phase composition	TiB_2	TiB_2	TiB_2
Chemical composition, %			
TiB_2	96.32	93.61	97.24
free boron	1.00	0.42	0.42
titanium free	1.15	-	-
titanium dioxide	-	3.84	-
carbon	-	1.12	-
magncsium	-	-	0.41
oxygen	0.61	1.54	0.82
nitrogen	0.92	1.01	0.41
Specific surface, m^2/kg	43000	38000	2600
The particle size range, μm	0.01-0.06	0.02-0.08	<5
Oxidation x10^7, kg O_2/m^2	1.42	4.05	31.5

3. Results and discussion

3.1. Determination of oxidation

The oxidation of titanium diboride samples contacting with air during 240 hours is shown in figure 1. During the first 2 days, the oxygen content in the samples was monitored every 4 hours, on the following day – every 12 hours.

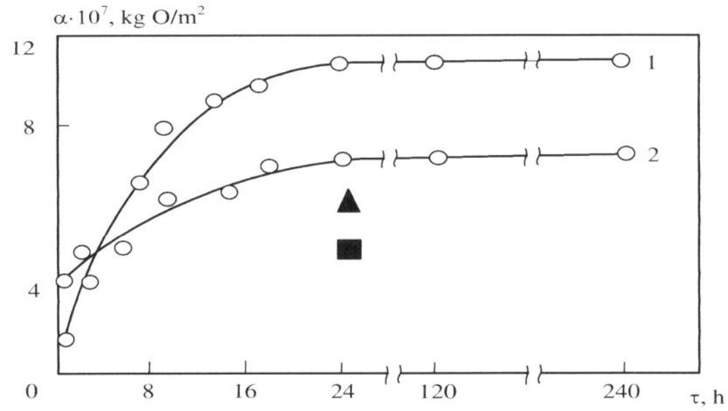

(1): (1.42 ÷ 11.06) ± (0.02 ÷ 0.18); (2): (4.05 ÷ 8.0) ± (0.06 ÷ 0.016)

Figure 1. Change in oxidation (α) of titanium diboride samples when contacting with air (▲, ■ – oxidation of samples 1 and 2, thermally desorbed in vacuum after storage in the air for 24 hours).

The analysis of the obtained results makes it possible to single out the following four scientific aspects for discussion:

1) The significant increase in oxidation occurs during the first 20-24 hours. In the subsequent period of time the oxidation practically does not change. At the same time the increase in oxidation is observed for TiB_2 (1) approximately in 8 times (from 1.42·10-7 to 11.06·10-7 kg O_2/m^2), for TiB_2 (2) – ~ in 2 times (from $4.05 \cdot 10^{-7}$ to $8.15 \cdot 10^{-7}$ kg O_2/m^2). The oxides of titanium and boron are not detected in the X-ray patterns of the samples. The increase in oxidation is most likely due to the adsorption of oxygen and water vapor. This fact is confirmed in a number of works devoted to the study of high-temperature nanosized compounds [9-12]. However, the mechanism of sorption – physical adsorption or chemical adsorption – has not yet been studied.

The obtained data made it possible to determine the dependence of the titanium diboride oxidation on the duration of its contact with the air, which is described for TiB_2 (1) by the equation of the form

$$a= [-0.0134\tau^2 + 0.7065\tau + 1.5175]\cdot 10^{-7} \quad (1)$$

for TiB_2 (2) –

$$a= [-0.0044\tau^2 + 0.2672\tau + 4.1023]\cdot 10^{-7} \quad (2)$$

where a – the oxidation, kg of oxygen·m^{-2};
τ – duration of contact of TiB_2 with air, h.
These dependencies are graphically represented in figure 2.

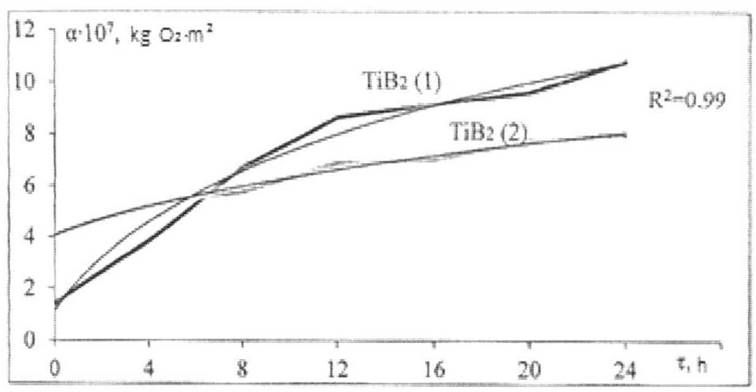

Figure 2. Graphical dependence of oxidation (α) of samples of titanium diboride TiB_2 (1) and TiB_2 (2) when they are in contact with the air (τ – the contact duration of TiB^2 with air (τ =0 ÷ 24 h), h, R^2 – the determination coefficient).

2) The sample of TiB_2 (2) is more resistant to the influence of atmospheric oxygen, which is due to its adsorption in the stage of carbon monoxide production having a high passivating power with respect to high-dispersed powders of transition metal and their compounds with carbon, boron and nitrogen. Indeed, the concentration of CO in the gas phase for option 2 is 3.2 – 5%, option 1 – 0.8 – 1.0%.

3) Vacuum annealing of samples (1) and (2) contacted with the air at a temperature of 973 K for desorption of gases ensures removal of only 40 % of oxygen (points ▲ and ■ in figure 1). The remaining amount of oxygen under these temperature conditions apparently participates in the surface oxidation of nanoparticles, with the formation of amorphous oxide films, which is confirmed by the results of determining the temperature of the beginning of samples oxidation described below.

4) The comparison of the oxidation of samples 1, 2, held in the air for 24 hours, and 3 ($11.06 \cdot 10^{-7}$, $8 \cdot 10^{-7}$ and $31.5 \cdot 10^{-7}$ kg O_2/m_2 surface, respectively) confirms that according to this characteristics nanopowders are not only inferior to micropowders, but also exceed them.

3.2. Determination of thermal and oxidative stability.

The oxidation process during heating in the air is usually investigated by two methods: high-temperature X-ray imaging and thermogravimetric analysis. In the first case, the temperature dependence of the intensity of one of the characteristic peaks of the studied material is considered, and therefore no special requirements concerning the content of the main phase and impurities are not imposed on the object of research. However, the method allows only the temperature interval to be stated in which the oxidation process begins. The temperature of the onset of oxidation can be determined much more accurately by thermogravimetric analysis, but with the condition that there is no or minimal content of impurities oxidized in the same temperature range as the main phase. Titanium diboride obtained according to options 1 and 2 is in fact a number of compositions, % $92.55TiB_2$+$1.05Ti_{free}$+$1.05B_{free}$ (1) and $91.25TiB_2$+$0.83B_{free}$+1.21 C (2). In this case, pyrolytic boron and carbon are oxidized in the temperature ranges 550-870 K and 671-790 K, respectively, which creates insurmountable difficulties for the correct thermogravimetric analysis of nanocrystalline TiB_2 (see figure 3).

It should be expected that the oxidation temperature of nanocrystalline TiB_2 can be 30-50° lower than TiB_2 micropowder [13], which oxidation begins at a temperature of 688±5 K. The X-ray thermal characteristics of TiB_2 (1) and TiB_2 (2) nanopowders and TiB_2 micropowder is shown in figure 4. When the samples are heated in the temperature range (273 – 623)±15 K, the change in the intensity of peaks is not observed. With further heating of samples TiB_2 (1) and (2) in the temperature range (623 – 673)±15 K and heating of TiB_2 micropowder sample of the in the temperature range (673 – 723)±15 K there is a significant decrease in the intensity of peaks, observed in the higher temperature range, which indicates the beginning and development of oxidation processes. In the temperature range under study, there are no peaks corresponding to titanium and boron oxides in the X-ray patterns, which suggests their formation in the amorphous state.

4. Conclusion

The oxidation and thermal-oxidative stability of titanium diboride nano- and micropowders in the air were studied. Titanium diboride, when stored in the air, actively adsorbs oxygen and moisture during the first 24 hours: the oxidation of TiB_2 (1) synthesized according to the option (Ti + B) varies within the range $(1.42\text{-}11.06) \cdot 10^{-7}$ kg O_2/m^2, TiB_2 (2) synthesized by the variant (TiO_2 + B) – $(4.05 - 8.15) \cdot 10^{-7}$ kg O_2/m^2 and is described by the equations $a(1) = [-0.0134\tau^2 + 0.7065\,\tau + 1.5175] \cdot 10^7$, $a(2) = [-0.0044\tau^2 + 0.2672\tau + 4.1023] \cdot 10^7$. Comparison of samples oxidation 1, 2, held in the air for 24 hours, and 3 ($11.06 \cdot 10^{-7}$, $8 \cdot 10^{-7}$ and $31.5 \cdot 10^{-7}$ kg O_2/m^2 of the surface, respectively) confirms that in this characteristics the nanocrystalline diboride exceeds the larger microcrystalline diboride.

Titanium diboride thermally desorbs in vacuum at a temperature of 973 K not more than 40% of the oxygen absorbed during its storage, which confirms the adsorption-diffusion mechanism of its interaction with atmospheric gases and the possibility of oxidizing of nanoparticles with the formation of amorphous oxide layers upon heating. Titanium diboride nanocrystals with heated in the air are oxidized in the temperature range (623 – 673)±15 K, which is 66 degrees below the temperature of the beginning of oxidation of its microcrystals 688±5 K. Pyrolytic boron and carbon accompanying diboride are oxidized in the temperature ranges (550 – 870)±12 K and (671 – 790)±17 K, respectively.

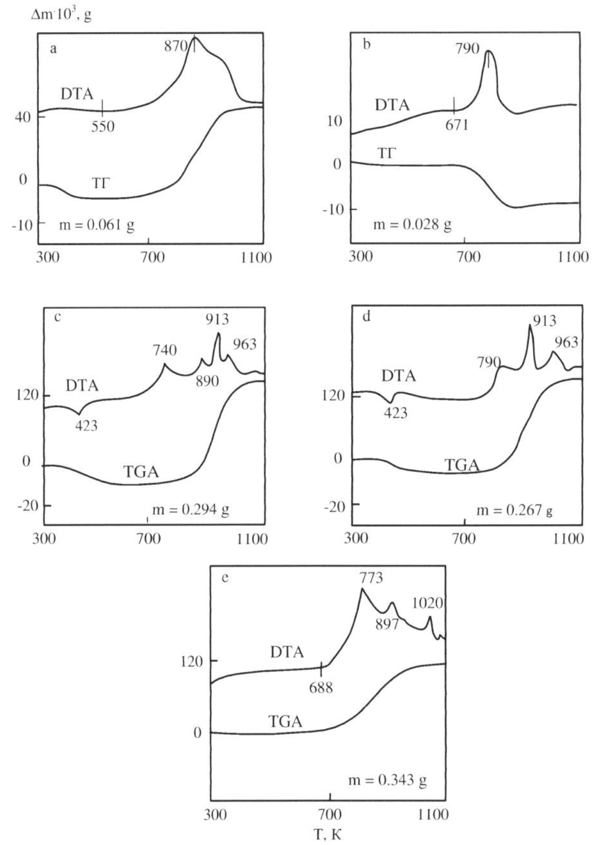

Figure 3. Thermograms of nanopowders with boron (a), carbon (b), titanium diboride (1) (c), (2) (d) and titanium diboride micropowder (e).

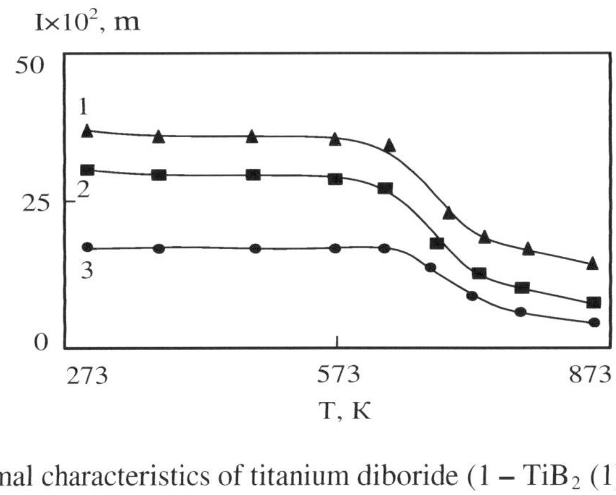

Figure 4. X-ray thermal characteristics of titanium diboride (1 – TiB_2 (1), 2 – TiB_2 (2), 3 – TiB_2 micropowder).

References

[1] Gusev A I 2005 *Nanomaterials, Nanostructures, Nanotechnologies* (Moscow: Fizmatlit) p 416
[2] Rudskoy A I 2007 *Nanotechnologies in Metallurgy* (St. Petersburg: Science) p 186
[3] Kolmakov A G, Barinov S M and Alymov M I 2012 *Fundamentals of Technology and Application of Nanomaterials* (Moscow: Fizmatlit) p 208
[4] Baloyan B M, Kolmakov A G, Alymov M I and Krotov A M 2007 *Classification, Features of Properties, Application and Production Technology* (Moscow: International University of Nature, Society and Man "Dubna", Branch of "Ugresh") p 125
[5] Gurov A A, Karmanov V I, Porozova S E and Shokov V O 2014 *Vestnik of the Perm National Research Polytechnic University. Engineering, Materials Science* **16 (1)** 23–29
[6] Rudneva V V 2006 *Izvestiya of the Higher Educational Institutions. Ferrous Metallurgy* **5** 16 – 19
[7] Shiryaeva L S, Nozdrin I V, Galevsky G V and Rudneva V V 2014 *Perspective Materials* **9** 40–46
[8] Karpov I V, Ushakov A V and Lepeshev A A 2013 *Technology of Mechanical Engineering* **2** pp 39–42
[9] Lepeshev A A, Ushakov A V and Karpov I V 2012 *Plasma-Chemical Synthesis of Nanodisperse Powders and Polymer Nanocomposites* (Krasnoyarsk: SFU Publishing House) p 310
[10] Nozdrin I V, Shiryaeva L S and Rudneva V V 2014 *Chromium Carbide – Nanotechnology, Properties, Application: Monograph* (Saarbrücken (Germany): LAP LAMBERT Academic Publishing) p 297
[11] Nozdrin I V, Rudnev V V and Galevsky G V 2014 *Scientific and Technical Bulletins of SPbSPU. Metallurgy and Material Science* **4(254)** p 233
[12] Ushakov A V, Lepeshev A A, Karpov I I and Krushenko G G 2012 *The Technology of Metals* **10** 27–32
[13] Ivanov V V, Blokhina I A, Kirik S D 2013 *Refractories and Technical Pottery* **4-5** 10–15
[14] Poilov V Z and Pryamilova E N 2016 *Journal of Inorganic Chemistry* **16(1)** 59–62

IOP Conf. Series: Materials Science and Engineering **411** (2018) 012021 doi:10.1088/1757-899X/411/1/012021

Melting of crude antimony in the low-tonnage production

G V Galevsky[1], V V Rudneva[1] and S G Galevsky[2]

[1]Siberian State Industrial University, 42 Kirova street, Novokuznetsk, 654007, Russia
[2]St. Petersburg Mining University, V.O., 21 Line, 2, St. Petersburg, 199106, Russia

E-mail: kafcmet@sibsiu.ru

Abstract. The project of low-tonnage production of crude antimony is proposed. The process scheme is based on the process of precipitation smelting. The possibility of combining the process of crude antimony smelting and its electrothermal refining from impurities in one melting unit is considered, the main technical and economic indicators are determined.

1. Introduction

Antimony, due to its fragility, is hardly used in industry in its pure form, but it is widely used in the form of alloys with lead, tin and copper. From 50 to 70% of all consumed antimony is in the form of metal for the alloys production, the rest amount of antimony is used in the form of various compounds. Antimony alloys with metals have an increased hardness, resistance to corrosion and wear resistance. The most known alloys of antimony are hard lead or antimony lead, printing metal, bearing metal. Hard lead containing from 5 to 15% of Sb is used for the manufacture of battery plates, sheets and pipes for the chemical industry, cable telegraph, telephone, electrical cable-sheaths, electrodes for certain types of electrolysis with insoluble anodes. Bearing metals or babbits are used in the manufacture of bearing liners for rail and motor transport and for other machines. Tin-based bearing metals contain 4-15% Sb, 60-80% Sn, 2-8% Cu, up to 25% Pb, have the required hardness, high abrasion resistance. Lead-based bearing metals containing 60-90% Pb serve as cheap substitutes for tin-based alloys. Metal antimony is also used for military purposes: it is added to lead, which is used to make shrapnel and bullets [1].

For a long time, the main supplier of antimony and its compounds to the regions of the former USSR remained the Kadamjay Antimony Plant (now the Kyrgyz Republic). But the current economic situation made production at this plant unprofitable. The Sakha-Ural Antimony Plant put into operation in 1996 is not able to fully meet the growing year by year demand for antimony and its compounds both in Russia and abroad. Thus, one of the ways out of the current situation is the development and implementation of a technology for a low-tonnage antimony production in the regions where heavy industry enterprises are concentrated. To date, the Sentachanskoye gold-antimony deposit in the Sakha Republic (Yakutia) is the most promising for the resource base of antimony production in Russia.

2. Smelting of crude antimony in the low-tonnage production

For the conditions of low-tonnage production the pyrometallurgical scheme for the antimony raw materials processing is most preferable. The technological scheme of antimony raw materials processing using the precipitation smelting process is shown in figure 1. The scheme includes precipitation melting, the products of which are matte, slag, gas phase and crude antimony. The latter, in its turn, is subjected to fire refining. The content of antimony in the feedstock is ~40%. The precipitation smelting of antimony ore is based on the great affinity of sulfur and oxygen to iron, rather than to antimony.

The main purpose of smelting the antimony raw materials is the transfer of the main substance of ore – antimony in a less contaminated by iron crude metal. During melting, sulfides react with iron, forming sulphurous iron and metallic antimony, droplets of which collect on the bottom of the furnace, and the empty rock is slagged. The addition of soda as a flux for antimony-sodium matte formation binds the iron sulphides of antimony to stable thiosalts, increasing the yield of matte and providing its good separation from the metallic and slag phases [2]. Iron turnings (steel or cast iron) are used in melting as a precipitator. During the precipitation smelting coal or coke are used as reducing agents, which prevent the oxidation of antimony and sulphide sulfur by the air oxygen. The coal consumption is 2-4% of the mass of ore raw materials. The raw materials and fluxing materials supplied to the electric smelter must be stored in the enclosed spaces. The moisture content of the charge components should not exceed 4-5%. To ensure better dosage and averaging of the composition before melting lump ore, coal, limestone are crushed to size of 20-30 mm, and the iron turnings are rattled to remove swarf strand from turnings. Melting of the concentrate is carried out at a temperature of 1250 ÷ 1400 ^{0}C. The resulting crude metal contains, wt%: 85 – Sb, 10 – Fe, 3.25 – S, 0.06 – As, 1.68 – others.

It is proposed to use an ore-thermal furnace with a capacity of 0.35 MW as the main unit for the antimony production. The electric furnace is devoid of a number of drawbacks inherent in reflective furnaces. Thus, the absence of a gas flare contributes to the release of less gases, which reduces the possibility of entrainment of antimony with them. During electric smelting the operate at a higher temperature without increasing the loss of antimony is possible. The use of more refractory slags can reduce the consumption of alkaline fluxes by replacing them in part with cheaper calciferous ones [3].

Slag and matte are transported to the dump because their processing is economically inexpedient in view of the small volume of production.

The furnace has the following parameters: casing diameter – 3400 mm; height of the furnace – 2600 mm; diameter of the bath – 1880 mm; height of the bathtub - 1800 mm; the area of the furnace bottom – 2.8 m^2; number of electrodes – 3; the diameter of the electrodes – 200 mm. Furnace round in section, three-phase. The bath of the furnace is lined with periclase-spinel refractory. The bottom of the furnace is laid out in the reverse arch; the vault is metal, water-cooled, lined with heat-resistant refractory concrete. Evacuation of gases is carried out through the opening in the roof of the furnace. Slag belt of the furnace is cooled by the applied caissons. Over the caissons are cast-iron plates, which carry the upper part of the lining. The current to the melt is fed by graphitized electrodes fixed in the electric holders and entering the arch of the furnace through the sealing rings [4, 5, 6]. Electromelting of antimony concentrate in the ore-thermal furnace is carried out in the regime of periodic charging of batch and periodic production of melting products. The technological cycle includes operations for preparation of melting, heating the melt, charging the batch, melting the charge, discharging the slag, releasing the crude metal or matte. The nominal load on the electrodes is maintained automatically by means of a power regulator.

The main materials used in the production and fire refining of antimony are: antimony concentrate KSUF – 4 (GOST 48 – 101 – 76); soda ash of grade B (GOST 11078 – 78); limestone flux grade 1 (TU 48-7-2-77); coke; iron turnings; sulfate of sodium technical grade 2 (GOST 6318-77); graphite electrodes with a diameter of 200 mm (GOST 4426-71). When preparing the charge, the fluxes are dosed according to the calculation of slag production of 60% SiO_2, 15% Na_2O, 15% CaO and 10% of others. The practical consumption of fluxes and reagents during the smelting of different types of concentrates is: soda ash 20 – 25%, limestone 15 – 20%, iron turnings 12 – 20%, coal 2 – 4% of the mass of the concentrate. The charge is loaded into the furnace by feeders through the arch. With this loading and nominal power selection the batch that enters the furnace melts without the formation of slopes, and the temperature under the roof is stably maintained at 600 – 700 ^{0}C. The rarefaction under the furnace roof is regulated remotely and does not exceed 2 mm of water in order to avoid a large suction of air into the furnace and the burnout of the electrodes.

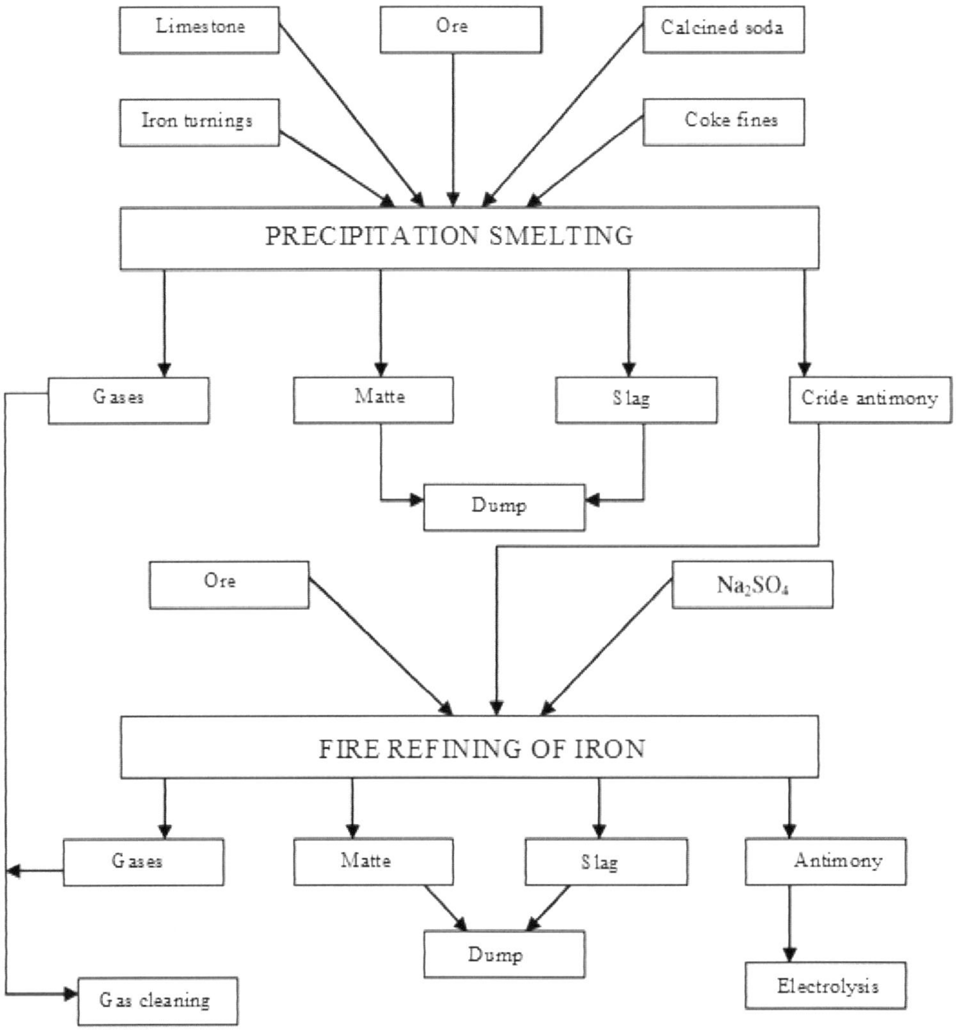

Figure 1. Technological scheme of antimony raw materials processing using the process of precipitation smelting.

The slag is discharged once every 8 hours through the slag tap into the cinder pot. Density of slag is 3.3 – 2.5 g/cm^3. At a time about 1.8 tonne is poured out. Before the release, the floating layer of the batch is melted, and the melt is heated to reduce the viscosity of the slag and improve the sludge. The slag has a temperature of 1250 – 1400 ^{0}C at discharge. The crude metal and matte are poured twice a day, i.e. in each shift either crude metal or matte is tapped. Matte is poured out through the matte tap-hole before the slag goes. Matte is poured into a special container (lined), made by the type of a cinder pot. At a time 2.2 tonnes is produced. When tapped, the matte has a temperature of the order of 1000 – 1250 ^{0}C. The crude metal is poured out through the tap-hole for the antimony tapping. At one time about 1800 kg of metal is discharged. The metal during discharging has a temperature of 900 – 1100 ^{0}C. From the tap-hole the metal enters the intermediate trough, from which it pours out onto the roll crystallizer. The melt, falling into the wedge-shaped space formed by the rolls, spreads along their entire length, simultaneously forming "crusts" that easily break down and slopes down into the box for crude antimony. Then this box is transported to the section for batch preparation, where the charge is prepared for refining. In the refining cycle, the metal is mainly discharged, and the matte and slag are discharged as they accumulate. The number of furnace gases varies from 600 to 1300 m^3/h; gases contain, % (volume): up to 20 - 24 – CO_2, 2 - 6 – CO and 1% – O_2. Melting of the concentrate in the electric furnace is 2.4 – 2.7 t/(m^2·day); electricity consumption 1050 – 1250 kW·h per 1 tonne of concentrate or 3700 - 4400 kW·h per 1 tonne of crude metal; the consumption of graphitized electrodes is 4-6 kg/t and 15-20 kg/t, respectively. The direct yield of antimony into the crude metal is 83 – 87 %, into dust 8 – 12%, into matte 2 – 3%, and into slag 1 – 1.5%. The annual capacity of the department is 625 tonnes of metal.

In order to save capital investments, the technology involves combining in one melting unit the processes of smelting of crude antimony and its refining from iron. This is achieved by successively carrying out the melting and refining operations in the same electric furnace.

The refining of antimony is based on the oxidation or sulphidation of impurities contained in it. The regime for performing the operation of antimony firing is determined by the processes of removal of iron and arsenic. Smelting of black antimony is conducted in succession for 20 days. Then for 8 days the fire refining is conducted. Single loading of crude metal into the electric furnace is 1.75 tonnes.

The number of refining cycles is 4 per day. The daily capacity of the furnace in the refining mode is 1.75·4 = 7 tonnes. The composition of the refined metal, %: 96 – Sb, 0.08 – Fe, 0.067 – As, 3.5 – S, 0.11 – Pb, 0.243 – others.

Purification of technological gases of ore-thermal furnaces is carried out in three stages: capture of coarse dust is carried out in the cyclone, dust of medium size is caught in the coolers, and for fine cleaning bag filters are used [7]. In connection with the small scale of production and low concentration of SO_2 in the gases, the organization of sulfuric acid production in this case is economically and technologically inexpedient. Therefore, the calciferous method is chosen for purification from small contents of sulfur dioxide. This method is distinguished by its simplicity and is carried out with the help of cheap reagents (limestone, calcite, chalk, dolomite, lime), and therefore requires a small initial capital investment, which predetermines the low operating costs at a sufficiently high degree of gas purification (85 – 98%).

3. Conclusion

The proposed technological scheme is based on modern efficient and environmentally safe processes that provide high technical and economic indicators of ore and industrial products processing, such as precipitation electric smelting, refining in the electric furnaces, chemical cleaning of waste gases. The technology of antimony production is practically non-waste. The refined slags, captured dust, partially matte are returned to the processes for the antimony extraction. The dump slags, calcareous (gypsum) sediments from wet cleaning are used in the construction industry as raw materials for the production of building materials.

References

[1] *Antimony* 1977 ed S M Melnikov (Moscow: Metallurgy) p 536
[2] Shiyanov A G 1961 *Production of Antimony* (Moscow: Metallurgizdat) p 176
[3] Strunsky B M 1972 *Rudnothermic Smelting Furnaces* (Moscow: Metallurgy) p 368
[4] Rozhikhina I D 2000 *Constructions and Designing of Electric Furnaces* (Novokuznetsk) p 148
[5] Strunsky B M 1982 *Calculations of Ore-thermal Furnaces* (Moscow: Metallurgy) p 192
[6] Gurova M I 1977 *Refractory Products and Materials* (Moscow: Metallurgy) p 216
[7] Gordon G M and Peikhasov I P 1977 *Dust Collection and Purification of Gases* (Moscow: Metallurgy) p 456

Thermodynamic modeling of processes in carbide-forming systems Ti–C–H–N and Ti–O–C–H–N

A K Garbuzova, G V Galevsky and V V Rudneva

Siberian State Industrial University, 42 Kirova street, Novokuznetsk, 654007, Russia

E-mail: kafcmet@sibsiu.ru

Abstract. Thermodynamic analysis of high-temperature interactions in Ti–C–H–N, Ti–O–C–H–N systems was carried out, corresponding to the processes of titanium carbide synthesis in the conditions of plasma flow using various titanium-containing raw materials. Modeling processes in these carbide-forming systems showed that the formation of titanium carbide is possible in equilibrium and quasi-equilibrium conditions, at temperatures below 3500 K, with stoichiometric and excess carbon according to gas-phase reaction involving titanium and cyanide vapour. The possibility of achieving a high yield of titanium carbide in real synthesis conditions is predicted.

1. Introduction

The objects of thermodynamic modeling are plasma-metallurgical jet processes of for the carbides production, which unlike traditional ones have some specific features [1, 2]:

- transience of the carbide formation processes which makes it possible to question the possibility of achieving equilibrium;
- high-temperature flow conditions for fairly rapid achievement of equilibrium;
- achieving a high yield of carbides exclusively in the conditions that ensure the gas-phase character of the reactions of carbide formation;
- the production of carbides in the systems consisting of several chemical elements, that form multiphase multicomponent mixtures, necessitates the performance of their thermodynamic analysis on a computer;
- change in the conditions for the existence of condensed phases due to their formation in the form of nano-sized particles (with a large surface curvature) for which the partial pressure above the condensed phase exceeds the saturation pressure above the flat surface. However, at present there are no unified opinion on the size threshold (1 or 2-10 nm) and the corresponding reference data;
- unconditional expediency of performing thermodynamic calculations of equilibrium compositions of carbide-forming systems at various parameters in connection with insufficient information to describe with the necessary accuracy and reliability the kinetics and mechanism of the carbides formation.

Taking into account these features, the results of thermodynamic modeling of carbide formation can be considered as predictive, requiring experimental confirmation and as technological benchmarks, perfection indicators of the reactor, as well as the level of organization in it of heat and mass exchange processes.

The purpose of thermodynamic modeling of synthesis processes is to predict the optimal parameters for the production of titanium carbide (the ratio of components and temperature), to determine the equilibrium parameters of the process (the degree of conversion of raw materials to carbide, the composition of gaseous and condensed products), to evaluate the contribution to the carbide formation of gas-phase reactions that ensure the efficient processing of dispersed raw materials in the conditions of plasma-metallurgical technology.

2. Results and discussion

The objects of the study were Ti–O–C–H–N and Ti–C–H–N systems, since the raw materials used in the synthesis processes were: fine powder of titanium and titanium dioxide TiO_2, methane and plasma-forming gas – nitrogen.

The calculation of the equilibrium compositions of gaseous and condensed products was carried out by the "const method" based on the joint solution of the following equations: the law of acting masses, the material balance, the total number of moles of the gas mixture, the existence of a condensed phase and the Dalton law [1].

The calculations were carried out using the computer simulation of high-temperature complex chemical equilibria PLASMA (Institute of Chemistry of Solids and Mechanical Chemistry of the RAS Siberian branch), which has a built-in database of interaction products for oxide, boride, carbide, and nitride-forming systems. In the calculations, the temperature range 1000-6000 K was considered at a total pressure in the system of 0.1 MPa. The initial data were the equilibrium constants of the reactions of compounds formation from the elements given in [3-9].

The compositions of the gas and condensed phases of the systems under study are given in table 1.

Table 1. Compositions of gas and condensed phases.

System	The composition of the gas phase	Composition of the condensed phase
Ti-O-C-H-N	H, H^+, H_2, N, N_2, NH, NH_2, NH_3, C, C^+, C_2, C_3, C_4, C_5, CH_2, CH_3, CH_4, C_2H, C_3H, C_4H, C_2H_2, C_2H_3, C_2H_4, CN, CN^{-1}, NCN, NCC, C_2N_2, C_4N_2, HCN, HCCN, C_3HN, O, O_2, NO, NO_2, H_2O, CO, CO_2, Ti, TiO, TiO_2, TiOH	Ti, TiC, TiN, TiO, Ti_2O_3, Ti_3O_5, Ti_4O_7, TiO_2
Ti-C-H-N	H, H^+, H_2, N, N_2, NH, NH_2, NH_3, C, C^+, C_2, C_3, C_4, C_5, CH_2, CH_3, CH_4, C_2H, C_3H, C_4H, C_2H_2, C_2H_3, C_2H_4, CN, CN^{-1}, NCN, NCC, C_2N_2, C_4N_2, HCN, HCCN, C_3HN, Ti	Ti, TiC, TiN

The choice of the ratios of the initial components in the investigated systems was determined by the stoichiometry of the reactions of the target products formation and the parameters of the technological equipment. The ratios chosen for the calculations are shown in table 2.

Table 2. The ratio of components in the studied systems.

Ratio of components, mole	Note
Ti:O:C:H:N=0.25:0.5:0.75:3:20	According to the stoichiometry for the reaction $TiO_2+3CH_4=TiC+2CO+6H_2$ (1)
Ti:O:C:H:N=0.25:0.5:0.78:3.12:20	104% of carbon content compared with the stoichiometry for reaction (1)
Ti:O:C:H:N=0.25:0.5:0.81:3.24:20	108% of carbon content compared with the stoichiometry for reaction (1)
Ti:O:C:H:N=0.25:0.5:0.84:3.36:20	112% of carbon content compared with the stoichiometry for reaction (1)
Ti:O:C:H:N=0.25:0.5:0.875:3.5:20	116% of carbon content compared with the stoichiometry for reaction (1)

Ti:O:C:H:N=0.25:0.5:0.91:3.64:20	120% of carbon content compared to the stoichiometry for reaction (1)
Ti:O:C:H:N=0.25:0.5:0.94:3.76:20	124% of carbon content amount compared to the stoichiometry for reaction (1)
Ti:C:H:N=0.25:0.25:1.0:20	According to the stoichiometry for the reaction $Ti+CH_4=TiC+2H_2$ (2)
Ti:C:H:N=0.25:0.375:1.0:20	150% of carbon content compared with the stoichiometry for reaction (2)

The results of thermodynamic calculations of equilibrium compositions of Ti–O–C–H–N system are shown in figure 1.

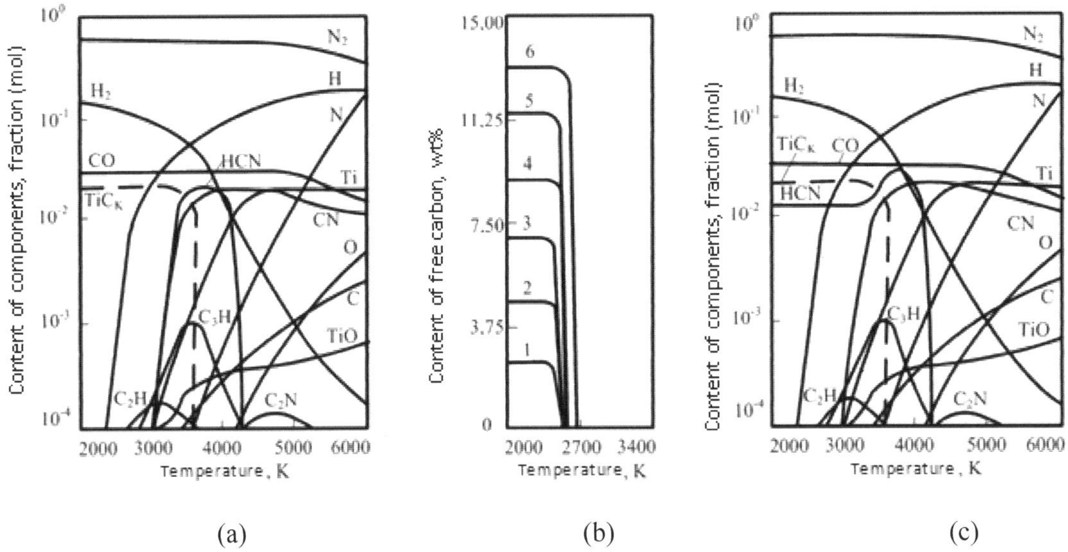

Figure 1. Results of thermodynamic calculations of Ti–O–C–H–N system: a – equilibrium compositions of the gas and condensed phases depending on the temperature with ratio Ti:O:C:H:N=0.25: 0.50: 0.75: 3: 20; b – dependence of the equilibrium concentration of free carbon in the condensed phase on the ratio TiC = 0.25:0.28 (1); 0.25:0.31 (2); 0.25:0.34 (3); 0.25:0.375 (4); 0.25:0.41 (5); 0.25:0.44 (6); c – quasi-equilibrium compositions of the gas and condensed phases of the Ti–O–C–H–N system with components ratio Ti:O:C:H:N=0.25:0.50:0.875:3.5:20.

In the system under consideration, the production of titanium carbide is possible at temperatures below 3500 K with all the component ratios considered in the calculations. The following gas-phase reactions of titanium carbide formation are thermodynamically possible:

$$Ti_{vapour} + 2HCN = TiC_s + H_2 + N_2 \quad (3)$$

$$Ti_{vapour} + 2CN = TiC_s + N_2 \quad (4)$$

$$2TiO_{vapour} + 4HCN = 2TiC_s + 2CO + 2H_2 + 2N_2 \quad (5)$$

$$6Ti_{vapour} + 2C_3H = 6TiC_s + H_2 \quad (6)$$

Of the reactions 3 and 6, the most probable from a thermodynamics point of view is reaction 3: the degree of conversion of Ti to TiC in this reaction is 0.94. The complete conversion of titanium to carbide is achieved in the temperature range 3200-3500 K at a stoichiometric ratio Ti:C (figure 1a) and carbon excess in the system (figure 1b). However, in this case the condensed products contain free carbon, the equilibrium concentration of which varies from 2.55 to 12.75 wt% when the Ti:C ratio is changed within 0.25:(0.28÷0.44).

In the conditions of quasiequilibrium with an excess of the carbidizer at the ratio Ti:C=0.25:0.375 (figure 1c), the cyanide is stable in a wider temperature range – (2000-3800) K, which makes it possible to obtain carbide with no free carbon.

In the Ti–C–H–N system, the formation of titanium carbide is possible in the conditions of equilibrium (a) and quasi-equilibrium conditions (b) (figure 2). In the first case, when the ratio Ti:C = 0.25:(0.0625-0.25) is changed, the degree of conversion of titanium to carbide varies from 0.25 to 1. And at the ratio Ti:C = 0.25:0.375 (excess carbide), the synthesis product is a carbide – carbon composition, containing 9.1 wt% free carbon. In the second case, no condensation of carbon occurs. In both cases, the formation of carbide is thermodynamically possible according to the gas-phase reaction.

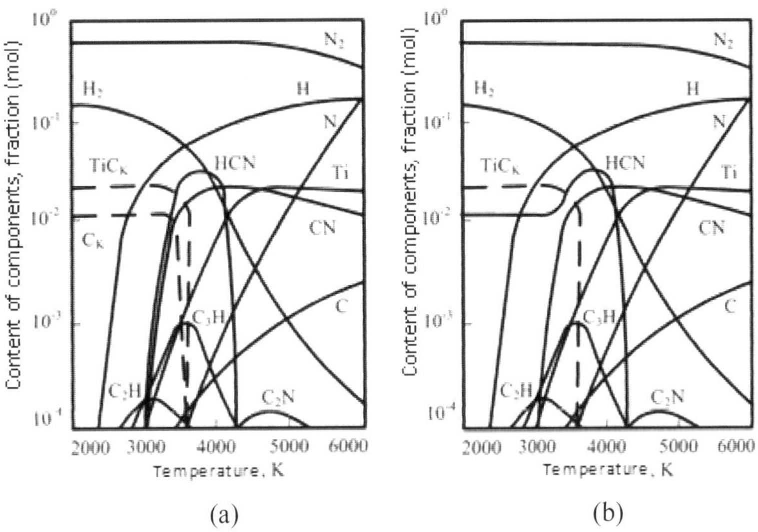

Figure 2. Equilibrium (a) and quasi-equilibrium (b) compositions of the gas and condensed phases of the Ti–C–H–N system with the ration of components Ti:C:H:N = 0.25:0.375:1.5:20.

The closeness of temperatures of titanium condensation (3533 K) and melting of the carbide formation (3530 K) does not allow an unambiguous assumption to be made about the possible mechanism of carbide formation, although the formation of carbide by the mechanism "vapour – crystal" is more likely.

3. Conclusion

Thermodynamic modeling of processes in carbide-forming systems Ti–C–H–N and Ti–O–C–H–N showed that in the Ti–O–C–H–N system titanium carbide formation is possible at a temperature below 3500 K with stoichiometric and excess amount of carbon. With an excess of carbon in the condensed products there is a free carbon, the equilibrium concentration of which varies from 2.55 to 12.75% by weight when the Ti:C ratio changes within 0.25:(0.28÷0.44). In the conditions of quasiequilibrium with an excess of antigraphitizer the carbide cyanide is stable in a wider temperature range – (2000-3800) K, which makes it possible to obtain carbide that does not contain free carbon.

In the Ti–C–H–N system the formation of titanium carbide is also possible in equilibrium and quasi-equilibrium conditions. In the condition of equilibrium, when the ratio Ti:C = 0.25:(0.0625-0.25) varies, the degree of conversion of titanium to carbide varies from 0.25 to 1, and at 1.5 times excess of the antigraphitizer, carbide contains 9.1 wt% free carbon. In the conditions of quasi-equilibrium no condensation of carbon occurs.

For each case the formation of titanium carbide according to the gas-phase reaction involving titanium and cyanide vapours is typical, that is, according to the "vapour – crystal" scheme, which makes it possible to predict the possibility of achieving its high yield in the real synthesis conditions.

References

[1] Suris A L 1985 *Thermodynamics of High Temperature Processes: Handbook* (Moscow: Metallurgy) p 568
[2] Garbuzova A K, Rudneva V V et al 2014 *Bulletin of the Mining and Metallurgical Section of the Russian Academy of Natural Sciences. Branch of Metallurgy: Proceedings* (Novokuznetsk: SibSIU) vol **32** p 196
[3] Glushko V P 1978-1982 *Thermodynamic Properties of Individual Substances: Handbook* ed. V Glushko (Moscow: Nauka)
[4] Spangehberg H J 1974 *Zeitschrift für Physikalische Chemie* **255(1)** 1–15
[5] Gurvich L V 1962 *Thermodynamic Properties of Individual Substances* vol 2 (Moscow: AN SSSR) p 775
[6] Barin J and Knacbe O 1973 *Thermochemical Properties of Inorganic Substances* (New York: Academicpress) p 647
[7] JANAF 1966 – 1975 *Thermochemical Tables* (Washington: American Insitute of Physics)
[8] Gerasimov Ya N et al 1973 *Chemical Thermodynamics in Nonferrous Metallurgy* vol 5 (Moscow: Metallurgy) p 296
[9] Weeks K E and Block F E 1965 *Thermodynamic Properties of 65 Elements and their Oxides, Halides, Carbides and Nitrides* (Moscow: Metallurgy) p 126

A study on changes in the properties of silumin surface layers modified by yttrium oxide

V E Gromov[1], Yu F Ivanov[2], D V Zagulyaev[1], O S Tolkachev[2], E A Petrikova[2] and V Konovalov[2]

[1]Siberian State Industrial University, 42 Kirova street, Novokuznetsk, 654007, Russia
[2]Institute of High Current Electronics SB RAS, 2/3Akademichesky avenue, Tomsk, 634055, Russia
[3]Samara National Research University, 34 Moskovskoye Shosse, Samara, 443086, Russia

E-mail: zagulyaev_dv@physics.sibsiu.ru

Abstract. In the present work, the analysis of the structure and properties of silumin, modified by deposition of yttrium oxide on the surface by electric-explosive alloying, was made, (alloying with a droplet-plasma flow produced by the electric explosion of a conductive foil with a sample of powder material). The structure and phase composition of silumin were studied by scanning and transmission diffraction electron microscopy. Formed coatings are characterized by a high level of roughness and porosity.

1. Introduction

It is known that aluminum is one of the most wide spread metals and occupies the first place in terms of content in the earth crust – 81.3 kg of aluminum per ton of the earth crust, which is 31.3 kg more than iron [1]. Due to aluminum occurrence the scientific community focus their attention on the studying of its properties and the possibilities of modifying them [2-4]. One of the most common methods for modifying aluminum is the addition of alloying elements, for example, silicon during the casting process, the resulting alloys are called silumins.

Silumins – the most common foundry alloys based on aluminum [5-8]. They are widely used as a structural material due to the combination of a complex of high operational and foundry-technological properties. In particular, the pistons of internal combustion engines and compressors are manufactured mainly from eutectic and hypereutectic silumins [5-10]. Despite this fact, taking into account continuously increasing requirements to the quality of cast products, efficiency and environmental friendliness of the technological process, it is still necessary to search for new reserves for improving the technological process and effective methods for improving the silumin properties, based on modification, microalloying and intensification of the solidification process (crystallization of the alloy) in the foundry form. One of these methods, combining the introduction of various elements and refractory particles into the melt with a high rate of crystallization of the material, is electric explosive alloying [10].

The purpose of this work is to study the structure and properties of silumin eutectic composition subjected to modification by yttrium oxide particles by the methods of electric explosive alloying.

2. Materials and methods

Eutectic silumin was used as the study material, the elemental composition of which is given in table 1.

Table 1. Results of x-ray spectral analysis of silumin samples (wt%, Al is the rest).

Si	Fe	Cu	Mn	Mg	Ni	Ti	Cr	Zn	Pb
11.1	0.25	2.19	0.029	0.58	0.92	0.047	0.005	нет	нет

Studies of the phase composition and the state of the silumin crystal lattice were carried out by X-ray diffraction analysis (XRD-7000s, Shimadzu, Japan). Investigation of the elemental and phase composition, the defective silumin substructure, was carried out by scanning methods (Philips SEM 515 equipped with EDAX ECON IV microanalyzer) and transmission diffraction (JEM-2100F, JEOL, Japan) electron microscopy. Foils (the research object of the material by the methods of transmission electron diffraction microscopy) were produced by ion thinning of thin (100 μm) plates on the Ion Slicer EM 09100IS instrument.

3. Research and discussion

Earlier, it was established in [9] that the electric explosive alloying (hereinafter referred to as EEA) of silumin of eutectic composition is accompanied by the formation of a layer with a highly developed topography of the modification surface. Indeed, the studies presented in this paper show that as a result of silumin EEA, a surface layer is formed, characterized by the high level of roughness, containing a large number of micropores, microcraters and microcracks (figure 1).

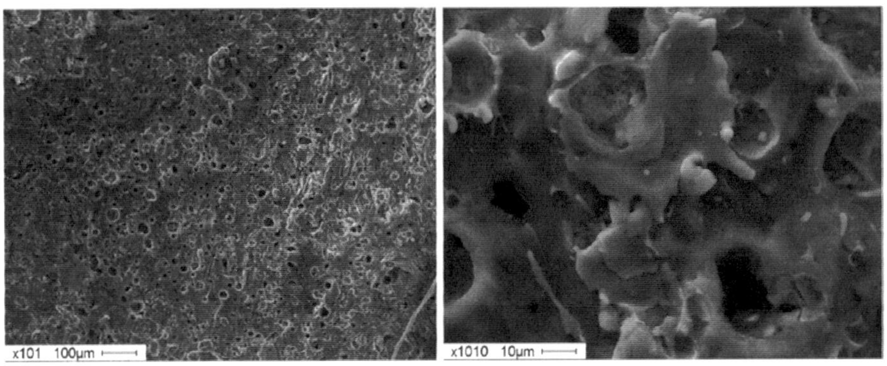

Figure 1. Silumin surface structure of eutectic composition subjected to EEA. Scanning electron microscopy.

Using the micro-X-ray spectral analysis, silumin elemental composition of the surface layer, subjected to EEA, was investigated. The results of the performed studies are presented in table 2.

Table 2. Elemental composition of the silumin surface layer, subjected to EEA, revealed by micro-X-ray spectral analysis of the zones shown in figure 2. Scanning electron microscopy. Results are presented in weight %.

Zone	Al	Si	Mg	Ti	Fe	Ni	Cu	Y	O	C
Figure 2, a	47.2	3.0	0.6	1.0	0.7	1.3	1.8	16.2	10.8	17.4
Figure 2, c	0.8	0.0	0.0	0.2	0.7	0.5	0.7	34.0	28.1	35.0

Analyzing the results presented in table 2, it can be noted that EEA leads to the formation of a surface layer with a high level of inhomogeneity in the distribution of alloying elements, which is most noticeable in the distribution of yttrium and oxygen atoms. These results confirm the fact that there is a presence of powder particles of alloying material in the plasma flow, previously identified in many studies [9, 10].

The structure of silumin volume subjected to EEA was analyzed by the cross-section method. A typical image of the alloyed layer structure revealed by scanning electron microscopy is shown in figure 2. Analyzing the results obtained, it can be noted, first, that the thickness of the modified layer varies within (30-50) μm. Second, a high level of porosity; pore sizes vary from units to tens of micrometers.

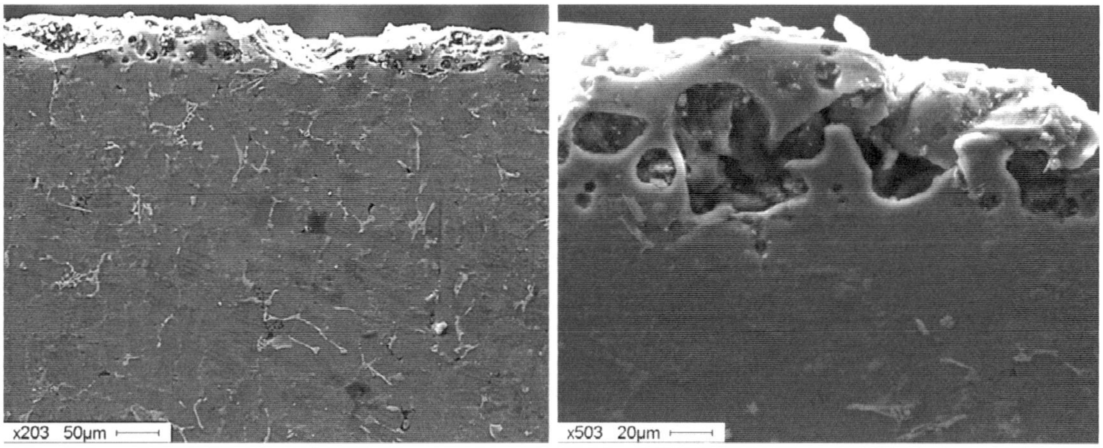

Figure 2. The structure of the silumin cross-section of eutectic composition subjected to EEA. Scanning electron microscopy.

Table 3. Results of X-ray diffraction analysis of silumin sample of eutectic composition subjected to EEA.

Phase	Content, rel. %	Parameter of a lattice, nm	D(CIR), nm	Δd/d, 10^{-3}
Al	68.2	0.40485	75.01	0.24
Si	30.1	0.54231	16.4	0.80
Y_2O_3	1.7	1.06010	16.6	7.88

The phase composition of silumin modified by EEA was investigated by X-ray diffraction analysis. The quantitative results of the phase analysis of the X-ray diffraction obtained from the studied material are presented in table 3.

Analyzing the results presented in table 3, one can note, first, an essentially high level of silicon in the surface layer of silumin, which can indicate the evaporation of some aluminum layer during EEA. Second, the presence of the phase Y_2O_3, which can be due to the introduction of particles of the initial yttrium oxide powder into the surface layer of silumin during EEA.

4. Conclusion

The modification of the silumin surface layer of eutectic composition by yttrium oxide particles was carried out by the method of electric explosive alloying. It was established that silumin EEA is accompanied by the formation of a highly porous surface layer with a thickness up to 50 μm, characterized by inhomogeneity in the distribution of alloying elements (silicon, yttrium and oxygen), a submicrocous and nanoscale multiphase structure, strengthening phases o which are silicon particles, Y_2O_3, YSi_2 and Y_2Si_2O.

Acknowledgements

The work was carried out with the financial support of the state task No. 3.1283.2017/4.6 and partial financial support of the grant from the Russian Foundation for Basic Research (project No. 16-58-00075-Bel_a).

References

[1] Drozdov A 2007 *Alyuminiy. Trinadtsatyy Element: Entsiklopediya* (Moscow: Biblioteka RUSALa) 239
[2] Zagulyayev D V, Konovalov S V and Gromov V Ye 2010 *Vestnik Yuzhno-Ural'skogo Gosudarstvennogo Universiteta* (Seriya: Matematika. Mekhanika. Fizika) **9** 53–56
[3] Zagulyaev D V, Konovalov S V, Shlyarov V V, Anuchina E A, Komissarova I A and Gromov V E 2016 *IOP Conf. Ser.: Mater. Sci. Eng.* **150** 012038
[4] Zagulyayev D V, Konovalov S V, Ponomareva M V and Gromov V Ye 2010 *Fundamental'nyye Problemy Sovremennogo Materialovedeniya* **1** 32–35
[5] Stroganov G B, Rotenberg V A and Gershman G B 1977 *Splavy Alyuminiya S Kremniyem* (Moscow: Metallurgiya) p 272
[6] Belov N A, Savchenko S V and Khvan A V 2008 *Fazovyy Sostav I Struktura Siluminov* (Moscow: MISIS) p 282
[7] Zolotorevskiy V S, Belov N A 2005 *Metallovedeniye Liteynykh Alyuminiyevykh Splavov* (Moscow: MISIS) p 376
[8] Belov N A 2009 *Fazovyy Sostav Alyuminiyevykh Splavov* (Moscow: MISIS) p 392
[9] Laskovnev A P, Ivanov Yu F, Petrikova Ye A et al 2013 *Modifikatsiya Struktury I Svoystv Evtekticheskogo Siluna Elektronno-Ionno-Plazmennoy Obrabotkoy* (Minsk: Belarus. Nauka) p 287
[10] Bagautdinov A Ya, Budovskikh Ye A, Ivanov Yu F and Gromov V Ye 2007 *Fizicheskiye Osnovy Elektrovzryvnogo Legirovaniya Metallov I Splavov* (Novokuznetsk: SibSIU) p 301

IOP Conf. Series: Materials Science and Engineering **411** (2018) 012024 doi:10.1088/1757-899X/411/1/012024

Microstructure and wear properties of Hardox 450 steel surface modified by Fe-C-Cr-Nb-W powder wire surfacing and electron beam treatment

V E Gromov[1], V E Kormyshev[1], A M Glezer[2], S V Konovalov[3], Yu F Ivanov[4,5] and A P Semin[1]

[1]Siberian State Industrial University, 42 Kirov street, Novokuznetsk, 654007, Russia
[2]The National University of Science and Technology MISIS, 4 Leninskiy ave, Moscow, 119049, Russia
[3]Academican S.P. Korolev Samara National Research University, 34 Moskovskoye Shosse, Samara, 443086, Russia
[4]Institute of High-Current Electronics of Siberian Branch of Russian Academy of Sciences, 2/3, Academicheskii ave, Tomsk, 654052, Russia
[5]Tomsk National Research State University, 36 Lenina ave, Tomsk, 634050, Russia

E-mail: gromov@physics.sibsiu.ru

Abstract. Structural phase states and tribological properties of the coating surfaced onto Hardox 450 martensite low-carbon steel with powder wire Fe-C-Cr-Nb-W and modified by subsequent electron-beam processing are studied by methods of modern physical material science. It is shown that irradiation of ~ 5 mm thick surfaced layer with high intensity pulsed electron beams results in the formation of ~ 20 μm thick surface layer with the master phases of α-Fe and NbC, Fe_3C and $M_6C(Fe_3W_3C)$ carbides. It is established that wear resistance of the surfaced layer after electron-beam processing increases more than 70-fold relative to wear resistance of Hardox 450 steel and friction coefficient decreases significantly (~3-fold).

1. Introduction

The important fundamental task is to obtain the coatings with high service properties ensuring the increase in operational life of products in the extreme conditions of high wear, corrosion, mechanical loads and temperatures [1]. Hardfacing using superior material coatings has been widely used to achieve longer service life [2-5]. Among different methods welding is considered as an economical choice as a variety of process can be utilized to deposit a desired coating [6].

Thorough analysis of «wear parameters – hardness microstructure» relation is necessary in research and practical application of surfacing of various types in the critical components and products [7-10]. Only in this case it is possible to obtain products with high operational parameters.

The purpose of the research is the analysis of structure and tribological properties of the layer formed on Hardox 450 steel with electrocontact surfacing of Fe-C-Cr-Nb-W wire and modified by high intensity pulsed electron beam irradiation.

2. Material and methods

Hardox 450 steel (0.19-0.26 C; 0.70 Si; 1.6 Mn;0.025 P; 0.010 S; 0.25 Cr; 0.25 Ni; 0.25 Mo; 0.004 B; balance – Fe, weight %) was used as a base material. The surfacing of the strengthening layer was done by with consumable metal electrode shielded by inert/active gas with automatic feeding of filler wire) inert gas shielded welding (Ar – 98%, CO2 – 2%), under welding current 250-300 A and voltage 30-35 V. The powder wire 1.6 mm in diameter of the following chemical composition (weight %): 1.3 C; 7.0 Cr; 8.5 Nb; 1.4 W; 0.9 Mn; 1.1 Si; balance – Fe was used as surfaced electrode. Surfacing results in the formation of high strength surface layer ~ 5 mm in thickness.

Modification of the surfaced layer for increasing in its tribological properties was done by surface irradiation with intensity electron beam at the facility «SOLO» [11] in the regime of meting and high speed crystallization. Electron-beam processing was done in two stages: parameters of electron beam at the first stage – density of electron beam energy in pulse Es = 30 J/sm2; pulse duration τ = 200 μs; quantity of pulses N = 20; at the second stage – Es = 30 J/cm2; τ = 50 μs; N = 1. The irradiation regimes were chosen by calculation results of temperature field forming in the surface layer of the material in irradiation align in one pulse regime [12]. Investigations of tribological properties, defect substructure, chemical and phase composition of modified surface layer are carried out using the methods of modern material science.

3. Results and discussion

The irradiation of the surfaced layer with high intensity pulsed electron beam results in the formation of the modified surface layer up to 20 μm thick. The modified layer differs from the major volume of the surfaced material by the degree of structure dispersion revealed in ion etching of the transverse metallographic section.

In the volume of the surfaced layer unirradiated with electron beam the dimensions of the etched structural elements reach 1.5 μm, and after electron-beam processing dimensions of etching elements (evidently, high melting compounds possessing a comparatively low level of etching with ion beam) vary within 150-750 nm (figure 1).

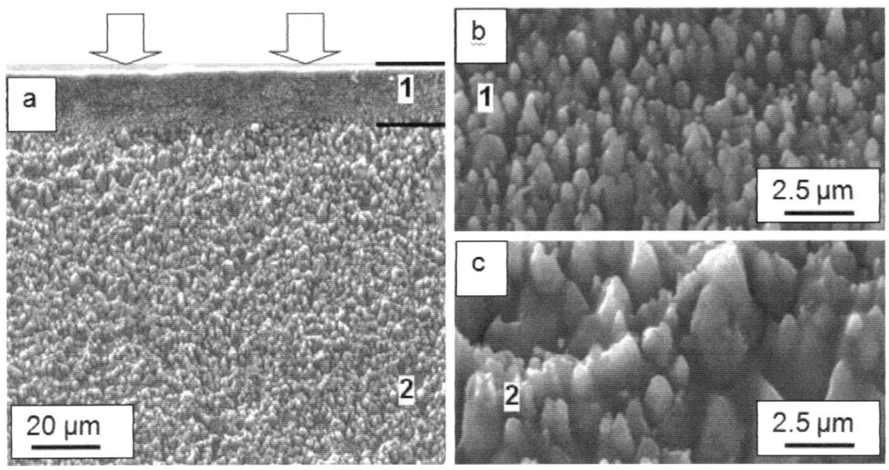

Figure 1. Structure of the surface layer. The transverse etched metallographic section. The arrows designate the irradiation surface of the layer with pulsed electron beams. Figures designate: 1 – the layer modified with electron-beam processing; 2 – the main volume of the surfacing.

When analyzing the results of X-ray phase analysis shown in figure 2 and in table 1 it can be revealed that the master phases of the studied surface layer of the surfacing are α-Fe (solution of Fe based bcc lattice) and niobium carbide NbC. A comparatively small parameter of crystal lattice of niobium carbide engages the attention. The niobium carbide NbC has a parameter of crystal lattice within 0.4429 to 0.4471 nm [13].

A comparatively small value of crystal lattice parameter of niobium carbide observed by us in the experiment (table 1) may be caused by a high level of concentration of vacant interstitial sites having the smaller linear dimension compared to the occupied interstitial sites. It is confirmed by the results of experimental works [14, 15].

Table 1. Results of X-ray phase analysis of the surfaced layer after the processing with intense pulsed electron beams.

Revealed phases	Phase content, mass. %	Lattice parameters, a, nm	Dimension of coherent scattering region, nm	$\Delta d/d*10^{-3}$
α-Fe	46.9	0.28553	61.8	2.36
NbC	53.1	0.43691	12.7	6.47

TEM image analysis shows that the second phase inclusions are located largely along the grain boundaries in the form of extended interlayers 100-150 nm in thickness. The second phase inclusions located in the grain junctions have the shape of the extended triple node and dimensions of such inclusions reach 1 μm.

Using the methods of microdiffraction analysis with dark field technique it is shown that the second phase inclusions located along grain boundaries in the form of interlayers are carbide $M_6C(Fe_3W_3C)$. In the volume and along the boundaries of martensite crystals the particles of iron carbide Fe_3C (possibly, M_3C) are revealed.

The surface layer structure of the surfacing irradiated with intense pulsed electron beams is characterized by the presence of the facet shape inclusions located chaotically in the grain volume (figure 2). The dimensions of such inclusions reach 2 μm. Indexing of microelectron diffraction pattern obtained from such inclusions is indicative of their being niobium carbide NbC.

Thus, by methods of electron diffraction microscopy it is shown that the surface layer of the surfacing modified with intense pulsed electron beam is a multi-phase aggregate with the master phases of α-iron based solid solution and carbides M_6C, NbC and Fe_3C.

The principal difference of the surface layer modified with intense pulsed electron beam from the nonmodified volume of the surfacing is the morphology and dimensions of the second phase inclusions. In the modified layer of the surfacing the inclusions have smaller dimensions (in comparison with the volume of the surfacing) and are located mainly in the form of comparatively thin interlayers along grain boundaries. In the volume of the surfacing nonmodified with electron-beam processing the main morphological type of the inclusions is the facet shape particles located chaotically in grain volume.

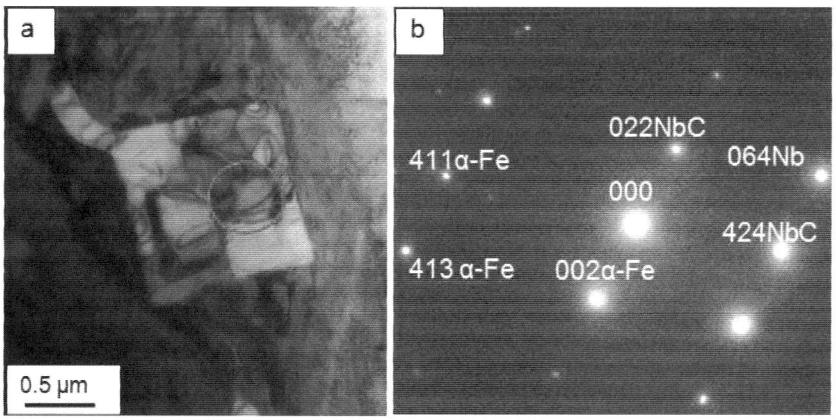

Figure 2. Structure of surfaced layer modified with electron beam. (a). Microelectron diffraction pattern (b) is obtained from the particle designated by the arrow.

The facet shape niobium carbides are also revealed at 5 mm depth from the electron beam modified surface. The performed tribological tests revealed 70-fold increase in wear resistance of the surfaced layer modified with intense pulsed electron beam in relation to wear resistance of steel.

Figure 3 shows the change in friction factor in tribological tests of the layer modified with electron beam. The two-stage change in friction factor is noticeable. In the first stage the friction factor value is ≈ 0.17, in the second stage ≈ 0.5. The friction factor of steel without the surfacing ≈ 0.26. When analyzing the change in fraction factor in tribological tests (figure 3) it may be supposed that modification of the surfaced layer with intense pulsed electron beams results in substantial ($\approx$ 3-fold) decrease in friction factor of the surfaced layer.

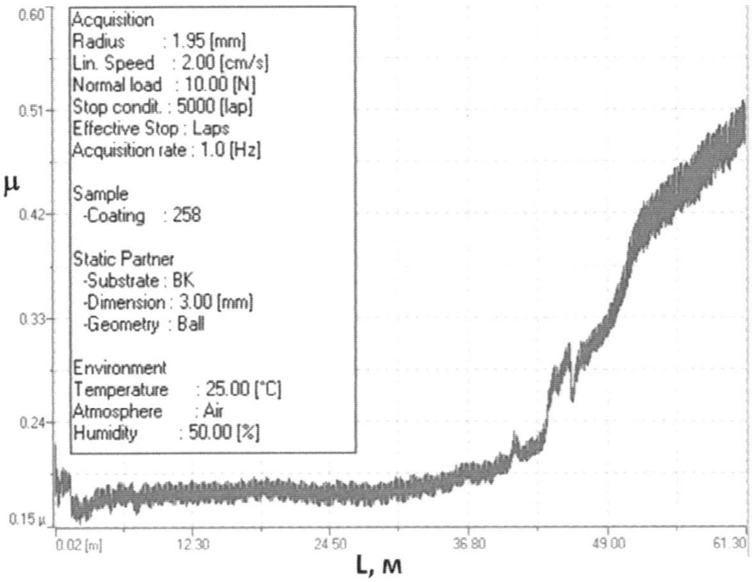

Figure 3. Dependence of friction factor on track length of tribological tests. The insert shows the conditions of tribological tests.

4. Conclusion

The studies of structure and tribological properties of the layer formed on Hardox 450 steel by electrocontact surfacing of Fe-C-Cr-Nb-W wire modified by irradiation with high intensity pulsed electron beam have been carried out. It has been shown that electron-beam processing of the welding surface results in the structure refinement and change in morphology of carbide phase of the layer as well as the substantial reduction in its friction factor.

Acknowledgements
The research is financially supported by grant RSF (project №15-19-00065).

References

[1] Kapralov E V, Raikov S V, Budovskikh E A, Gromov V E, Vaschuk E S and Ivanov Yu F 2014 *Physics* **78** 1015–1021
[2] Yüksel N and Sahin S 2014 *Materials & Design* **58** 491–498
[3] Venkatesh B, Sriker K and Prabhakar V S V 2015 *Procedia Materials Science* **10** 527–532
[4] Gualco A, Marini C, Svoboda H and Surian E 2015 *Procedia Materials Science* **8** 934–943
[5] Teker T, Karatas S and Osman Yilmaz S 2014 *Protection of Metals and Physical Chemistry of Surfaces* **50** 94–103
[6] Zahiri R, Sundaramoorthy R and Lysz P 2014 *Surface and Coatings Technology* **260** 220–229

[7] Chang C-M, Chen Ye-C and Wu W 2010 *Tribology International* **43** 929–934
[8] Fan C, Chen M-C, Chang C-M and Wu W 2006 *Surface and Coatings Technology* **201** 908–912
[9] Berns H and Fischer A 1997 *Materials Characterization* **39** 499–527
[10] Coronado J J, Caicedo H F and Gomez A L 2009 *Tribology International* **42** 745–749
[11] Koval' N N and Ivanov Yu F 2008 *Physics* **5** 60–70
[12] Koval' N N and Ivanov Yu F 2016 *Structural Evolution of Steel Surface Layer Subjected to Electron – Ion – Plasma Methods of Processing* (Tomsk: NTL)
[13] Bannykh O A, Budberg P B, Alisova S P et al 1986 *Constitutional Diagrams of Binary and Multi-component Systems on Iron Base* (Metallurgia, Moscow)
[14] Lipatnikov V N, Rempel A A and Gusev A I 1990 *Inorganic Materials* **26** 252–256
[15] Rempel' A A 1996 *Successes of Physical Sciences* **166** 33–62

Experimental estimation of drilling energy efficiency and life time of new rotary bits equipped with combined inserts on the basis of superhard composite materials

L T Dvornikov[1], S M Nikitenko[2], V A Korneyev[1] and P A Korneyev[1]

[1]Siberian State Industrial University, 42 Kirova Street, Novokuznetsk, 654007, Russia
[2]Federal Research Centre of Coal and Coal Chemistry, Siberian Branch of the Russian Academy of Sciences, 18 Sovetskiy Ave., Kemerovo, 650000, Russia

E-mail: korneev_va@list.ru

Abstract. The article is devoted to the drilling energy consumption and new design of rotary bits, its life time estimation. The feature of the bits is the use of combined cutting inserts based on cubic boron nitride. Tests of the experimental design of mining tools were carried out on a specially designed drilling bench, as well as by Fletcher and Wombat standard drilling rigs.

1. Introduction

Rapid development of technologies in the mining industry, success in the development of materials with unique properties actualize scientific research on the practical application of scientific and technological progress results. Investigations aimed at creating new designs of drill bits with increased wear resistance were carried out by the team of the Federal Research Center of Coal and Coal Chemistry of the SB RAS (FRC CCC SB RAS). FRC CCC SB RAS within the Federal Program framework on the theme: "Development of experimental designs of a combined tool using superhard composite materials for effective rock destruction" [1]. The results of this work are presented in this article.

2. Materials and methods

The most technologically advanced materials capable of replacing tungsten-cobalt alloys in the cutting inserts of rotary bits in the near future are superhard composites based on cubic boron nitride (CBN) and synthetic diamond [2]. And if the experience of using technical diamonds in a drilling tool has been accumulated quite extensively, then the development of designs based on cubic boron nitride has not at present been implemented in principle or was carried out on a non-significant scale [3]. In this regard, the experience of designing drilling bits with cutting inserts based on CBN, as well as their testing represent actual scientific and technical information.

In accordance with the requirement specification, 3 designs of rotary bits with combined cutting inserts were developed, which is a hard alloy coated with a CBN layer. The cutting inserts had circular and semi-circular shape (figure 1). Further, on their basis, cutters were manufactured for 28 mm diameter borehole rotary drilling with two-and three-point designs.

To evaluate the characteristics of the developed rotary bits in comparison with the existing ones, the tests were also carried out for two-point bits with a diameter of 30 mm with plates made of a standard tungsten-cobalt alloy. Photos of the bits and their designations in the experiment are shown in figures 2 and 3. Geometric and physico-chemical properties of the bit designs are presented in table 1.

The tests were carried out on a specially developed drilling bench, which allows the cutting tool movement, advance force, rotation speed and torque to be recorded. To simulate the rock during the experiments sandy-cement blocks (SCB) were used. Photos of the drilling bench and SCB after the tests are shown in figure 4.

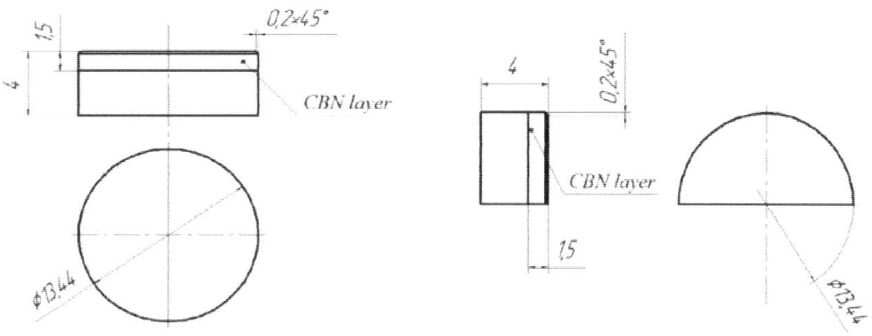

(a) (b)

Figure 1. Combined cutting plates used in experimental designs of rotary bits: round shape (a); semicircular shape (b).

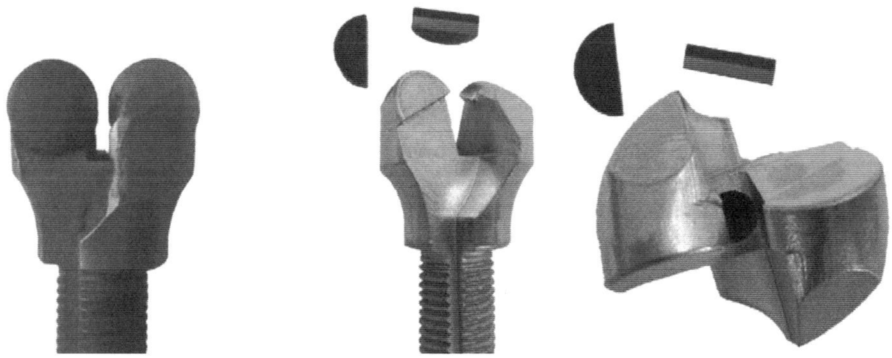

(a) (b)

Figure 2. Experimental designs of two-point rotary bits with a round insert (Type 1) (a); with a semicircular insert (Type 2) (b).

To evaluate the physics-mechanical properties of the SCB, samples were taken from each block, which were subsequently subjected to research in the profile laboratory. Each SCB used for testing had storage time of at least 48 days. The results of laboratory studies showed that the average strength coefficient value according to Protodyakonov scale of hardness was f = 8-9.

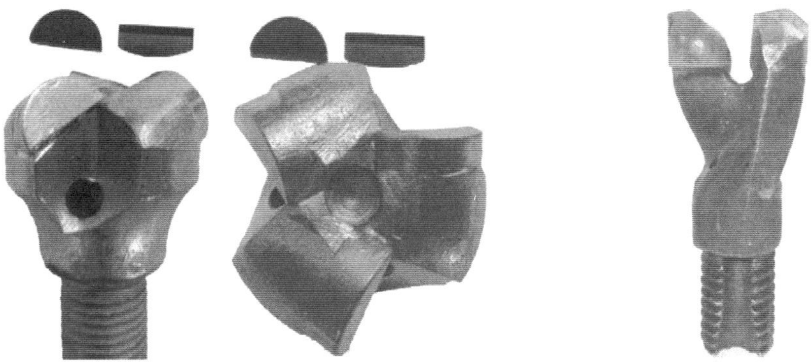

(a) (b)

Figure 3. Rotary bits: an experimental three-point with a semicircular insert (Type 3) (a); standard with cutting inserts made of tungsten-cobalt alloy (Type 4) (b).

Table 1. Geometrical parameters of cutting inserts and physics-chemical properties of cutter bodies.

Sample type	Cutting edge shape	Front angle	Rear angle	Chemical body composition /body hardness	Cutting plate chemical composition
No.1	Symmetric	-15°	18°	35KhGSA/350-390NV	CBN
No.2	Symmetric	-15°	18°	35KhGSA/350-390NV	CBN
No.3	Symmetric	-15°	16°	35KhGSA/350-390NV	CBN
No.4	Asymmetric	-2°	18°	35KhGSL/240-270NV	VC8V

Parameters of the drilling mode, the number of tested cutters of each design, as well as the number of blastholes drilled with each sample, are given in table 2.

After drilling 15 blastholes with each experimental design drill bits (Type 1, Type 2, Type 3), the bench tests were completed in connection with obtaining the cutting tool stable characteristics. As a result of the conducted studies, drilling diagrams, recording in time the advance force, the tool movement, the speed of its rotation, and the torque were obtained. Drilling diagrams were recorded for each drilled blasthole. Figure 5 shows one of the diagrams obtained when sample No. 1 of the Type 1 rotary bit was used.

The failure of experimental cutters during drilling at the stand was not registered, and in this connection they were transferred for further rock drilling tests to standard Fletcher and Wombat drilling rigs in the conditions of limited liability company “Alardinskaya Mine” and open joint stock company “UK "Neryungriugol” mine Denisovskaia. Drilling of SCB with cutters with reinforcing hard alloy inserts (Type 4) was stopped after the cutters were worn out.

(a) (b)

Figure 4. Drilling bench (a) and SCB after tests (b).

Further experiments on drilling rock with Fletcher and Wombat rigs were carried out in 2 stages. During the drilling process, the total length of drilled blastholes was registered, as well as the drilling speed.

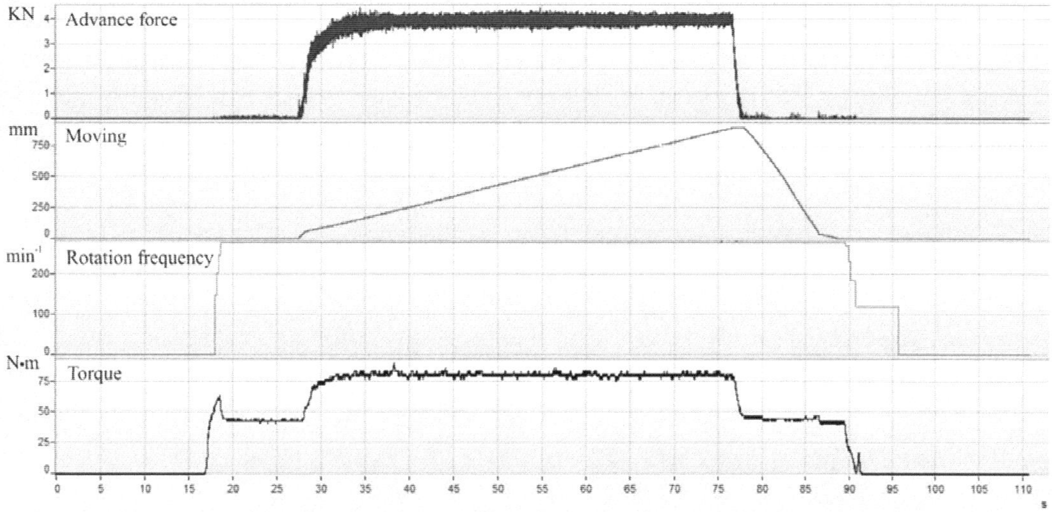

Figure 5. The drilling diagram obtained at the drilling bench when the sample No.1 of the rotary bit Type 1 was working.

Table 2. Drilling mode parameters.

Drilling mode parameters	Sample No. 1	Sample No. 2	Sample No. 3	Sample No. 4	Sample No. 5
		Type 1			
Advance force, kN	3.7	3.7	3.7	3.7	3.7
System pressure, MPa	0.2	0.2	0.2	0.2	0.2
Rotation frequency, min $^{-1}$	280	280	280	280	280
Specific advance, mm/rev	3.8	3.8	3.9	3.7	3.8
Advance speed, mm/s	17.6	17.7	18.0	17.4	17.5
Meters of drilled blasthole	13.2	13.4	13.4	13.2	13.3
		Type 2			
Advance force, kN	3.7	3.7	3.7	3.7	3.7
System pressure, MPa	0.2	0.2	0.2	0.2	0.2
Rotation frequency, min $^{-1}$	280	280	280	280	280
Specific advance, mm/rev	4.8	4.5	4.4	4.5	4.6
Advance speed, mm/s	22.3	21.1	20.7	20.8	21.3
Meters of drilled blasthole	13.5	13.7	13.7	13.6	13.7
		Type 3			
Advance force, kN	7	7	7	-	-
System pressure, MPa	0.5	0.5	0.5	-	-
Rotation frequency, min $^{-1}$	280	280	280	-	-
Specific advance, mm/rev	2.6	2.5	2.5	-	-
Advance speed, mm/s	12.1	11.6	11.7	-	-
Meters of drilled blasthole	13.6	13.7	13.7	-	-
		Type 4			
Advance force, kN	3.7	3.7	3.7	3.7	3.7
System pressure, MPa	0.2	0.2	0.2	0.2	0.2
Rotation frequency, min $^{-1}$	280	280	280	280	280
Specific advance, mm/rev	5.7	5.2	4.5	4.5	5.5
Advance speed, mm/s	26.6	24.2	21.0	20.8	25.5
Meters of drilled blasthole	7.2	7.1	6.2	7.3	6.3

At the first stage, the sandstone of strength f = 8-9 according to Protodyakonov scale of hardness was drilled by the Wombat pneumatic drilling rig with the use of Type 1, Type 2, Type 3 of rotary bits. At the second stage drilling of sandstone with strength f = 8-9 of Protodyakonov scale of hardness was performed with a Fletcher hydraulic rig using rotary bits Type1, Type 2.

Testing of the Type 3 rotary bit was not carried out due to the rig advance force lack, revealed at the first stage of the rock drilling experiment. Drilling of blastholes in both cases was carried out with water washing.

Photos of Type 1, Type 2 and Type 3 rotary bits after the experiments are shown in figures 6, 7.

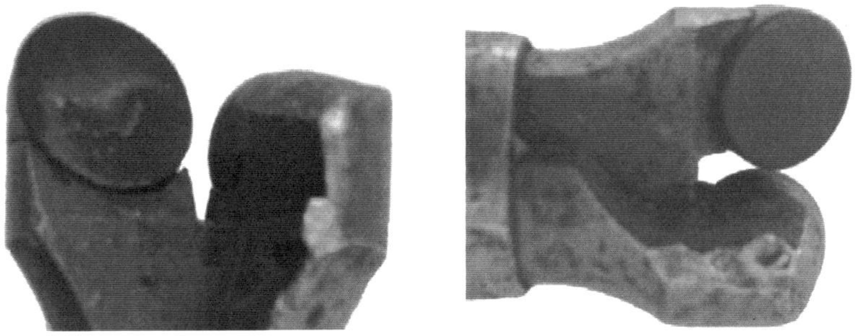

Figure 6. Experimental Type 1 rotary bit after tests.

(a) (b)

Figure 7. Experimental designs of rotary bits after tests: Type 2 (a), Type 3 (b).

3. Results and discussion

The experimental data obtained during the research were statistically processed and allow us to draw the following conclusions.

Figure 8 compares the resource of the developed experimental designs of the rotary bits (Type 1, Type 2, Type 3) with reinforcing inserts made of CBN and a tool with tungsten-cobalt inserts (Type 4), which is widely used in practice.

The drilling tool resource was determined as the average value obtained from the tests results at the drilling bench when drilling SCB and drilling rock with rock hardness ratio f = 8..10 units at industrial drilling rigs Wombat and Fletcher with pneumatic and hydraulic supply.

It should be noted that in connection with the full-scale resource tests lack, as well as the use of SCB at the bench, the results obtained are of an evaluation nature and are considered preliminary.

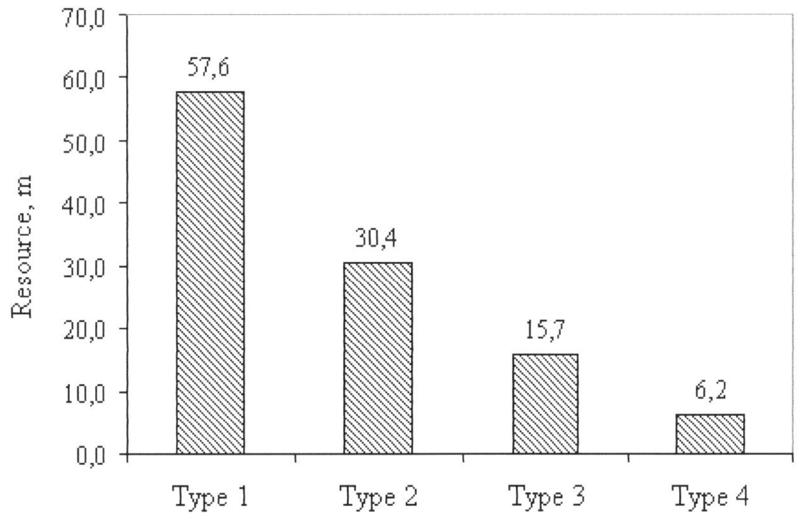

Figure 8. Evaluation of rotary bit life time.

It can be seen from the diagram that the life time of the most effective Type 1 experimental design cutter exceeds the life time of the standard rotary bit by more than 9 times.

Other experimental designs (Type 2, Type 3) also have high wear resistance, which indicates the significant prospects of using superhard composite materials for reinforcing rotary bits cutting inserts.

Figure 9 shows the values of the average specific advance rate, which is achieved when drilling with bits of new designs. As can be seen from table 2, the set advance force, as well as the number of revolutions when drilling with Type 1, Type 2, Type 4 cutters were the same. Experimental Type 3 rotary bit in connection with the three-point design required a greater advance force, while the specific advance of such a cutter per revolution was the least.

It should be noted that the specific advance rate value when drilling with the Type 2 cutter is intermediate between the advance values achieved when drilling with Type 1 and Type 4 cutters, indicating that there are potential opportunities for the performance of the Type 1 cutter improvement. The smaller value of the specific advance rate and, correspondingly, the drilling speed of the Type 1 rotary bit are related to its blades design, which can be optimized through additional studies.

Specific advance rate during experimental data processing was determined by the following expression:

$$\Delta = \frac{L_{av.} \times 60}{n \times t_{av.}}, \text{ mm/rev.} \quad (1)$$

Where $L_{av.}$ – average blasthole depth, mm; n – rotation frequency, rev/min.; $t_{av.}$ – average blasthole drilling time, s.

Figure 10 shows the values of the average drilling speed for bench tests, which was calculated by the expression:

$$v = \frac{L_{av.}}{t_{av.}}, \text{ mm/s.} \quad (2)$$

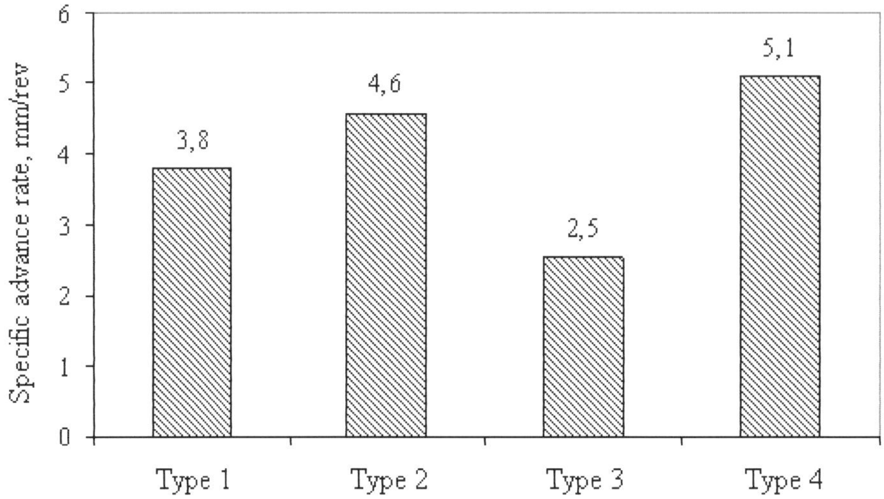

Figure 9. Specific advance rate at rotary bits bench test.

The drilling speed during rotary bits testing Type 1, Type 2 and Type 4 in the working conditions at various rigs was practically the same and reached about 26 mm/s. At the same time drilling with Type 3 rotary bit was not carried out due to the rig used in researches low advance force.

Figure 11 gives information on the test cutters drilling energy consumption. To calculate the energy consumption, the results of experiments carried out on the drilling bench were used. The value of drilling energy consumption was determined by the work spent on drilling to the destroyed rock volume ratio, while the consumption for the tool advance was not taken into account due to its insignificant share in the total energy consumption.

The following dependencies were used in the calculations:

$$P = M_{CR.} \cdot \omega \text{, W.} \tag{3}$$

Where P – the power spent for drilling; $M_{CR.}$ – the torque, N•m; ω is the angular velocity, rad/s.

$$A = P \cdot t \text{, J.} \tag{4}$$

Where A is work; t – drilling time, s.

$$V = \pi r^2 \cdot L \text{, m}^3. \tag{5}$$

Where V is the volume of destroyed rock; r – radius of the blasthole, m; L – blasthole, m.

The drilling energy consumption by means of the rotary bit developed designs Type 1 and Type 2 differs from the similar characteristic of the standard rotary bit Type 4 within the statistical error. At the same time, the blasthole drilling energy consumption with a 3-point experimental rotary bit Type 3 exceeds the specific power consumption for drilling by more than 2 times for each of the tested bits. The most likely reason for the situation is the difficulties associated with the removal of the cuttings. The third point presence in the bit design reduces the free space, which leads to densification and re-destruction of the drill cuttings, as a result the specific power consumption for drilling significantly increases.

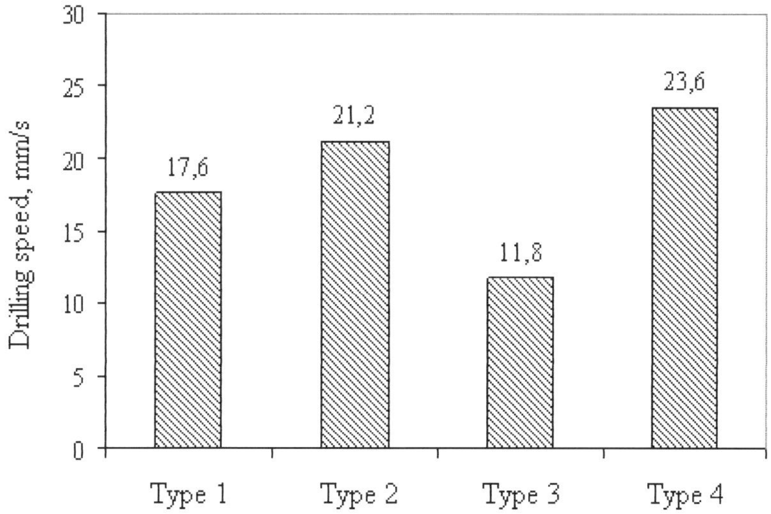

Figure 10. Drilling speed at rotary bit bench tests.

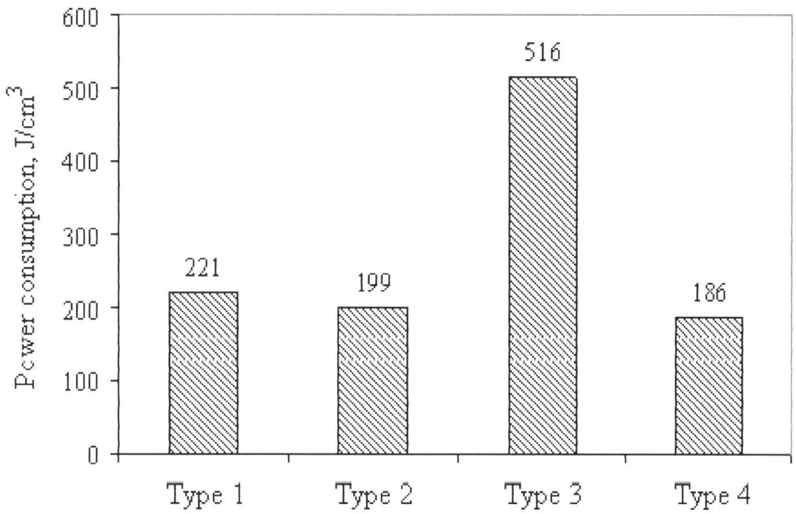

Figure 11. Drilling energy consumption at rotary bits bench tests.

4. Conclusion

The experimental study results have shown that the use of CBN for reinforcing blades of rotary bits can significantly increase the life of a mining tool without its other technological characteristics, such as drilling speed and power consumption worsening. At the same time, the high thermal stability of the CBN allows the based on it mining tool to be operated in combination with high-performance drilling modes at temperatures up to 900 °C in the cutting zone.

In connection with the lack of full-scale resource tests, the data obtained are indicative and intend to assess the prospects of using the mining tool with reinforcing inserts made of CBN. Further studies will allow us to refine and supplement the results obtained.

References

[1] Dvornikov L T, Klishin V I, Nikitenko S M, Korneyev V A and Korneyev P A 2016 *IOP Conf.*

Series: Earth and Environmental Science **45** 33–42
[2] Nikitenko S M, Kol'ba A V, Anokhin A S, and Kukueva E V 2016 *Glass and Ceramics* (New York: Springer US) pp 458–464
[3] Nikitenko S M, Anokhin A S, Strel'nikova S S, Kukueva E V, Shipkov A N and Tkachenko V V 2015 *Glass and Ceramics* (New York: Springer US) pp 290–293

Simulation of crystallization parameters of casting alloys during modifying treatment by physical impact

V B Deev[1,2], O G Prikhodko[3], K V Ponomareva[3], A I Kutsenko[3], E S Prusov[4], E Kh Ri[5] and S V Smetanyuk[2]

[1]Wuhan Textile University, 1 Textile Road, Wuhan, 430073, China
[2]National University of Science and Technology "MISIS", 4 Leninskiy ave., Moscow, 119049, Russia
[3] Siberian State Industrial University, 42 Kirova street, Novokuznetsk, 654007, Russia
[4]Vladimir State University n.a. Alexander and Nikolay Stoletovs, 87 Gorky Street, Vladimir, 600000, Russian Federation
[5]Pacific National University, 136 Tikhookeanskaya street, Khabarovsk, 680042, Russia

E-mail: deev.vb@mail.ru

Abstract. The paper presents the developed mathematical model for calculating the crystallization parameters of engineering casting alloys processed by external physical impacts, allowing the critical size of the crystallizing nuclei to be predicted and the amount of nuclei for evaluating the effectiveness of the melt modifying treatment. To automate the calculations for the proposed model, the application software has been created using modern methods of visual software design on the basis of modular and object-oriented approaches. Using A356 aluminum alloy as an example, it was shown that the melt treatment by physical impacts helps to reduce the critical size of the crystallizing nuclei and increase their amount, with the greatest effect being achieved for alloys with a higher proportion of secondary materials in the charge. Prospects of using the calculation technique for the proposed mathematical model in real production conditions during development of technological processes for the production of foundry alloys and castings from them are indicated.

1. Introduction

It is known that the quality of foundry engineering alloys is determined by their structure, which affects the level of mechanical, technological and operational properties of the parts.

At present, methods of physical (external) modifying effects on melts such as ultrasound, vibration, high-temperature superheating, electric current, a magnetic field, etc., are being introduced in the industrial foundry and metallurgical technologies [1-5]. These effects promote the formation of a fine-grained structure and increase the mechanical and operational properties of castings without the introduction of special modifying additives, in some cases having a rather high cost. The main advantage of physical influences is that they do not change the chemical composition of the melt and do not lead to the accumulation of undesirable impurities in casting alloys during further remelting [6-8]. In this regard, the use of physical effects conditions the prospects of using an increased number of secondary materials during melting.

The aim of the work consists in the development and approbation of a computational model for calculating the critical size of the nucleus and the number of nuclei in the melt volume during the modifying treatment of melts using physical methods of action.

2. Methods

The influence of physical effects on the melt is proposed to be interpreted from the position of the cluster model [8]. In accordance with the cluster model of the structure of melts, the main structural components of the liquid are considered separate clusters surrounded by a three-dimensional grid of a disordered zone with a chaotic atomic structure [9]. Clusters in microvolumes can be of different size, composition and density. The internal atomic structure of clusters is close to the structure of the original solid. The characteristic cluster size has a nanoscale that does not exceed the dimensions of the critical nucleus (r_c).

Taking into account the modern concepts of the cluster model, the following expressions were obtained by mathematical transformations for spherically formed nuclei:

$$r_c = \frac{3J\frac{\Delta W}{e}T_0}{\rho L}\left(\frac{\Delta T}{\Delta \tau}\right)^{-1}, \tag{1}$$

$$n_0 = \frac{3c\Delta T_S}{4\pi L}\frac{1}{r_c^3}, \tag{2}$$

where J – current pulse due to the transition of electrons from the ionization level of the free atom in the disordered zone to the level Fermi nucleus (cluster); $\frac{\Delta W}{e}$ –jump of a potential at the boundary of cluster with a disordered zone; $\Delta W = (J_1 - W_{lib})$ – the work expended by the valence electron upon transition to the level Fermi cluster when a nucleus of critical dimensions is formed; J_1 – the first potential of ionization; W_{lib} – the work of the electron liberation; T_0 – the equilibrium temperature of crystallization; $\frac{\Delta T}{\Delta \tau}$ – rate of melt cooling in the period of crystals nucleation; n_0 – the number of nuclei per unit volume of the melt; ρ – density; L and c – the specific heat of crystallization and the heat capacity of the melt; ΔT_S – supercooling of the melt.

This mathematical model is universal and allows the main parameters of melts crystallization (the critical size of the nucleus r_c and the number of nuclei per unit volume of the melt n_0) to be predicted after their treatment by various physical effects.

For the automation of calculations based on the proposed model, the original software "Calculation of melt crystallization parameters after treatment external actions" was developed using modern methods of visual design of computer software based on modular and object-oriented approaches. Delphi integrated development environment was used to create the calculation modules. The thermophysical properties of the substances incorporated in the program database and open for additions or changes serve as the initial data for the calculations. The program makes it possible to calculate the critical radius of the nuclei r_c, the number of nuclei per unit volume of the melt n_0, depending on the experimentally determined crystallization parameters – supercooling of the melt and the nucleation time interval. It is possible to simultaneously calculate both r_c and n_0 for the initial melt; the melt after physical action (magnetic field, electric current, vibration, inert gas); the melt with a complex processing technology, as well as visualization of the calculation results.

Calculations can be carried out for the following alloys: carbon steels, cast irons, bronzes, aluminum alloys of various compositions. In the case under consideration, the object of the study was alloy A356 (equivalent to the brand EN AB-42100 in accordance with standard EN 1676-2010, the chemical composition according to table 1), which is now widely used in various industries, including for casting parts of metallurgical engineering (instrument parts, pump cases, carburettors operating at temperatures up to +200 °C, etc.). The initial data for determining the crystallization parameters are given in table 2.

Table 1. Chemical composition of alloy A356 according to EN AB-42100, wt%.

Al	Si	Mg	Fe	Cu	Mn	Zn	Ti	Other impurities	
								each	total
base	6.5-7.5	0.30-0.45	≤0.15	≤0.03	≤0.10	≤0.07	≤0.18	≤0.03	≤0.10

Table 2. Reference data for calculating the crystallization parameters of A356 alloy using the proposed model.

Material	The first ionization potential, eV	Current impulse, A/m^2	Work function of electron, eV	Density, kg/m^3	Specific heat of crystallization, J/kg	Heat capacity, J/kg K
A356	5.984	8.134	4.23	2500	281 000	1085

The necessary values for the calculation of melt supercooling and the time interval of nucleation during the treatment of A356 alloy by physical actions were determined experimentally by thermal cooling curves. The melting of the experimental alloy and its processing were carried out in the induction furnace IF-0.06. Refining of melts before casting was carried out by manganese chloride (0.2% of the melt mass). The physical effects applied were thermal treatment (T = 970…980 °C, τ = 7…10 min – with 100 % pig materials in the charge, T = 1000…1020 °C, τ = 10…12 min – with 50…100 % of pig materials in the charge), a constant magnetic field with a gradient $\Delta B/\Delta x$ = 0.886 T/m when casting into a mold, a constant electric current during crystallization ($j = 0.92 \cdot 10^5$ A/m^2), and their combinations. The secondary materials in the charge in all options contained in their volume: 50 ... 55% of fine scrap and waste alloys, 45 ... 50% of briquetted shavings of the alloy with the same chemical composition.

3. Results and discussion

Of great interest is the study of the parameters r_c and n_0 depending on the quality of the charge used and various methods of physical effects on melts. The results showed a significant influence on the crystallization process of such methods of physical actions as the temperature-time treatment (TTT) and the magnetic field.

Table 3 presents the results of calculations for the developed model using a software application for A356 alloy, obtained from different charge.

Table 3. Calculations results of crystallization parameters of A356 alloy using the software application "Calculation of melt crystallization parameters after treatment external actions".

Composition of charge	Method of melt treatment	$r_c \cdot 10^{-7}$, m	$n_0 \cdot 10^{14}$, m^{-3}
A356 (pig metals)	Initial (without treatment)	0.9253	0.1083
	TTT	0.3322	0.3071
	Magnetic field	0.4056	0.2453
	TTT + magnetic field	0.2098	0.4669
	Electric current	0.4815	0.2173
	TTT + electric current	0.3071	0.3291
A356 (50 % of pig metals + 50 % of secondary	Initial (without treatment)	0.9706	0.1029
	TTT	0.3689	0.2688
	Magnetic field	0.4588	0.2114

materials)	TTT + magnetic field	0.2271	0.4043
	Electric current	0.5058	0.2011
	TTT + electric current	0.3100	0.3196
A356 (100 % of secondary materials)	Initial (without treatment)	0.9997	0.1005
	TTT	0.4234	0.2095
	Magnetic field	0.4979	0.2096
	TTT + magnetic field	0.2658	0.3854
	Electric current	0.5431	0.1842
	TTT + electric current	0.3239	0.3087

It follows from the data in table 3 that the melts treatment by physical actions (both individually and in a complex) leads to a decrease in the critical radius of the nuclei (r_c) and an increase in the number of nuclei per unit volume of the melt (n_0). As a result, because of the bulk character of the crystallization, a fine-grained structure is formed in the alloys. It should be noted that physical actions had a greater effect on the alloy from the charge with an increased amount of secondary materials (scrap and waste).

The presented mathematical model realized in the developed program allows the crystallization parameters (r_c, n_0) to be predicted after applying various physical actions and the effectiveness of a particular technology of modifying melt processing in the production of thin-walled castings of various purposes from ferrous and non-ferrous alloys to be evaluated. For calculation the reference data on the thermophysical properties of materials and the experimentally determined crystallization characteristics of melts are required.

A promising direction for development of the above indicated approaches to predicting the parameters of crystallization can be their adaptation to foundry composite materials on a metal basis [10]. Modifying treatment of metal matrix composites reinforced with endogenous phases (TiB_2, TiC, Al_3Ti, Mg_2Si, etc.) is not only advisable from the standpoint of improving the micro- and macrostructure of the matrix, but also for controlling the morphological characteristics of the forming reinforcing particles. Considering the significant potential of using physical methods of melts treatment for modifying composites, the development of reliable calculation models for evaluating the efficiency of such treatment will allow the most rational technological modes for obtaining composite alloys with specified properties to be choosen.

4. Conclusion

A mathematical model is developed for calculating the crystallization parameters of foundry machine-building alloys treated by physical actions, which allows the critical size and the number of crystallization nuclei for evaluating the effectiveness of the melt modifying treatment to be predicted. The proposed methodology for calculating crystallization parameters can be applied in production and research laboratories and centers dealing with the issues of obtaining high-quality castings for machine-building purposes.

Acknowledgments

The work was carried out within the framework of the state work "Organization of Scientific Research" of the state task of the Ministry of Education and Science of Russia in the field of scientific activity for 2017-2019 (task No.11.5684.2017/6.7).

References

[1] Komarov S, Ishiwata Y and Mikhailov I 2015 *Metallurgical and Materials Transactions* **46** 2876–83

[2] Vorozhtsov S, Kudryashova O, Promakhov V, Dammer V and Vorozhtsov A 2016 *JOM* **68(12)** 3094–100

[3] Rabiger D, Zhang Y, Galindo V, Franke S, Willers B and Eckert S 2014 *Acta Materialia*

79 327–38
[4] [4]Zhang Y, Cheng X, Zhong H, Xu Z, Li L, Gong Y, Miao X, Song C and Zhai Q 2016 *Metal* **6** 170
[5] Li Q L, Xia T D, Lan Y F and Li P F 2014 *Materials Science and Technology* **30(7)** 835–41
[6] Deev V B, Selyanin I F, Ponomareva K V, Yudin A S and Tsetsorina S A 2014 *Steel in Translation* **44(4)** 253–4
[7] Timoshkin I Y, Nikitin K V, Nikitin V I and Deev V B 2016 *Russian Journal of Non-Ferrous Metals* **57(5)** 419–23
[8] Deev V B, Selyanin I F and Tsetsorina S A 2008 *Steel in Translation* **38(8)** 623–4
[9] Mudry S, Shtablavyi I and Shcherba I 2008 *Archives of Materials Science and Engineering* **34(1)** 14–18
[10] Prusov E S, Panfilov A A and Kechin V A 2017 *Russian Journal of Non-Ferrous Metals* **58(3)** 308–16

A study on the extrusion processes of hollow products in the die with a movable matrix

V V Evstifeev[1], A A Aleksandrov[1], A V Evstifeev[2], A I Kovalchuk[3] and V N Peretyatko[4]

[1]Siberian State Automobile and Highway University, 5 Mira ave, Omsk, 644080, Russia
[2]JSC Omsk Production Association “Irtysh”, 18 Gurtyeva street, Omsk, 644060, Russia
[3]JSC “Omsktransmash”, 2 Krasniy per., Omsk, 644020, Russia;
[4] Siberian State Industrial University, 42 Kirova street, Novokuznetsk, 654007, Russia

E-mail: aleksandrov.omsk@mail.ru

Abstract. Calculation formulas of general type and the developed mathematical model of the direct extrusion process are presented, which makes it possible to determine the force mode of deformation. The examples of calculations and graphs showing the possibility of determining the optimal tool geometry and determining the amount of decrease in deformation force when using a die with a movable matrix are given.

1. Introduction

The hollow parts having conical surfaces find wide application in various designs and products. The manufacture of such parts by cold extrusion forging provides an improvement in performance due to the possibility of obtaining parts with the oriented fiber structure, and also makes it possible to reduce or completely eliminate the products finishing by cutting, which ensures their competitiveness.

The use of dies with a moving matrix (figure 1) for the manufacture of parts with conical cavities by direct extrusion from annular or cylindrical workpieces allows the deforming force to be reduced, since in this case the frictional forces acting on the surface between the workpiece and the matrix play an active role.

2. Methods of research

The magnitude of the deforming extrusion force can be determined by the theoretical upper bound method [1, 2] using the process scheme and the velocity hodograph, adopting a planar deformation scheme.

The velocity hodograph (figure 2b), where the velocity vectors are constructed from the pole point O, with the numbering corresponding to the blocks, will be the same for both forming schemes shown in figure 2a and figure 2c.

At the same time, when extruded in a die with a movable matrix (figure 2c), the costs for overcoming the frictional forces acting on the contact surface between the deformed workpiece and the matrix are substantially lower due to the decrease in velocity by the value V_1.

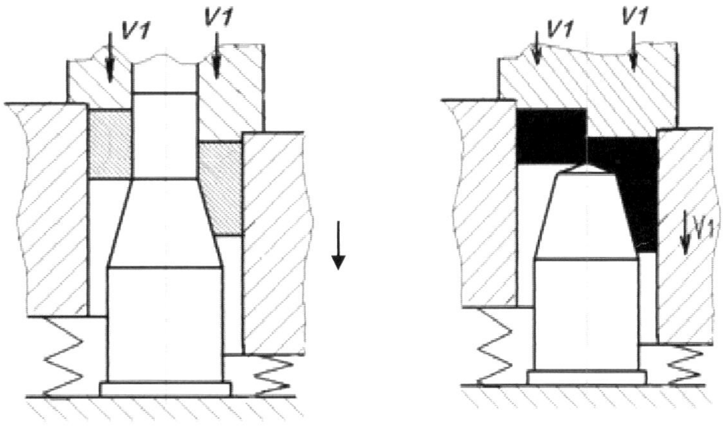

Figure 1. Scheme of workpiece deformation between the fixed counterpunch and the movable matrix: (a) extruding the annular preform; (b) extruding a continuous blank.

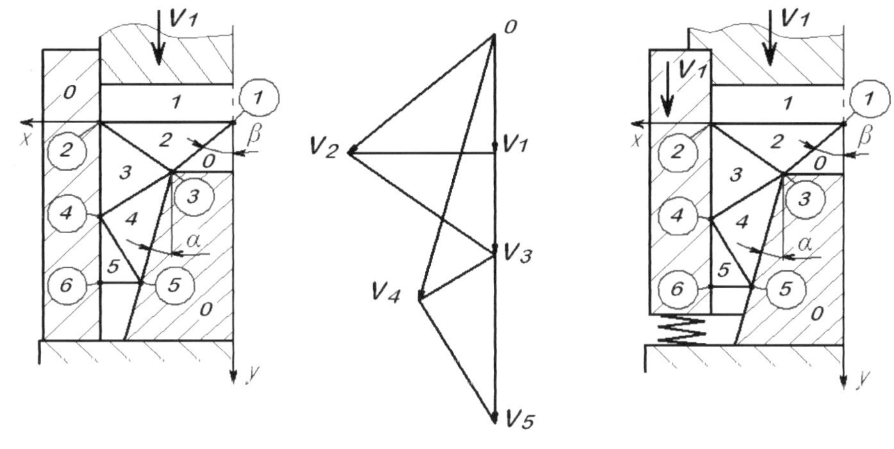

Figure 2. Calculation schemes: a – extrusion in the die with a fixed matrix; b – velocity hodograph corresponding to this scheme; c – extrusion in the die with a movable matrix.

Note that the design of the hodograph is the most time-consuming calculation stage of the known theoretical method [1]. However, it is possible to determine the magnitude of the deforming force of extrusion even without the velocity hodograph construction, using only the design scheme of the process [2]. In this case, the relative specific force q acting on the punch is represented by the sum calculated from equations (1) and (2), the fractions of this force expended on shear and friction along the boundaries of the blocks in accordance with the scheme of the plastic deformation process.

$$q_i = \mu \frac{l_i}{h_i}, \tag{1}$$

$$q_{ij} = \frac{l_{ij}^2 \cdot \sin\gamma}{h_i \cdot h_j}, \tag{2}$$

where $\mu = \tau_k / \tau_s$; τ_s – the metal yield strength on shear; τ_k – the contact shear stress, $l_{i,j}$ – the lengths of boundaries between blocks i and j; l_i and l_j – the lengths of the sides of the i-th and j-th blocks parallel to the current lines in these triangular blocks, h_i and h_j – the block heights; γ – is the angle of current lines rotation on the adjacent boundary of the blocks i and j.

For the considered schemes, the lengths of the boundaries of triangular blocks (with regard to the accepted notations) can be expressed through coordinates of six nodes (table 1) and trigonometric functions of two angles α and β.

Table 1. Coordinates of nodes.

Node numbers and node coordinates					
1	2	3	4	5	6
$x_1 = 0$	$x_2 = R$	$x_3 = r$	$x_4 = R$	$x_5 = r_5$	$x_6 = R$
$y_1 = 0$	$y_2 = 0$	$y_3 = h_0$	$y_4 = l_3$	$y_5 = L$	$y_6 = L$

Equation for calculating the power mode of the process:

$$q = q_1 + q_2 + q_3 + q_4 + q_5 + q_{12} + q_{23} + q_{34} + q_{45}. \qquad (3)$$

According to the calculation scheme: the total height of the forging (l_1+L), as well as the coordinates of the 5th and 6th nodes, correspond to the required forging dimensions at the final extrusion time; $l_{12} = R$ and parallel to X axis; the origin of coordinates coincides with the first node; h_0 – height of the 0th hard block of the workpiece; l_1 – length of the contact boundary of the 1st rigid block of the workpiece and the matrix. Given this, we find the length of the blocks boundaries:

$$l_2 = \frac{r}{\sin\beta},\ l_{23} = [h_0^2 + (R-r)^2]^{1/2},\ l_{34} = [(l_3 - h_0)^2 + (R-r)^2]^{1/2},\ l_4 = \frac{r_5 - r}{\sin\alpha},$$

$$l_{45} = [l_5^2 + (R - r_5)^2]^{1/2},\ l_5 = L - l_3.$$

Let us calculate the block heights:

$$h_2 = \frac{R \cdot h_0}{l_2},\ h_3 = R - r,\ h_4 = [R - r - (l_3 - h_0) \cdot tg\alpha]\cos\alpha,\ h_5 = R - r_5.$$

Further, using equations (1) and (2), we define the components of the right-hand side of equation (3) that make up:

$$q_1 = \frac{\mu \cdot l_1}{R},\ q_2 = \frac{\mu \cdot l_2}{h_2},\ q_3 = \frac{\mu \cdot l_3}{h_3},\ q_4 = \frac{\mu \cdot l_4}{h_4},\ q_5 = \frac{\mu \cdot l_5}{h_5},$$

$$q_{12} = \frac{R \cdot \sin\beta}{h_2},\ q_{23} = \frac{l_{23}^2 \sin\beta}{h_2 h_3},\ q_{34} = \frac{l_{34}^2 \sin\alpha}{h_3 h_4},\ q_{45} = \frac{l_{45}^2 \sin\alpha}{h_4 h_5}.$$

Equation (3), after substituting all found components, is a mathematical model designed to calculate the process force mode of the direct extrusion in the die with a fixed matrix (figure 2a). It allows us to find the theoretical value of the deformation force and optimize the process, determining the geometric relationships of the forming tool, in which extrusion is carried out with the lowest energy costs. Note that varying the coordinates of the 1st, 2nd and 4th nodes increases the accuracy of calculations in determining the optimal ratios.

When extruded in the die with a movable matrix (figure 2c), due to the decrease (by value V_1) in the relative speed of movement of the formed metal on the surface of its contact with the matrix, the relative specific force acting on the punch decreases, which improves the resistance of the deforming tool. When using the developed mathematical model, this decrease in force is taken into account as follows: $q_1 = 0$; q_3 and q_5 are multiplied by $(R-h_3)\cdot h_3^{-1}$ and by $(R-h_5)\cdot h_5^{-1}$, respectively.

The constructed mathematical model of the process, taking into account several variable parameters, allows their optimal values to be calculated and the optimum tool ratios to be found for which the minimum deformation power is ensured, which is shown in figures 3-6, where the coefficient μ, taking into account friction conditions on the tool, changed within 0 to 0.5.

3. Results and discussion

The dependencies in figure 3 show that when the ordinate node 4 is varied in the deformation area, expressed through l_3, it is always possible to find its position (for $r = 5$, $\alpha = 30^o$), in which the minimum deformation power is reached. However, for different instrument ratios this position can vary significantly. For example, figures 4 and 5 show that for $R = 30$, $r_5 = 20$, $l_1 = 15$, $\beta = 45^o$, the optimal values of l_3 and the cone angle α, depending on the radius r determining the initial gap, and the coefficient of friction μ, vary much.

Figure 6 shows the theoretical values of the relative specific deformation force at the optimum values of the variable parameters.

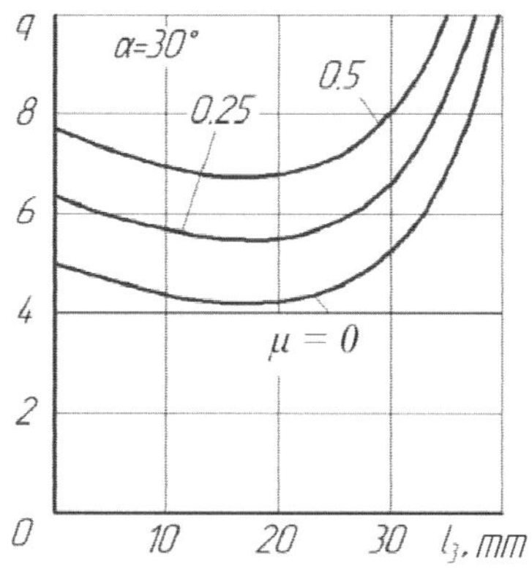

Figure 3. Graphs of the change in the relative specific force q as a function of l_3 and μ.

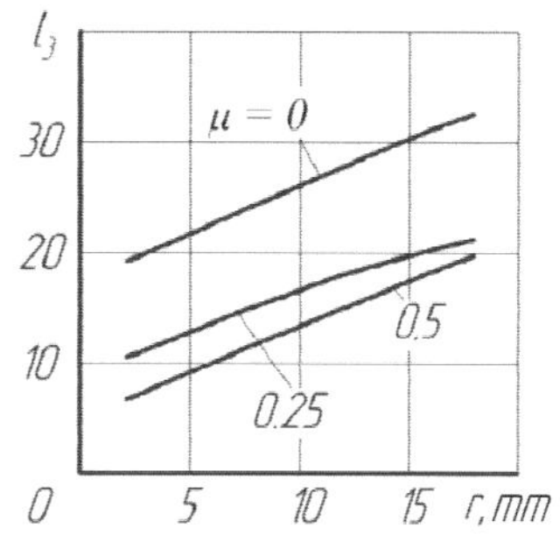

Figure 4. Graphs of the change in the optimal value l_3 as a function of r and μ.

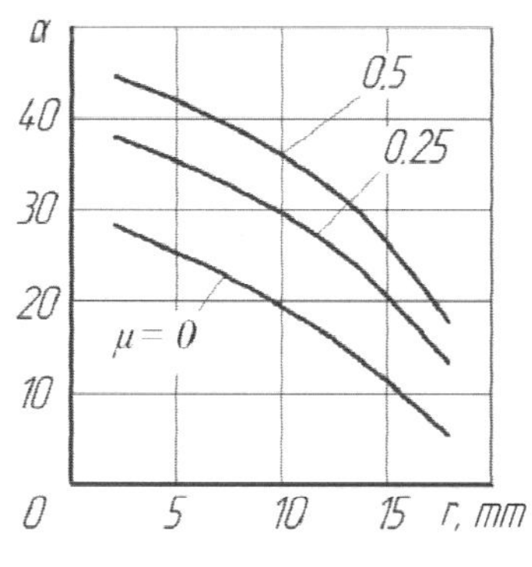

Figure 5. Graphs of the change in the optimal angle α (deg.) as a function of r and μ.

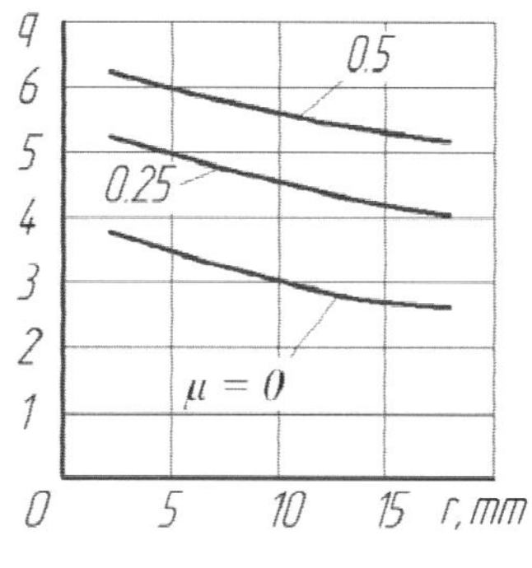

Figure 6. Graphs of the change in the specific force q at the optimal l_3 and α.

IOP Conf. Series: Materials Science and Engineering **411** (2018) 012027 doi:10.1088/1757-899X/411/1/012027

4. Conclusion

The obtained results show that the direct extrusion of the workpiece in the die with the movable matrix helps to reduce the technological load improving the resistance of the deforming tool, in comparison not only with direct extrusion in the fixed matrix, but also with the reverse extrusion, due to the reduction of the prevailing influence of frictional forces acting on the workpiece contact surface with the matrix.

The method of calculation, which does not require the construction of a velocity hodograph, shown in the example of the extrusion process in the die with a movable matrix, reduces the complexity of developing mathematical models of various forming processes and can be used to calculate the technological load and optimize the geometry of the forming tool.

References

[1] Unksov E P et al 1992 *Theory of Forging and Forming* (Moscow: Mashinostroyeniye) p 720
[2] Evstifeev V V, Aleksandrov A A 2014 *Analysis of Forming Processes based on Energy methods and Standardized Calculation Schemes: Monograph* (Omsk: SibADI)

IOP Conf. Series: Materials Science and Engineering **411** (2018) 012028 doi:10.1088/1757-899X/411/1/012028

Technologies, which allow to reduce an impact of metal silicon production on the environment

K S Yolkin[1], D K Yolkin[2] A D Kolosov[3], N A Ivanov[3] and M G Shtayger[4]

[1]UC RUSAL, RUSAL Engineering Technology Centre, 40 Pogranichnikov street, Krasnoyarsk, 660111, Russia
[2]RUSAL, JS Company "Kremniy", 26 Yuzhnaya street, Shelehov, 666034, Russia
[3]Irkutsk National Research State Technical University, 83 Lermontov street, Irkutsk, 664074, Russia
[4]MC Mechel Steel, 1 Krasnoarmeysky street, Moscow, 125167, Russia

E-mail: karlinat@mail.ru

Abstract. It is shown that silicon carbide can be used as a reducing agent in the production of metallic silicon. The results of tests of silicon carbide are presented. The possibility of reducing the amount of furnace gases is shown. The calculated amount of decrease in the amount of waste furnace gases is confirmed by the test results.

1. Introduction

Organic-silicon compounds combine the best properties of silicates; on the basis of silicones, heat-resistant lubricating oils, liquid dielectrics, varnishes, enamels, rubbers, plastics, fiberglass, effective sorbents, catalysts are produced.

High purity silicon is a typical semiconductor and is widely used in modern electrical engineering, radio engineering and electronics. A special place is occupied by solar power on the basis of crystalline silicon, share of which increases significantly to obtain cheap electricity.

Modern technology of silicon production, used in world and domestic practice, is associated with a risk of negative impact on the environment, because in addition to the final product, other reaction products are formed, including dust from incomplete use of charge materials.

Metallic silicon is produced by high-temperature reduction of silica with the help of reducing agents in arc ore thermal electric furnaces. A feature of the industrial production of silicon is the use of silica-containing raw materials and carbonaceous reducing agents, characterized by a high content of useful components and a low content of slag-forming impurities, and therefore, silicon smelting reduction is practically a slag-free process [1]. The gases, released by smelting in ore-thermal furnaces, are characterized by the content of a large amount of fine dust. 94-96% of dust content is silicon dioxide [2]. As a result of using the furnace gases, which are entered to be cleaned, as sulfur-containing raw materials, sulfur compounds in the form of SO_2 are present in the these furnace gases, in addition, there are also nitrogen oxides.

The reduction of silicon occurs from the total reaction

$$SiO_2 + 2C = Si + 2CO \qquad (1)$$

Modern theory considers that the recovery of silicon in ore-thermal furnaces occurs in two stages:

1 stage (low-temperature zone of the furnace) - formation of silicon carbide by the reaction

$$SiO_2 + 3C = SiC + 2CO \quad (2)$$

2nd stage (high-temperature zone of the furnace) - production of silicon by the reaction

$$SiO_2 + 2SiC = 3Si + 2CO \quad (3)$$

Considering the coefficients, total overall reaction has the following form:

$$3SiO_2 + 6C = 3Si + 6CO \quad (4)$$

As can be seen from this reaction, amount of reaction gases formed by mass is twice that the amount of the final product. To reduce the negative impact on the environment was one of the main aims of this work.

If to compare the above mentioned reactions 1, 3, and 4, it can be seen that in the production of three moles of silicon in the first case, reactions 1, 4, six moles of CO are formed, as received the same amount of silicon by reaction 3, two moles of CO are formed. This relationship formed the basis for industrial tests on the use of silicon carbide in the production of technical silicon in powerful industrial electric furnaces according to the technology developed by the authors [2]. During the tests, assessment of the furnace productivity and quality of the smelted silicon was made, amount of waste furnace gases was measured, presence of contaminants in them, and comparison with the operation of the furnace on a typical charge for the production were made.

The tests were carried out on a 25 MVA furnace.

The furnace operation performances, in stages, are summarized in table 1, change in furnace operation, productivity, specific electricity consumption, which depends on the amount of used silicon carbide, are shown in figure 1.

Table 1. Furnace operation performances with the use of silicon carbide materials.

	Consumption per 1 ton of silicon, t/t				
	Typical charge	1 stage	2 stage	3 stage	4 stage
Quartzite	3.046	2.787	2.658	2.654	2.324
Silicon carbide material	-	0.323	0.409	0.508	0.627
Electricity, mWh/t	16.28	16.200	14.623	14.164	13.223
Productivity, t/day	29.40	30.96	34.47	35.35	37.96

During the tests, calculations of the amount of reaction gases, occurring in the silicon smelting reduction process, on typical and experimental charges were carried out (table 2).

Table 2. The amount of occurring gases per 1 ton of silicon.

gases	typical charge	replacement 10 % C
CO	2414 nm^3	1985 nm^3
CO_2	1134 nm^3	803 nm^3
H_2O	3041 nm^3	2079 nm^3
Total	6589 nm^3	4867 nm^3

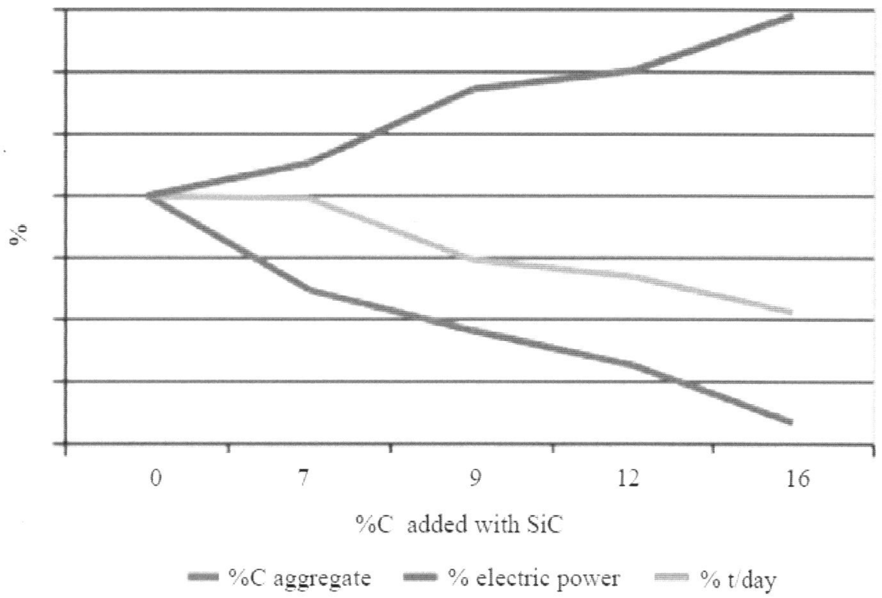

Figure 1. Change of the furnace operation performances using silicon carbide materials.

During the tests, measurements of the amount and composition of gases, leaving the furnace, during the operation on a typical charge and in case of replacing a part of the reducing agent with silicon carbide materials. Changes in the number and types of contaminants in the composition of waste gases were estimated. All measurement results are summarized in table 3.

Table 3. Results of instrumental determination of the composition of gases and the amount of contaminants of the silicon furnace.

Used charge	Typical charge	Charge with the use of SiC			
		replacement 7% C		replacement 10% C	
Parameter	Emission value	Emission value	Change in comparison with the typical one,%	Emission value	Change in comparison with the typical one,%
Gas volume, nm^3/h	236625	214200	-9.5	205600	-13.1
Dust, kg/h	1252.97	991.27	-20.1	801.098	-36
kg/t Si	1019.5	678.95		549.3	
CO, kg/h	31.2	22.465	-28.1	20.07	-35.7
kg/t Si	25.47	15.387		13.74	
NO, kg/h	64.6	29.16	-55	9.5	-64.5
kg/t Si	52.6	19.975		6.51	
NO_2, kg/h	4.99	1.836	-63	0.54	-0.89
kg/t Si	4.06	1.257		0.32	
SO_2, kg/h	5.66	5.32	-6	1.7	-70
kg/t Si	4.6	3.6		1.15	

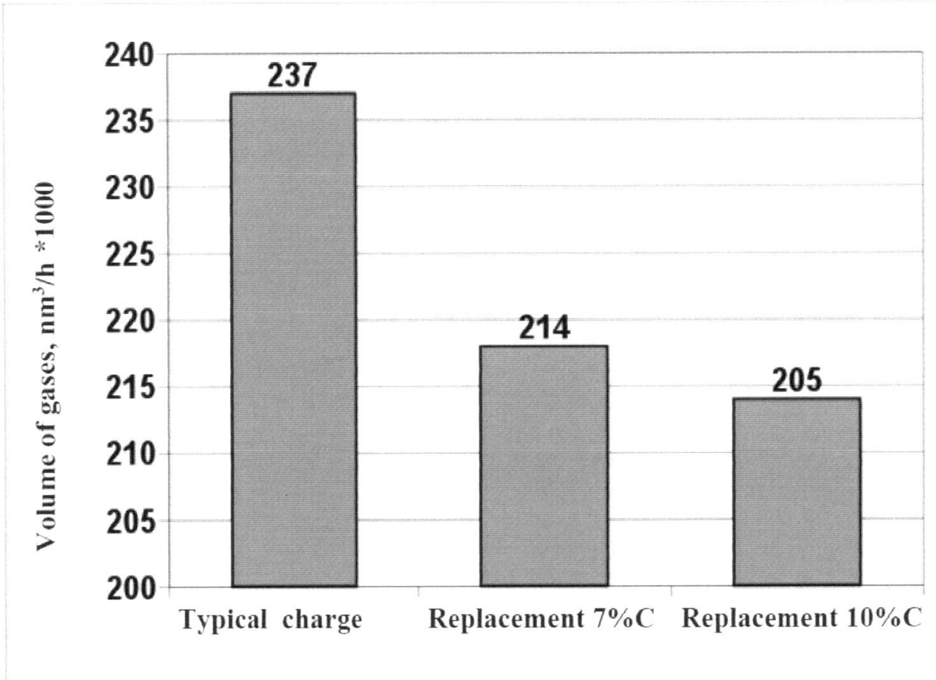

Figure 2. Change in quantity of furnace gases.

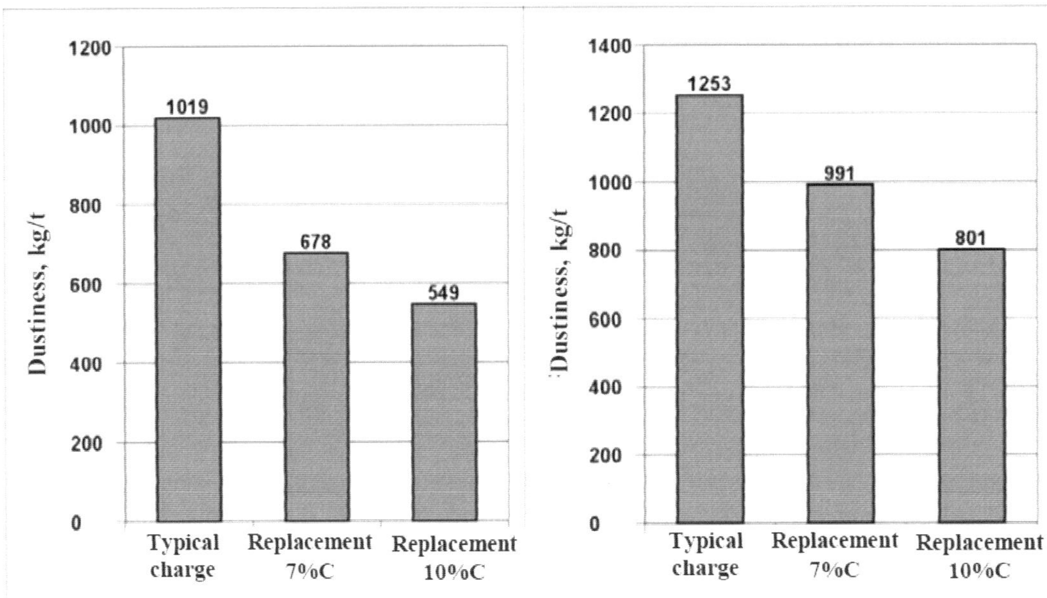

Figure 3. Change in dust content of furnace gases.

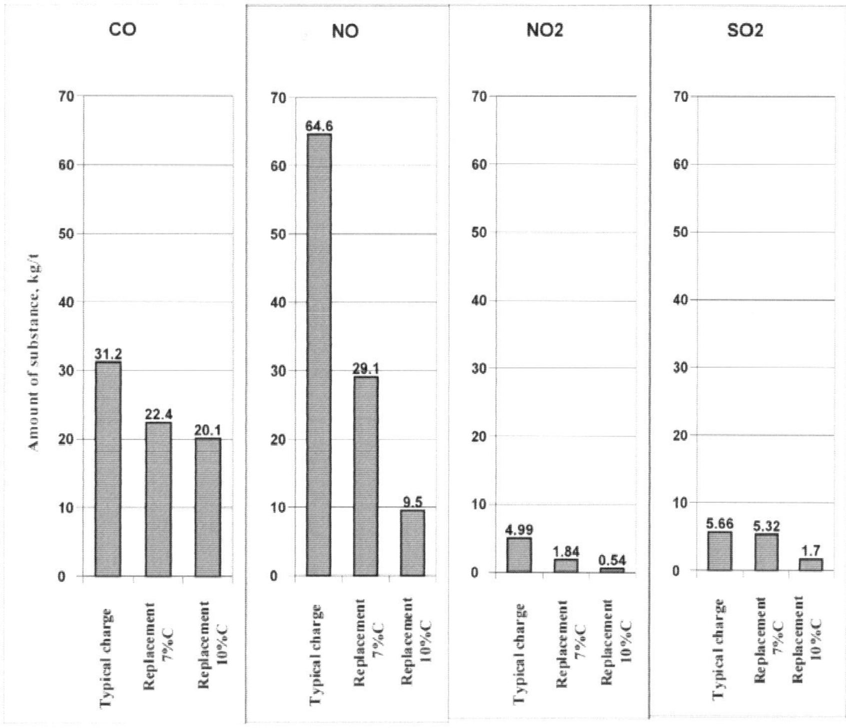

Figure 4. Change in composition of furnace gases.

The conducted studies showed high efficiency of silicon carbide materials application as a reducing agent in the production of metallic silicon. Efficiency was achieved due to increased productivity of the furnace, decrease in the consumption of raw materials and materials.

A significant ecological effect was achieved due to a decrease in the amount of waste gases and a decrease in the dust content of the gases. The composition of the gases also changes significantly, which leads to a reduction in the unfavorable environmental load in the area of the location of the enterprise.

2. Conclusions

1. The use of silicon carbide materials in the silicon reduction smelting increases the productivity of the furnace, reduces the consumption of process electricity in proportion to the amount of carbon replaced by silicon carbide.

2. Replacement of carbon with silicon carbide material in the composition of reducing agents reduces dust content, amount and composition of waste furnace gases, by reducing the amount of contaminants, which reduces the environmental load.

Acknowledgements

This paper has been prepared with financial support from the Ministry for Education and Science of Russian Federation (Order No.218 adopted on the 9th of April 2010 by the Government of the Russian Federation.) within the framework of the project "Development of a complex resource-saving technology and organization of high-tech production of nanostructures based on carbon and silicon dioxide to improve the properties of building and construction materials" (Agreement No. 02.G25.31.0174; code 2015-218-06-7159) using the equipment kindly provided by the Joint Instrumentation Centre "Baikal Center of Nanotechnologies".

References

[1] Zelberg B I, Chernykh A E and Yolkin K S 1997 *Metallurgy* 170 p
[2] Yolkin K S, Zelberg B I, Barantsev A G, Kryuchkov V K and Yolkin D K 2013 *MANEB* 364 p
[3] Yolkin K S, Yolkin D K, Ivanova O B and Kirillov M A. patent RU 2570153, C 01 B 33/025, filed on August 29, 2014, published on 10.12.1015.
[4] Yolkin K S, Yolkin D K, Koshkin S V, Peganov D V, Molyavko A A and Ferefyorova T T 2015 *Non-Ferrous Metals and Meneurs* pp 613-6
[5] Ivanchik N, Kondratiev V and Chesnokova A 2016 *Procedia Engineering* **150** 1567-73
[6] Kondratiev V V, Ivanov N A, Balanovskiy A E, Ivanchik N N and Karlina A I 2016 *Journal of the Siberian Federal University. Series: Engineering and Technology* vol 9 **5** 671-685
[7] Ivanchik N N, Balanovsky A E, Kondratiev V V, Tyutrin A A and Kuzmin M P 2016 *Bulletin of the Irkutsk State Technical University* vol 20 **12(119)** 165-172
[8] Kuz'min M P, Kondrat'ev V V, Larionov L M, Kuz'mina M Yu and Ivanchik N N 2017 *Metallurgist* vol 61 **1-2** 86-91
[9] Kondratiev V V, Govorkov A S, Kolosov A D, Gorovoy V O and Karlina A I 2017 *International Journal of Applied Engineering Research (IJAER)* vol 12 **22** 12373-7
[10] Kondratiev V V, Nebogin S A, Sysoev I A, Gorovoy V O and Karlina A I 2017 *International Journal of Applied Engineering Research (IJAER)* vol 12 **22(2017)** 12809-13

Conduct of reduction smelting of metallic silicon: theory and practice

K S Yolkin[1], D K Yolkin[2], A A Nemarov[3], I A Sysoev[3] and A I Karlina[3]

[1]UC RUSAL, RUSAL Engineering Technology Centre, 40 Pogranichnikov street, Krasnoyarsk, 660111, Russia
[2]RUSAL, JS Company «Kremniy», 26 Yuzhnaya street, Shelehov, 666034, Russia
[3]Irkutsk National Research State Technical University, 83 Lermontov street, Irkutsk, 664074, Russia

E-mail: karlinat@mail.ru

Abstract. Analysis of reduction smelting of silicon showed the formation of a secondary stable silicon carbide. The possibility of lading fresh charge to the lower horizons of the furnace tank was shown, which excludes the formation of stable silicon carbide. The initial temperatures of the interaction of silica and silicon carbide, obtained using various carbonaceous reducing agents, are determined.

1. Introduction

Complexity of the reduction of silicon by carbon is related to the fact that the conditions for the realization of both basic and accompanying reactions sharply differ from each other: some of the basic and concurrent reactions occur most intensively at high temperatures, while the other part – at low temperatures.

$$SiO_2 + 3C = SiC + 2CO \quad (1)$$

$$2SiO_2 + SiC = 3\ SiO + CO \quad (2)$$

$$SiC + SiO = 2Si + CO \quad (3)$$

It means that the processes, described by the reactions mentioned above, can occur in different zones of the tank of the ore-thermal furnace. This leads to the fact that in the course of the reduction of silicon both stable condensed and gaseous intermediate products are formed, which move in the tank of the furnace towards each other: condensed substances, as they heat and spend in the lower horizons of the tank under the action of gravity, – from the top to the bottom, from the cold sections of the tank into the hot ones, gaseous, on the contrary, – from the bottom to the top, from the hot sections of the tank into the cold ones.

Silicon smelting reduction characteristics depend both on the completeness of the use of SiO (d) in the high-temperature zone, and on the completeness of trapping of its residues, which can, according to the considered scheme, occur with the participation of carbon by the reaction and only in the low-temperature zone of the furnace tank [1,2,3].

2. Loading the furnace with the charge

Loading the furnace top with the charge is carried out by charging pipes and with the help of loading machines. Loading is carried out uniformly in time, in small portions 2-3 times per hour, so that the level of the furnace top is constant.

A cone of charge should be maintained around each electrode. The cone disposition of the charge around the electrodes creates resistance to the escape of gases along the electrodes and improves the conditions for uniform gas release throughout the active surface of the furnace top, prevents the narrowing of the reaction crucibles. The height of the cones should be 400-500 mm.

Melting of silicon in ore-thermal furnaces is conducted in a continuous manner, with a constant loading of the proportioned charge on the surface of the furnace top. In the first stage of silicon reduction, silicon carbide is formed (table 1).

Since silicon carbide in the furnace tank can be in different polymorphic states (figure 1), it is important to obtain a primary silicon carbide of cubic syngony, since the hexagonal silicon carbide is more stable and its interaction with silicon monoxide to form silicon requires higher temperatures, which adversely affects the efficiency of the silicon production process [4].

Table 1. Reaction capacity of silicon carbides.

Reducing agent	Kind SiC	Reacting capacity			Temperature of interaction start, K		
		SiO_2+3C	50% SiC	95% SiC	SiO_2+3C	50% SiC	95% SiC
Charcoal	3C	0.292	0.270	0.261	1788	1805	1834
Petroleum coke	2H	0.190	0.158	0.153	1898	1933	1943
Black coal	3C	0.219	0.204	0.202	1839	1866	–
Silicon carbide	6H	–	–	0.166	–	–	1963

It is important, during the melting of silicon and siliceous ferroalloys, to use full section of the furnace tank without forming cavities, since in these places two gas phases interact by the reaction.

$$SiO + CO = SiC + SiO_2 \quad (4)$$

Uncontrolled consumption of SiO, which often leads to a reduction in the technical and economic performances of the reduction smelting.

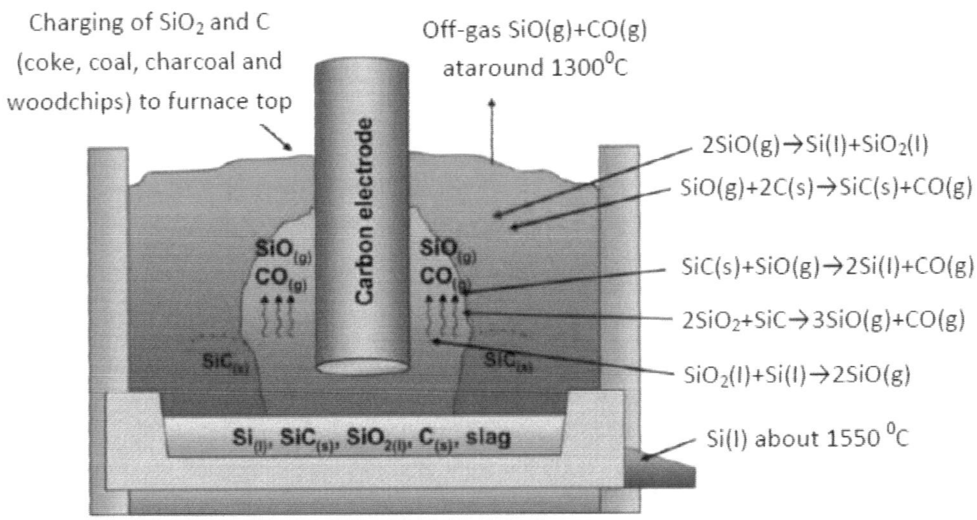

Figure 1. Silicon production states.

In order to prevent the formation of stable hexagonal silicon carbide, it is necessary to have a supply of carbon to the surface and the deeper horizons of the furnace tank for trapping the gaseous intermediate products of SiO during melting. Continuity of mass transfer is supported by periodic processing of the furnace top (destruction of the crushed charge on separate sections of the surface of the furnace tank).

Equipment operation with certain functions for processing the furnace top, and for making the profile of the charge on the furnace top and in the furnace tank, necessary for maximum mass transfer is very important.

Rotation of the furnace tank about the vertical axis has an auxiliary function for loosening the furnace top. Possible variants of rotation: reversible or circular, where the circular rotation of the tank is preferred. Circular rotation gives a certain profile to the smelting products on the bottom of the tank, which facilitates the free escape of silicon from the furnace.

3. Processing of the furnace top

In the silicon smelting reduction process, intermediate reduction products are in the solid state (SiC) and gaseous state (SiO). With a continuous supply of reduction products to the high-temperature reaction zone, where the temperature is higher than 1817 °C, silicon carbide is consumed to form silicon by the reaction (4):

$$SiC + SiO = 2\,Si + CO \quad (5)$$

When the supply of silicon carbide to the high-temperature zone of the tank slows down, heat of electric arcs is consumed to overheat the charge, part of silicon monoxide in the temperature range 1710-1940 C disproportionates by the reaction (5):

$$2SiO_{(g)} = Si + SiO_2. \quad (6)$$

This reaction is exothermic, which leads to further overheating of the furnace top. In addition, disproportionation products in this temperature range are: silicon in the liquid phase, silica in the plastic state (quartz glass), and this significantly reduces the gas permeability of the furnace top. These processes occur at a depth of 300-500 mm from the surface of the furnace top, so gases that were previously filtered through a loose charge, in this case tend to leave the reaction zone in the hottest places (along the electrodes), which leads to local gas release in the form of "fistulas" and losses of silicon.

High chemical activity of fresh charge leads to the formation of unstable silicon carbides with a high rate of interaction with other reaction products, prevents the formation of stable silicon carbide with low reactivity, maintains a constant rate of charge loss.

To prevent losses of silicon, simultaneously with the loading of fresh charge on the furnace top, this is processed by the machine for its processing. Basically, at the Russian smelters, processing is carried out by pneumatic machines. The periodicity of processing of the furnace top must be at least 2-3 times per hour.

The firm "Dango & Dienental" has developed a hydraulic loading-stirring machine for processing of the furnace top. The processing is carried out by immersing the pole of the machine in the charge, which makes it possible to move the charge during the processing of the furnace top, breaking the sintered sections of the furnace top. The processing of the top is conducted along the electrode, then between the electrodes in the center of the furnace.Charge is dragged forwards by the movement of the machine itself. The machine has removable devices (troughs) for loading separate components of the charge in the furnace pol in the form of additives which adjust the course of the furnace. The machines of the "Dango & Dienental" company occupy 65-70% of the market of this equipment (data of "Tekma" and "Dango & Dienental" companies).

Particular attention should be paid to the processing of the furnace top between the electrodes and the center of the furnace tank. In these places, level of the furnace top should be much lower than that of the electrodes, and this will lead to an increase in the active cross-sectional area of the furnace top.

4. Conclusion

Almost all of the reduced silicon accumulates in a circular trough and freely leaves the furnace through a tap-hole under the electrode in a place equipped for the release of silicon.

Correspondence of technological, energetic and geometric parameters of the entire furnace unit is important for high-performance operation of the furnace in the production of silicon.

5. References

[1] Tolstoguzov N V 1992 *Metallurgy* p 283
[2] Kulikov I S 1986 *Metallurgy* p 344
[3] Yolkin K S, Zelberg B I, Bartsev A G, Kryuchkov V K and Yolkin D K 2013 *St. Petersburg, ManEB* p 364
[4] Yolkin K S and Yolkin D K 2012 *Perfection of Technology and Equipment in the Production of Silicon and Siliceous Ferroalloys* pp 16–18
[5] Kondrat'Ev V V, Nemchinova N V, Ivanov N A, Ershov V A and Sysoev I A 2013 *Metallurgist* vol 57 **5-6** 455–9
[6] Yolkin K S, Ivanov N A, Karlina A I and Ivanov N N 2015 *Non-Ferrous Metals and Minerals* pp 224–5
[7] Rassokhin A S, Ponomarev A N and Figovsky O L 2018 *Magazine of Civil Engineering* **02(78)**
[8] Kondratiev V V, Ivanov N A, Balanovskiy A E, Ivanchik N N and Karlina A I 2016 *Journal of the Siberian Federal University. Series: Engineering and Technology* vol 9 **5** 671–685

IOP Conf. Series: Materials Science and Engineering **411** (2018) 012030 doi:10.1088/1757-899X/411/1/012030

Energy-efficient technologies for gas cleaning systems of aluminum smelters

V A Ershov, T I Zimina, I V Komogortsev, V O Gorovoy and A S Govorkov

Irkutsk National Research Technical University, 83 Lermontov street, Irkutsk, 664074, Russia

E-mail: ziminati@istu.edu

Abstract. The article presents the relevancy of creating a shell-and-tube heat exchanger for utilizing the heat of process gases from an aluminum electrolyser. On the basis of the results of virtual modeling, a heat exchange unit was created, which is currently being tested in industrial conditions under various operating conditions. In the future, recovered heat will be used to increase the energy efficiency of production, by using heat energy to generate electrical energy, preheat anodes and alumina.

1. Introduction

According to [1], up to 3% of heat is carried away from electrolysers with a self-baking anode with anode gases that are removed into the system of gas cleaning units (GCU), and from electrolyzers with burned anodes - 12.5%. At the same time, the volume of the dust-gas-air mixture, withdrawn from the electrolyzer with a self-baking anode, is, on average, 600 ÷ 1000 m^3/h, and with the burned anodes – 7000 ÷ 14000 m^3/h [2]. For example, more than 2 million m^3/h are withdrawn from two buildings of V-series of electrolysis of the Irkutsk Aluminum Smelter.

In connection with the requirements of the Russian legislation in the field of ecology [3] and energy [4], studies on increasing the degree of catching of pollutants and increasing energy efficiency by utilizing the heat of waste process gases are relevant.

The authors of [5, 6, 7] noted that when the temperature of the gases, that are being cleaned, decreases, efficiency of gas cleaning units increases (Figure 1), due to the intensive desorption of the trapped hydrogen fluoride (HF) by alumina.

2. Modelling of the process

Therewith, a decrease in the temperature of process gases has a beneficial effect on the operability and longevity of gas cleaning equipment. In accordance with figure 2, when electrolytic gases are cooled, their physical volumes decrease, which, in turn, reduces the operating costs of gas cleaning units.

According to [2], capital expenditures are from 15 to 40 US $ per m^3 of capacity of gas that are being cleaned.

The need for cooling is due to the fact that in the warm season the temperature of gases at the entrance to the gas cleaning units can reach 190 °C, and the bag filters made of polyester, used in gas cleaning technology in recent years, have service limit values of 140 ÷ 145 °C. The use of filters made of heat-resistant fabrics, that withstand temperatures above 200 °C, will result in their 3 ÷ 4-fold increase in cost [8].

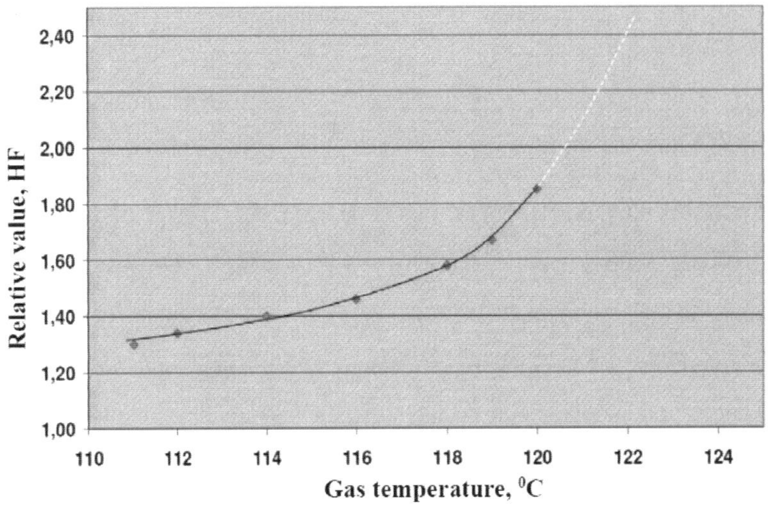

Figure 1. Interrelation between temperature of gases and desorption of hydrogen fluoride.

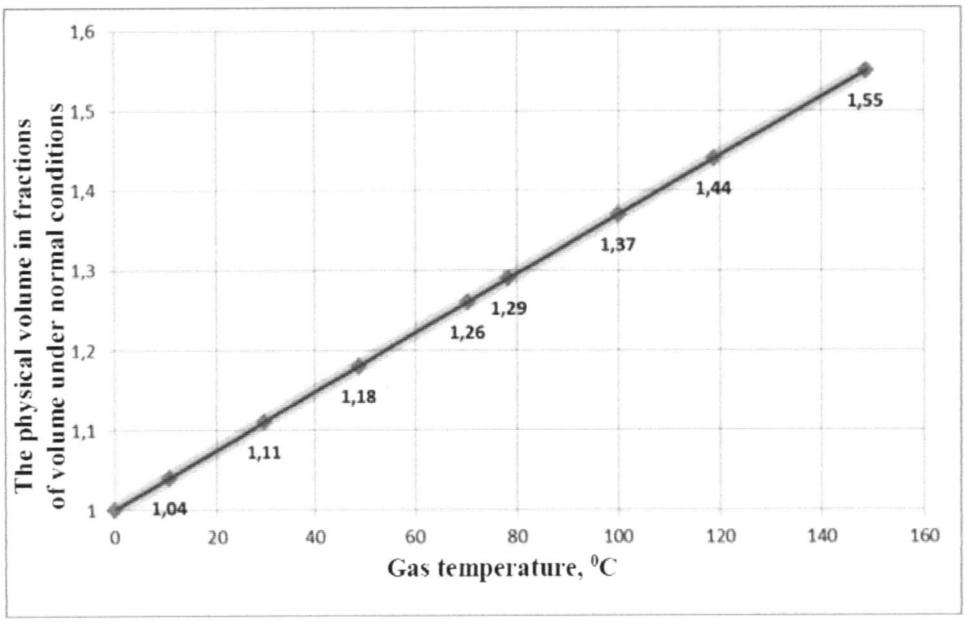

Figure 2. Physical volume of gases-temperature relationship.

At present time, various approaches to solve the problem of cooling technological gases are proposed. The simplest solution to the problem of high process gas temperatures is the addition of air to the gas ducts through adjustable dampers. [9] However, this entails an increase in the energy consumption of the fans, need for additional filter cloth areas will increase, and the maintenance costs of the gas cleaning units will be significantly increased [10].

Another way of cooling is to spray water in the gas stream before entering the gas cleaning facilities [1]. But it additionally requires the consumption of compressed air and purified water and, as a result, constant monitoring and maintenance. There is also a risk of appearance of corrosion on the equipment, increased hydrogen fluoride emissions due to the hydrolysis of solid fluorides, as well as an increased water content in enriched alumina [2].

It is possible to achieve gas cooling as far as they move along the gas duct due to the core type cooler, developed and patented by Solios [3]. The cooler has vertical metal rods, installed inside the U-shaped section of the gas duct that connects the electrolysers to the collecting gas duct. The pressure losses in such a cooler are compensated by a reduction in the physical volume of evacuated gases due to their cooling by, on the average, 10 °C.

Additional cooling can be achieved by equipping the gas duct grid with heat-sinking fins, located spirally on the outer surfaces of the gas ducts, which allows partial winding of the air flushes, which are surrounding the gas duct, increasing the heat transfer to the environment. [2]. The installed fins are able to increase the area of the heat-transfer surface of the gas-duct grid by 10 ÷ 25% and cool the anode gases by 20-25 °C, but at the same time they increase the dimensions of the gas ducts grids.

According to some authors, one of the most effective solutions for heat utilization is the use of heat exchangers [1]. This method has its drawbacks, because it requires maintenance due to the tendency of the pipe still to clog, and the use of additional equipment (water pipe, heat exchange circuit, measuring equipment, etc.). The advantage of this technology is the recovery of heat energy, which allows to achive greater energy efficiency of production by: heating water for production needs; for heating, or cooling; obtaining distilled water; power generation; [5, 9, 10] preheating of alumina [1; 5] or baked anodes. Ultimately, efficiency of the use of heat exchangers should be evaluated from the standpoint of capital and operating expenditures.

In the course of work, it was determined that the most optimal, in terms of weight, overall dimensions, strength and performance characteristics. The body of the experimental heat exchanger is made in the form of a "pipe" of rectangular cross-section. 13 (thirteen) cooling blocks, consisting of heat-exchange elements, are inserted inside the body. The blocks are fixed to the supporting structure by means of a bolted connection on the gaskets at the top and bottom of the body by the process windows made in the outer walls. The windows are closed with lids on the gaskets. Cold water is supplied through the bottom cover via the piping system, pressure and rate of water supply to each heat exchange unit is regulated by means of a ball valve mounted on the water supply pipes.

On the bottom cover, there are branch pipes through which one coolant passes through the space between the parts of the supporting structure. To increase the heat transfer, fining of the heat exchange tubes is used, which represents a tape soldered along the entire length of the tube.

The heat exchange part, through which the coolant passes, consists of 169 pipes. Pipe still of heat exchangers and fins were made of copper sheet, there are holes in the fins to increase heat transfer and gas flow.

3. Conclusion

The article presents the relevancy of creating a shell-and-tube heat exchanger for utilizing the heat of process gases from an aluminum electrolyser. On the basis of the results of virtual modeling and optimization of the design, an experimental heat exchanger was created, which is planned to be tested in industrial conditions under various operating conditions.

Subsequently, recovered heat will be used to increase the energy efficiency of production by using heat energy to transform energy into the electric one, preheat the anodes and alumina.

The use of accumulated heat energy will allow to increase the technical and economic performance of aluminum smelters, reduce capital expenditures in operation and designing of gas cleaning systems and rationally solve the problem of utilization of process gases.

Acknowledgement

This paper has been prepared with financial support from the Ministry for Education and Science of Russian Federation (Order No.218 adopted on the 9th of April 2010 by the Government of the Russian Federation.) within the framework of the project 02.G25.31.0181 "Development of heavy duty energy efficient technology for aluminum RA-550" in the framework of realization of complex projects in high-tech production, approved by the RF Government decree № 218 from April 9, 2010.

References

[1] Shakhrai S G, Budnik E V and Kondratiev V V 2013 *Metallurgist* vol 56 **9-10** 700–4
[2] Shakhrai S G, Gron V A, Kondrat'ev V V, Nikolaev V N and Belyanin A V 2015 *Metallurgist* vol 59 **1-2** 126–130
[3] Shakhrai S G, Nemchinova N V, Kondrat'ev V V, Mazurenko V V and Shcheglov E L 2017 *Metallurgist* vol 60 **9-10** 973–7
[4] Sysoev I A, Ershov V A and Kondrat'ev V V 2015 *Metallurgist* vol 59 **5-6** 518–525
[5] Kuz'min M P, Kondrat'ev V V, Larionov L M, Kuz'mina M Y and Ivanchik N N 2017 *Metallurgist* vol 61 **1-2** 86–91
[6] Kondrat'ev V V, Rzhechitskii E P, Karlina A I, Sysoev I A and Shakhrai S G 2016 *Metallurgist* vol 60 **5-6** 571–5
[7] Shakhrai S G, Korostovenko V V, Gron' V A, Kondrat'ev V V and Belyanin A V 2014 *Metallurgist* vol 58 **1-2** 138–140
[8] Kondrat'ev V V, Ershov V A, Shakhrai S G, Ivanov N A and Karlina A I 2016 *Metallurgist* vol 60 **7-8** 877–882
[9] Ershov V A, Sysoev I A and Kondrat'Ev V V 2013 *Metallurgist* vol 57 **3-4** 346–351
[10] Kondrat'Ev V V, Nikolaev V N, Rzhechitskii E P, Kornyakov M V and Afanas'Ev A D 2014 *Metallurgist* vol 57 **9-10** 779–782

IOP Conf. Series: Materials Science and Engineering **411** (2018) 012031 doi:10.1088/1757-899X/411/1/012031

Results of development of an experimental heat exchanger

V A Ershov, T I Zimina, A S Govorkov, N A Ivanov and N N Ivanov

Irkutsk National Research Technical University, 83 Lermontov street, Irkutsk, 664074, Russia

E-mail: ziminati@istu.edu

Abstract. For stable technology of electrolysis with an increase of current, it requires efficient heat removal from elements of construction, including from the waste process gases. Reasonability of cooling the electrolytic gases before cleaning is due to the desire to reduce physical volume of cleaned gases, that, in turn, reduces the operation expenditures of gas cleaning units. The article describes feasibility of waste gases cooling in the production of aluminium with the help of the electrolytic method.

1. Introduction

It is known that in the production of aluminum with anodic gases, removed into the system of gas cleaning units (GCU), up to 3% of heat is carried away from electrolysers with a self-baking anode, and from electrolysers with calcined anodes - 12.5% [1, 2]. At the same time, volume of the dust-gas-air mixture withdrawn from the electrolyser with the self-baking anode is on the average 600 ÷ 1000 m3/h, and with the calcined anodes - 7000 ÷ 14000 m3/h [3].

One of the most effective solutions for utilizing heat from gases is the use of heat exchangers [4, 5, 6]. This method has its draw-backs, because it requires maintenance due to the tendency of the tubular to clog, and it requires additional equipment (water pipe, heat exchange circuit, measuring equipment, etc.).

2. Development of the thermoelectric converter

The advantage of this technology is the recovery of heat energy, which allows to achive greater energy efficiency of production thanks to: heating water for production needs; for heating, power generation, preheating of alumina or calcined anodes [7]. Ultimately, efficiency of the use of heat exchangers should be assessed from the standpoint of capital and operating expenditures[8, 9].

According to the design, heat exchangers are of different types, and shell-and-tube design is much more compact and efficient in terms of heat transfer and less expensive to be constructed (there is no need for a rigid body due to water pressure). Using the method of mathematical modeling and with the help of laboratory studies [1, 10], it was determined that the most optimal, in terms of mass, overall dimensions, strength and performance characteristics is the design shown in figure 1.

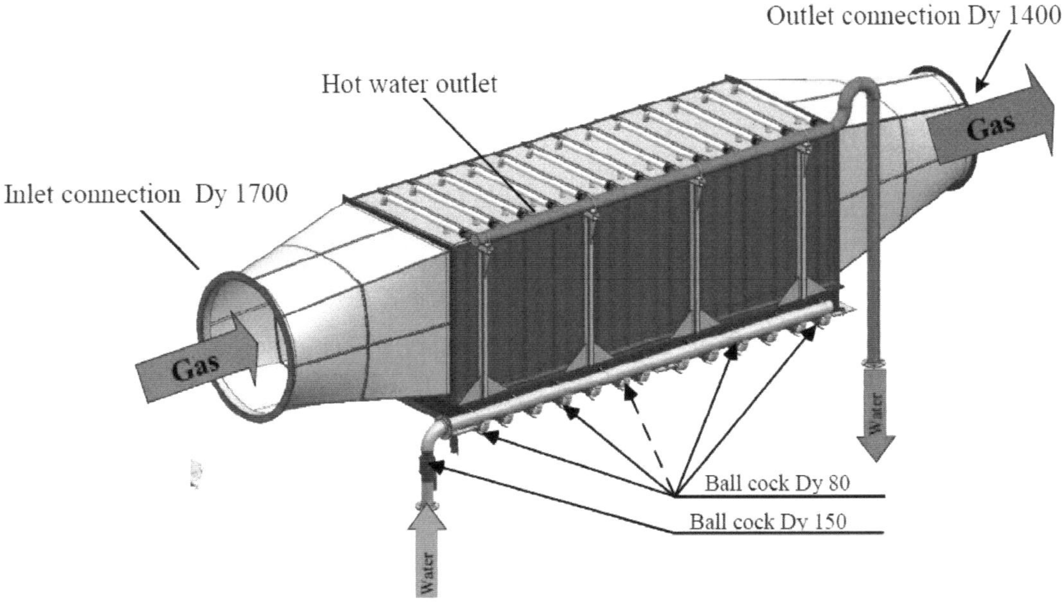

Figure 1. General view of the experimental heat exchanger.

The body of the experimental heat exchanger (EHE) is made in the form of a "pipe" of rectangular cross-section. 13 (thirteen) cooling blocks, consisting of heat-exchange elements, are inserted inside the body. The blocks are fixed to the frame by means of a bolted connection on the gaskets at the top and bottom of the body by the process windows made in the outer walls. The windows are closed with lids on the gaskets (Figure 2).

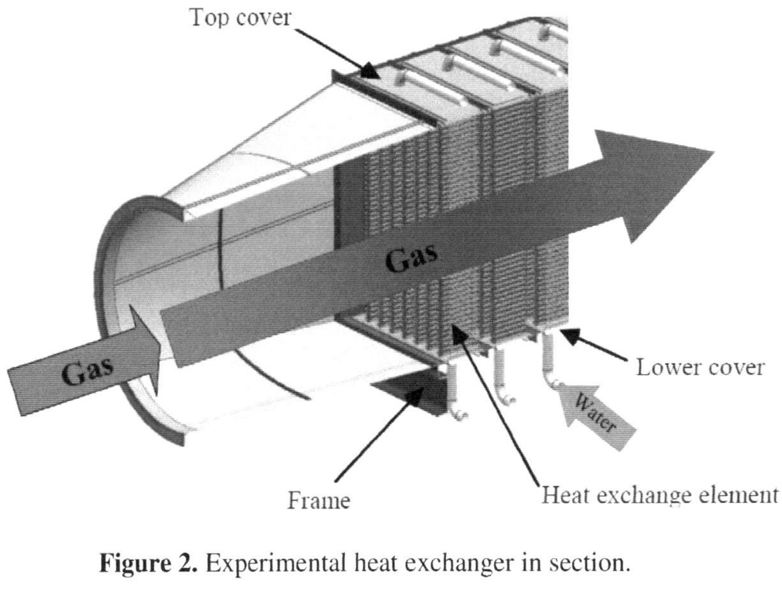

Figure 2. Experimental heat exchanger in section.

Cold water is supplied through the bottom cover via the piping system, pressure and rate of water supply to each heat exchange unit is regulated by means of a ball valve mounted on the water supply pipes.

On the bottom cover there are branch pipes through which one coolant passes through the space between supporting structure. To increase the heat transfer, fining of the heat exchange tubes is used, which represents a tape soldered along the entire length of the tube.

Diagram of gas and water flow in the heat exchanger, which is being developed, is shown in Figure 3.

Conical junctions, that connect the heat exchanger with the gas ducts, are mounted on the ends by the bolted connection to the body and to the shell.

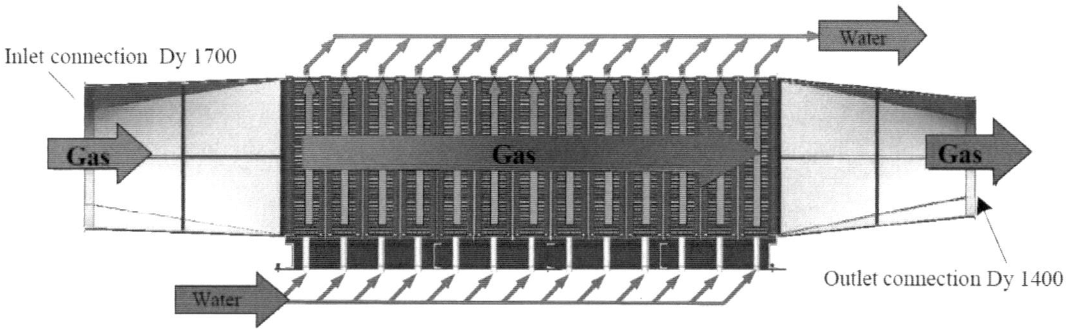

Figure 3. Diagram of heat exchanger (longitudinal section).

The heat exchange part, through which the coolant passes, consists of 169 pipes. Pipe still of heat exchangers and fins were made of copper sheet, there are holes in the fins to increase heat transfer and gas flow.

Heat exchange elements are made in the form of copper tubes and mounted in a block (Figure 4), dimensions and weight of which were selected taking into account the convenience of installation and maintenance during operation.

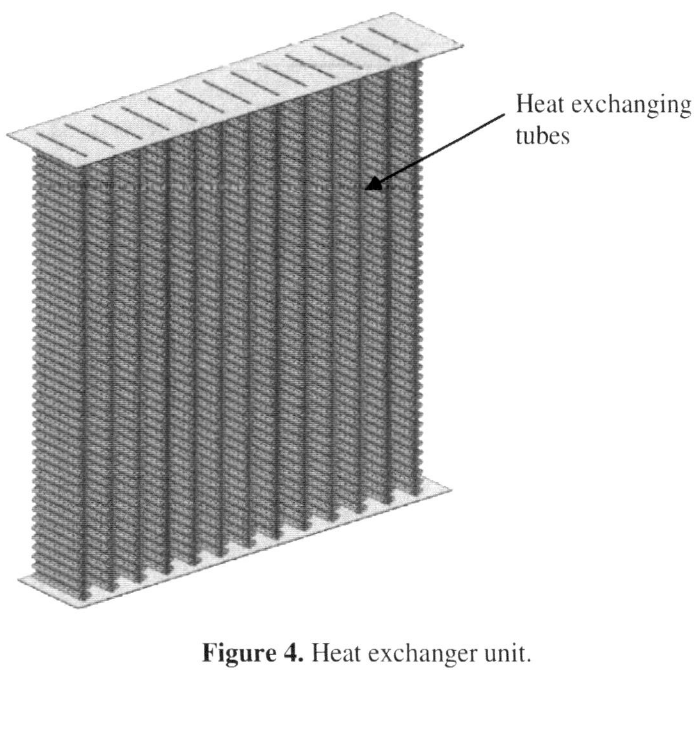

Figure 4. Heat exchanger unit.

Body of the heat exchanger is a shell welded from 2 flanges of a closed section joined together by longitudinal beams of the angle section. There are steel sheets reinforced with vertical stiffeners on the inner side. Thickness of the wall of the shell is determined by the pressure of the production environment and the size of the shell, considering the data on the calculation of the plastic deformations of metal.

Supporting structure of the heat exchanger shell is supported by a rigid frame made of a C-beam. The structure of the rigid frame was selected by modeling with the calculation of plastic deformations as a result of load stresses considering the safety factor. On the support platforms there are adjusting bolts for adjusting the position of the rigid frame during structural fabrication.

3. Conclusion

On the basis of the results of virtual modeling and optimization of the design, an experimental heat exchanger was created, which is currently being tested in industrial conditions. In the future, recovered heat will be used to increase the energy efficiency of production by using thermal energy to preheat the anodes and alumina.

The use of accumulated thermal energy will allow to increase the technical and economic performance of aluminum smelters, reduce capital expenditures in operation and design of gas cleaning systems and rationally solve the problem of utilization of process gases.

Acknowledgement

This paper has been prepared with financial support from the Ministry for Education and Science of Russian Federation (Order No.218 adopted on the 9th of April 2010 by the Government of the Russian Federation.) within the framework of the project 02.G25.31.0181 “Development of heavy duty energy efficient technology for aluminum RA-550” in the framework of realization of complex projects in high-tech production, approved by the RF Government decree No. 218 from April 9, 2010.

References

[1] Kondrat’ev V V, Rzhechitskii E P, Karlina A I, Sysoev I A and Shakhrai S G 2016 *Metallurgist* vol 60 **5-6** 571–5
[2] Kuz’min M P, Kondrat’ev V V, Larionov L M, Kuz’mina M Y and Ivanchik N N 2017 *Metallurgist* vol 61 **1-2** 86–91
[3] Kondrat'ev V V, Ershov V A, Shakhrai S G, Ivanov N A and Karlina A I 2016 *Metallurgist* vol 60 **7-8** 877–82
[4] Ershov V A, Sysoev I A and Kondrat'ev V V 2013 *Metallurgist* vol 57 **3-4** 346-351
[5] Shakhrai S G, Nemchinova N V, Kondrat’ev V V, Mazurenko V V and Shcheglov E L 2017 *Metallurgist* vol 60 **9-10** 973–7
[6] Sysoev I A, Ershov V A and Kondrat'ev V V 2015 *Metallurgist* vol 59 **5-6** 518–525
[7] Shakhrai S G, Budnik E V and Kondratiev V V 2013 *Metallurgist* vol 56 **9-10** 700–4
[8] Shakhrai S G, Gron V A, Kondrat'ev V V, Nikolaev V N and Belyanin A V 2015 *Metallurgist* vol 59 **1-2** 126–130
[9] Shakhrai S G, Korostovenko V V, Gron' V A, Kondrat'ev V V and Belyanin A V 2014 *Metallurgist* vol 58 **1-2** 138–140
[10] Kondrat'ev V V, Nikolaev V N, Rzhechitskii E P, Kornyakov M V and Afanas'ev A D 2014 *Metallurgist* vol 57 **9-10** 779–782

Gas-phase reduction of iron and chrome from chromites

O V Zayakin and V I Zhuchkov

Institute of Metallurgy of Ural Branch of Russian Academy of Science, 101 Amundsena street, Ekaterinburg, 620016, Russia

E-mail: zferro@mail.ru

Abstract. This paper focuses on studies of iron and partially chromium gas-phase reduction from chrome ore raw material of the Kempirsay Massif. The studies were carried out using a thermogravimetric method in a CO flow followed by chemical analysis of the samples. It is shown that: depending on the time of the samples exposure at the CO consumption rate 75 m^3 / h and the temperature of 1200 ° C, basic reduction processes take place in the first 20 minutes of exposure; with temperature growth from 1000 to 1400 ° C and an exposure time of 20 minutes, the degree of Fe and Cr reduction from the considered samples increases from 13 to 37%. Gaseous reducing agent and low temperatures contribute to a greater reduction of iron rather than chromium. Gaseous CO is a good iron reducing agent (with a small excess it allows to obtain a high degree of iron reduction) and a weak chromium reducing agent (the degree of chromium reduction is 14% with the multiple excess of CO and on condition of the chromium carbides formation).

1. Introduction

Gaseous-phase reduction of chromite ore elements can be considered as one of components preliminary reduction methods - metallization of charge materials. The issues of obtaining and using metallized chrome materials (pellets, briquettes) are considered in a number of works [1-4]. The firm "Outokumpu" in Finland carried out industrial research on the chromium ores preparation. According to the Daavittila J. data (Outotec Oyj, Finland), selective reduction of iron (more than 80%) and chromium (10%) was performed, however, technological difficulties were noted in the process for the chromium ores of the Kemi deposit, which contain a large number of silicates with a low melting point. Nevertheless, it is known that preliminary reduction and heating of chrome-ore materials allow to save 10-20% of electric power and to substantially increase the productivity of ore-reduction electric furnaces.

2. The materials studied

In this paper during studying the preliminary reduction of iron and partially chromium from chromium ore, CO was used as the gaseous reducing agent. According to the reaction:

$$MeO(s) + CO\ (s) \rightarrow Me\ (s) + CO_2\ (gas)$$

reduction process is accompanied by samples mass decrease due to the transition of oxygen from the solid (iron and chromium oxides) to the gas phase, therefore a thermogravimetric method was used to study the kinetics of gas-phase reduction. Chemical composition of the initial chromic ore materials is presented in table 1.

Table 1. Chemical composition of chromium ore materials

No.	Origin of chromium ore materials	Chemical composition, wt. %					
		Cr_2O_3	FeO	Al_2O_3	SiO_2	MgO	CaO
1	Kazakhstan, Kempirsay Massif	52.0	13.3	8.2	5.3	18.3	1.4
2	Russia, Saranovskoye field	40.9	29.0	19.8	3.1	15.5	-

3. Experiment and discussion

Influence of the exposure time from 0 to 40 min on the sample mass loss of 1.5-2.0 mm Kempirsay Massif chromium ore fraction during CO gas-phase reduction with a flow rate of 75 m^3 / h and a temperature of 1200° C was studied. Results of the experiments are presented in figure 1. It is shown that main mass change in the sample, and, consequently, the reduction processes, takes place in the first 20 minutes of exposure. With subsequent exposure to 40 minutes, the mass of the sample remains almost unchanged.

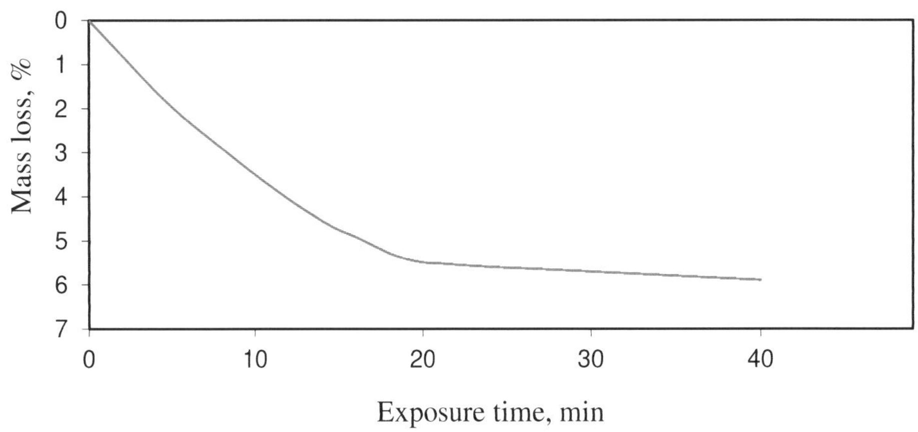

Figure 1. Influence of the exposure time on the mass loss of chromium ore sample from the Kempirsay Massif at CO consumption rate 75 m^3 / h and temperature 1200 ° C.

Under similar conditions, the temperature effect on the mass loss of a chromium ore sample with an exposure time of 20 min was studied at each temperature. The results of experiments are shown in figure 2. Total reducibility of Fe and Cr is calculated from the mass loss. It was found that temperature growth from 1000 to 1400 °C, with CO consumption of 75 m^3 / h and an exposure of 20 minutes, increases the reducibility of Fe and Cr samples from Kempirsai massif from 13 to 37%.

Similar studies on the components reducibility of the Saranovskoye deposit chromium ore samples were carried out in [5] under comparable conditions. In the high-temperature region ($\geq$ 1200 ° C), experimental data on Fe and Cr reducibility from Kempirsay ore samples are consistent with literature data on the reducibility of the same elements from Saranovo ore. In the low temperature range (1000-1200 ° C), Fe and Cr reducibility of the Saranovo ore is much higher, which can be justified by the low ratio of Cr / Fe (1.2) in the Saranov ore, compared to the Kempirsai massif sample (Cr / Fe = 3, 4). Gaseous reducing agent and low temperatures contribute to a better reduction of iron than chromium. Gaseous CO is a good iron reducing agent from FeO (with a slight excess allows to completely reduce iron) and a weak chromium reducing agent (the degree of chromium reduction is 14% with a multiple excess of CO and the condition of the carbide Cr_7C_3 formation).

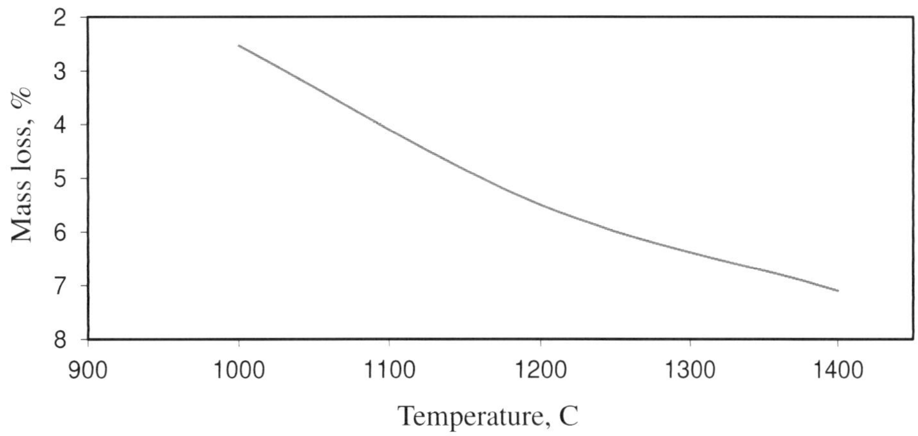

Figure 2. Temperature effect on the mass loss of chromium ore sample from the Kempirsay Massif at CO consumption rate 75 m^3 / h and at 20 min exposure time.

The influence of atmosphere (reducing, weakly reducing and oxidizing) on the reducibility of chromium ore elements by a solid carbon reductant at 1000-1400 ° C with a calculated amount of reducing agent on obtaining metal with 20% Cr was studied. A coke with 73% of C_s content was used as a solid reducing agent. It is shown that during the transition from reducing to oxidizing atmosphere, the rate and completeness of the elements reduction from chromium ore decreases. Thus, at 1400 ° C and 60 min exposure during the transition from oxidizing atmosphere to the reducing, the loss of the sample mass increases by 25% (rel.).

4. Conclusion

Thus, when considering the processes of chromium and iron gas-phase reduction from chromium ore materials:

- the effect of temperature and exposure time on the mass loss of samples and the degree of iron and chromium reduction have been studied;
- it is shown that temperature growth from 1000 to 1400 °C, at CO consumption rate of 75 m^3 / h and an exposure of 20 minutes, increases Fe and Cr reducibility of Kempirsai massif samples from 13 to 37%;
- it was found that, depending on the time of samples exposure there is a significant change in the mass of the samples at the 75 m^3 / h CO consumption rate and 1200 ° C, and therefore basic reduction processes occur in the first 20 minutes of exposure.

Acknowledgements
The work was carried out within the framework of the State Task No. State Registration 0396-2015-0084 and with the support of the RFBR Project No. 16-03-00092.

References

[1] Topilsky S P et al 1989 *Conf. Proceedings: Problems of Scientific and Technical Progress of Electrothermy of Inorganic Materials* (Dnepropetrovsk: DmetI) p 55
[2] Kudryavtsev V S 1975 *Metals* **6** 3–9
[3] Kadarmetov Kh N and Povolotsky V D 1987 *Metals* **3** 19–21
[4] Chernobrovin V P et al 2004 *Theoretical Foundations of Carbon Ferrochrome Production Processes from the Ural Ores* (Chelyabinsk: SUSU) p 346
[5] Sokolov I A 1938 *Blast Furnace Process* (Sverdlovsk: "Polygraphbook") p 398

IOP Conf. Series: Materials Science and Engineering **411** (2018) 012033 doi:10.1088/1757-899X/411/1/012033

Results of development of the test prototype of a thermoelectric converter

T I Zimina, N N Ivanov, V A Ershov, A S Govorkov and A M Pankov

Irkutsk National Research Technical University, 83 Lermontov street, Irkutsk, 664074, Russia

E-mail: ziminati@istu.edu

Abstract. Earlier studies have confirmed the reasonability of using thermoelectric converter (TEC) for heat recovery of electrolytic gases. The use of process flow diagram, which consists of a heat exchange unit along with thermoelectric converters is an innovative technology and will allow to increase the energy efficiency of production and achieve economic advantages of the aluminium production process.

1. Introduction

One of the ways to increase the energy efficiency of production is the recovery and return of thermal energy, carried away by process gases into the system of gas exhaust, into the technological process [1, 2, 3, 4].

So, for example, with anodic gases, removed to the system of gas cleaning plants (GCP), up to 12.5% of heat is carried away from electrolysers with burned anodes [5].

Previous studies, presented in [6, 7, 8], confirmed the efficiency of using a shell-and-tube heat exchanger for cooling gases with a heat-transfer fluid.

As a result of working out the design features of the heat exchanger, it was determined that it is expedient to use a thermoelectric conversion device (TCD) to recover thermal energy. Herewith, the efficiency of the use of heat exchangers should be evaluated from the standpoint of capital and operating expenditure [9, 10]

2. Development of the thermoelectric converter

The developed thermoelectric converter (TEC) is a semiconductor oscillator that generates electric current due to the temperature difference on its opposite surfaces. The basis for the calculations was the calculated model for the distribution of temperatures on the surface of the adapter sleeve of the heat exchanger in the ANSYS application program package, developed by the specialists of the INRTU [1].

Inside the TEC, there is a number of thermoelectric modules, whose operation is based on the Seebeck effect, the essence of which is the appearance of an electromotive force due to the appearance of a potential difference in a chain of heterogeneous semiconductor materials when a temperature gradient is created.

One side of the TEC (heat sink) is made in the form of a flat heat-conducting surface, which is fixed on the hot plane of the system with a coolant (gas, liquid). The second side of the TEC (radiator) is a cooling module into which a cooling liquid is supplied under pressure.

General view of the thermoelectric converter is shown in figure 1.

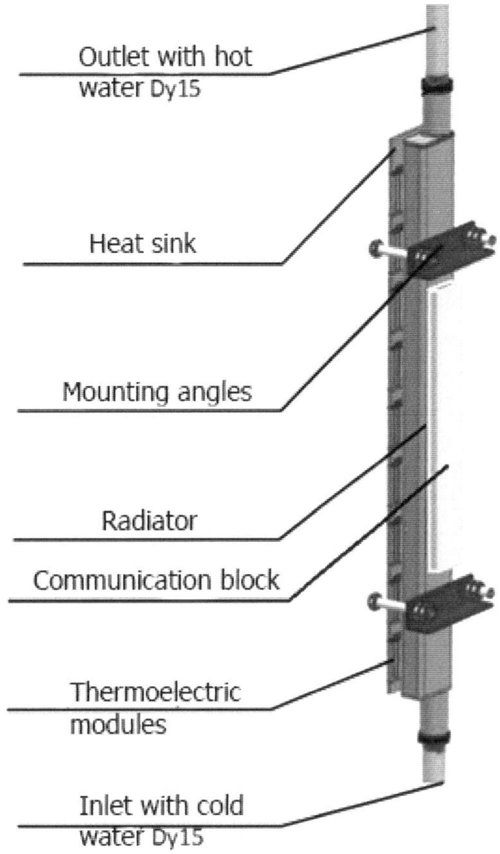

Figure 1. General view of the TEC for the section of the gas inlet duct into the gas cooling system.

The structure of one TEC includes 12 thermoelectric modules, which are mounted on a heat-removable plate (heat sink) by fixing on the silicon-organic paste КПТ-8 and pressing the cooling radiator with a mounting corner with bolts that can monitor the tightening torque.

Electrical connection of thermoelectric modules to each other can be carried out in parallel, in series or in parallel-series in the switching unit, depending on the required output parameters.

The switching unit with overall dimensions: 250x20x40 mm is located on the lateral side of the cooling radiator and serves to accommodate the connections of the input contacts of thermoelectric modules and the output terminals to connect the power consumption devices, or for further connection with switching units of other TECs (in case the system consists of several TECs).

The switching unit has a removable cover on 4 bolted connections to protect from moisture ingress, short circuit, accidental mechanical damage of the contacts and the risk of electric shock.

To ensure the necessary temperature difference and heat removal from thermoelectric modules, a liquid cooling radiator was used. In case of heat exchange between a solid body and a liquid (gas), occurring when they are in contact and are simultaneously transferring heat by heat conduction and convection, convective heat exchange occurs.

An important factor which determines the choice of the module is its location. Since the sample of the thermoelectric converter will be located in close proximity to the source of water supply with a coolant temperature of 30 °C, it was decided to use a liquid cooling radiator.

In the initial structural modeling, aluminum and bimetallic liquid radiators with different heat transfer coefficients were used as the liquid cooling radiator. These liquid radiators, with the help of a hardware connection, were attached to a steel gas pipe for surface matching of thermoelectric modules. Despite the good coefficient of heat conduction of aluminum (λ = 200 W/m * °C), compared to steel (λ = 55 W/m * °C), such radiators did not manage to cool the surface of thermoelectric modules to the required temperature. This is due to the fact that the surface of the radiator was being heated at the expense of thermal radiation from the steel pipe of the gas pipe and the small area of the heat transferring element (fin) of the radiator directly to the wall with the cooling liquid. It is possible that the material, from which the radiator is made, has a porous internal structure and the presence of a large amount of impurities in the composition, which could have a significant effect on the heat conduction coefficient.

In view of the fact that it was not possible to produce or purchase an aluminum radiator that would be suitable for solving this problem, it was decided to produce a steel cooling radiator where the cooled surface would directly contact with the wall surface with the cooling liquid. That is why, samples of steel water radiators were made, which, despite poor heat conduction compared to aluminum, provided cooling of the surfaces of the thermoelectric modules to the required temperature (figure 2).

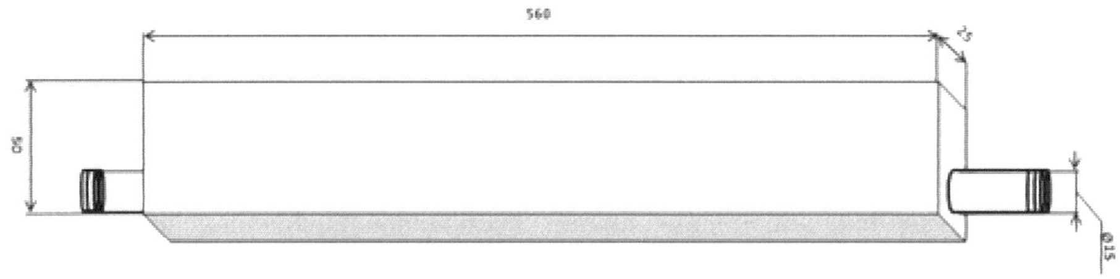

Figure 2. Steel water cooling radiator for the TEC.

Connection of steel liquid radiators to the water supply system was carried out with the help of plumbing of the required length.

Thus, the main characteristics of the produced prototype of TEC:

- generated average power 24 W;
- maximum operating temperature 200 °C;
- recommended coolant temperature 30 °C;
- pressure in the cooling system is not more than 10 bar.

3. Conclusion

Previous studies have confirmed the reasonability of using TEC to recover heat of electrolytic gases. The main functions of the TEC are:

- reduction of the temperature and physical volume of the gases which are being cleaned at the gas inlet duct into the gas cooling system;
- conversion of the thermal energy of electrolytic gases into electrical energy;
- DC power supply of power consumption devices in the operational area and other technological needs.

The use of a process flow diagram which consists of a heat exchanger along with thermoelectric converters is an innovative technology, and will allow to increase the energy efficiency of production and achieve economic advantages of aluminum production technology.

Acknowledgements

This paper has been prepared with financial support from the Ministry for Education and Science of Russian Federation (Order No.218 adopted on the 9th of April 2010 by the Government of the Russian Federation.) within the framework of the project 02.G25.31.0181 “Development of heavy duty energy efficient technology for aluminum RA-550" in the framework of realization of complex projects in high-tech production, approved by the RF Government decree № 218 from April 9, 2010.

References

[1] Kuz'min M P, Kondrat'ev V V, Larionov L M, Kuz'mina M Y and Ivanchik N N 2017 *Metallurgist* vol 611 **2** 86–91
[2] Kondrat'ev V V, Rzhechitskii E P, Karlina A I, Sysoev I A and Shakhrai S G 2016 *Metallurgist* vol 60 **5-6** 571–575
[3] Kondrat'ev V V, Ershov V A, Shakhrai S G, Ivanov N A and Karlina A I 2016 *Metallurgist* vol 60 **7-8** 877–882
[4] Ershov V A, Sysoev I A and Kondrat'ev V V 2013 *Metallurgist* vol 57 **3-4** 346–351
[5] Shakhrai S G, Budnik E V and Kondratiev V V 2013 *Metallurgist* vol 56 **9-10** 700–704
[6] Shakhrai S G, Gron V A, Kondrat'ev V V, Nikolaev V N and Belyanin A V 2015 *Metallurgist* vol 59 **1-2** 126–130
[7] Shakhrai S G, Nemchinova N V, Kondrat'ev V V, Mazurenko V V and Shcheglov E L 2017 *Metallurgist* vol 60 **9-10** 973–977
[8] Sysoev I A, Ershov V A and Kondrat'ev V V 2015 *Metallurgist* vol 59 **5-6** 518–525
[9] Kondrat'ev V V, Nikolaev V N, Rzhechitskii E P, Kornyakov M V and Afanas'ev A D 2014 *Metallurgist* vol 57 **9-10** 779–782
[10] Shakhrai S G, Korostovenko V V, Gron' V A, Kondrat'ev V V and Belyanin A V 2014 *Metallurgist* vol 58 **1-2** 138–140

Heat treatment of steel 30KhGSA on the specialized equipment

A A Ivanov and T N Oskolkova

Siberian State Industrial University, 42 Kirova street, Novokuznetsk, 654007, Russia

E-mail: Ivanov.aa@termomet.com

Abstract. This article presents the results of the developed technology for the production of the part "striker" made of steel 30KhGSA with the use of vacuum heat treatment. It was established that application of vacuum heat treatment for critical parts can significantly increase the percentage of yield of good products, reduce the number of technological operations by 2.5 times and the cost of manufactured products by 30%, and also obtain high repeatability of the result, which is an important indicator in the production of serial products .

1. Introduction

Modern industry in Russia includes many dynamically developing directions of production of metal products. At many stages of production a large number of defective products are produced. If the part is critical and needs to meet strict requirements for mechanical properties, in most cases thermal treatment should be applied to it. At this stage, a fairly large percentage of the output of defective products is observed.

If we consider a modern plant, then it will have quite a lot of modern high-precision equipment for mechanical treatment, i.e. automated lines, machining complexes with CNC, etc. In the heat treatment section, old thermal furnaces or salt baths are usually in a very worn condition.

Differences in temperature in these units reach 70 °C, which is unacceptable for the manufacture of critical parts. In most cases, enterprises take into account the high percentage of rejects in the production of critical parts and simply overstate the prices for them without conducting any work to improve the technological process of production. In this paper we will consider the method of vacuum heat treatment on modern high-precision equipment, describe the advantages of using innovative technologies, as well as the economic effect of their use.

Heat treatment of steel is a very important final operation in the manufacture of parts and tools. It is responsible for giving the final mechanical properties to the material, on which the performance of the part will depend [1, 2]. The vacuum heat treatment, although described in the literature of the 80s, was mainly used at enterprises that produce parts for the defense industry, thus, vacuum heat treatment for Russia is a new direction, and enterprises are not fully aware of the possibilities and the advantages of using this process in modern production [3].

One of the main advantages of vacuum heat treatment is the absence of decarburization (oxidation) of the surface during the heating and cooling of the part, as well as the reduction of shrinkages and deformations due to uniform heating and cooling. The vacuum furnace B63 used in this work is shown in figure 1.

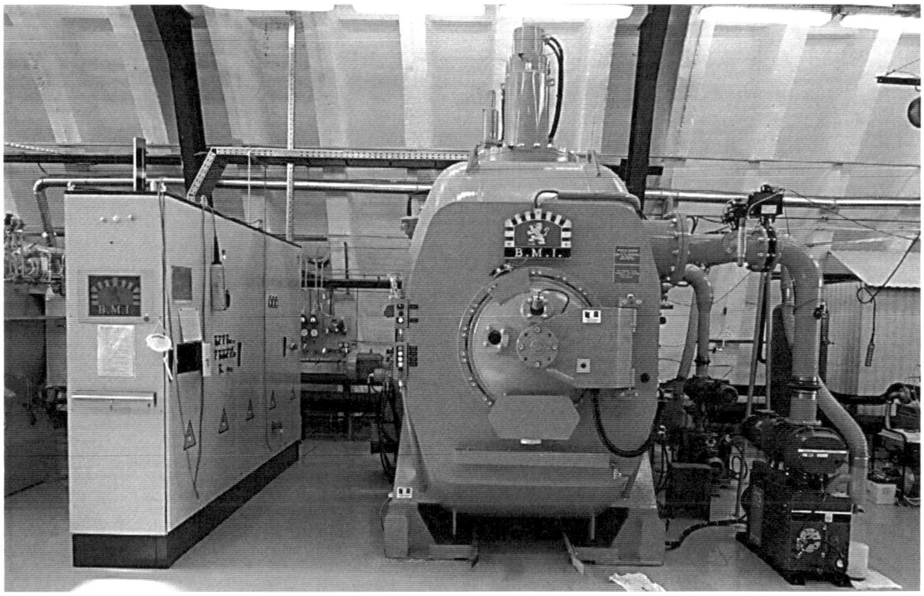

Figure 1. Vacuum furnace BMI B63 for heat treatment.

2. Results and discussion

The paper considers the use of vacuum heat treatment for steel 30KhGSA. Structural alloy steel 30KHGSA is one of the numerous steels used in the aviation industry. The main alloying elements in the steel are chromium, manganese, silicon. The chemical elements manganese and chromium sharply increase the hardenability. A feature of this high-quality steel is its relatively inexpensive cost and good weldability [4].

Analyzing the results of scientists in the field of aviation materials, it can be concluded that the standard thermal treatment for the material in question is the improvement: hardening from the temperature of 900±10 °C in the oil with a subsequent high-temperature tempering. This mode of heat treatment is designed to obtain certain mechanical properties, namely, high fatigue strength [5].

The technology was tested on cylindrical samples 25 mm in diameter made of steel 30KhGSA, the chemical composition of which meets the requirements of GOST 4543- 71. We used ARL optical emission spectrometer. The chemical composition of the steel is shown in table 1.

Table 1. Analysis of the chemical composition of two batches made of steel 30KhGSA, %.

Batch	C	Mn	Si	P	S	Ni	Cr	Cu	Fe
1	0.28	0.89	0.95	0.01	0.01	0.13	0.89	0.08	res.
2	0.31	0.89	0.96	0.01	0.01	0.13	0.98	0.09	res.
GOST	0.28-0.34	0.8-1.1	0.9-1.2	0.025	0.025	0.3	0.8-1.1	0.3	res.

The hardening was carried out in the vacuum thermal furnace BMI B63. The hardening temperature was 900±5 °C, the heating was carried out in a vacuum medium with partial nitrogen pressure. The supply of partial pressure prevents the oxidation of the parts surface when heated. The hardening was carried out in the vacuum oil heated to 40 °C. The cooling rate is controlled by the oil temperature parameters, the pressure in the hardening chamber, and the intensity of the oil mixing. After hardening to confirm the quality of heat treatment, hardness measurement and metallographic analysis were carried out.

Measurement of hardness according to Vickers (GOST 2999-75) was carried out on the Tukon 2500 microhardness testing instrument. The Rockwell hardness test (GOST 9013-59) was performed on TK-2M hardness testing instrument. The results of hardness measurements are presented in table 2.

Table 2. Hardness of the samples after hardening 900±5 °C

Batch No.	$HV5_{CENT.}$	$HRC_{SURF.}$	$HV5_{SURF.}$	$HRC_{SURF.}$
Batch 1 Ø 20 mm	490	48	500	48
Batch 1 Ø 25 mm	480	47	480	47
Batch 2 Ø 20 mm	490	48	500	48
Batch 2 Ø 25 mm	490	48	490	48

Analyzing the data of table 2, it can be concluded that samples from batch 1 and batch 2 with a diameter up to 25 mm have a uniform hardness throughout the section. The obtained hardness parameters satisfy the data described in the specialized literature [3].

Metallographic analysis after hardening showed a homogeneous martensite structure throughout the cross section of the samples. The microstructure of samples 25 mm in diameter from batch 1 is shown in figure 2.

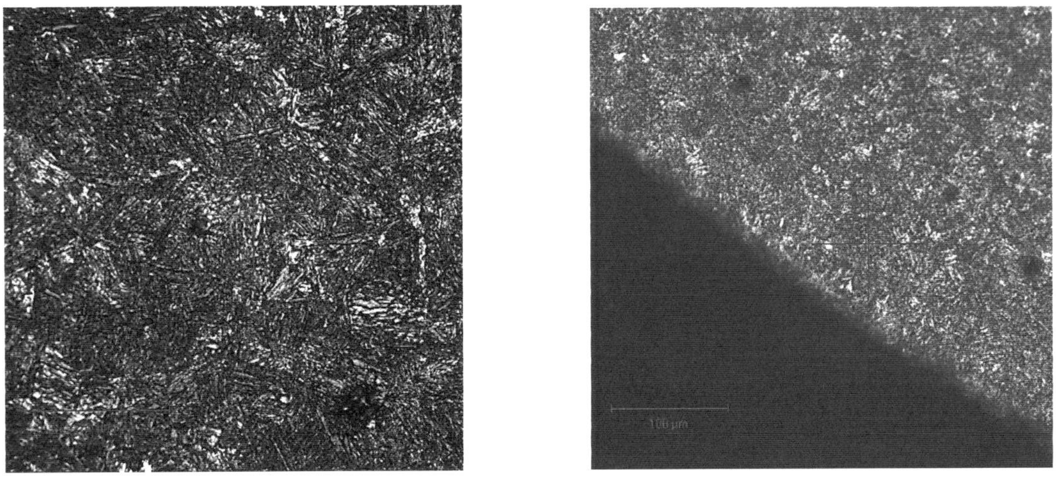

Center x 200 Surface x 200

Figure 2. Microstructure of the sample with Ø25mm (batch 1) after hardening.

After hardening, high-temperature tempering was carried out in the B-53RNC+ vacuum furnace in an inert nitrogen atmosphere (99.999%) at a temperature of 510±5 °C and a pressure of 300mbar. Steel grade 30KhGSA tends to temper brittleness of the second kind, so when it is tempered, it needs to be cooled quickly. In the furnace for uniformity of temperature during heat treatment (± 5 °C), as well as accelerated cooling, a turbine is installed. This turbine provides a high speed of cooling parts after tempering, which prevents the formation of temper brittleness.

After carrying out the tempering operation, hardness testing and metallographic analysis were performed to confirm the quality of the heat treatment. The results of hardness measurements are presented in table 3.

Analyzing the data in table 3, it can be concluded that samples from batch 1 and batch 2 with a diameter up to 25 mm have a uniform hardness throughout the section. The obtained hardness indicators satisfy the data described in the specialized literature.

Table 3. Hardness of samples after high-temperature tempering 510 ± 5 °C.

Batch No.	$HV5_{CENT.}$	$HRC_{SURF.}$	$HV5_{SURF.}$	$HRC_{SURF.}$
Batch 1 Ø 20 mm	355	35	365	35
Batch 1 Ø 25 mm	360	35	370	35
Batch 2 Ø 20 mm	380	37	400	38
Batch 2 Ø 25 mm	380	37	390	37

The metallographic analysis after high-temperature tempering showed a uniform structure of sorbitol in the entire sample section. The microstructure of samples 25 mm in diameter from batch 1 and batch 2 is shown in figure 3.

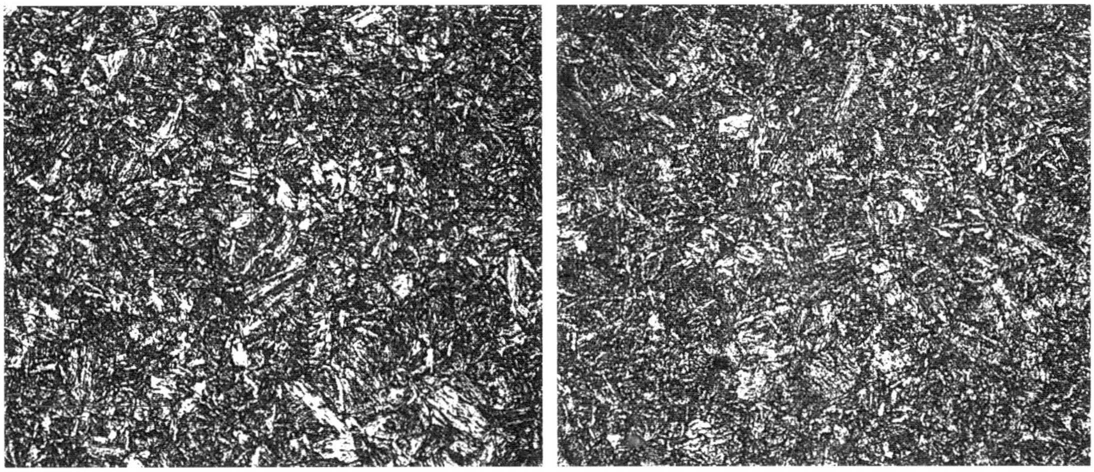

Batch 1 center x 500 Batch 2 center x 500

Figure 3. Microstructure of samples with Ø25mm after high-temperature tempering.

In hardened and tempered samples from batch 1 and batch 2, which have uniform hardness over the entire section, a homogeneous structure of sorbitol is observed (figure 3).

For parts used in aircraft construction, it is required to obtain a temporary tensile strength 1200±10 MPa. Tensile tests were carried out on INSTRON 300DX tensile testing machine. The test results are shown in tables 4 and 5.

Analyzing the data of tables 4 and 5, it can be concluded that the samples from batch 1 and batch 2, after carrying out final heat treatment on vacuum equipment, fully comply with the requirements for critical aviation components.

After working out the technology, vacuum heat treatment was applied to the "striker" part.

At the factory that produces these parts, the following problems were observed: the output of a good product on average was 60%. The current production technology of the "striker" component consisted of 22 operations. The thermal treatment was carried out in accordance with the production instructions. The main part of technological operations (17) was carried out after heat treatment, since atmospheric furnaces were used in the heat treatment process. In the process of heat treatment in atmospheric furnaces, the surface of the part reacts with the furnace gases inside the furnace, as a result of which the part is oxidized. Oxidation of the part reduces the number of alloying elements in the surface layer. This process adversely affects the mechanical properties of the part. As a result of oxidation of the surface after heat treatment, a large number of operations are required to smoothing the parts in size to remove the formed oxidized layer. These operations significantly lengthen the process of creating a part. Most of the defective products are detected after the heat treatment operation, therefore, the technology of manufacturing the "striker" with the use of vacuum heat treatment was developed in cooperation with "TermoMet" JSC.

Table 4. Results of tensile tests of samples (Batch 1).

Sample No.	Conventional yield strength, MPa	Ultimate strength, MPa	Modulus of elongation, %	Reduction of area, %
1	1003	1115	16.3	47.1
2	1008	1107	17.5	48.7
3	1011	1112	15.1	48.5
4	1009	1115	15.2	47.1

Table 5. Results of tensile tests of samples (Batch 2).

Sample No.	Conventional yield strength, MPa	Ultimate strength, MPa	Modulus of elongation, %	Reduction of area, %
1	1043	1221	14.2	43.8
2	1001	1207	16.8	45.2
3	1058	1222	14.0	46.7
4	1042	1221	15.6	51.0

The entire technological process of creating a "striker" with the use of vacuum heat treatment includes 9 operations (4 operations before heat treatment + heat treatment + 4 operations after heat treatment).

To test the developed technology, a pilot batch of 40 pieces was produced. After passing through the whole technological process, the output of the good product was 100%, which is almost twice as much as for the parts manufactured using the old technology.

The main results of calculations of technical and economic indicators are given in table 6.

As a result of the work, the quality of the vacuum heat treatment of samples with a diameter up to 25 mm made of steel 30KhGSA was confirmed.

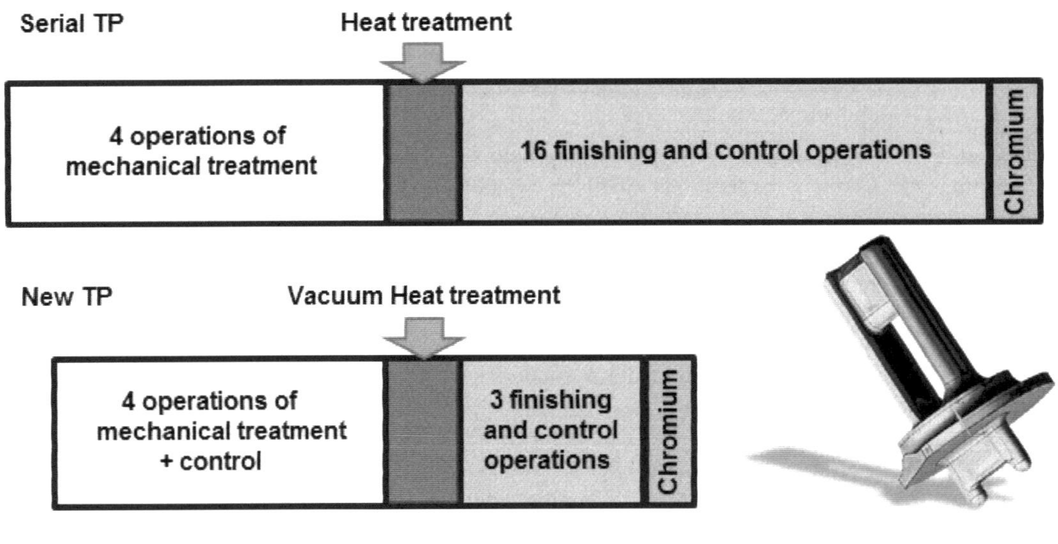

Figure 4. Technological scheme of manufacturing the "striker" part.

IOP Conf. Series: Materials Science and Engineering **411** (2018) 012034 doi:10.1088/1757-899X/411/1/012034

Table 6. Results of technical and economic indicators of the "striker" part.

Name	Old technology	New technology	Efficiency
Number of operations	21	8	2.6 times
Labour-output ratio, (standard hour)	6	4	33
Cost, rub.	5000	3500	30%

3. Conclusion

As a result of the work carried out, the technology of manufacturing the "striker" part, made of steel 30KhGSA, with the use of vacuum heat treatment, was developed. Using vacuum heat treatment for critical parts, it is possible to significantly increase the percentage of yield of good products, as well as to obtain high repeatability of the result, which is an important indicator in the production of serial products. The use of vacuum heat treatment has great prospects for implementation in modern manufacturing plants in Russia.

References

[1] Lakhtin Yu M 2015 *Metallurgy and Heat Treatment of Metals* (Moscow: Alliance) p 446
[2] Lakhtin Yu M, Leontieva V P 2013 *Materials Science* (Moscow: Alliance) p 528
[3] Kablov D E 2002 *Aviation Materials* (Moscow: VIAM, MISIS) p 422
[4] Gulyaev A P, Gulyaev A A 2012 *Metal Science* (Moscow: Alliance) p 643
[5] Elagina O Yu 2009 *Technological Methods of Increasing Wear Resistance of Machine Parts* (Moscow: Logos) p 488

Capability enhancement of production of activating fluxes for arc welding using ultradispersed products of silicon waste processing

N N Ivanchik[1], A E Balanovsky[1], M G Shtayger[2], I A Sysoev[1] and A I Karlina[1]

[1]Irkutsk National Research Technical University, 83 Lermontov street, Irkutsk, 664074, Russia
[2]MC Mechel Steel, 1 Krasnoarmeysky street, 125167, Moscow, Russia

E-mail: kvv@istu.edu

Abstract. Experimental results on welding with the help of activating flux are presented. It includes nanosilica obtained from metallurgical wastes. As a result of the use of the developed composition of the activating flux, melting capacity of the welding arc increased by 1.5-3 times, energy intensity of the process decreased by 30-50%.

1. Introduction

To achieve an increase in welding productivity the melting capacity of the arc by reducing the diameter of the electrode wire (increasing the current density and, as a consequence, melting rate of metal increases) [1], or using activating fluxes (activation of the arc-A TIG welding) [2-10].

2. Equipment and materials

To conduct researches, a DC welding arc of direct polarity with a current strength of 70-400 A, burned vertically between a non-consumable tungsten electrode (2.4-4 mm in diameter) and a metal surface in an argon shielding gas (TIG welding). Welding experiments were performed on plates with dimensions of 150*50*(2.4.6) mm from low-carbon steel 3. Before A-TIG welding, a uniform controlled layer of activating flux (SiO_2-silicon dioxide) with a width of 5 mm and a thickness of 20 μm was applied.

For the production of the activating flux, gas cleaning dust of the electrothermal silicon production was used. Dust contained components in the following ratio, %: SiO_2 75÷95, SiC 4.0÷11.0, C_{free} 3.0÷7.0, MgO 0.4÷1.2, Al_2O_3 0.5÷1.0, Fe_2O_3 0.3÷0.9, CaO 1.0÷2.0, rest 0.8÷2.3.

Several activating granulometric compositions of the charge (table 1) were prepared for the study. The first composition had a particle size of 150-300 μm (macro silica) [2-10].

3. Results and discussion

Figure 1 shows the results of photographing the welding process without activating flux (figure 1, a) and with the activating flux of figure 1, b. Photographing the process shows that the front and tail parts of the welding pool surface at A-TIG welding with an activating flux have a shape with a clearly defined characteristic depression (cup) near the center of the crater (see figure 1, b). Such a significant depression is not observed at TIG welding without the activating flux (figure 1, a). On the contrary, we see a plain surface of the liquid pool in the area of the heating spot. With increase of the arc current during welding with activating fluxes, the depth of this cup increases using all the investigated thicknesses of the samples.

Table 1. Physical properties of the activating flux.

Properties	Activating flux		
	Macroparticles SiO_2	Microparticles SiO_2	Nanoparticles SiO_2
Particle size	200-300, μm	30-70, μm	40-80, nm
Bulk density (g/cm^3)	2.64	2.42	2.10

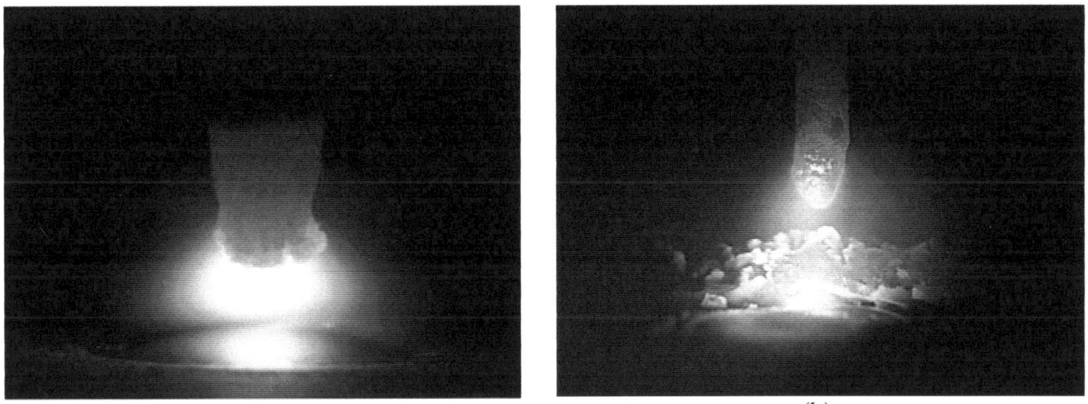

(a) (b)

Figure 1. Change in the structure of the welding arc during argon-arc welding of plates of steel 3 with a thickness of 4 mm. a – without flux; b, c – with the activating flux – SiO_2. DC power of direct polarity – 100 A. Arc length – 2 mm. Welding speed – 14 cm/min.

When the flux is melting, a screening slag phase is formed on the surface of the weld pool (figure 1, b) with low electrical conductivity, which reduces the surface tension of the molten metal and increases the deflection of the weld pool. Figure 2 shows the oscillograms which show an increase in the voltage during the welding with an activating flux at equal current values. These results indicate an increase in the effective thermal power of the welding arc during the welding with the use of an activating flux.

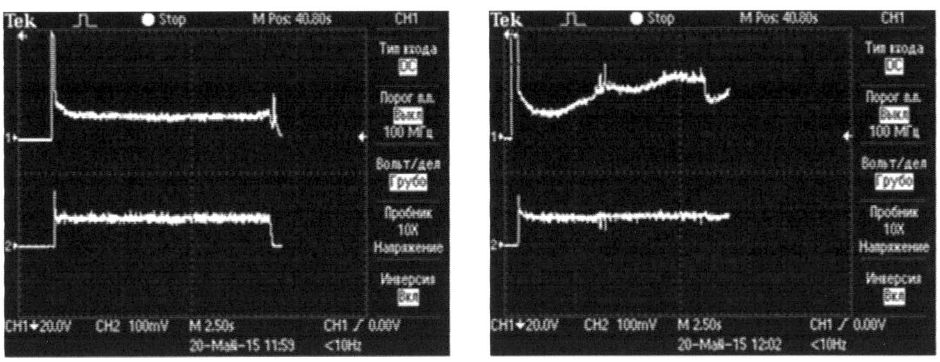

Figure 2. Oscillograms of current and voltage of the welding arc: a – without flux; b – with flux.

According to the thermal power, welding arc in argon with an activating flux approaches to the welding arc in a mixture of gases, but surpasses it by its melting capacity. The increase of the melting capacity of the arc during the welding with an activating flux changes the conditions for the formation of welds. The welds are formed narrow and deep, wedge-shaped (figure 3).

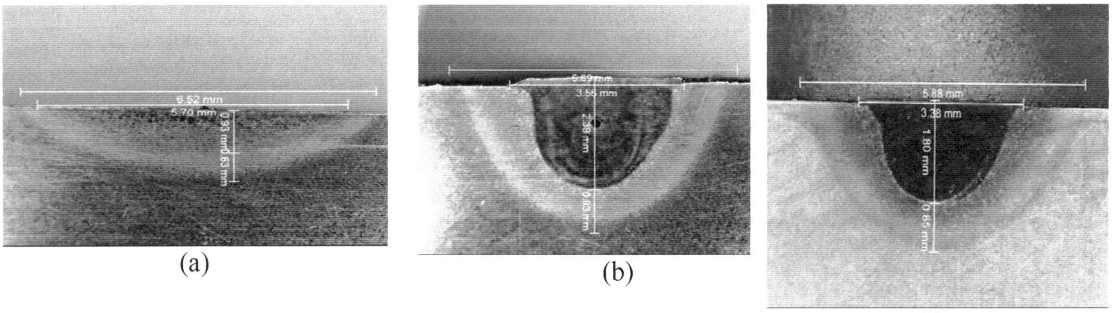

Figure 3. Cross section of the weld penetration zone during the welding in argon and current strength 80A Arc length 2 mm. Welding speed 18 cm/min. (a) without activating flux, (b) with activating flux (nanosilica), binder - ethyl alcohol, (c) with an activating flux (nanosilica), binder - heat-resistant varnish.

During the experiments it was established (figure 4) that the particle size of the activating flux plays a significant role in the melting capacity of the welding arc. All three studied compositions of activating fluxes show a positive effect on the melting capacity of the welding arc. Compared with macro and micro particles of SiO2, the SiO2 nanoparticle has a greater effect in melting capacity during the welding with an activating flux (figure 4, c). In the regime of arc heating and evaporation, thermal dissociation and decomposition of nanoparticles of silicone dioxide SiO2 occur much faster than thermal dissociation and decomposition of microparticles of silicone dioxide SiO2.

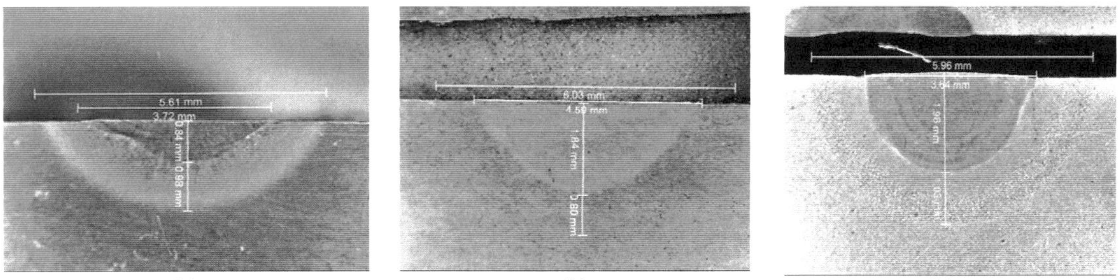

Figure 4. Cross section of the weld penetration zone during the welding in argon and current strength 90A Arc length 2 mm. Welding speed 25 cm/min. (a) activating flux (macrosilica) binder - heat-resistant varnish; (b) activating flux (microsilica); binder - heat-resistant varnish; (c) activating flux (nanosilica) binder - heat-resistant varnish.

As a result, nanoparticles of silicon dioxide SiO2 have a great influence on the arc melting properties during the welding in comparison with microparticles of silicon dioxide SiO2, which causes an increase in the weld penetration depth by 50-70%. Figure 5 shows the summarizing results of the positive effect of using an activating flux in welding low-carbon steel in argon service as compared to welding without an activating flux, both in argon and in a mixture of gases service (argon + helium). The positive effect of using an activating flux at high currents and with increasing thickness of the metal appeared considerably. From the standpoint of welding technology, we see (figure 5) that for welding metal with a thickness of 2.4 mm using an activating flux, it is possible to significantly reduce the electric power of the welding arc. The values obtained in figure 5 are in good qualitative relationship with the results of the work of other authors who considered activating fluxes where the main component was metal oxides.

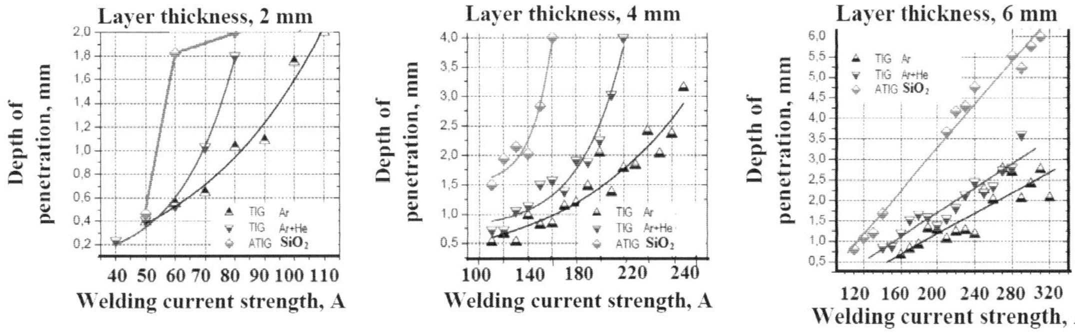

Figure 5. Experimental dependences of the depth of weld penetration of steel 3 of different thicknesses from the strength of the welding current at different welding regimes (welding in argon, in a mixture of gases, in argon with an activating flux).

4. Conclusions

1. The technology of obtaining the target final product of the activating flux for welding from secondary raw materials - wastes of metallurgical production

2. Compared with the traditional method, the use of activating fluxes provides a reduction in heat input for single-pass welding of metal of equal thickness up to 3-5 times in comparison with the traditional method.

3 It is shown that in the mode of arc welding heating, evaporation, thermal dissociation and decomposition of nanoparticles of silicone dioxide SiO2 occur much faster than thermal dissociation and decomposition of microparticles of silicone dioxide SiO2. This leads to the maximum melting capacity of the welding arc by increasing the effective thermal power of the welding heat source.

4. The use of an activating flux which has in its composition nano particles of silicon dioxide SiO2 during the welding in shielding gases, allows to increase the arc melting capacity by 2-4 times and increase the productivity of welding at low currents.

Acknowledgements

This paper has been prepared with financial support from the Ministry for Education and Science of Russian Federation (Order No.218 adopted on the 9th of April 2010 by the Government of the Russian Federation.) within the framework of the project "Development of a complex resource-saving technology and organization of high-tech production of nanostructures based on carbon and silicon dioxide to improve the properties of building and construction materials" (Agreement No. 02.G25.31.0174; code 2015-218-06-7159) using the equipment kindly provided by the Joint Instrumentation Centre "Baikal Center of Nanotechnologies".

References

[1] Akulov A I 1977 *Mechanical Engineering* p 432
[2] Savitsky M M 1999 *Automatic Welding* **12** 20–28
[3] Parshin S G 2003*Welding Production* **6** 40–43
[4] Parshin S G 2014 *Metalurgija* **2** 167–170
[5] Niagaj J 2008 *Biuletun Instytutu Spawalnictwa w Gliwicach* **1** 46-50
[6] Lenivkin V A 2010 *Welding Production* **10** p 38
[7] Parshin S G 2010 *Welding* **2** p 37
[8] Balanovsky A E 2015 *Strengthening Technologies and Coatings* **6 (126)** 27–32
[9] Gyu V V, Balanovsky A E and Kondratiev V V 2017 *Metallurgy of Machine Building* **1** 9–15

Resource-saving technology of obtaining nanosilica

V V Kondratiev[1], A D Kolosov[1], V O Gorovoy[1], S A Nebogin[1] and K S Elkin[2]

[1]National Research Irkutsk Technical University, 83 Lermontova street, Irkutsk, 664074, Russia
[2]LLC RUSAL Engineering and Technological Centre, 37 Pogranichnikov street, Krasnoyarsk, 660111, Russia

E-mail: vog@istu.edu

Abstract. Possibility of obtaining nanoscale silica from the dust of dry gas cleaning of silicon production by thermal methods is considered in the report. Analysis of the chemical and granulometric compositions of the initial dust, dust after annealing in the muffle furnace is given. A new technology for the thermal-vortex enrichment of dust of dry gas cleaning of silicon to obtain nanosilica is proposed, the results of modeling and laboratory studies are presented.

1. Introduction

Nanosilica is one of the most popular nanomaterials in modern industry. Different types of amorphous nanosilica are used in construction, in the rubber, chemical, tire, food and other industries. There are prospects for its use in metallurgy [1-2]. High-purity nanosilica is used in medicine and is necessary for many scientific studies.

High demand for nanosilica is explained by its unique physico-chemical properties:

- nanosilica is chemically neutral with respect to many known mineral and organic compounds, but it reacts well, for example, with hydrofluoric acid with the creation of hydrofluoric acid;
- has a large specific surface, due to which it is widely used as sorbents, modifiers, etc.

Depending on the purity, granulometric composition, nature of origin and method of production, cost of different varieties of nanoilica can vary thousands of times.

Gas cleaning dust of silicon production. In the production of silicon in ore-thermal furnaces, a large amount of dust is released, which today is one of the accumulating wastes.

2. Gas cleaning dust of silicon production

In the production of silicon in ore-thermal furnaces, a large amount of dust is released, which today is one of the accumulating wastes. Russian enterprises, which produce technical silicon, switch to dry gas cleaning technology, one of the stages of which is dust trapping with the help of bag filters. It is the dust captured by bag filters (hereinafter - dust of bag filters) that is of greatest interest for the producing of amorphous nanosilica.

For researches within the scope of the report, dust samples of bag filters from ZAO "Kremniy" in Shelekhov were selected. Dust of bag filters of ZAO "Kremniy" has the chemical composition given in table 1.

Table 1. Chemical composition of dust of bag filters of ZAO "Kremniy".

S No.	Substance	Content (%)
1	SiO_2	> 92
2	C	< 8
3	Possible admixtures	< 1

Granulometric content of dust of bag filters is given in figure 1 and table 2.

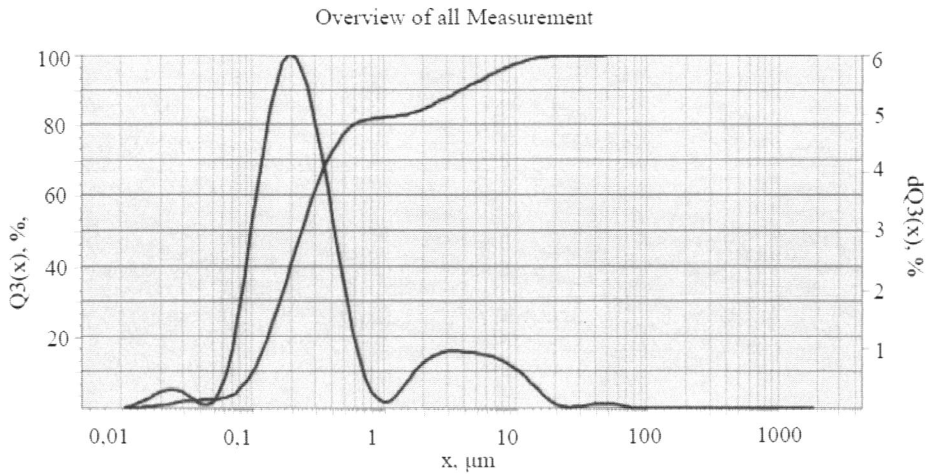

Figure 1. Granulometric content of dust of bag filters of ZAO "Kremniy".

Table 2. Granulometric content of dust of bag filters of ZAO "Kremniy".

x less than, μm	Q (x), %
0.1	10.1
0.2	29.4
0.3	58.6
0.5	77.7
0.8	81.9
1.3	82.6
2.3	85.4
3.8	89.8
6.5	94.3
10	97.3
18	99.4
30	99.6

From the chemical and granulometric analyses it can be seen that the dust of bag filters is a mixture of nanostructures based on silicon dioxide and carbon. Isolation of the nanosilica is a promising direction for researches [3-10].

Heat treatment of gas cleaning dust. Based on the chemical composition of dust, it seems appropriate to use heat treatment of dust of bag filters to oxidize (burn out) carbon and obtain silica of a certain purity.

Within the framework of the laboratory experiment, a series of experiments was conducted to burn carbon from dust of bag filters using the laboratory muffle furnace ЭКПС-10. Cuvets with dust of bag filters were placed in the furnace chamber for a time of 5-45 minutes, at different temperatures 450-800 °C.

A series of experiments showed that the best mode of annealing is holding for 15 minutes at a temperature of 650 °C. A further increase in the annealing time does not lead to a more complete burnout of carbon, and an increase in temperature greater than 650 °C leads to considerable sintering of the particles and the formation of large stable conglomerates.

A visual comparison of the initial and annealed dust of bag filters shows that the annealed dust is much whiter than the original one, which indicates a decrease in the amount of carbon.

Table 3 shows the chemical composition of the dust of bag filters, annealed at 650 °C for 15 minutes.

Table 3. Chemical composition of dust of bag filters, annealed in muffle furnace.

S No.	Substance	Content (%)
1	SiO_2	> 98
2	C	< 2
3	Possible admixtures	< 1

Granulometric composition of dust of bag filters, annealed in a muffle furnace, is shown in figure 3 and in table 2.

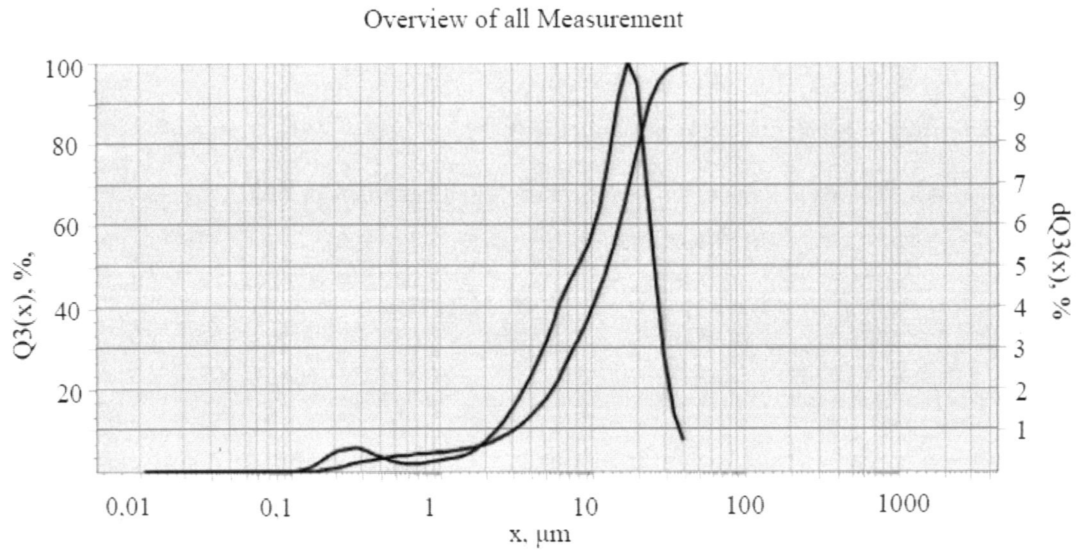

Figure 2. Granulometric composition of dust of bag filters, annealed in a muffle furnace.

The results of analyses of dust of bag filters annealed in the muffle furnace show that part of carbon is successfully oxidized, increasing the percentage of silica. However, a significant increase in particle size is observed. These particles can not be decomposed by ultrasound i.e. particle sintering occurred due to prolonged exposure to temperature. Method of annealing the dust of bag filters in the muffle furnace showed its efficiency, however, the following drawbacks were revealed:

- Insufficient burn of carbon;
- Sintering of silica particles into large conglomerates;
- High energy consumption per kilogram of the output.

Table 4. Granulometric composition of dust of bag filters, annealed in a muffle furnace.

x in μm	Q3(x) in %
0.1	0
0.2	0.3
0.3	2.1
0.5	3.7
0.8	4.3
1.3	5.2
2.3	7.4
3.8	12.8
6.5	24.8
10	39.8
18	73.4
30	96.7

As a result of studying the process of heat treatment of dust of bag filters by the example of annealing in a muffle furnace, the authors proposed a method for burning carbon that does not have the disadvantages mentioned above.

3. Conclusion

Laboratory studies have shown the efficiency of the method of thermo-vortex dust enrichment of bag filters of silicon production. It was possible to achieve a significant reduction in the amount of carbon, as well as to avoid excessive sintering of silica particles. A greater degree of oxidation of carbon without increase of the temperature can be achieved by increasing the time particles spent in the furnace chamber. One way to increase the residence time in the working zone is to increase the size of the working chamber.

A further line of research is the development and construction of a thermal vortex enrichment installation with a large size of working chamber, which will allow make more complete burn of carbon, and also generate more product to study its consumer qualities.

Acknowledgements

This paper has been prepared with financial support from the Ministry for Education and Science of Russian Federation (Order No.218 adopted on the 9th of April 2010 by the Government of the Russian Federation.) within the framework of the project "Development of a complex resource-saving technology and organization of high-tech production of nanostructures based on carbon and silicon dioxide to improve the properties of building and construction materials" (Agreement No. 02.G25.31.0174; code 2015-218-06-7159) using the equipment kindly provided by the Joint Instrumentation Centre "Baikal Center of Nanotechnologies".

References

[1] Yolkin K S, Ivanov N A, Karlina A I and Ivanov N N 2015 *Non-Ferrous Metals and Minerals* pp 224–5
[2] Nemarov A, Lebedev N, Kondrat'ev V, Kornyakov M and Karlina A 2016 *International Journal of Applied Engineering Research* vol 11 **20** 10222–6
[3] Rassokhin A S, Ponomarev A N and Figovsky O L 2018 *Magazine of Civil Engineering* **02(78)**
[4] Kondratiev V V, Ivanov N A, Balanovskiy A E, Ivanchik N N and Karlina A I 2016 *Journal of the Siberian Federal University. Series: Engineering and Technology* vol 9 **5** pp 671–85
[5] Kondrat'Ev V V, Nemchinova N V, Ivanov N A, Ershov V A and Sysoev I A 2013 *Metallurgist* vol 57 **5-6** 455–9
[6] Ivanchik N, Kondratiev V and Chesnokova A 2016 *Procedia Engineering 2nd International*

Conference on Industrial Engineering pp 1567–73

[7] Kondratiev V V, Govorkov A S, Kolosov A D, Gorovoy V O and Karlina A I 2017 *International Journal of Applied Engineering Research (IJAER)* vol 12 **22** 12373–7

[8] Kondratiev V V, Nebogin S A, Sysoev I A, Gorovoy V O and Karlina A I 2017 *International Journal of Applied Engineering Research (IJAER)* vol 12 **22** 12809–13

[9] Ershov V A, Gorovoy V O and Karlina A I 2016 *Modern Technology. System Analysis. Modeling* **4 (52)** 114–121

[10] Ivanchik N N, Kondratiev V V, Ivanov N A and Karlina A I 2015 *Non-Ferrous Metals and Minerals* pp 234–5

Capabilities of use of dry separation of microsilica to obtain target products

V V Kondratiev, S A, Nebogin, A D Kolosov, V O Gorovoy and A A Nemarov

Irkutsk National Research Technical University, 83 Lermontov street, Irkutsk, 664074, Russia

E-mail: s.a.nebo@yandex.ru

Abstract. This article is devoted to methods of processing of the dust entrained from the ore-thermal furnaces for silicon production. Chemical and granulometric analysis of dust entrainment of silicon production was carried out. Possibility of using gravitational-centrifugal separation to obtain target products, such as spherical silica and nano-structured carbon was considered.

1. Introduction

As of today, many factories in the metallurgical industry do not subject their waste to further processing capable of turning wastes into useful material, but prefer to recover wastes by transporting them to mud storage areas. This method of waste disposal leads to environmental problems, as one ton of finished products of the plant comes with a ton of dust products. For example, the amount of trapped dust in a silicon plant may amount to 42,000 tons per year, which inevitably leads to an increase in the size of mud storage areas and ecosystem damage.

2. Granulometric composition of dust captured by a preliminary purification system

The gas cleaning system of silicon plants provides for two stages of purification. The first stage is a pre-cleaning system and serves to capture large, heavy dust that can disrupt the operation of the fine cleaning system. Most often this system is a cyclone battery [1-3]. A typical granulometric composition of the dust collected at the 1 stage of purification is shown in figure 1 and table 1.

Table 1. Disperse composition of pre-cleaning dust.

Number of particles in the sample (%)	5	10	25	50	75	90	95	99	100
Size (μm)	1.8	3.6	11.2	25.2	48.1	71.9	85.9	111.2	115.3

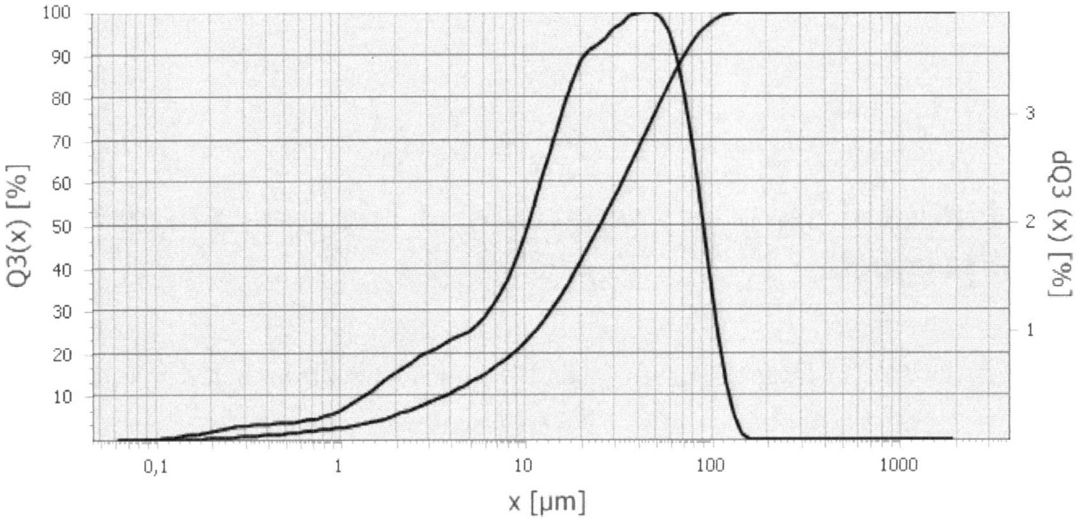

Figure 1. Granulometric composition of dust captured by a preliminary purification system, measured on "Analysette 22 NanoTec" particle size analyzer.

3. Granulometric composition of fine dust

Dust trapped by the fine cleaning system is of great interest. Studies carried out by different groups of scientists show that the main component of dust (at least 92%) trapped by the fine cleaning system is finely dispersed spherical silicon dioxide. Figure 2, obtained with the help of JEOL JIB-Z4500 scanning electron microscope, shows the spherical shape of the silica particles. The second largest component of the mixture is highly dispersed, nanostructured carbon. Granulometric analysis of dust of fine cleaning is shown in figure 3 and table 2.

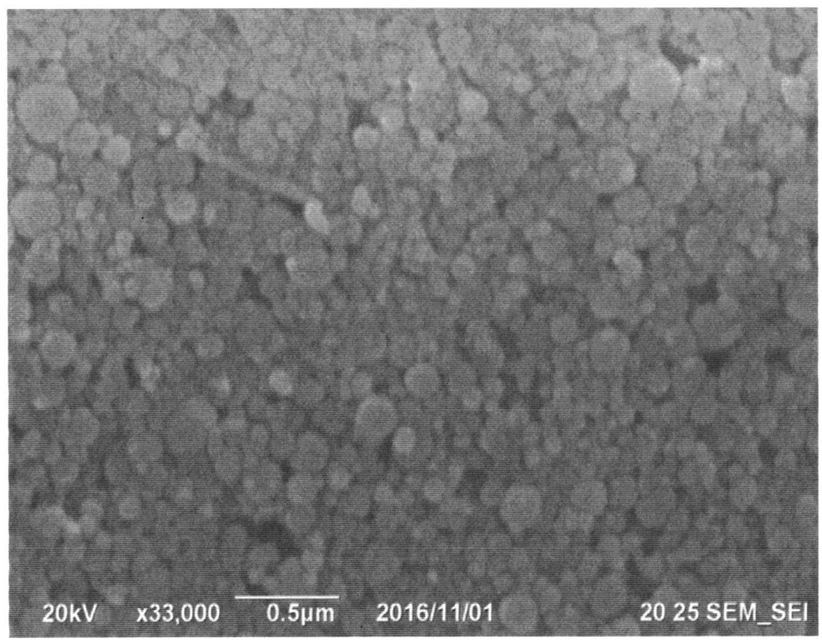

Figure 2. Image of fine dust obtained with the help of JEOL JIB-Z4500 scanning electron microscope.

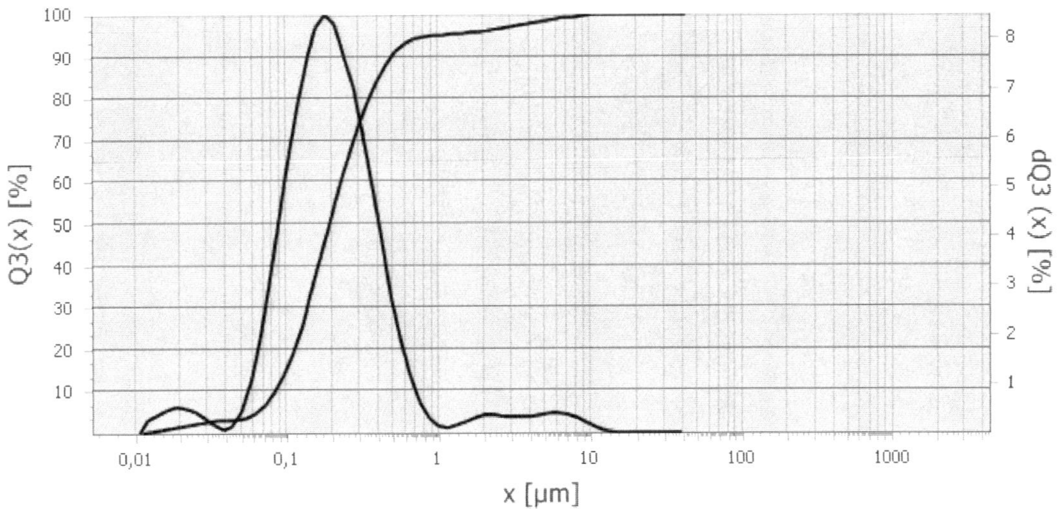

Figure 3. Granulometric composition of fine dust, measured on a laser particle size analyzer "Analysette 22 NanoTec".

Table 2. Disperse composition of fine dust.

Number of particles in the sample (%)	5	10	25	50	75	90	95	99	100
Size, less than µm	0.06	0.09	0.1	0.2	0.3	0.5	1	6.6	103

Nanodisperse spherical silica with a high mass fraction of dioxide has become widely used as a modifying additive in the production of construction concretes and high-strength composite alloyed metal alloys. In addition, it is used in the production of rubber products, and it is a filler in the production of paints, ceramics, etc. [4-7]. Consequently, for further use of dust of fine cleaning of silicon production furnaces, its enrichment to a SiO_2 content of not less than 98% is required. In the process of processing fine dust, it is equally important to provide for the possibility of obtaining a fraction of a product with a high content of carbon nanostructures. The carbon nanostructured material itself is an interesting and promising material, and is also widely used as modifying additives.

There is a flotation method for separating dust from the removal of silicium production furnaces [3, 8]. In detail, this method was considered in articles [1, 9, 10]. The flotation method makes it possible to separate with high efficiency the microsilica remaining in the chamber product from the carbon fraction, separated out as a foam product. However, flotation method has a number of disadvantages associated with the progress of the separation process in the liquid medium. This circumstance entails further drying and disintegration of the obtained product. This increases energy costs and, as a consequence, the cost of the final product.

We considered the possibility of dry separation of microsilica on devices of gravitational centrifugal dust collection (cyclones). Carbon and silicon dioxide have a different density, that´s why, provided that the particles have the same dimensions, they will be deposited differently in the cyclones. To test the hypothesis, a stand consisting of 5 successively mounted cyclones (figure 4) was constructed. Dusty gas stream was created with the help of a compressor and a dust-collecting device, which activated a slotted shutter. As raw materials, dust of furnaces, caught in bag filters, taken from a pipeline that carries a dust and gas stream from silicon furnaces was used. The separation process on gravitational-centrifugal dust-collecting devices was mathematically modeled, and predicted the possibility of gravity-centrifugal separation of particles of dioxide and carbon.

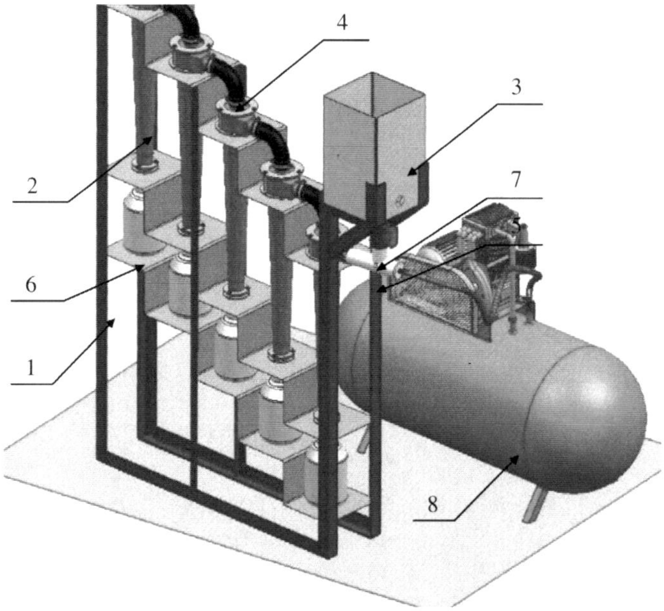

Figure 4. Installation of dry dust separation of silicon production furnaces: 1 – frame; 2 – cyclone; 3 – loading hopper; 4 – elbow flexible; 5 – slit dispenser; 6 – capacity; 7 – disk shutter with the electric drive; 8 – compressor.

4. Conclusion

As a result of a series of tests, it was possible to achieve an increase in the concentration of microsilica from 93% to 98% with a yield of up to 35% by weight. It was possible to achieve an increase in the concentration of the carbon fraction from 6% to 24% with a yield of 10%. Mathematical calculations and results of mathematical modeling testify to the possibility of increasing the separation characteristics of the installation, in case of a change in the geometric dimensions of the dust-collecting devices, and the operating conditions of the draft equipment of the stand.

Acknowledgements

This paper has been prepared with financial support from the Ministry for Education and Science of Russian Federation (Order No.218 adopted on the 9th of April 2010 by the Government of the Russian Federation.) within the framework of the project "Development of a complex resource-saving technology and organization of high-tech production of nanostructures based on carbon and silicon dioxide to improve the properties of building and construction materials" (Agreement No. 02.G25.31.0174; code 2015-218-06-7159) using the equipment kindly provided by the Joint Instrumentation Centre "Baikal Center of Nanotechnologies".

References

[1] Ivanchik N, Kondratiev V and Chesnokova A 2016 *Procedia Engineering "2nd International Conference on Industrial Engineering, ICIE 2016"* pp 1567–73
[2] Kondratiev V V, Govorkov A S, Kolosov A D, Gorovoy V O and Karlina A I 2017 *International Journal of Applied Engineering Research (IJAER)* vol 12 **22** 12373–7
[3] Kondratiev V V, Nebogin S A, Sysoev I A, Gorovoy V O and Karlina A I 2017 *International Journal of Applied Engineering Research (IJAER)* vol 12 **22** 12809–13
[4] Kondrat'Ev V V, Nemchinova N V, Ivanov N A, Ershov V A and Sysoev I A 2013 *Metallurgist*

vol 57 **5-6** 455–9
[5] Ershov V A, Gorovoy V O and Karlina A I 2016 *Modern Technology. System Analysis. Modeling* **4 (52)** 114–121
[6] Ivanchik N N, Kondratiev V V, Ivanov N A and Karlina A I 2015 *Non-Ferrous Metals and Minerals* pp 234–5
[7] Yolkin K S, Ivanov N A, Karlina A I and Ivanov N N 2015 *Non-Ferrous Metals and Minerals* pp 224–5
[8] Nemarov A, Lebedev N, Kondrat'ev V, Kornyakov M and Karlina A 2016 *International Journal of Applied Engineering Research* vol 11 **20** 10222–6
[9] Rassokhin A S, Ponomarev A N and Figovsky O L 2018 *Magazine of Civil Engineering* **02(78)**
[10] Kondratiev V V, Ivanov N A, Balanovskiy A E, Ivanchik N N and Karlina A I 2016 *Journal of the Siberian Federal University. Series: Engineering and Technology* vol 9 **5** 671–685

Study of combined packaging materials for steel products

L G Kolyada, E V Tarasyuk, N L Medyanik, A P Ponomarev and A V Smirnova

G.I. Nosov Magnitogorsk State Technical University, 38 Lenina street, Magnitogorsk, 455000, Russia

E-mail: gepod@inbox.ru

Abstract. Metal products are subject to atmospheric corrosion during transportation, long-term storage and operation. According to experts, corrosion destroys up to a third of the annual production of ferrous metals that is the reason for the annual economic losses of large metallurgical enterprises. An important area in the fight against corrosion is the search for new packaging materials containing volatile corrosion inhibitors. The urgency of the problem under investigation lies in the fact that the export of metal products is mainly effectuated by sea through areas with a humid tropical climate. The paper investigates structural, physical, deformation, strength, barrier and corrosion resistance properties of domestic packaging papers in comparison with Fislage's foreign equivalent. It studies packaging paper protection of cold-rolled steel under periodic humidity concentration. In order to increase the competitiveness of domestic packaging composite materials for metal products, it is necessary to improve their production technology, for example, by increasing the inhibitor content in packaging materials and the creping of the paper base, or by increasing the deformation-strength properties due to reinforcement.

1. Introduction

In the Russian Federation the annual loss of metals due to corrosion reaches 12% of the total mass of the country's metal reserves. Innovative methods of protecting metal products from corrosion during transportation and storage are the use of packaging materials containing corrosion inhibitors. Currently, some manufacturers offer combined paper-based materials containing corrosion inhibitors that can favorably compete with standard corrosion protection aids (oils, lubricants, varnishes, paints enamels, etc.) [1-6]. The issue of metal protection against atmospheric corrosion has become particularly acute due to expanded metal export volumes as metal products are often transported by open top wagons or by sea through the regions with wet tropic climate.

Up-to-date package for steel products shall fully exclude intrusion of corrosive water steams and aggressive gases to metal surface and have the required mechanical properties providing its own integrity and preservation of the packed steel products against mechanical damage during handling and transportation over long distances.

The above conditions are met by combined packaging materials with volatile corrosion inhibitors [2-6] that are made of a layer of crepe paper laminated with polyethylene and/or reinforced with polymeric threads and cloth. Reinforcement with polymeric threads contributes to better material strength. Similar packaging materials are widely used for packaging steel rolls, steel sheet packs, wire and other metal products. These materials may be used at power-driven and automated packaging units. The following de-preservation at the customer site is reduced to package removal.

In recent years, foreign combined materials were mostly available in the market. However, after Russian export sanctions have been imposed, domestic packaging papers are more widely used.

The aim of this investigation is to estimate deformation, strength and corrosion protection properties of combined packaging materials containing corrosion inhibitors.

To achieve the target goal, the following tasks were solved: investigation of deformation and strength properties of packaging papers; investigation of barrier properties of packaging papers; estimate of corrosion protection properties of packaging papers under atmospheric corrosion condition.

2. Materials and methods

Research targets are combined packaging materials manufactured by JSC PP Technokhim (Magnitogorsk); crepe inhibited paper laminated with polyethylene foil EuroBum manufactured by Production Commercial Association "Europack-Lebedyan" (Moscow); crepe laminated inhibited paper reinforced with polypropylene grid by FISLAGE Flexibles GmbH (Germany). Samples of packaging papers and their properties are listed in table 1.

Table 1. Survey objects.

Sample designation	General view	Manufacturer	Packaging material characteristics
BLIK		JSC PP Tekhnokhim	Crepe packaging paper with polyethylene coating
BLIK-1	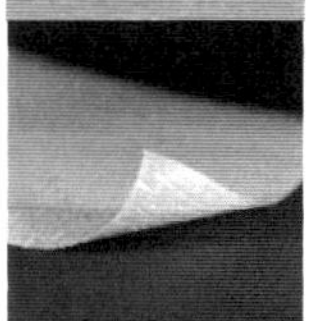	JSC PP Tekhnokhim	Crepe inhibited laminated paper
BLIK-3		JSC PP Tekhnokhim	Crepe inhibited laminated paper BLIK reinforced with polypropylene threads
EuroBoom	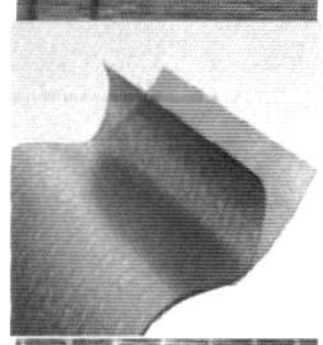	PLC Production Commercial Association Evropak-Lebedyan	Crepe inhibited paper laminated with polyethylene foil
Fislage	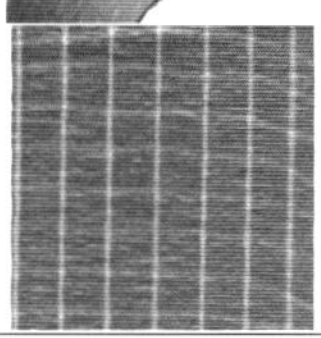	FISLAGE Flexibles GmbH	Crepe inhibited laminated paper reinforced with polypropylene grid

Deformation and strength properties of materials were tested at digitally controlled IP 5158-0,5 test bench as per GOST 30436-96 (ISO 1924-2-85). Based on test results, the strength and tensile elongation were calculated.

The weight of 1 m^2 paper was computed according to GOST 13199-88.

Inhibitor residuals in anti-rust paper were determined according to GOST 16295-93.

Vapor permeability of combined materials was calculated as per GOST 9.507-88. The test issue is in determination of the weight of moisture vapor passing through 1 m^2 combined material for 24 h at 20 °C and (95±5) % relative humidity.

Water adsorbency at one-side wetting was determined according to GOST 12605-97 (Cobb's method). The method includes determination of the water weight (g) absorbed by paper surface at wetting one side of the sample for 60 sec.

Samples of 08-Yu cold rolled products were used for study of corrosion protection properties.

Protective properties of inhibitor paper sheets were tested with the gravimetrical method in the heat and moisture chamber according to GOST 9.054-75. Corrosion damage was visually checked. The index and rate of uniform corrosion were calculated based on the test results [7-8].

3. Investigation results and their discussion

Results of calculation of structural and physical properties of packaging papers are provided in table 2.

Table 2. Structural and physical properties of packaging papers.

Specimen designation	Thickness (mm)	1 m^2 paper weight (g/m^2)	Inhibitor weight (g/m^2)	Inhibitor content
BLIK	0.27	122	-	without an inhibitor
BLIK 1	0.21	112	12	Sodium nitrite and urotropin (1:1)
BLIK 3	0.21	117	9	
EuroBoom	0.54	166	5.4	Contains nitrites[a]
Fislage	0.35	157	9.5	

[a]Note: calculated according to GOST 16295-93

The weight of 1 m^2 packaging papers correlates to their thickness. Content of volatile inhibitor in 1 m^2 papers under study is nearly the same and amounts 9-12 g/m^2, which is sufficient for metal corrosion protection. The inhibitor content in the EuroBoom paper is low– 5.4 g/m^2.

Packaging papers of JSC PP Technokhim contain UNII volatile inhibitor specified in the reference documents for these products. Inhibitor content of the Fislage's and EuroBoom papers is confidential, but identification tests in water extracts based on packaging papers detected nitrite ions [8].

Deformation and strength properties of packaging papers are provided in table 3.

Table 3. Deformation and strength properties of packaging papers.

Specimen designation	Breaking force (N)	Relative elongation (%)	Tensile strength (kN/m)
BLIK	40	38	2.7
BLIK 1	63	5	4.2
BLIK 3	86	22	5.7
EuroBoom	55	55	3.7
Fislage	88	36	5.9

According to the strength properties, domestic packaging paper is inferior to foreign analogues, except for paper laminated with polypropylene fabric: the tensile strength of EuroBoom packaging paper is 1.6 times lower. Due to the inadequate strength properties of EuroBoom packaging paper, problems can arise in the mechanized packing of steel rolls.

Investigation of barrier properties of packaging materials is essential for metal protection against atmospheric corrosion, as the package should fully prevent water vapor access to metal surface because condensing moisture is a corrosion source and enhancer. Results of vapor permeability computing at 20 °C are shown in table 4.

Table 4. Barrier properties of combined packaging materials.

Specimen designation	Thickness (mm)	Vapor permeability ($g/m^2 \cdot /day$)	Adsorbency (g/m^2)
BLIK	0.27	64	60
BLIK 1	0.21	76	64
BLIK 3	0.21	66	61
EuroBoom	0.54	42	58
Fislage	0.35	81	95

Relatively law vapor permeability is specific for EuroBoom specimen as this packaging material has the greatest thickness. A high adsorbency of Fislage's packaging paper is conditioned by the greater creeping ratio.

Metal plates packed in corrosion protection papers were exposed to 95 % relative humidity and at 40 ºC for 7 h and for 17 h at room temperature and moisture condensation. Test total duration was 20 cycles/480 h. The area damaged by corrosion was determined based on samples. The mass of corrosion products and the rate of uniform (general) corrosion were gravimetrically determined [6-7]. The test results are shown in table 5.

Table 5. Results of corrosion protection tests at cold-rolled steel samples under conditions of periodic humidity condensation.

Specimen designation	Corrosion product weight (g)	Corrosion area (cm^2)	Corrosion index (g/m^2)	Corrosion rate ($g/m^2/h$)
BLIK	0.003	1.60	0.24	$5.0 \cdot 10^{-4}$
BLIK 1	0	0	0	0
BLIK 3	0.0002	0.02	0.02	$3.3 \cdot 10^{-5}$
EuroBoom	0.02	1.38	0.96	$3.9 \ 10^{-3}$
Fislage	0.002	0.04	0.17	$3.5 \cdot 10^{-4}$

Figure 1 shows appearance of samples after corrosion tests under conditions of periodic humidity concentration.

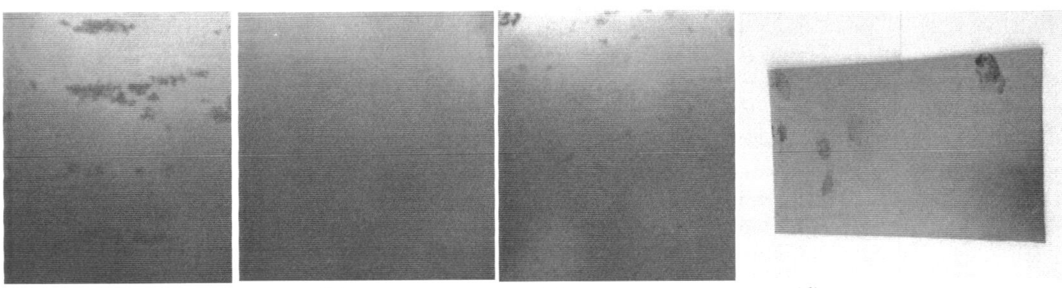

(a) (b) (c) (d)

Figure 1. Appearance of samples after corrosion tests under conditions of periodic humidity concentration: (a) – BLIK; (b) – BLIK 1; (c) – Fislage; (d) – EuroBoom.

Conducted tests confirm corrosion protection efficiency of combined packaging materials of JSC PP TechnoKhim. All studied materials of JSC PP TechnoKhim have better protective properties than those of foreign products. The EuroBoom paper does not have any protective action. It has poorer corrosion protection properties than those of the Fislage's paper: the corrosion index is by 5.6 times higher, which can be explained by a lower content of corrosion inhibitor.

4. Conclusion

The issue of selection of packaging materials for metal products is especially relevant under condition of import substitution. Comparison of domestic packaging materials manufactured by JSC PP TechnoKhim and EuroBoom paper with the foreign counterpart has proven that products offered by JSC PP TechnoKhim provide the best protection against atmospheric corrosion than that of previously used foreign analogues. These combined materials may be recommended for packaging cold-rolled steels.

In conclusion it should be noted that to improve the competitiveness of domestic composite packaging materials for metal products it is necessary to improve their production technology. This can be achieved by increasing the content of the inhibitor in packaging materials, by increasing the creping of the paper base, by increasing the deformation-strength properties due to reinforcement, and by developing new packaging materials [5].

References

[1] Kolyada L G, Tarasyuk E V, Ponomarev A P and Vafin V R 2017 *Iron and Steel Industry* **6 (1410)** 61–73
[2] Antropov L I, Makushin E M and Panasenko V F 1981 *Metal Corrosion Inhibitors* (Technika: Kiev) p 183
[3] Rosenfeld I L and Parsiantseva V P 1985 *Atmospheric Corrosion Inhibitors* (Nauka: Moscow) p 278
[4] Rogova A N and Rasumkov A V 2002 *Transport and Consumer Packaging* **6** 44–47
[5] Medyanik N L, Shadrunova I V, Kolyada L G, Tarasyuk E V and Chekushina T V 2016 *Metallurgical and Mining Industry* **8** 14–20
[6] Kichenko A B and Kushnarenko V M 2005 *Practice of Anti-corrosive Protection* **4 (38)** 17–22
[7] Panchenko Yu I and Strekalov PV 2005 *Metal Protection* **41(4)** 402–16
[8] Girevaya Kh Ya, Kolyada L G and Ponomarev A V 2015 *Current High-End Technologies* **7** 19–22

IOP Conf. Series: Materials Science and Engineering **411** (2018) 012039 doi:10.1088/1757-899X/411/1/012039

Development of innovative production technology of automotive high quality in the conditions of cold rolling mill

S A Kondrashov, E M Golubchik and Yu N Kondrashova

Nosov Magnitogorsk State Technical University, 38 Lenina street, Magnitogorsk, 455000, Russia

E-mail: Soulsnack@mail.ru

Abstract. The main tendency of ferrous metallurgy development is improvement of automotive sheet quality and production of the new steel grades which meet the requirements of the international standards. The studied subject is continuous cold strip mill 2500. Analysis of the defects for emulsion spots and spotted carburization showed that their distribution is irregular across the strip width. In the world practice, there are various ways to remove the lubricating and cooling liquid from the rolled strip during the production of automotive sheet by a cold rolling mill. Special modern systems are developed by both foreign and domestic firms are used to remove cooling lubricant from the strip. But significant disadvantage of such solutions is high cost. So, the rational solution of this problem is the development of an innovative technology for removing the cutting compound based on the principle of blowing with minimum amount of compressed air. The article describes an innovative system for cleaning the strip from the residues of the lubricating-cooling liquid in the production of an automotive sheet on the four-stand cold rolling mill 2500 and calculation of parameters. The implementation of improved automotive sheet production technology will reduce the compressed airflow more than 2.5 times, compared with the current system and significantly to expand the range of products.

1. Introduction

At the present time one of the main tendency of metallurgy development is improvement of metal product quality by the implementation of innovative methods of the metal production, and production of the modern steel grades. The problem of quality improving of cold-rolled sheets is a special meaning at achievement of high efficiency of its use in a market economy [1].

2. Main part

The studied subject is the continuous 2500 cold strip mill, which is a major power consumer of the industrial enterprise of ferrous metallurgy [3,4,5,6,7,8].

The analysis of defects was carried out and circle charts were constructed for emulsion spots, oil, electrolyte and spotted carburization (figure 1a, 1b). These diagrams show that the distribution of these defects is irregularly across the width of a cold-rolled strip.

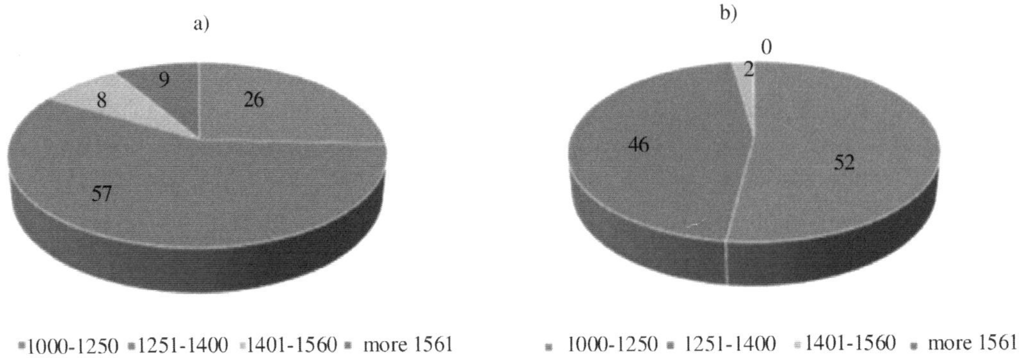

Figure 1. Distribution of defects in width a) emulsion spots, b) spotted carburization.

There are various ways to remove coolant remains from the strip in a cold rolling mill. One way is to use felt or calico cleaners which are in gripping unit and the strip are rolled through it. Weakness of this method is that the part of draft force of coiling device is lost in the process of friction. Felt or textile fiber quickly absorb the erasable liquid, at the same time gradually lose absorbing properties, and thus the quality of erasure decreases with time. Besides, these materials gradually lose their fibers, which are remained on the sheet and cause surface defects of the rolled sheet during further processing. One else problem is that facings which are stuck in a felt or calico cleanser can cause further scratches on the sheet [2].

Removal of the coolant remains is possible by using of a wiping roll. However, final result of erasure depends on the surface of the work rollers and on the erasing rollers. But this method isn't effective, because of contaminants pick up to the rollers and subsequently to the rolled strip and in which case the quality of removal of contaminants from the surface depends on the flatness of the sheet, its thickness and rolling speed.

Special modern systems for blowing of a strip are also used for cleaning of a strip, which are developed by both foreign, and domestic-owned firms, including Siemens, SMS Demag, Spraying System Co. and etc. As an example we will consider using of the perspective VacuRoll device by Spraying System Co and the principle of its work. The system consists of two horizontally arranged rollers (figure 2).

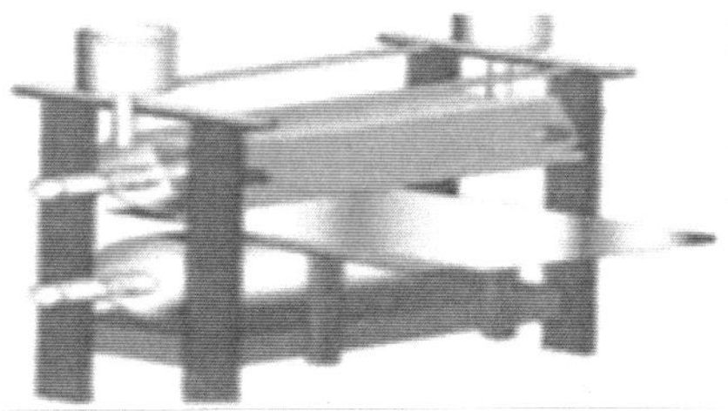

Figure 2. VakuRoll system with two horizontally arranged rollers

The principle of the system is that the system creates a uniform vacuum along the entire length of the roller, including its edges. As the coolant is removed from the strip, it passes through the roller shell. Thus, the emulsion and oils are drawn into the surface of the roller shell, and solid inclusions are gripped by the discs and remain between them to prevent scratches on the strip. The emulsion passes through the hollow center of the roller housing and then enters into the emulsion system for reuse. But a significant disadvantage of this solution is the overprice, the equipment reconstruction and the location choice for this strip cleaning system in the output part of the cold rolling mill [3,4].

Based on the above methods of cleaning the strip from the emulsion, a rational solution is the development of innovative production technology, parameter determination and the introduction of a new strip cleaning system based on the principle of blowing with compressed air. This method has a significant disadvantage in increased compressed air flow, so this system is designed in such a way as to take into account all the features of the cold rolled strip production with a minimum compressed air flow. At introduction this system will reduce the air flow more than by 2.5 times in comparison with the current system [2,5].

The strip cleaning system based on the compressed air supply principle consists of two parts:

1. Development and installation of a new plow for the compressed air supply, in order to clean the strip behind the IV stand from cutting compound remains. It is necessary to place 2 independent compressed air supply zones on this plow for effective metal cleaning of different widths from cutting compound (figure 3).

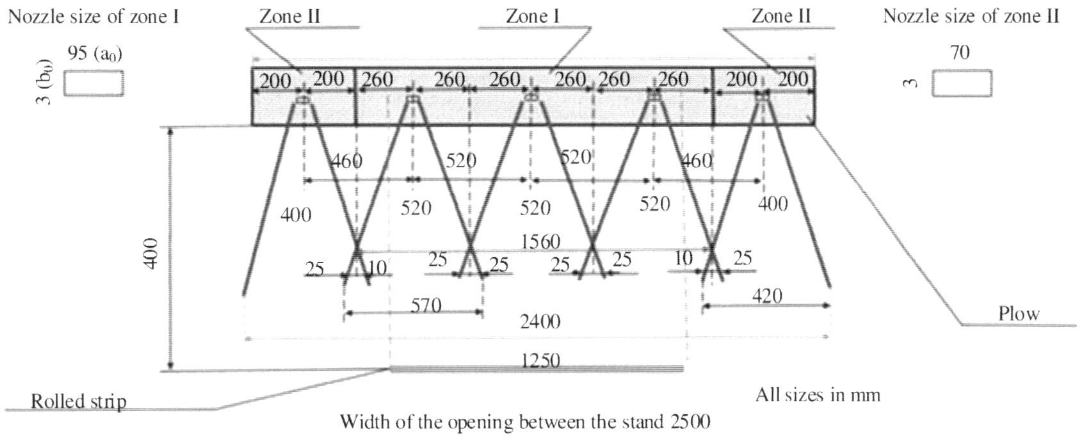

Figure 3. Scheme of a new plow for supplying compressed air.

This version of the plow will not only effectively remove cutting compound remains from the strip, but also save the compressed air consumption.

2. Production and installation of a new sticker box construction behind the IV stand (figure 4).

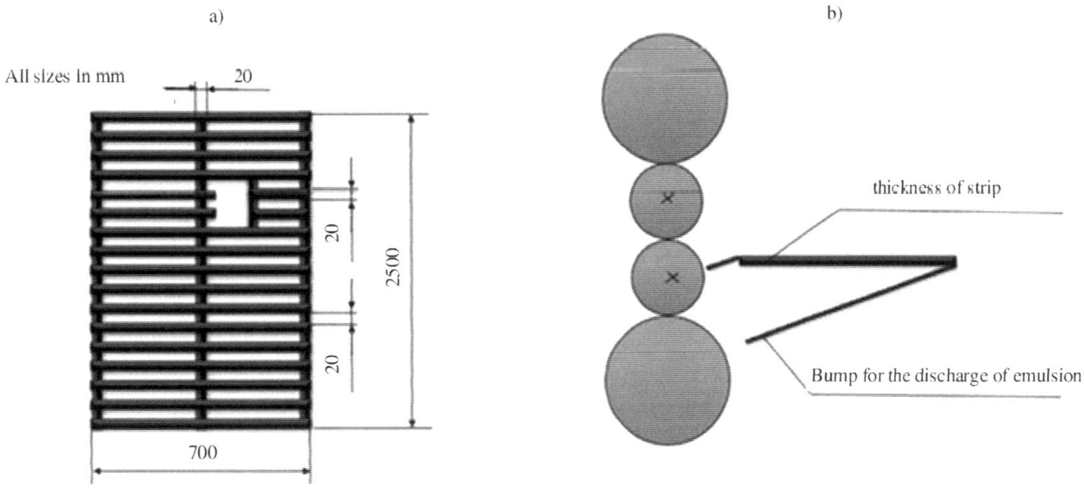

Figure 4. Advanced sticker a) box design b) installation scheme.

This version of the sticker box will allow to flow of emulsion with no hassle into the crankcase of the IV stand, then into the emulsion system, and also it will eliminate possibility of emulsion accumulation on the sticker box. The structural feature of this version of the sticker box will prevent the emulsion from getting into the coiler crankcase by installing a bump under the sticker box (figure 4).

The system for the strip cleaning from the cutting compound behind the IV stand will work according to figure 5 as a result of such implementation.

The emulsion falling on the sticker box by another methods does not accumulate on it and it gets with no hassle from the table on the bump into the crankcase of the IV stand, then into the emulsion system. Consequently, the cutting compound does not fall on the rolled strip during the blowing, with the formation of swirling air currents [2].

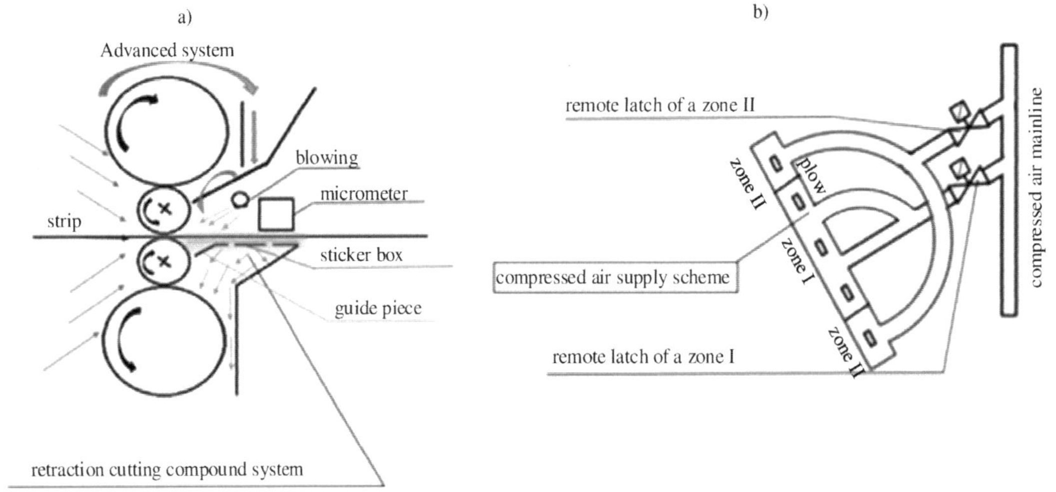

Figure 5. The scheme of the improved a) system for cleaning the strip behind the IV stand from the cutting compound b) the scheme for connecting the plow to the air line.

At installation of a new plow design (fig. 5) it is planned to eliminate the risks of the arising swirling air currents during the blowing for cleaning the rolled strip from the cutting compound. Such plow design is achieved due to its calculated length of 2360 mm and a blow-out area of 2400 mm, which is less than the width of the aperture between the base box, which is 2500 mm. It is important to divide the plow into 2 zones for saving of compressed air and reducing the risk of getting cutting compound to the rolled strip. The first zone (center) is designed for blowing with a width of the rolled strip in the range from 1000-1560 mm, the second zone (edges) is jointed for rolling a strip with a width of more than 1561 mm. For the on-line activation and de-activation of the compressed air supply are provided a remote opening and closing of the valve of the 1st and 2nd blow zones, the control unit of which is installed in the main post of the 4-stand mill (figure 5).

3. Calculation

Holes of rectangular cross-section are made on the plow proceeding from conditions and features at which the blowing will work.

Slot jets when flowing out of elongated rectangular holes with a ratio of sides $a_0/b_0>5$. The jet of compressed air formed at the outflow from the elongated rectangular hole is calculated as flat at a distance $x<6a_0$, where a_0 is the size of the larger side of the rectangular hole, and x is the distance from the nozzle to the strip [1,2].

The plow parameters are calculated.

For zone I:

$a_0 = 95$ mm, $b_0 = 3$ mm, $x = 400$ mm.

$$x<6a_0, \tag{1}$$

where a_0 is the size of the larger side of the rectangular hole; x is the distance from the nozzle to the strip.

Based on the formula the area of the compressed air coating of one nozzle is 570 mm (6*95).In order to prevent the emulsion from entering on the strip between the nozzle cover zones it is necessary to place the nozzles in Zone I 520 mm apart to provide a 25 mm overlap zone on each side. Based on the formula $x<6a_0$ (400<570) the distance from a plow nozzle to the rolled strip was accepted 400 mm for ensuring necessary parameters.

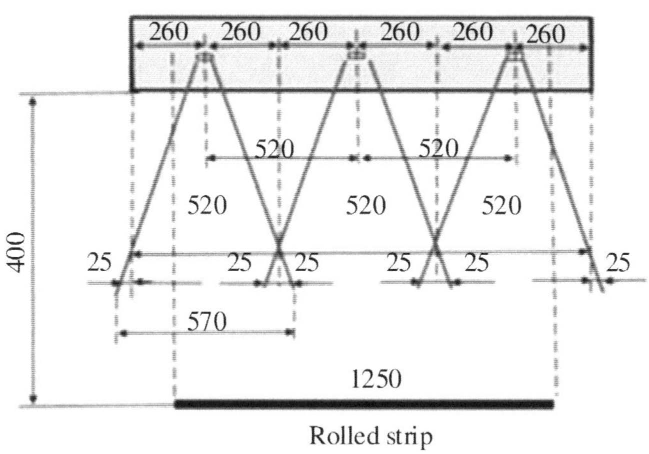

Figure 6. Distribution of nozzles on the plow in zone I

For zone II:

a_0 = 70 mm, b_0 = 3 mm, x = 400 mm.

Based on the formula (1) the area of the compressed air coating of one nozzle is 420 mm (6*70). In order to prevent the emulsion from entering on the strip between the nozzle cover zones it is necessary to place the nozzle in Zone II 460 mm from the edge of the nozzle of zone I to provide a 10 mm overlap zone.

Based on the formula (1) 400<420 the distance from a plow nozzle to the rolled strip was accepted 400 mm for ensuring necessary parameters.

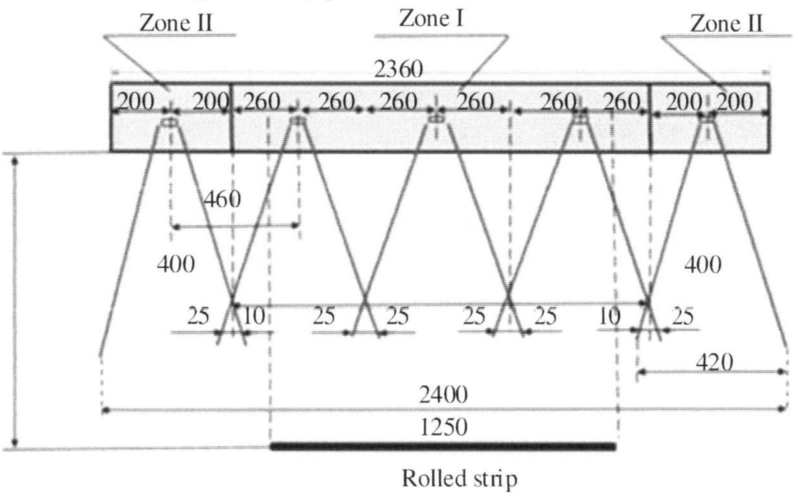

Figure 7. Distribution of nozzles on the plow in zone II

Calculation of compressed air flow is given below.

The main parameters for the calculation of compressed air flow are adopted:

Pressure is 5 atmospheres (0.51 MПа) and temperature. The available hours of mill.

For zone I:

a_0 = 95 mm, b_0 = 3 mm, x = 400 mm. 95/5>5, 400<6*95.

The flow capacity of the nozzle is calculated by the formula:

$$G = 3.16B_3\alpha_1F\sqrt{(P_1 + 0,1)\rho}, \quad (2)$$

where G is flow capacity, kg/h;

B_3 is coefficient that takes into account the physical and chemical properties of gases and vapors at operating parameters (for compressed air 0.770);

α_1 is the flow coefficient corresponding to the area F for atmosphere (0.66);
F is nozzle cross-sectional area, mm^2;
P_1 is pressure before the nozzle, MPa;
ρ is density of compressed air, kg/m^3.

$$\rho = \frac{(P_1+0.1)*10^6}{B_4RT_1}, \tag{3}$$

where B4 is compressibility factor of real gas (for air 1);
R is gas constant (for air 287);
T1 is temperature of the working medium before the nozzle, K.

$$\rho = \frac{(0.51 + 0,1) * 10^6}{1 * 287 * 283} = 7,51\frac{kg}{m^3},$$

$$G = 3.16 * 0.77 * 0.66 * (95 * 3)\sqrt{(0.51 + 0.1) * 7.51} = 979.65\ kg/h$$

Calculated by the formula the mass air flow is converted into a volumetric:

$$V = 979.63 \div 1.2 = 816.36\ m^3/h$$

(When the temperature of the compressed air in the main line decreases the density increases and the flow rate increases. At 0° C the flow in zone I is 831 m^3/h).

The volume of air passing through one nozzle for 5996 operating hours of the mill is calculated by the formula:

$$V^I = 816.36 * 5996 = 4894886.77\ m^3 = 4894.89\ thousand\ m^3$$

Volume of air passing at full functional of zone I (3 nozzles in operation) for 5996 hours is calculated by the formula:

$$V_{f.f.} = 4894.89 * 3 = 14684.66\ thousand\ m^3$$

For zone II:
a_0 = 70 mm, b_0 = 3 mm.
The width of the coating for blowing one nozzle is 420 mm (6a_0). For a covering of width over 1561 mm, two nozzles must be distributed on the plow at a distance of 460 mm from the edge nozzle of zone I. To avoid the formation of a thin jet of coolant on the strip the zone of overlapping air flows will be 10 mm (figure 7).

$$G = 3.16 * 0.77 * 0.66 * (70 * 3)\sqrt{(0.51 + 0.1) * 7.51} = 721.83\frac{kg}{h}$$

$$V = 721.83 \div 1.2 = 601.53\ m^3$$

Hour age for the production of metal with a width of more than 1561 mm in 2015 is calculated based on the ratio of the total number of tons for the average hot hour for this range of width:

$$t = 50826.96 \div 288.37 = 176.3\ hours$$

During the work of 2 nozzles the air consumption on production of wide metal will be:

$$G_{II} = 3601.53 * 2 * 176.3 = 212098.6\ m^3 = 212.1\ thousand\ m^3$$

The air consumption for the constant operation of II zones during 5996 hours will be:

$$G_{total} = 14684.66 + 212.1 = 14896.76\ thousand\ m^3$$

The air consumption for the constant operation of II zones during 5996 hours will be:

$$G'_{total} = 14684.66 + \left(\left(\frac{601.53}{1000}\right) * 2 * 5996\right) = 14684.66 + 7213.52 = 21898.18\ thousand\ m^3$$

According to economists the air flow at the current system was 38235 thousand m^3 in 2015.

Consequently for the production of wide metal (when zone II is connected) the savings in compressed air will be:

$$G_{sav} = 38235 - 14896.76 = 23338.24\ thousand\ m^3$$

According to the economist the cost of 1 thousand m^3 is 359.8 rubles.

Saving during the work of zone I for 5996 hours and connection of zone II for 176.3 hours in rubles will be:

$$E = 23338.24 * 359.8 = 8397099.15\ rub = 8.4\ million\ rubles$$

Reduction of compressed air consumption will be:

$$G_{reduction} = \frac{38235}{14896.76} = 2.57$$

During the continuous work of zone I and zone II for 5996 hours of the mill operation:

$$G'_e = 38235 - 21898.18 = 16336.83\ thousand\ m^3$$

Reduction of compressed air consumption during operation of zone I and zone II during 5996 hours:

$$G^{I} = 38235/21898.18 = 1.75$$

Saving during the continuous work of zone I and zone II for 5996 hours in rubles will be:

$$E'_{c.a.} = 16336.82 * 359.8 = 5877988.67\ rubles = 5.9\ million\ rubles$$

4. Conclusion

In the conclusion, the implementation of innovative production technology of an automotive sheet of an improved quality of a 4-stand cold-rolling mill will make it possible to reduce the consumption of compressed air by more than 2.5 times compared with the current system and significantly expand the range of products.

References

[1] Kondrashov S A, Golubchik E M and Martynova T Y 2017 *Proc. Int. Conf. Metallurgy Technologies Innovations Quality* vol 1 (Novokuznetsk: Siberian State Industrial University) p 144

[2] Kondrashov S A and Mamaev I N 2017 *The Proceedings Of The Distributors* vol 1, ed Spirin S A and Chuvikova L K (Magnitogorsk) p 116

[3] Polyakova M A and Golubchik E M 2015 *Conference On Structural Mechanical And Material Engineering* vol 19 p 17

[4] Polyakova M A, Gulin A E and Golubchik E M 2014 *Applied Mechanics and Materials* **656** 497

[5] Golubchik E M 2014 *Nosov Magnitogorsk State Technical University Bulletin* **1** 63–9

[6] Malapheev A V, Zaslavets B I, Igumenshev V A, Bulanova O V and Rotanova Yu N 2008 *South Ural State University Bulletin, Power Engineering Series* **9** 3–8

[7] Bulanova O V, Zaslavets B I, Igumenshev V A, Malapheev A V and Rotanova Yu N 2009 *Higher School Bulletin, Electro mechanics* **1** 60–5

[8] Karandaev A S, Rotanova Yu N, Kornilov G P, Karandaeva O I, Rovneyko V V and Galyamov R R 2009 *South Ural State University Bulletin, Power Engineering Series* **12** 16

IOP Conf. Series: Materials Science and Engineering **411** (2018) 012040 doi:10.1088/1757-899X/411/1/012040

Cr_3C_2 and VC influence on the structure and properties of WC-Co solid alloys

Yu Krutskii, S Veselov, N Cherkasova, V Chushenkov and R Vorobiev

Novosibirsk State Technical University, 20, K. Marks ave, Novosibirsk, 630073, Russia

E-mail: j_krutskii@rambler.ru

Abstract. In this paper, the effect of inhibiting additives of highly dispersed vanadium and chromium carbides on the structure and properties of the VC6 hard alloy was studied. It has been established that a correctly selected amount of additives (up to 1.5% by weight) can significantly reduce the grain size (from 2.5 to 1.5 μm) of the strengthening phase.

1. Introduction

Carbides are known to be the alloys of metals with carbon and represent the most important and the broadest class of inorganic compounds with high mechanical and physical-chemical characteristics. Now we should focus the attention on tungsten carbide (WC) and the alloys with cobalt on its base, for instance, WC-Co. Hardness parameters for tungsten carbide are relatively stable and in comparison with other carbides, they change slightly at the temperature increase up to ~ 1000 °C. Moreover, in contrast with carbides of other transition metals tungsten carbide has higher Young's modulus (~ 700GPa) and a lower coefficient of thermal expansion ($5.5 \cdot 10^{-6} K^{-1}$). It is the set of mentioned above characteristics and thermal stability that influence the use of WC as a base material in production of solid alloys suitable for metal working and drilling operations. Sintering is considered to be the most important stage of a conventional technique of solid alloys fabrication. The product is sintered from a pressed powder mixture with a plasticizer additive. After sintering, an average grain size of a carbide phase noticeably increases when compared with the grain size of the initial WC powder. Since fine-grained alloys display higher serviceability, the use of a fine-grained mixture for sintering will yield higher quality alloys. Alloying additives of other carbides are used to retard the grain growth in carbide phase of WC-C alloys. Growth inhibitors mainly find industrial application for sintering of WC-Co fine-grained solid alloys. The inhibitors could be arranged in the following order with regard to their relative efficiency: VC>Cr_3C_2>NbC>TaC [1].

Consequently, vanadium carbide (VC) and chromium carbide (Cr_3C_2) have the best effect on the grain growth. It is important to note that combination of VC with other additives ensures the most efficient growth control [2]. In spite of its best ability to inhibit grain growth and to increase the hardness of solid alloys significantly, some decrease in product strength and crack resistance could be observed. A negligible quantity of alloying additives (0.5-1.0 wt%) is added into the initial charge, but it's quite sufficient to improve the properties of solid alloys belonging to WC-Co system. The decrease in growth of tungsten carbide grains is explained by the fact that VC or other solid alloying additives segregate on WC crystal surfaces, that leads to reduction in WC solubility in Co by 2 – 4 times [3]. In turn, it impedes coalescence and formation of agglomerates during sintering [4]. Such an approach to inhibition of WC-Co grain growth appears to be the most perspective. It should be recognized that the studied solid alloying additives were used in nanodisperse state. The information about the use of growth inhibiting additives with micron size particles is practically not available. The objective of this research is to study how the inhibitors of tungsten carbide (vanadium and chromium carbides) grain growth can affect the structure and properties of sintered solid alloys.

2. Materials and experimental procedure

Vanadium ($VC_{0.88}$) and chromium (Cr_3C_2) carbides were synthesized by the carbothermal method via reactions:

$$V_2O_3+4.76C = 2VC_{0.88} + 3CO \quad (1)$$

$$3Cr_2O_3+13C = 2Cr_3C_2 + 9CO \quad (2)$$

The synthesis technique is characterized by the use of nanofibrous carbon (NFC), which is a new kind of a carbon material. NFC consists of 4-8 mm granules formed by closely interwoven carbon nanofibers with the diameters of 30-100 nm. NFC is quite pure: its impurities are the residues of the starting catalyst (90 wt % Ni/10 wt % Al_2O_3) and their content does not exceed 1.0 wt % [5]. To carry out synthesis, the NFC granules were ground in an agate mortar and sifted through a sieve with a mesh size of 100 μm. The specific surface area of the powdered NFC is approximately equal to150 000 m^2/kg, which is much higher compared with the specific surface area of lampblack (~ 50 000 m^2/ kg) [5, 6]. The other reagent was vanadium oxide (V_2O_3) (6-09-02-390-85 Specification, purity 99 wt. %, the average particle size is 0.92 μm). All of the starting reagents were dried in the oven at 100 °C, to remove moisture before the experiment.

The synthesis of the above mentioned carbides was conducted in a crucible-type induction furnace under the inert gas (argon) atmosphere, thus eliminating the probability of undesirable nitriding of the reaction products. The starting reagents were measured out in doses in accordance with the stoichiometry of the reactions (1) and (2). The starting powders were mixed by passing them through a sieve with a mesh size of 100 μm. One-phase solid alloys were obtained at the holding time of 20 minutes and at 1500 °C for vanadium carbide and 1300 °C for chromium carbide. The impurity content did not exceed 2 wt % [7, 8].

These carbides were used as alloying additives to fabricate products from WC-Co (6 wt. % of Co) solid alloy. Sedimentation analysis was carried out with the use of the MicroSizer 201 facility to determine the average sizes of the powder particles, the degree of dispersion, values of standard deviations and to obtain asymmetry indicator (skewness) histograms. The above mentioned parameters are determined by means of the «geometric» method described in [9]. The obtained data are presented in table 1.

Table 1. The results of sedimentation analysis.

Sample	The average size of particles, μm	Standard deviation	Skewness
WC-Co	1.68	1.71	0.03
vanadium carbide	4.07	2.5	-0.01
chromium carbide	7.10	2.1	-0.18

The values of the standard deviations for the powder of WC-Co solid alloy are indicative of the fact that the particles are «moderately sorted», i.e. they are distinguished by a relatively wide range of particle size distribution. The values of the standard deviations for the powders of carbides give evidence of a wide range of particle size distribution (they are «poorly sorted» or polydisperse). Comparatively low values of the asymmetry indicators must be a proof of the histograms symmetry.

The first stage of the powder mixture study presupposes preparation of suspension on the ethanol base. The powder content in the mixture equaled 60 wt %.Then, grain growth inhibitors were added into the WC-Co powder suspension. Vanadium and chromium carbides were chosen as the inhibitors in weight ratio of 1:1. Their content against WC-Co powder content was measured in wt %: 0.30 (0.15 + 0.15); 0.60 (0.30 + 0.30); 0.90 (0.45 + 0.45); 1.20 (0.60 + 0.60) и 1.50 (0.75 + 0.75). To achieve deagglomeration of WC-Co solid-alloyed mixtures and homogeneous distribution of inhibitors in WC-Co suspensions, they were exposed to ultrasound for 30 minutes with the use of 2kW generator.

The next stage presupposes granulation process to improve quality of the compacted samples (density increase). To increase strength of the pressed material, we added the PVAC100 bonding additive in proportion of 2wt % from the mass content of the powder. The suspension was dried at 110° C for 24 hours. Fritsch Analysette 3 vibrosieves were applied at the final stage of pressed-powder preparation. The size of the granules for molding was 100-250 µm. Mold pressing of the powder was done by uniaxial pressing with the «Instron 3369» universal electromechanical test machine, Great Britain. A metal press-mold was used for compacting the powder. The prepared compacts had the dimensions of 5x5x40mm The compacts were sintered in the electric vacuum furnace with resistance of СГВ- Liquid-phase sintering was carried out in vacuum of 2·10-4 mm Mercury, at 1380°C, for 40-minute holding time. The heating rate was equal to 5°C/min. The sintered samples were cooled inside the furnace.

The density of the sintered samples was determined by hydrostatic weighing method (Archimedean method). Bending strength is considered to be one the main mechanical characteristics of metal ceramics. Three-point bending tests were conducted with the «Instron 3369» universal electromechanical testing machine. Vickers hardness was measured with the use of the 402 MVD hardness meter under 200g load. Fabrication of polished sections from solid-alloys includes the following stages: 1.The samples were cut with «Discotom-65» (StruersA/S) machine. 2. Filling of the material was completed on the SimpliMet press. 3. The sample sections were ground for surfacing on each abrasive disk of the «Buehler, Ltd» grinding/polishing machine. 4. The sections were polished on the hand machine with each of the polishing pastes (60/40, 28/20, 14/10 and 1/0 dispersiveness) for ten minutes. The polished sections were etched for 5 minutes in the fresh mixture of potassium ferrocyanide ($K_3[Fe(CN)_6]$), potassium hydroxide (KOH) or NaOH 20% solutions used in equal proportions. Microstructure studies of the prepared samples were conducted with the CarlZeiss EVO 50 scanning electron microscope (SEM). The sintered sample from WC-Co solid alloy without additives was studied to make correlation between the results of the analogous investigations.

3. Results and discussion

Research results for all the series of experimental and reference samples made only from WC-Co solid alloy and from WC-Co solid alloy containing inhibiting additives are given in table 2. The experiments have proved that the materials containing vanadium and chromium carbides have lower strength and hardness values than WC-Co used as a reference material. At the same time, the addition of vanadium and chromium carbides results in significant microstructure refinement and, consequently, the additives act as grain growth inhibitors. Degradation of mechanical (strength and hardness) and physical characteristics (density) for the studied experimental samples series is related to the volume ratio of defects. Wedge-like discontinuities are thought to be the most common defects, which have greater impact on the material strength.

The analysis of changes in grain sizes revealed the efficiency of vanadium and chromium carbide mixtures used as grain growth inhibitors (figure 1). The size of the strengthening phase decreased from ≈2.5 µm up to 1.5 µm with the increase of the additives content. In these conditions, the maximum change in the grain size was registered for the material with 1.5 wt% of additives. At the same time, the greatest amount of macro-defects are observed (figure 2).

Table 2. Physical-mechanical properties and microstructure characteristics of WC-Co solid alloy with different content of alloying additives.

Sample	Relative density of a sintered sample, %	Hardness, HV	Flexure strength, MPa	Grain size,μm
WC	99.76	1810	760	2.5
WC + 0,15% VC + % 0,15 Cr_3C_2	83.65	1330	-	1.93
WC+0.3%VC + 0.3%Cr_3C_2	94.36	1510	-	1.91
WC+0.45%VC +0.45%Cr_3C_2	86.23	1510	562	1.76
WC +0.6 %VC + 0.6 %Cr_3C_2	84.79	1380	-	1.64
WC +0.75%VC + 0.75%Cr_3C_2	89.04	1070	400	1.5

(a) (b)

Figure 1. WC-Co structure with out (a) and with 0.75% VC + 0.75% Cr_3C_2 (b).

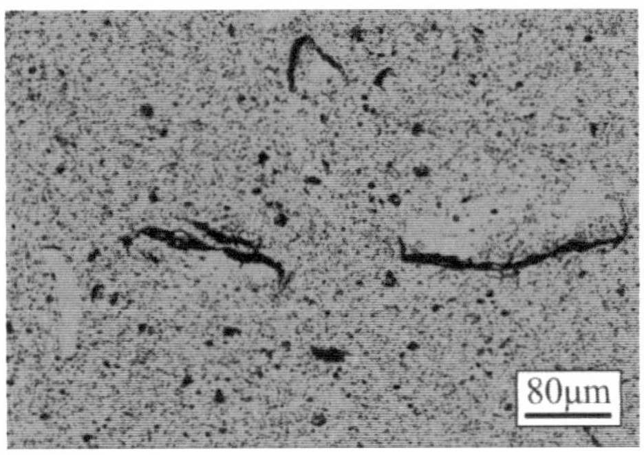

Figure 2. Typical defects of structures.

The morphology character of large defects is indicative of their formation just at the stage, when the experimental samples are being pressed. The dependence of samples micro-hardness on the content of a modifying additive is quite monotonous with the extremum for vanadium and chromium carbide content from 0.6 up to 0.9 % (1510 HV).

We managed to determine strength of materials only for the samples from series 4 and 6, and tensile strength at bending was equal to 562 and 400MPa, respectively. According to Hall-Petch relationship, the carbide phase refinement has to increase both strength and hardness of the material. However, the quality of the processing procedures has the greatest impact on mechanical and physical properties of brittle materials fabricated by powder metallurgy techniques. The defect structure formed in the material at the stage of pressing tends to affect the resulting properties of sintered materials.

The same results were obtained when vanadium carbide was used as an alloying additive.

4. Conclusion

The obtained characteristics of the sintered samples microstructure give evidence of the positive effect of highly dispersed vanadium and chromium carbides used as additives inhibiting tungsten carbide grain growth. Linear reduction of the grain size is detected due to the increase in alloying additives content. A minimum size of the major carbide phase is detected for the material with 1.5 wt% of vanadium and chromium carbide mixture.

Low rates of mechanical characteristics (strength, hardness) are indicative of high defect concentration in the sample material. It should be noted that high defect concentration of the samples depends on the material preparation procedures and prevents us from giving the correct estimation of the effect which the addition of chromium and vanadium carbides has on the mechanical-physical material properties. The objective of further investigations should be optimization of material preparation procedures aimed at the use of inhibiting additives from the transition metals carbides.

References

[1] Kurlov A S and Gusev A I 2013 *Physics and Chemistry of Tungsten Carbides* (Moscow: Fizmatlit)
[2] Morton C V, Wills D J and Stjernberg K 2005 *Int. J. of Refract. Met. and Hard Mater.* **23** 287–293
[3] Dai-hong X, Yue-hui H, Wei-Hong L and Min S 2009 *Transactions of Nonferrous Metals Society of China* **19** 1096–1105
[4] Fang Z Z, Wang X, Ryu T, Hwang K S and Sohn H Y 2009 *Int. J. of Refract. Met. and Hard Mater.* **27(2)** 288–299.
[5] Kuvshinov G G, Mogilnykh Y I, Kuvshinov D G, Yermakov D Y, Yermakova M A, Salanov A N and Rudina N A 1999 *Carbon* **37(8)** 1239–46
[6] Shumilova R G and Kosolapova T Y 1968 *Soviet Powder Metallurgy and Metal Ceramics* **7(11)** 906–909
[7] Krutskii Yu L, Kvashina T S, Chushenkov V I and Lozhkina E A 2017 *Materials of VIII International Scientific and practical conference «Innovations in mechanical engineering»* (Novosibirsk, Russia) 361–364
[8] Krutskii Y L, Bannov A G, Sokolov V V, Dyukova K D, Shinkarev V V, Ukhina A V and Kuvshinov G G 2013 *Nanotechnologies in Russia* **8(3-4)** 191–198
[9] Blott S J and Pye K 2001 *Earth surface processes and Landforms* **26(11)** 1237–48

IOP Conf. Series: Materials Science and Engineering **411** (2018) 012041 doi:10.1088/1757-899X/411/1/012041

Structure and properties of boron carbide ceramics fabricated by hot pressing

Yu Krutskii, A Tyurin , A Pel, N Cherkasova and R Kuzmin

Novosibirsk State Technical University, 20, K. Marks ave, Novosibirsk, 630073, Russia

E-mail: j_krutskii@rambler.ru

Abstract. The hot-pressing production technique of high-density ceramics on polydisperse boron carbide base with an average particle size of 2.1μm was studied. In some cases, a highly dispersed powder of chromium carbide was used as a sintering additive. Features of the structure were analyzed. The presence of pores between the grains was identified. Morphology of grains characteristic of highly dispersed boron carbide is observed.

1. Introduction

Boron carbide ($B_{12}C_3$ or B_4C) shows promise as the material with the high melting point, hardness, specific strength and chemical stability in aggressive media [1]. The coatings containing B4C are widely used and investigated [2, 3]. However its use in manufacture is limited by the complexities of high-density ceramics preparation.

The production technique of high-density ceramics on boron carbide base could be improved via sintering additives. Refractory oxygen-free compounds could find application as the sintering additives, such as chromium diboride (CrB_2) [4], titanium diboride (TiB_2) [5], chromium carbide (Cr_3C_2) [6]. Reaction sintering was carried out for an hour under an argon atmosphere at different temperatures [6]. The density of sintered products tended to grow with the increase of chromium carbide content and the temperature of sintering. The sample density reached 95% from the theoretical value at 2150 °C and Cr_3C_2 content in mixture of 25 wt%. The research objective is to study the sintering of polydisperse boron carbide by hot pressing and to examine some properties of the obtained ceramics.

2. Materials and experimental procedure

In some cases, a powder of boron carbide with the purity of ~ 98.5 % [7] and a sintering additive of Cr_3C_2 powder with ~ 98.0 % purity [8] were used as starting materials. Particle size distribution of boron and chromium carbides are shown in figure 1.

In compliance with the procedure described in [9], the average particle sizes, standard deviations and asymmetry indicators (skewness) were determined by means of the «geometric» method. The data are presented below in the table 1.

IOP Conf. Series: Materials Science and Engineering **411** (2018) 012041 doi:10.1088/1757-899X/411/1/012041

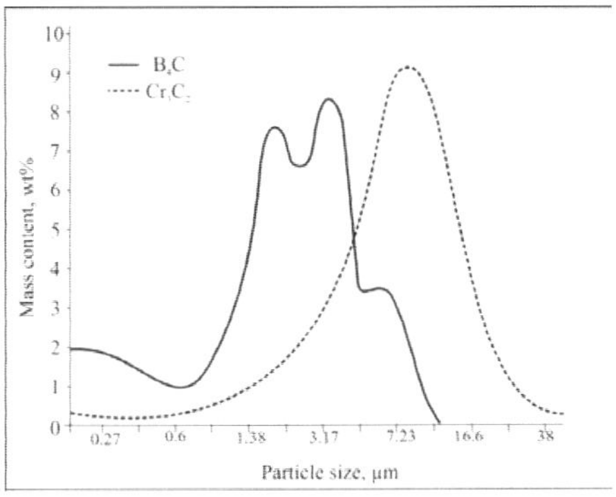

Figure 1. Particle size distribution of boron carbide and chromium carbide samples.

Table 1. The results of sedimentation analysis.

Sample	The average size of particles, μm	Standard deviation	Skewness
boron carbide	2.1	2.47	-0.20
chromium carbide	7.1	2.10	-0.18

The values of the standard deviation are indicative of a wide range of particle size distribution (the particles are "poorly sorted" or polydisperse). We consider that it could have a positive effect on the process of high-density ceramics fabrication. Low values of the asymmetry indicator (skewness) turn out to be the proof of the histograms symmetry.

To determine the values of bending and compressive strength, the work-pieces from the powder of boron carbide were exposed to hot pressing at the temperature of 2100 °C, under pressure of 35MPa, and 45minute-holding time. Volumetric porosity and water absorption capacity were also estimated for the samples obtained. The work-pieces were made from the samples in order to define the material mechanical characteristics (i.e. bending strength and compressive strength). Bending /compressive strength tests were carried out with the use of Instron machines 3369.

The other series of the tests was done to estimate density and microhardness of hot-pressed ceramics. Its microstructure was also examined. For comparison, we exposed to sintering not only boron carbide, but the mixtures with chromium carbide additives of 6 wt% and 20 wt%. Hot pressing was conducted in the argon flow, under pressure of 35MPa, at ~1950 °C. The temperature was set when taking into consideration the fact that boron carbide starts melting at this heat level [1] and, consequently it results in liquid-phase sintering. Microstructure studies of cleavage faces and polished sections were conducted with the use of CarlZeiss EVO50 scanning electron microscope (SEM). Microhardness was further determined on the polished sections. Microhardness was measured T with the use of 402 MVD (WolpertGroup) semiautomatic hardness meter.

3. Results and discussion

The values of volumetric porosity (0.01-0.02 %) and water absorption capacity (0.01 %) were extremely low for the sintered samples. The averaged values of bending and compressive strength were equal to 406 MPa and 1553 MPa for the samples under test, respectively. These averaged values are rather high, because the bending strength of boron carbide amounted to 340 MPa, and the compressive strength was in the range of 1100…1660 MPa.

While being sintered, boron carbide samples were found to shrink at ~ 1700 °C, but shrinkage of the samples with a sintering additive was observed at ~ 1500 °C. It's important to note that in accordance with [4], carbon oxide densification started at the temperature of 1800-1830 °C, but this process started at 1750 °C with the use of 10 wt% Cr_3C_2 sintering additive. Polydispersity (i.e. wide-range particle size distribution) and high dispersiveness of carbide powders are certain to be the reason of low-temperature shrinkage start. The values of the samples density ratio were estimated in percent (%): № 1 – 95.0; № 2 – 96.0; № 3 – 96.5. Consequently, the density ratio is slightly affected by the use of a sintering additive. Micrographs of samples with cleavage surfaces are shown in figure 2.

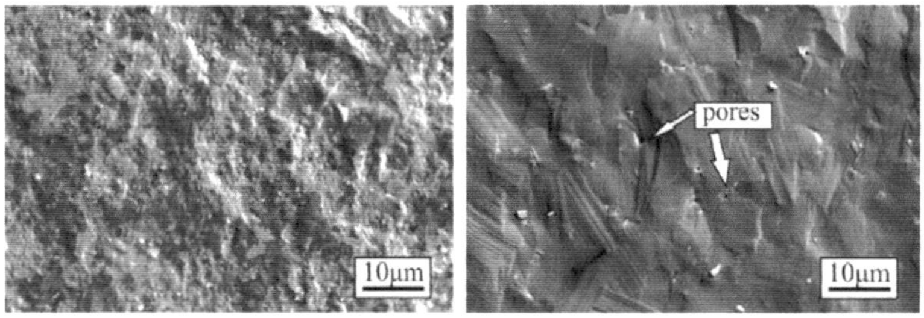

(a) (b)

Figure 2. Failure surface B_4C (a) and B_4C+6% Cr_3C_2 (b).

The analysis of fractured cleavage faces gives evidence of dense structure formation and the lack of large pores and discontinuities. Minor quantities of closed pores are found in the samples containing boron carbide. The pores or voids have a spherical shape and their volume concentration is quite small, therefore they don't have any negative impact on the material properties. Volumetric porosity tends to decrease considerably with the increase of carbon carbide content. A cast structure is also observed owing to the addition of alloying elements at the time of sintering. Grain boundaries are practically invisible.

Sample structure analysis allowed us to determine the grain size of sample №3 containing 20 wt% of chromium carbide. Its grain size equals 0.5 μm (figure 3). The grain size for the rest samples is nearly the same.

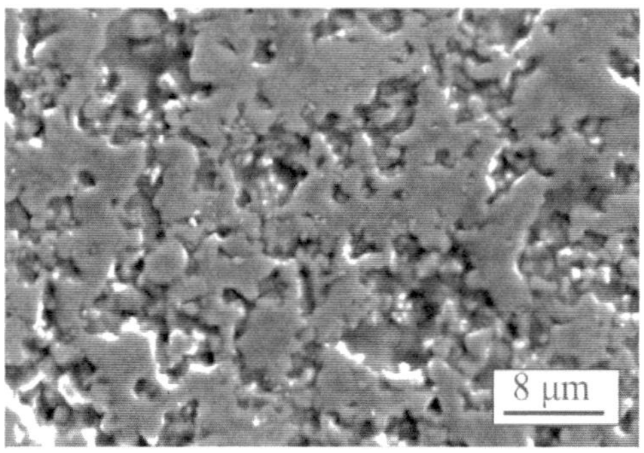

Figure 3. The structure of B_4C+ 20 % Cr_3C_2.

Microhardness reached the values of 41 ± 3.2GPa for sample №1, and 46 ± 2.8GPa for sample №2, and 45 ± 2.4GPa for sample № 3. The obtained results illustrate that the use of a sintering additive (in this case polydisperse chromium carbide) results in the increase of the sample microhardness. The microhardness values are comparatively high. They fall in the range of 29.7…47.9 GPa for the samples from boron carbide [1].

4. Conclusion

The article presents the study of the hot-pressing technique applied for polydisperse boron carbide and for chromium carbide used as a sintering additive. Some characteristics of the obtained ceramics are also reviewed. At relatively low pressure of hot-pressing (~ 35 MPa), the samples with the following quality attributes have been fabricated: volumetric porosity and water absorption capacity – not more than 0.02 %; the average value of bending strength – 406 MPa; the average value of compressive strength – 1553 MPa; relative density – 95-96%, microhardness – 42 GPa. The use of chromium carbide sintering additive resulted in the increase of microhardness up to 45-46 GPa.

References

[1] Kosolapova T 1986 *Properties, Use and Fabrication of High-melt Compounds* (Moscow: Metallurgy)
[2] Zimogliadova T A, Drobyaz E A, Golkovskii M G, Bataev V A, Durakov, V G and Cherkasova N Y 2016 *IOP Conf. Series: Materials Sci. and Eng.* **156(1)** 012017
[3] Kornienko E E, Kuz'min V I, Lozhkin V S, Gulyaev I P, Sivkov A A, Ivashutenko A S, Rahmatullin I A, Sergachev D V and Bezrukova V 2017 *Metal Working and Material Science* **3(76)** 42–50
[4] Yamada S, Hirao K, Yamauchi Y and Kanzaki S 2003 *Ceram. Int.* **29(3)** 299–304
[5] Nikzad L, Licheri R, Ebadzadeh T, Orrù R and Cao G 2012 *Ceram. Int.* **38(8)** 6469–80
[6] Li X, Jiang D, Lin Q, Chen Z Huang Z 2014 *J. Eur. Ceram Soc.* **34** 1073–81
[7] Krutskii Y L, Bannov A G, Sokolov V V, Dyukova K D, Shinkarev V V, Ukhina A V, Maksimovskii E A, Pichugin A Yu, Solov'ev E A, Krutskaya T M and Kuvshinov G G 2013 *Nanotechnologies* **8(3-4)** 191–198
[8] Krutskii Yu L, Dyukova K D, Bannov A G, Ykhina A V, Sokolov V V, Pichugin A Yu, Krutskaya T M, Netzkina O V, Samoylenko V V 2014 *Powder Metallurgy & Functional Coatings* **3** 3–8
[9] Blott S J and Pye K 2001 *Earth Surface Processes and Landforms* **26(11)** 1237-48

IOP Conf. Series: Materials Science and Engineering **411** (2018) 012042 doi:10.1088/1757-899X/411/1/012042

Metal loss reduction in the continuous furnace with a mechanized hearth due to the increase in uniformity of billet heating

O V Kuznetsova, K S Konoz, M V Temlyantsev, N V Temlyantsev, E Ya Zhivago and V V Dorofeev

Siberian State Industrial University, 42 Kirova street, Novokuznetsk, 654007, Russia

E-mail: uchebn_otdel@sibsiu.ru

Abstract. The analysis of the negative influence of the uniform heating of billets along the length in the continuous furnaces with a mechanized hearth on the work of heating furnaces, rolling mill and the quality of metal products is given. The influence of the non-uniformity of heating on metal loss is investigated. Based on the experimental data analysis of temperatures measurements of billets surface heated in the walking-hearth furnace, a statistical analysis was carried out and the equation of regression of the temperature distribution along the length of the billet was obtained using the least squares method. It is established that an increase in the non-uniformity of surface billet heating leads to an increase in the metal loss, and this dependence has a nonlinear character.

1. Introduction

Slabs and square billets are heated in the continuous furnaces of various designs before rolling. The most promising of them are continuous furnaces with a mechanized hearth (furnaces with a walking hearth FWH and furnaces with walking beams FWB). One of the important indicators of the quality of metal heating [1] in furnaces, along with the absence of cracks from temperature stresses [2], is the uniformity of temperature distribution along the thickness, length and perimeter of billets. A high heating non-uniformity of metal along the length of the billet is typical for continuous furnaces with a mechanized hearth, which is caused by gaps between the elements of the walking hearth or water-cooled walking beams [3, 4].

In FWH the appearance of local areas on the workpiece with a reduced temperature (“dark spots”) may be due to imperfection of the design or technical condition of the water closures. For example, in case of unsatisfactory operation of hydraulic locks, depending on the pressure regime in the furnace, atmospheric air can get inside through the gaps between the walking elements of the hearth, which causes cooling of the part of the workpiece located above the gap.

In FWB dark spots appear at the contact points of billets with water-cooled beams, and the magnitude of the uniformity largely depends on the design and material of the riders installed on the beams.

It is typical that the non-uniform heating of billets also appears during the rolling process. When fixing the currents I of the rolling mill engines, low-frequency oscillations are clearly visible, the number of which corresponds to the number of local sections of rolled products with a lowered temperature, and the magnitude of the load fluctuation depends on the magnitude of the temperature fluctuation.

High temperature non-uniformity along the length of billets has a number of negative consequences, negatively affecting the operation of heating furnaces, rolling mill and the quality of metal products:

- the non-uniformity of geometric dimensions and the mass of the running meter along the length of the workpiece;
- periodic increase in rolling forces and, as a result, an increase in the probability of rolls breakage;
- increase in the power consumption for rolling profiles;
- elimination of the possibility of expanding the assortment due to rolling high-precision and lightweight profiles, stable performance of finished rolled products in the field of negative tolerances;
- creates fluctuations in mechanical characteristics of rolled products along the length of the workpiece;
- complicates the setting of the mill, which leads to an increase in rejects and second-class quality in the geometric dimensions of profiles;
- increased metal loss and fuel consumption during heating.

The lower the metal temperature in the area of the dark spot, the higher the metal resistance to the plastic deformation, the force P of the mill and the energy consumption for rolling. A decrease in the temperature, for example, of 3ps steel from 1300 to 1150 °C, leads to an increase in the resistance of the steel plastic deformation and, accordingly, the rolling forces by 1.55 times on average by approximately 3% per 10 °C. In this regard, the need to obtain the required mass-average final temperature of metal (to roll dark spots) leads to the inevitability of raising the heating temperature of the rest of the workpiece with no dark spots. This leads to a general increase in the metal loss, steel decarburization, worsens the operating conditions of the furnace hearth lining, reduces the service life of the furnace.

Figure 1 shows schematically the dependence of the rolling forces and load of the stand motors on the workpiece temperature for various temperature non-uniformity along the workpiece length. At the same average metal temperature a more uniform heating (figure 1, c) leads to a lesser probability of grain growth, overheating, steel burning and melting of scale. It allows the cobbing to be intensified and the productivity of the rolling process to be increased. High non-uniformity of heating (figure 1, a), on the contrary, increases the probability of heating failure, leads to considerable fluctuations in the forces of the mill and the load of electric motors.

2. Methods of research

In the present work, the effect of non-uniformity of workpiece surface heating on the metal loss is investigated. Based on the experimental data of temperature measurements of the workpieces surface heated in the furnace with a walking hearth of one of the Russian metallurgical enterprises, statistical analysis was carried out and the equation of regression of the temperature distribution t over the length l of the workpiece was obtained using the least squares method:

$$t = t^{\circ} + A \cdot \cos(B \cdot l) \quad (1)$$

where $\overset{\circ}{t}$ – the average surface temperature of the workpiece, °C; A and B – empirical coefficients depending on the heating mode and the number of gaps between the elements of the hearth (heat dissipation sector).

3. Results and discussion

For example, figure 2 shows the results of one of the statistical processing options. In equation (1), t° = 1214 °C; A=60; B=3.76. The maximum value of the surface temperature is 1274 °C, the minimum is 1154 °C, the temperature difference Δt is 120 °C, respectively. The dependence has the form of a cosine wave. The ends of the billet are characterized by a higher heating temperature and the number

IOP Conf. Series: Materials Science and Engineering **411** (2018) 012042 doi:10.1088/1757-899X/411/1/012042

of temperature minima (dark spots) corresponds to the number and location of gaps between the elements of the hearth that have a cooling effect.

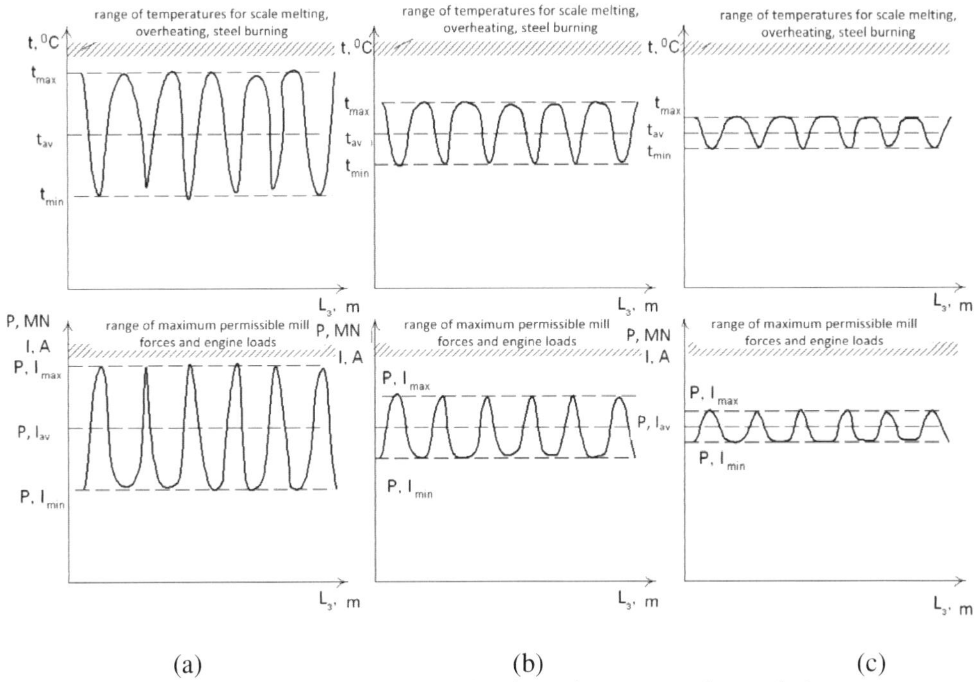

Figure 1. Dependence of rolling forces and loads of stand motors on the workpiece temperature for various temperature non-uniformity along the workpiece length.

Using equation (1) and the data on kinetics of oxidation of siliceous spring steels [5 – 10], an assessment of the effect of heating non-uniformity on metal loss was made. Comparative calculations were carried out holding steel 60S2KhA at the appropriate temperature along the length of the billet for 1 minute.

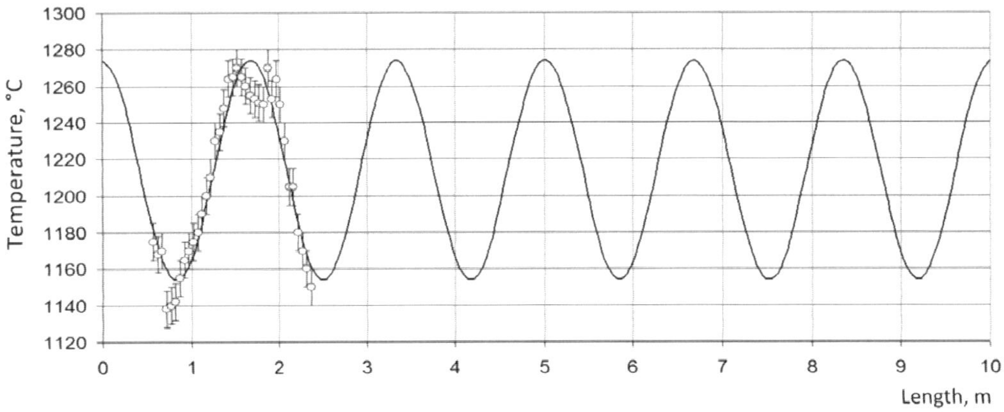

Figure 2. Temperature distribution along the length of the workpiece (points – experimental data, solid line – calculated).

Figure 3 shows the results of the study. It is established that an increase in the heating non-uniformity of the billet surface leads to an increase in metal loss, and this dependence has a nonlinear character. With an average temperature 1214 °C and a temperature difference 100 °C, the metal loss increases by almost 2% compared to non-uniformity heating, and at a drop of 200 °C by 7%.

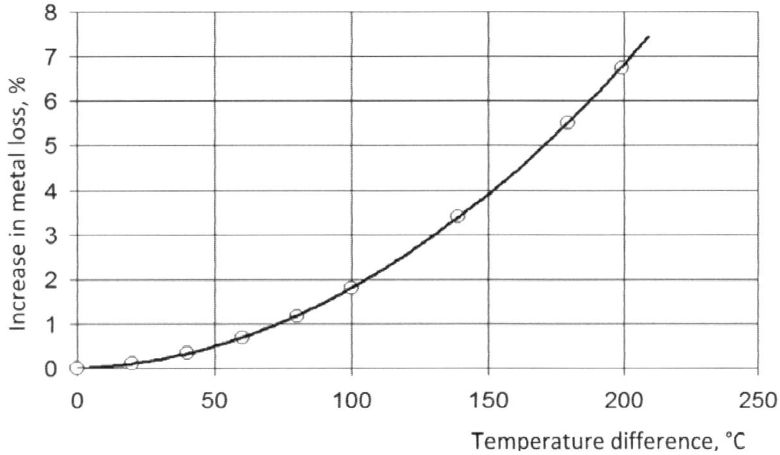

Figure 3. Dependence of the increase in metal loss on non-uniformity of heating of the surface workpiece.

4. Conclusion

It is established that an increase in the heating non-uniformity of the billet surface leads to an increase in metal loss, and this dependence has a nonlinear character.

References

[1] Temlyantsev M V et al 2005 *Izv. Vuzov. Ferrous Metallurgy* **6** 69–70
[2] Temlyantsev M V et al 2003 *Izv. Vuzov. Ferrous Metallurgy* **4** 56–58
[3] Kugushin A A et al 1980 *Steel* **10** 881–884
[4] Gusovsky V L et al 1970 *Steel* **9** 849–853
[5] Temlyantsev M V and Temlyantsev N V 2005 *Procurement Industries in Mechanical Engineering* **3** 50–52
[6] Temlyantsev M V and Temlyantsev N V 2005 *Izv. Vuzov. Ferrous Metallurgy* **2** 75–76
[7] Temlyantsev M V et al 2014 *Bulletin of the Mining and Metallurgical Section of the Russian Academy of Natural Sciences. Department of Metallurgy: Proceedings* (Novokuznetsk: SibSIU) 33 pp 55–63
[8] Temlyantsev M V and Mikhailenko Yu E 2006 *Oxidation and Decarburization of Steel in the Processes of Heating under Pressure Treatment* (Moscow: Teplotekhnik) p 200
[9] Peretyatko V N et al 2008 *Heating of Steel Slabs* (Moscow: Heat Engineer) p 192
[10] Filippova M V et al 2016 *Development and Implementation of Energy and Resource-saving Technologies for Metal Forming* (Novosibirsk: SB RAS) p 269

IOP Conf. Series: Materials Science and Engineering **411** (2018) 012043 doi:10.1088/1757-899X/411/1/012043

About the graphical modeling of production systems in metallurgy

S M Kulakov, A I Musatova, V N Kadykov, V I Bazaykin and A G Nikitin

Siberian State Industrial University, 42 Kirova street, Novokuznetsk, 654007, Russia

E-mail: kulakov-ais@mail.ru

Abstract. To create a regulatory basis for technical and economic indicators in situational planning and forecasting the operation of the production system, it is offered to use a set of interrelated graphic and mathematical models. Mathematical models include: formula-algorithmic normative models of time characteristics of production processes; table models of situational standards of process time characteristics by product types and mathematical normative models of situational technical and economic indicators of production systems. The main attention in this article is given to the task of constructing graphic models that visually reflect the spatial organization of material flows, the operational structure of production processes and the decomposition of production operations. The necessary stages for construction of a graphic model of spatial organization of material flows are determined. The technique for constructing a graphic model of production operations is given, which includes the decomposition of complex operations into elements and microelements. Each graphic model is accompanied by the description, purpose, justification and specific examples.

1. Introduction

For the adoption of rational management decisions in the conditions of operating automated production systems, regulatory models of their functioning are needed that can be presented in the form of a set of interrelated graphical and mathematical models that includes the following components:

1. Graphic models of production systems functioning, including the models of spatial organization of material flows, operational structure of production processes, decomposition models of production operations.

2. Structured multifactorial mathematical models of production systems functioning, to which the following are assigned:

- formula-algorithmic normative models of production process time characteristics (including the duration of technological, transport, labour, control operations, cycles);
- table models of situational standards of process time characteristics of by product types (assortment), for example [1, 2].
- mathematical models of technological processes, for example [3].

3. Graphical aggregate-temporal normative models of production systems functioning (complex graphs reflecting the sequence and parallel operation of equipment, labour actions of personnel, organization of material flows) [2].

4. Mathematical normative models of situational technical and economic indicators of production systems (PS), including: assessment of the PS performance per unit of time, taking into account the

assortment, the model of the PS work time fund, the standard production volumes, profit forecasts, profitability, etc., and also table models of normative technical and economic indicators.

In this paper, the main attention is paid to the procedure of constructing graphic models that are necessary for the development of normative mathematical models used for planning and forecasting the technical and economic indicators of PS operation.

There are known graphic models of different purposes, containing a set of interrelated elements [5 - 8]. The graphic model of the spatial organization of PS material flows is a conditionally graphic representation of the flow of materials (raw materials, semi-finished products, products), combined with a schematic layout of the equipment (units, machines, vehicles) of the specific PS (technological complex, shop, devisions, technological line).

The graphical model presented in the form of a scheme of material flows organization is necessary for the construction and rationale of a rational production structure with a different specification degree of the system (enterprise, subdivisions, units of subdivision) at all stages of its life cycle.

2. Stages of the graphic model construction

The construction of the graphic model includes the following steps:

a) preparation of background information:

- determination of the boundaries of the considered (investigated, simulated) PS; study of the PS plan (drawing) with the specification of equipment and technological instructions;
- PS decomposition into specialized divisions (departments, lines, sections, sites), taking into account their role in the production process, types and amount of equipment used (the example is given in the form of table 1);
- analysis of the range of products (planned products), indicating the share of each type for a certain period of time (year, quarter or month); study of technological routes (maps) for manufacturing all types of products by aggregates.

Table 1. The list and characteristics of PS production sections (fragment).

Name of sections	Purpoes of sections	Equipment of sections	
		name	quantity
1. Warehouse of workpieces	Receiving, unloading, storage, control, preparation and supply of blanks for heating	1. Electric overhead type crane	2
		2. Transfer trolley	2
		3. Racks for inspection and stripping blanks	4
		4. Table	2
2. Heating section	Heating of workpieces to the rolling temperature	1. Continuous furnace with walking beams	3
		2. Loading grate	3
		3. Furnace table	1
....	....	....	.....
8. Finished goods warehouse	....	....	.....

b) development of the main components of the organization scheme of PS material flows:

- construction of the enlarged diagram of interconnections of sections and PS units;
- performance of visual (monitoring) observations over the flow of materials at specific facilities (sections, PS units);
- verbal description of structural features and purpose of the main, auxiliary and transport equipment of PS and technological routes for all types of products;
- formation of the material flow scheme [4, 9] by integrating the technological routes by units taking into account the assortment (examples are shown in figures 1, 2);

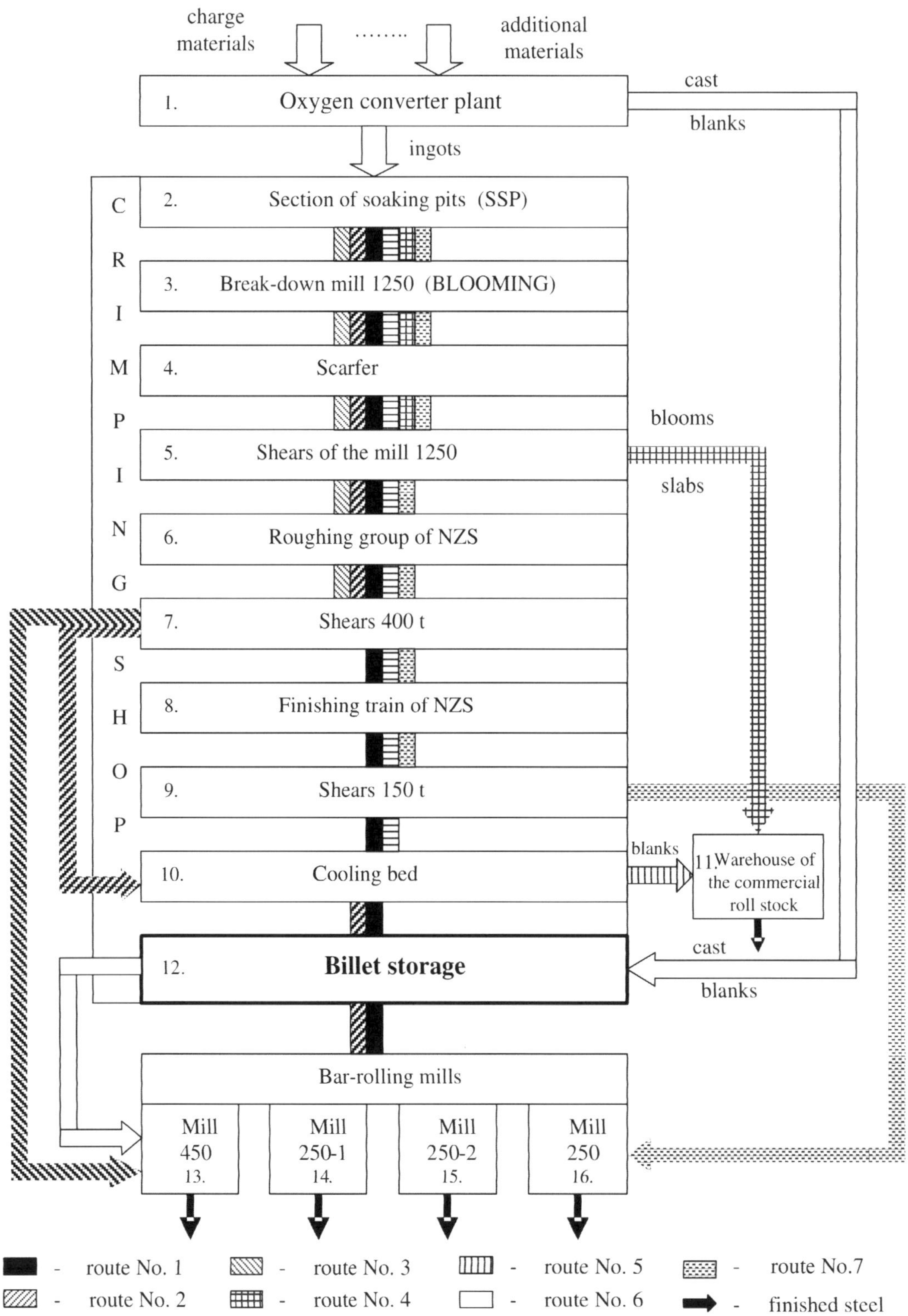

Figure 1. Integrated scheme of technological routes in rolling production.

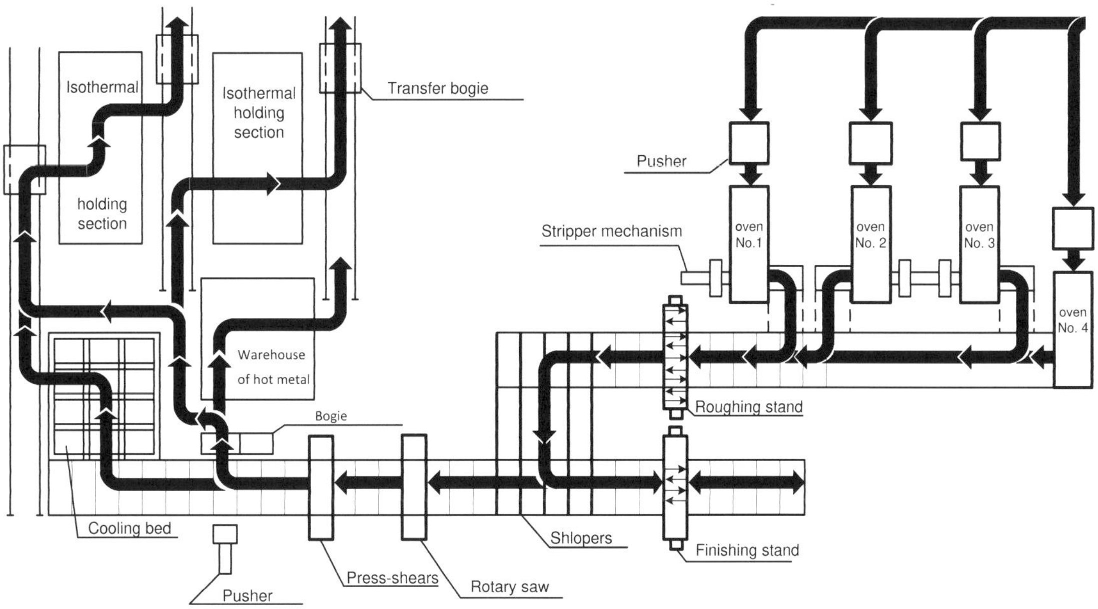

Figure 2. Scheme of the organization of material flows in the rolling shop (fragment).

3. Graphical model of the operational structure processes of PS

The graphical model of processes is a schematic representation of consecutive and parallel production operations and their connections on the basis of technological production routes and the model of organization of material flows. In accordance with the theory of graphs, this model can be called a directional production graph. For its construction, it is necessary to perform passive experiments (observations) in advance in order to study and represent the operations of the production process by sections and system units in the form of a tabular classification model of a set of operations, the example is presented in table 2.

Table 2. Classification of production operations by PS sections (fragment).

Index of operation[a]	The name of the operation	Type of the equipment	Borders of operation (moment)	Unit flow	Character of operation[b]
M	Stripping of the wire end	Wire pointer	The moment of turning on the machine – The moment of turning it off	Wire bundle	*A*
Ma	Preparation of a new coil for drawing	–	Starting time of restringing - The moment of its end	Wire	*A*
....	....	....	....	......	....
M	Drawing	Drawing station	The moment of turning on the machine – The moment of turning it off	Coiled wire	*T*

[a]*M* – mechanical, *M-Ma* – mechanical-manual, *Ma* – manual
[b]*A* – auxiliary, *T* – technological, *TP* – transportation, *K* – control

In the schematic representation of the operational graphic model, the following symbols of production operations are proposed:

IOP Conf. Series: Materials Science and Engineering **411** (2018) 012043 doi:10.1088/1757-899X/411/1/012043

 – *the main technological operation*, which is the main part of the production process, carried out in the leading unit, associated with the transformation of the raw materials (raw materials) into a semi-product (or product) with the required properties and characteristics (chemical composition, mechanical parameters, internal structure, shape, etc.). For example: *rolling a workpiece* in the work stand to produce a roll of a given cross-sectional shape; *melting of charge materials* in the furnace (converter) for the production of cast iron, steel or other alloy.

 – *auxiliary technological operation*, preceding or following the main operation. For example, preheating of billets in the furnace before rolling and operations following rolling: *cooling, cutting, baling*; preparation of charge materials for steel smelting in the converter and subsequent operations: *out-of-furnace processing of steel* (purging of argon steel in a special installation), *casting of steel* in a continuous casting machine for obtaining blanks, *slow cooling in chambers*.

 – *transport operation* associated with the movement of raw materials (materials, semi-finished products, packages, packs) inside the system, in particular: *unloading*, *loading*, *feeding*, *transfer* by electric bridge crane, conveyor, shlpeper, roller conveyor, transfer devices.

 – *control operation*, which includes an assessment of the characteristics of materials, semi-finished products, finished products and their verification for compliance with standards.

 – *element* of the operation (*refueling*, *preparation*, *installation*, etc.).

Each symbol contains: the name of the operation, reflecting the process (heating, rolling, smelting), type of the material (billet, ingot, cast iron, alloy), the name of the equipment used (methodic furnace with walking beams, continuous steel casting machine). For the transport operation the symbol indicates: the type of operation, the equipment used, the transportation object, the departure and destination points.

Solving problems of rating characteristics (indicators) of the production process requires a multilevel graphic description of each operation, based on its decomposition into elements and trace elements.

4. Decomposition graphical model of production operations

PS is a symbolic representation of the components of operations and their relationships. The decomposition makes it possible to investigate and describe the internal structure of operations, that is, to define and classify its elements and microelements, the sequence and parallelness of their fulfillment, to establish spatio-temporal *fixation points* corresponding to the moments of the beginning and ending of each element and microelement (actions, movements) and visualize the whole process as a whole. Examples of graphical models (the operational organization of PS production processes) in combination with the decomposition of operations are presented fragmentarily in figures 3 - 5, [1].

In the decomposition depiction of operations, additional symbols can be included that reflect the nature of their execution (the actions of the mechanisms, the workers – executors), figure 5. A table model is compiled that characterizes the elements and microelements of the PS operations, which in its form corresponds to table 2.

It should be noted that the decomposition of the operation is carried out on the basis of detailed timing (monitoring) observations with the obligatory fixation of the interaction of three components: the operation of the equipment (including individual mechanisms), the processing of materials (flow), the actions of the working personnel. For example, the study of the stroke of the break-down mill was carried out using the decomposition of technological operations of the rolling process.

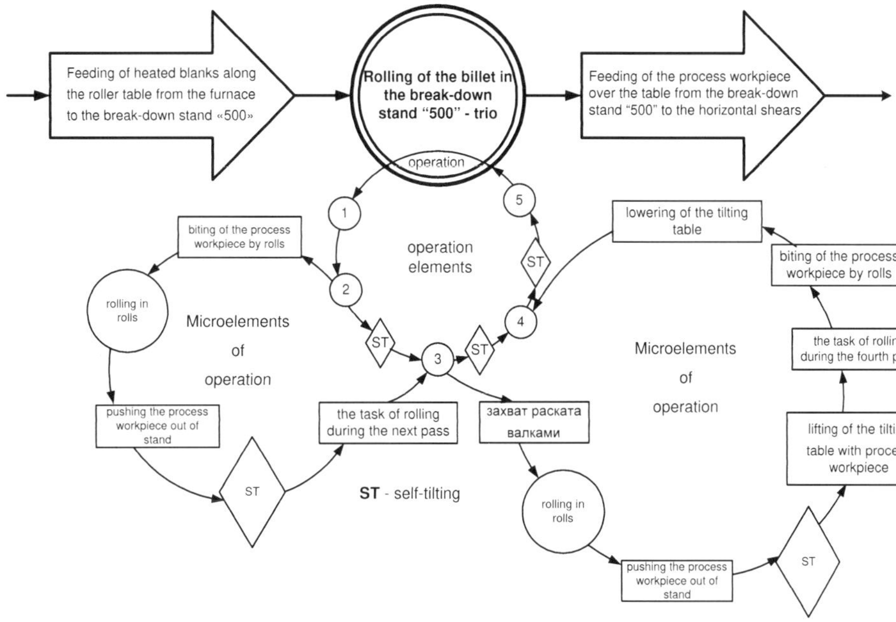

Figure 3. Decomposition of the production operation in the rolling shop.

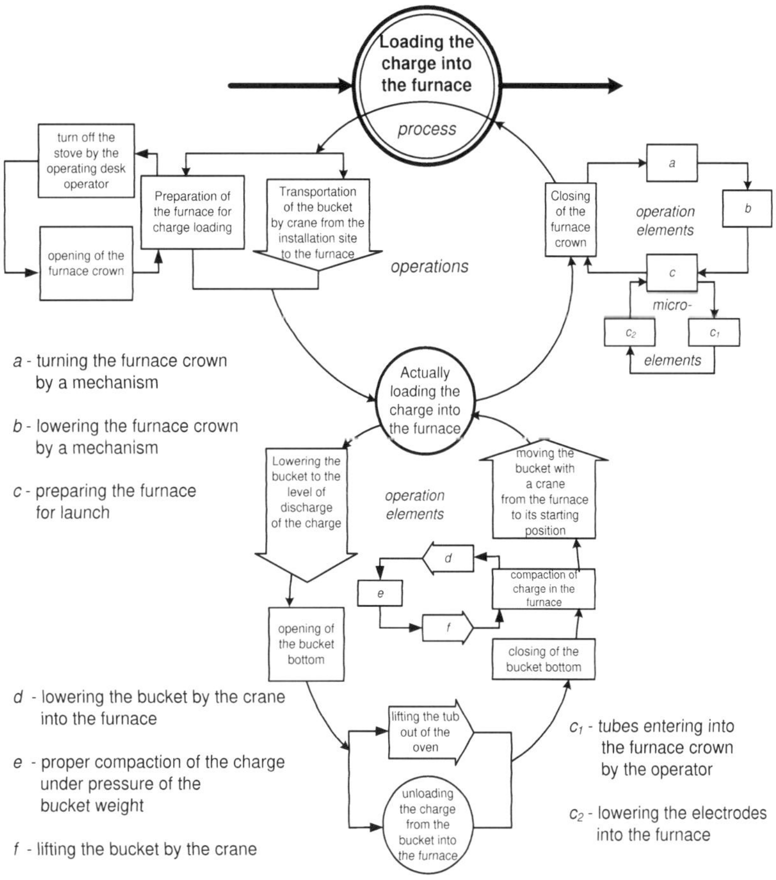

Figure 4. Decomposition of the process of charge loading into the arc steelmaking furnace.

IOP Conf. Series: Materials Science and Engineering **411** (2018) 012043 doi:10.1088/1757-899X/411/1/012043

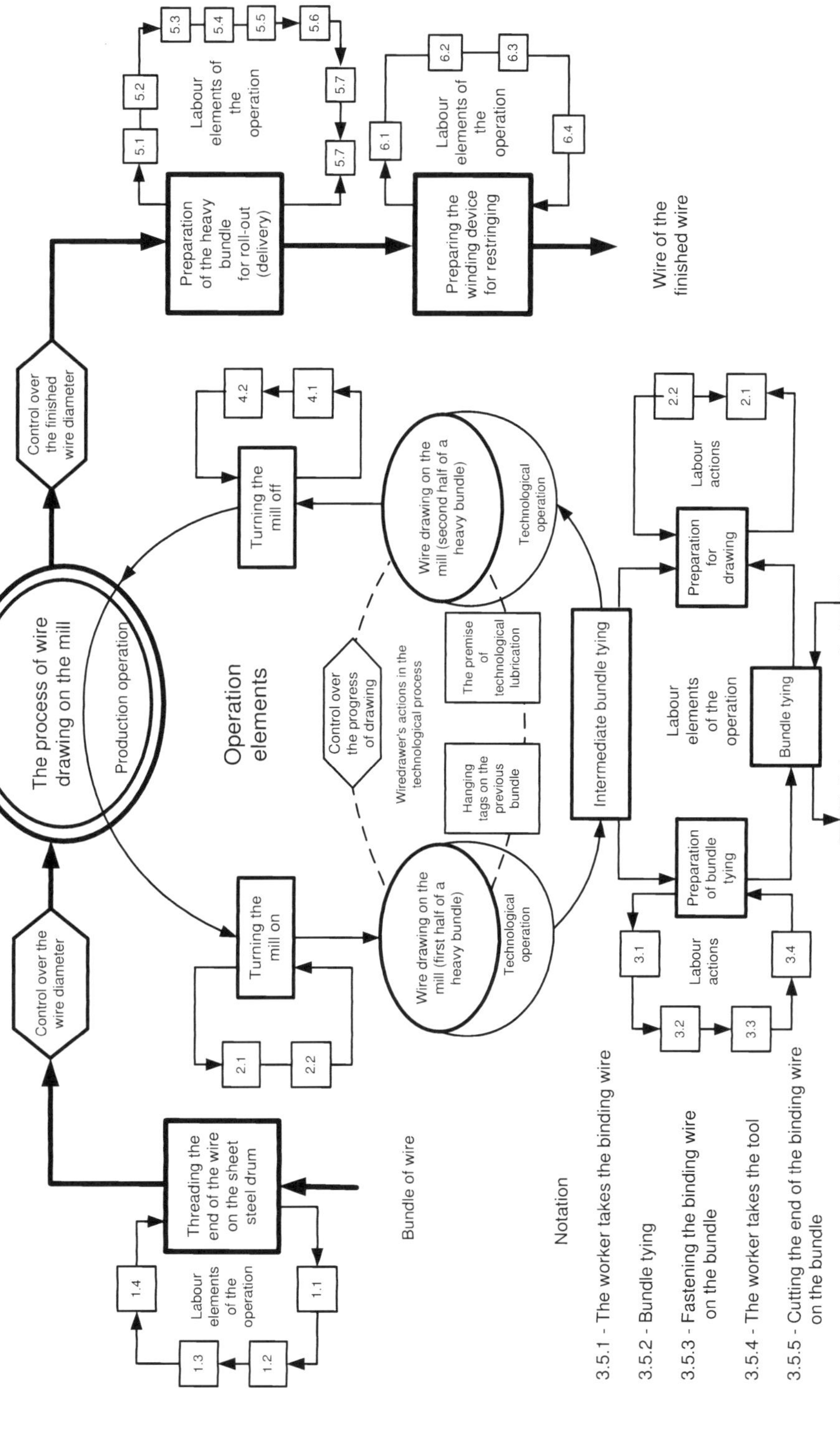

Figure 5. Decomposition of production operations of a drawing mill.

IOP Conf. Series: Materials Science and Engineering **411** (2018) 012043 doi:10.1088/1757-899X/411/1/012043

Fixation points were established in advance to visually determine the moments of beginning and end of the elements and microelements of operations. The two-variant definition of the *fixation points* of the elements of operations is presented in table 3.

Based on the decomposition (figure 6) and the selected methods for estimating the duration of the elements of operations, a model for the functioning of crimping stand was formed, then the regulatory durations of operations were determined, and the algorithm and calculations are given in [3].

Table 3. Determination of the fixation points of the operation elements.

For a unit of material flow			For a unit of equipment		
Name of operation elemnets	Fixation points of operation elements (moments)		Name of operation elemnets	Fixation points of operation elements (moments)	
	initial	final		initial	final
Initial pause	Stopping of the ingot before the stand	Beginning of the ingot biting by rolls	Stopping the ingot in before the stand. The time to prepare the work stand for rolling the ingot	Beginning of lifting of the upper stand roll	Setting the gap between the rollers for the first feed of the ingot (pass 1)
Deformation time of the ingot or rolling	Biting the front end of the ingot (rolling) with rolls	Exit of the rear end of the rolling from the rolls	Machine time rolling (skipping)	The beginning of the acceleration of the rolls' rotation during ingot capture (rolling)	End of deceleration of rolls after the release of the rolling stock
Pause between passes	Exit of the rear end of the rolling from the rolls	Biting the rear end of the roll with rolls after reverse	The preparation time the mechanisms for the next pass (the establishment of a given gap between the rollers)	Start slowing engine speed after the release of the rolling	The beginning of acceleration of rolls revolutions at capture of roll
The time for the tilting of the roll before the next pass	Contacting the roll with tilting finger	The end of the roll installation on the table	Running time of the tilting device	Beginning of the tilting device lifting from initial position	Start of the rotation of the working table rollers

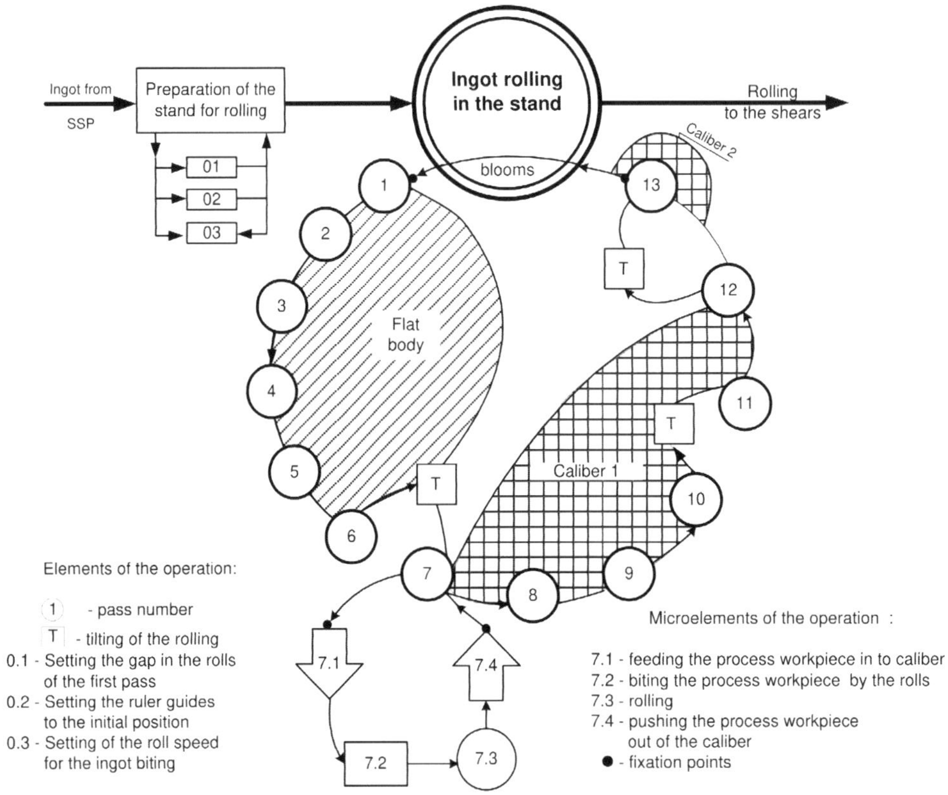

Figure 6. Decomposition scheme of the ingot rolling process.

5. Conclusion

In order to create a regulatory basis for technical and economic indicators for situational planning and forecasting the PS work, it was offered to use a set of interrelated graphic and mathematical models. In the paper the technique of formation of graphic models is described. It should be noted that the normative multivariant mathematical models for the PS functioning developed by the authors are applied in the management systems of objects of metallurgical enterprises and are published in papers [1 - 3, etc.]. Graphic models, which are the basis for creating the corresponding PS mathematical models, are not sufficiently described. Therefore, in this paper a step-by-step procedure for constructing graphic models, illustrated by concrete examples, is considered separately.

References

[1] Musatova A I and Kulakov S V 2014 *Scientific Review* **3** 139–149
[2] Musatova A I, Kulakov S M and Kadykov V N 2011 *Steel in Translation* **41(4)** 294–300
[3] Leushin I O 2013 *Modeling of Processes and Objects in Metallurgy: Textbook* (Moscow: Forum SIC INFRA-M) p 208
[4] Kulakov S M, Kadykov V N and Musatova A I 2003 *Izv. Vuzov. Ferrous Metallurgy* **12** 43–50
[5] Okunev L Ya 2016 *Modeling of Components and Elements of Integrated Circuits: Textbook* (St. Petersburg: Lan P) p 464
[6] Khripunov N V 2010 *Almanac of Modern Science and Education* **5** 144–147
[7] Fomina V V 2003 *Proc. on Education, Science, Production* (Ufa 2003) pp 42–44
[8] Rakov V S 2011 *Almanac of Modern Science and Education* **9** 37–41
[9] Zagorodnev D I 2010 *Scientific and Technical Bulletin of SPbGPU. Series "Science and Education"* **3(106)** 101–107

IOP Conf. Series: Materials Science and Engineering **411** (2018) 012044 doi:10.1088/1757-899X/411/1/012044

Influence of boron and nitrogen on tempered martensite embrittlement

A N Maznichevsky[1,2], R V Sprikut[1], A Ya Zaslavsky[1,2] and Yu N Goikhenberg[2]

[1] Lasmet Co, 2nd Paveletskaya street 18, Chelyabinsk, 454047, Russia
[2] South Ural State University, 76 Lenin ave, Chelyabinsk, 454080, Russia

E-mail: chiefteh@lasmet.ru

Abstract. The propensity of boron-containing and molybdenum-containing steels to tempered martensite embrittlement was carried out. Investigation conducted by serial curves of impact strength and fractographical studying of samples fracture obtained with a scanning electron microscope (SEM). Fracture of samples depending on conditions of their tempering (temperature and cooling rate) is determined. The study has shown that commercial boron-containing steel does not show a tendency to tempered martensite embrittlement. It was found that addition of boron into steel with a low concentration of nitrogen (0.004 mass%) leads to a decrease of impact strength, but does not embrittle steel. An increase of the nitrogen concentration to 0.016 wt. % leads to embrittlement, regardless of the boron concentration introduced up to 0.025% by weight.

1. Introduction

The phenomenon of tempered martensite embrittlement is known since middle of the last century. A considerable number of papers have been devoted to this problem [1 6]. At the same time, influence on tempered martensite embrittlement of such a microimpurity as boron introduced into steel in an amount of 0.0001 to 0.005 wh. %, i.e. much lower than phosphorus in the technical literature is mentioned little, and data given often contradict each other. For example, it is noted [7, 8] that boron promotes the development of tempered martensite embrittlement, and in [4, 9], on the contrary, boron suppresses it. The aim of this study is to clarify the effect of boron on the propensity of boron-containing steels to cold shortness and tempered martensite embrittlement. This investigation is carried out in comparison with molybdenum-containing steels (molybdenum suppresses tempered martensite embrittlement).

2. Materials and methods

A series of commercial (30Mn1B, 35MnTiB, 38CrVB, 40CrVTiB, 30CrMo, 40CrMo, 38CrMnMo, 40CrMnNiMo) and laboratory (40MnV-VI) meltings was investigated.

Commercial steel (one heat of each grade) was obtained in electric arc-furnace. Laboratory heats were obtained in a vacuum induction furnace VSG-30A with a crucible of 25 kg capacity. The melt was cast into a conical ingot with a diameter of 113 (upper) / 110 (bottom) × 220 mm and a mass of 20 kg. A total of 6 heats of 40MnV-VI steel with different microalloying were melted and investigated.

The ingots were rolled into square-section bars with a square side of 40 mm.

As a method of estimating propensity of steel to tempered martensite embrittlement, serial curves of impact strength tests carrying out at different temperatures was used. In our opinion and the authors' opinion [10], this method is more accurate than the method of evaluation of impact strength after tempering at different temperatures with subsequent tests only at room temperature.

Thermal treatment of the samples consisted of quenching in water at a temperature of 880 °C and tempering at temperature of the greatest propensity to embrittlement (cooling after tempering in water and with furnace).

Microfractograms were carried out on a scanning electron microscope JEOL JSM-6460 LV (South Ural State University). Impact strength were controlled in according to GOST 9454-78 on samples with a U-shaped concentrator (type 1).

The chemical composition of investigated steels is given in table 1.

Table 1. Chemical composition of investigated steels, wt. %.

Grade / Heat	C	Si	Mn	S	P	Cr	Ni	Mo	Cu	V	Ti	Al	B	N
Commercial boron-containing steel														
30Mn1B	0.31	0.14	1.20	0.009	0.015	0.27	0.06	0.01	0.06	0.006	0.017	0.026	0.005	0.006
35MnTiB	0.33	0.23	1.10	0.027	0.014	0.09	0.09	0.02	0.13	0.005	0.044	0.040	0.003	0.007
38CrVB	0.38	0.28	0.68	0.026	0.007	0.88	0.07	0.04	0.10	0.081	0.042	0.036	0.002	0.010
40CrVTiB	0.43	0.25	0.66	0.028	0.008	0.85	0.05	0.05	0.08	0.080	0.042	0.032	0.003	0.010
Commercial molybdenum-containing steel														
30CrMo	0.34	0.27	0.76	0.031	0.017	0.99	0.18	0.15	0.18	0.005	0.007	0.030	0.030	0.008
40CrMo	0.42	0.31	0.79	0.031	0.020	0.97	0.24	0.16	0.14	0.006	0.001	0.010	0.010	0.010
38CrMnMo	0.37	0.27	0.77	0.031	0.017	0.99	0.17	0.15	0.18	0.005	0.007	0.032	0.032	0.010
40CrMnNiMo	0.42	0.32	0.56	0.027	0.027	0.69	0.84	0.18	0.15	0.004	0.001	0.004	0.026	0.010
Laboratory boron-containing steel (40MnV-VI)														
43-1	0.36	0.18	0.96	0.015	0.014	0.28	0.02	0.02	0.05	0.07	tr.	0.018	0.001	0.004
43-2	0.35	-"-	-"-	-"-	-"-	-"-	-"-	-"-	-"-	-"-	tr.	-"-	0.002	-"-
43-3	0.35	-"-	-"-	-"-	-"-	-"-	-"-	-"-	-"-	-"-	0.027	-"-	0.002	-"-
44-1	0.35	0.19	0.99	0.015	0.014	0.36	0.02	0.02	0.08	0.07	tr.	0.013	0.001	0.016
44-2	0.36	-"-	-"-	-"-	-"-	-"-	-"-	-"-	-"-	-"-	tr.	-"-	0.002	-"-
44-3	0.36	-"-	-"-	-"-	-"-	-"-	-"-	-"-	-"-	-"-	0.028	-"-	0.002	-"-

-"- – the same concentration of chemical element in heat.
tr. – traces of element

3. Results and discussion

A comparative analysis propensity of boron-containing and molybdenum-containing steels to tempered martensite embrittlement was carried out in the following pairs: 30Mn1B - 30CrMo; 35MnTiB - 40CrMo; 38CrVB - 38CrMnMo; 40CrVTiB - 40CrMnNiMo. The results of impact strength tests are shown in figure 1.

After the tests, fractures of the samples were subjected to microfractographical studies. According to data obtained, 30Mn1B steel samples exhibited a fiber fracture at all test temperatures (figure 2a). At the same time, samples of steel 30CrMo (after cooling in water) observed a quasi-cleavage fracture (~ 15% of cleavage fracture) after testing at minus 20 ° C, and in the samples cooled with a furnace, at the same test temperature already 85-90% of cleavage fracture. The data obtained show that 30Mn1B and 30CrMo steels are not prone to tempered martensite embrittlement. However, along with this, steel 30CrMo was prone to cold shortness.

Cooling after tempering (in air or with furnace) does not have a significant effect on the appearance of 35MnTiB steel samples fractures, which in all cases remain shear (figure 2c). However, impact strength of samples were cooled with a furnace and tested at minus 75 °C, a lowered toughness was recorded (figure 1b), which is due to a sharp increase in the brittle component in fracture. Nevertheless, retaining a high proportion of the ductile component in all fractures of 35MnTiB steel samples. A significant deformation of specimens and shear fractures allows steel 35MnTiB that are not prone to tempered martensite embrittlement.

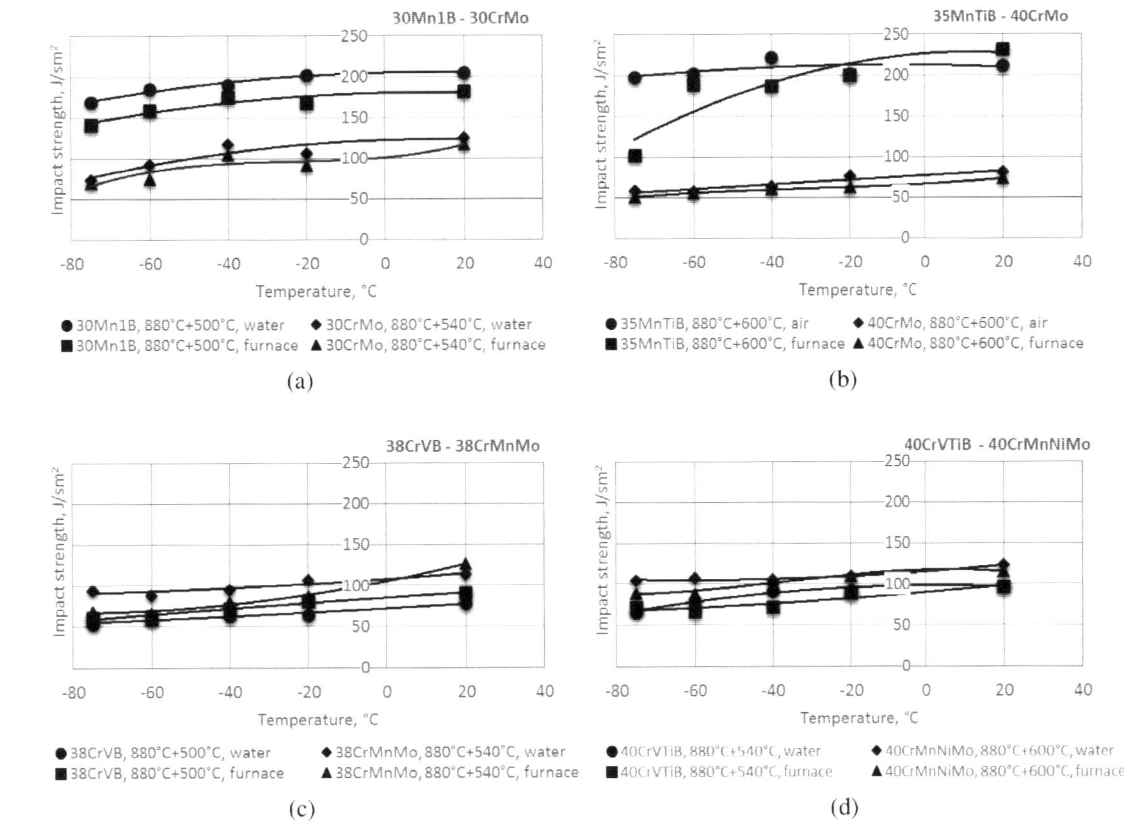

Figure 1. Serial curves of impact strength.

The serial curves of change impact strength for steel 40CrMo practically merge (see figure 1b). Magnitude of toughness decreases very insignificantly with decreasing temperature and without any sharp changes in test curves. Consequently, cooling rate after a two hour tempering at 600 °C does not affect on cold shortness or embrittlement of this steel. Despite alloying with molybdenum and a high tempering temperature, fractures of this steel had quasi-cleavage fracture after testing at negative temperatures (cleavage component predominates).

Microfractographical studies of steels 38CrVB, 38CrMnMo, 40CrVTiB and 40CrMnNiMo make it possible to draw similar conclusions about samples fracture at low temperatures occurred crystalline, mainly along the grain boundaries. However, it is important to note that boron-containing steels are not inferior to high cost nickel- and molybdenum-containing steels by impact strength. This circumstance makes it possible to consider boron-containing steels as an alternative to their replacement.

Combined effect analysis of boron, titanium and nitrogen on the tendency to tempered martensite embrittlement was carried out on laboratory fractional smelting (heat 43 and heat 44). Introduction of boron into steel without titanium (heat 43-2) and with titanium (heat 43-3) at a low concentration of nitrogen (0.004 wh. %) leads to a decrease of toughness (figure 4). Microfractographical study of samples fractures showed a large part of surface fracture has a shear component (table 2). Steel microalloyed with boron and with nitrogen of 0.014-0.020 wt. % (heat 44-2), characteristic for the metal of arc melting, retention of toughness value at the same level was observed (see figure 4).

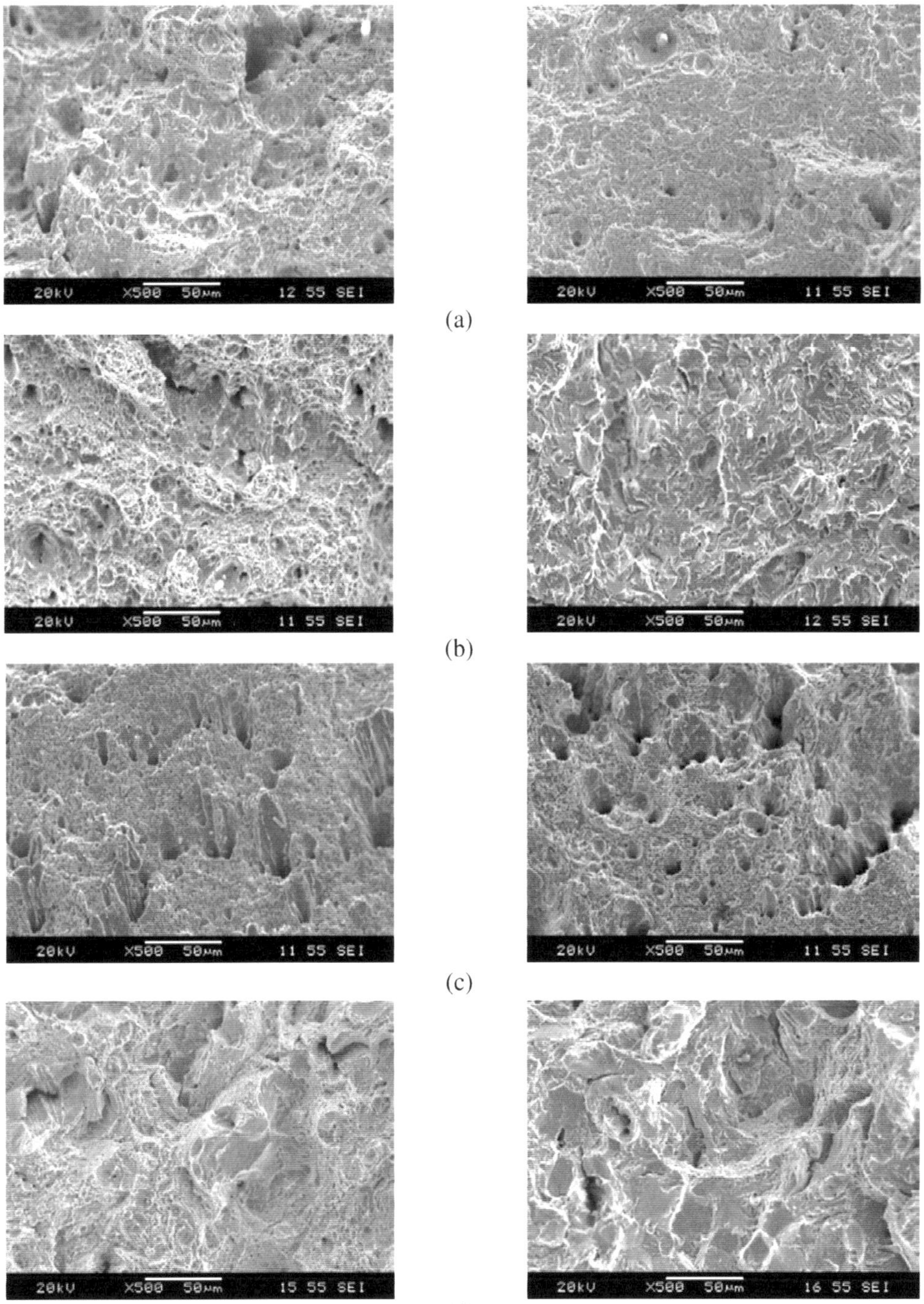

+20 °C -75 °C

a – 30Mn1B; b – 30CrMo; c – 35MnTiB; d – 40CrMo

Figure 2. Microfractograms of samples fracture at +20 °C (left) and -75 °C (right).

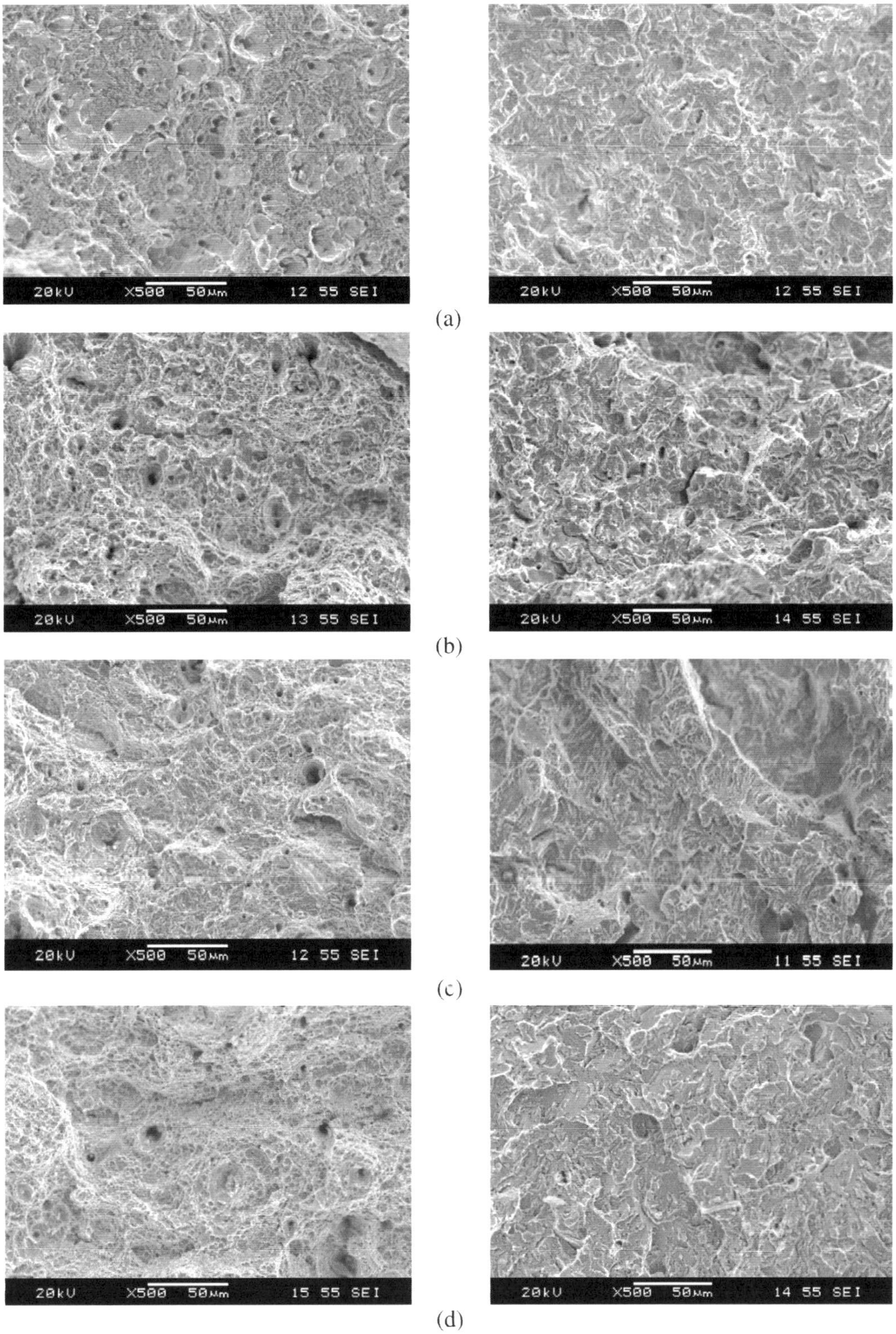

+20 °C -75 °C

a – 38CrVB; b – 38CrMnMo; c – 40CrVTiB; d – 40CrMnNiMo

Figure 3. Microfractograms of samples fracture at +20 °C (left) and -75 °C (right).

Table 2. Influence of microalloying on samples fracture.

Heat	Chemical composition, wt. %				shear component, %	cleavage component, %
	N	Al	B	Ti		
Laboratory boron-containing steel (40MnV-VI)						
43-1	0.004	0.018	< 0.0005	tr.	73	27
43-2	0.004	0.018	0.0016	tr.	70	30
43-3	0.004	0.018	0.0013	0.027	75	25
44-1	0.016	0.013	< 0.0005	tr.	25	75
44-2	0.016	0.013	0.0017	tr.	25	75
44-3	0.016	0.013	0.0021	0.028	25	75

tr. – traces of element

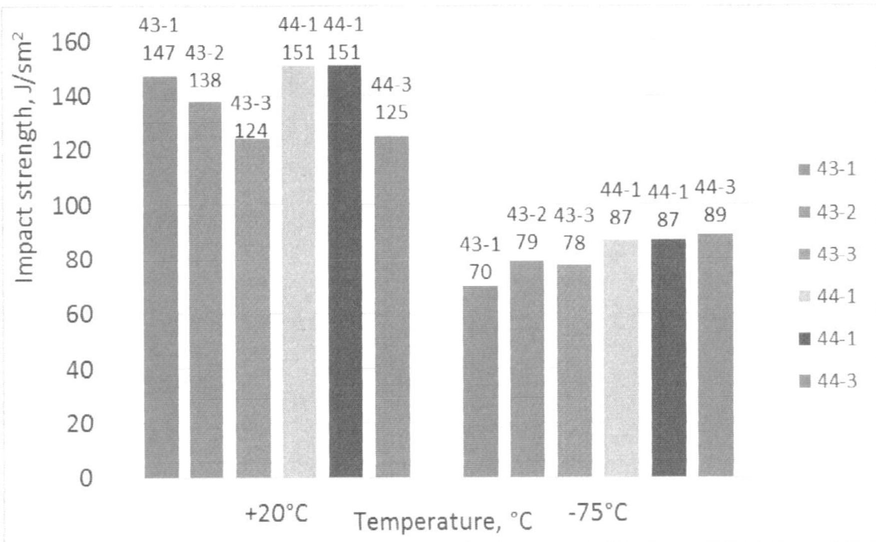

Figure 4. Microalloying influence on impact strength of 40MnV-VI.

Complex microalloying with boron and titanium (heat 44-3) showed a decrease value of impact strength by 15-20%. It is important to note that all three fractions (heat 44) have, in the main, a brittle fracture and only small areas of shear fracture (table 2).

4. Conclusion

The data given on the cold-shortness and sensitivity of steels to tempered martensite embrittlement indicate a significant superiority of steels microalloyed with boron over molybdenum-containing steels.

Boron-containing steels are not inclined to tempered martensite embrittlement and are not inferior to alloyed nickel- and molybdenum-containing steels under equal test conditions.

Introduction of boron into steel with a low concentration of nitrogen leads to a decrease of toughness without embrittlement of steel. Additional microalloying with titanium at a greater extent lowers the toughness, which is most likely due to formation of special titanium carbides along the grain boundaries.

It is noted that steel with a normal (for arc melting) nitrogen concentration (0.014-0.016 wt. %) without microalloying with boron and titanium has a significant effect on samples fracture (the samples have a predominantly brittle fracture) with a sufficiently high toughness. This circumstance indicates the embrittlement of nitrogen, which, when combined with nitride-forming elements (for example, chromium, vanadium or aluminum), is released along the grain boundaries. The introduction of boron into such a steel without microalloying with titanium does not lead to a change in the toughness or the type of fracture. And the introduction of boron into steel in the presence of titanium reduces the impact strength by 15-20% without changing the appearance of the fracture.

Acknowledgements
The work was supported by Act 211 Government of the Russian Federation, contract № 02.A03.21.0011.

Referemces

[1] Potak Ya M 1955 *Fragile Destruction of Steels and Steel Parts* (Moscow: Oborongiz) p 390
[2] Ustinovshchikov Yu I 1984 *Nature of Tempering Brittleness of Steels* (Moscow: Science) p 240
[3] Goldstein M I, Grachev S and Veksler Yu G 1985 *Special Steels* (Moscow: Metallurgy) p 408
[4] Utevsky L M, Glikman E E and Kark G S 1987 *Tempered Martensite Embrittlement of Steel and Iron Alloys* (Moscow: Metallurgy) p 222
[5] Bryant K L and Benerjee S K 1988 *Embrittlement of Constructional Steels and Alloys* (Moscow: Metallurgy) p 552
[6] Stolyarov V I and Laz'ko N V 1998 *High Energy Physics and Nuclear Physics* **22(11)** 25–8
[7] Zhang T, Hou H-X, Yi H-L, Chen J-P, Wang X-Q and Liu M 2011 *Jinshu Rechuli/Heat Treatment of Metals* **36(11)** 76–80
[8] Dudova N, Mishnev R and Kaibyshev R 2011 *ISIJ Int.* **51(11)** 1912–18
[9] Ovchinnikov D V, Sofrygina O A, Zhukova S Y, Pyshmintsev I Y and Bityukov S M 2011 *Steel in Translation* **41(4)** 356–60
[10] Smirnov M A, Schastlivtsev V M, Zhuravlev L G 1999 *The Fundamentals of Heat Treatment of Steel* (Ekaterinburg: UrO RAN) p 496

Features of roughness formation during production of the hot-rolled temper-rolled band

E M Golubchik, E M Medvedeva, A E Gulin and M A Polyakova

Nosov Magnitogorsk State Technical University, 38 Lenin Ave., Magnitogorsk, 455000, Russia

E-mail: fekla_med@mail.ru

Abstract. Active implementation of new technology processes is applied as one of way to increase the competitive ability of metal products. Today one of the modern trends is transfer of requirements of intermediate to final technological operations. The new production technology of a hot-rolled pickled temper-rolled band of a wide dimensional and branded range is developed by scientist of Nosov State Technical University jointly with specialists of PJSC Iron & Steel Works. Field of use this band is automobile wheel disk production at JSC AVTOVAZ. Researches on forming of band final properties showed a basic possibility of ensuring the rate band parameters at an exception of operation of cold rolling.

1. Introduction

The hot-rolled strip, inclusive pickled, annealed and temper-rolled is one of the mainstream metal product. This type of metal products is used for the production of welded building structures, pipes, car parts (in particular, wheel disks, rear suspension link connectors). This is due to the reduction in the number of technological process stage, in particular due to operation exception of cold rolling, and also the transformation of some mechanical properties and performance parameters of cold-rolled products to pickled hot-rolled products [1].

Studies on the development of a new production technology of the pickled hot-rolled strip with increased quality are conducted by scientists of the Nosov Magnitogorsk State Technical University and specialists of PJSC Iron & Steel Works [2,3].

2. Research objective

The developed technology (figure 1) makes it possible to produce a pickled tempered strip at the of PJSC Iron & Steel Works with using a continuous two-stage temper-rolling mill 630 (CTRM 630) to give uniform surface roughness.

The strip of 07GBYu steel in the size 3.5×410 mm has been made according to this technology. Its chemistry is presented in the table 1.

Further, according to presented scheme the hot-rolled steel with the dimensions of 3.5×1270 mm was subject to cutting to 410 mm length, hydrochloric pickling and temper rolling.

Process conditions of temper rolling are given in table 2.

3. Methodology

The anisotropy of mechanical properties and, in particular, roughness, was studied to establish the features of the final properties formation of hot-rolled pickled strip [4].

Measurements on the roll of the pilot lot were made every 50 meters along its length of 282 m in order to estimate the changes in the anisotropy of the roughness.

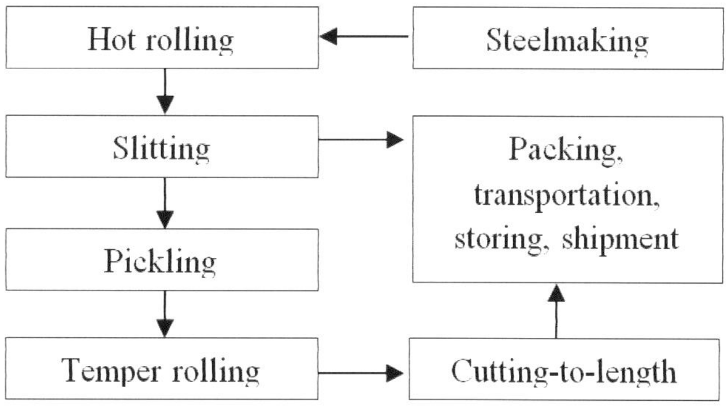

Figure 1. Scheme of hot-rolled strip production with temper rolling.

Table 1. Chemistry of 07GBYu steel.

Chemical elements content, %											
C	Si	Mn	S	P	Cr	Ni	Cu	N2	Al	Nb	Ti
0.08	0.28	0.69	0.003	0.012	0.04	0.02	0.04	0.005	0.038	0.020	0.015

Table 2. Process conditions of temper rolling for 07GBYu steel.

Thickness, mm	Roughness of working rolls Ra, μm	Linear speed of rolling, m/s	Percent reduction, %	Coiler tension	Decoiler tension
3.5	3.3-3.5	10.1	1.3	7.0	9.0
		10.2	1.2		
		10.1	1.4		
		10.3	1.4		
		10.4	1.3		
		10.7	1.3		
		10.5	1.5		
		10.5	1.3		

Surface roughness measurement was performed on a roughness integrator HOMMEL-TESTER T8000, as well as on a portable Surtronic 25. According to modern customer requirements the measurement was carried out at a base length of 2.5 mm, in directions oriented at 0 °, 45 ° and 90 ° to the rolling direction. The measurements were made at 5 points. The mean value was taken from the arithmetic mean of 5 measurements. The results of the average values of Ra are given in table 3.

It can be seen from the presented results that the anisotropy of the surface roughness before temper rolling between different angles to the rolling direction with respect to the averaged value of the Ra index is of the order of 0.15-0.25 μm and after temper rolling 0.07-0.15 μm. In addition to reducing the roughness anisotropy, temper rolling increases the degree of surface homogeneity ($\Delta Ra_{befortemperrolling} \approx$ 0.4 μm, $\Delta Ra_{aftertemperrolling} \approx$ 0.3 μm) and reduces the general level of roughness by 0.2 μm.

Besides, according to ISO 4287:1997, the roughness estimate can be performed in 3D tasks, evaluating the Primary profile as a rough profile (Roughness profile) and a Waviness profile. The ISO 4287:1997 standard defines three sets of parameters: P-parameters calculated on the basis of the primary profile, R-parameters calculated from the roughness profile, and W-parameters calculated on the basis of the ripple profile.

Table 3. Ra2.5 roughness measurement results.

Sampling site	Sample orientation	Measuring point along the length of the strip, m					
		0	60	110	160	220	270
Before temper rolling	0° (along)	1.71	1.78	1.71	1.8	1.63	1.71
	90° (across)	2	1.89	2.1	1.97	1.92	1.88
	45°	1.96	1.85	1.9	1.86	1.72	1.73
After temper rolling	0° (along)	1.6	1.52	1.56	1.5	1.47	1.5
	90° (across)	1.69	1.77	1.71	1.59	1.68	1.62
	45°	1.72	1.64	1.56	1.52	1.57	1.61

The study of these parameters was carried out using a modern scanning probe microscope (SPM) at the research institute "Nanosteels" of the Nosov Magnitogorsk State Technical University. The measurements were carried out on specially prepared samples, which were taken from the initial preform before and after temper rolling. The sample's surface has features characteristic of the entire strip. The maximum area of the base of study sample is 1.5 cm^2, the maximum height is 10 mm. When scanning samples the microscope probe was positioned at a right angle to the investigated surface.

To estimate the roughness parameters of the investigated samples the diagonal segment was selected as the sample length according to figure 2. The quantitative altitude distribution is based on the entire area of the image.

As a result of scanning SPM images were obtained which were visualized using computer graphics mainly in the form of two-dimensional brightness (2D) images of roughness profiles (figure 3). With 2D visualization each point of the surface Z = f(x, y) is associated with a tone of a certain color (or position in the scale of a multicolor palette) depending on the height (z-coordinate) of the surface point.

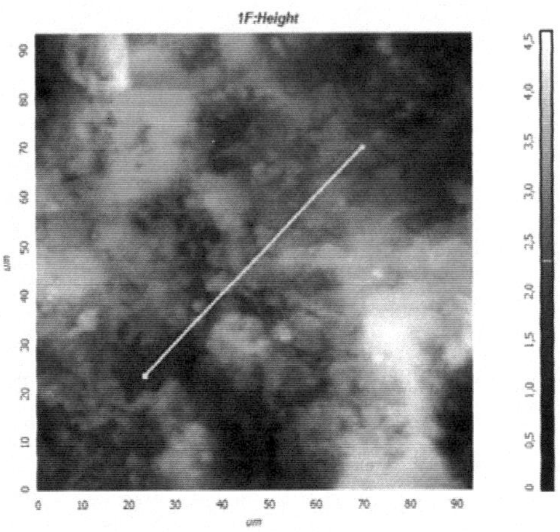

Figure 2. The sample length location for the roughness parameters estimating.

IOP Conf. Series: Materials Science and Engineering **411** (2018) 012045 doi:10.1088/1757-899X/411/1/012045

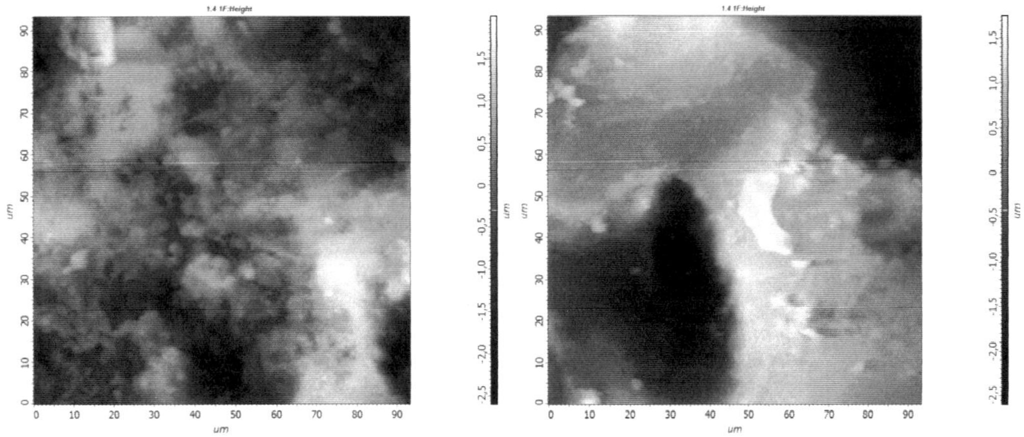

(a) (b)

Figure 3. SPM images of interested area (a – before temper rolling, b – after temper rolling).

Thus, from the presented graphical results, we can see a visual decrease in the roughness level and uniformity of the surface, which is confirmed by the numerical calculations presented above.

4. Conclusion

Figure 4 presents a comparative analysis of the experimental strip (a) with the serial tape (b). The results of processing the rolled stock into the finished product are shown in figure 5.

(a) (b)

Figure 4. Tempered strip surface of pilot run (a) and non-tempered strip surface of batch (b).

Based on the visual estimate, the tempered strip has a higher uniformity of roughness without the line of fracture, which is favorable for the process of paint coating applying.

Figure 5. Stamped wheel from pilot run of pickled temper-rolled strip of 07GBYu steel for the LADA automobile of JSC Avtovaz.

Thus, the carried out researches have established the formation features of the finished pickled hot-rolled strip and also ways of its management, for example by gage interference reducing, on-line management of the reduction process, more detailed processing of the strip ends with cutting out of off-grade pieces.

References

[1] Golubchik E M, Medvedeva E M and Martynova T Yu 2017 *Processing of solid and laminate materials* **1(46)** 59–4
[2] Golubchik E M, Medvedeva E M, Telegin V E and Vasil'ev I S 2015 *Proceedings of the 12th all-russian scientific and practical conference "Modern problems of the mining and metallurgical complex. science and production"* **1** 208–12
[3] Golubchik E M, Medvedeva E M and Telegin V E 2016 Proceedings of *the International scientific and practical conference "New solutions in the field of strengthening technologies: the view of young professionals"* **1** 254–58
[4] Golubchik E M and Medvedeva E M 2016 *Materials of the V International scientific and technical conference of young scientists, post-graduates and students "High technologies in modern science and technology"* 138–9

IOP Conf. Series: Materials Science and Engineering **411** (2018) 012046 doi:10.1088/1757-899X/411/1/012046

Formation and emission of sulfur dioxide in aluminum production

M Ya Mintsis[1], G V Galevsky[1], V V Rudneva[1] and S G Galevsky[2]

[1]Siberian State Industrial University, 42 Kirova street, Novokuznetsk, 654007, Russia
[2]St. Petersburg Mining University, V. O., 21 Line, 2, St. Petersburg, 199106, Russia

E-mail: kafcmet@sibsiu.ru

Abstract. The sources of formation of sulfur dioxide during the production of anodic mass (baked anodes) and aluminum by electrolysis are analyzed; the method for calculation of the formed sulfur dioxide anhydride depending on the quality of the used raw materials is presented. Total emissions of sulfur dioxide are estimated for different methods of anode mass production, electrolysis cell structures and methods for purifying waste gases.

1. Introduction

In the production of anode mass or baked anodes for aluminum electrolysis cells, electrode (pitch or petroleum) coke and coal tar pitch are used. Fuel oil or natural gas is used for baking or drying coke. This raw material and fuel contains sulfur in its composition, which is the source for sulfur dioxide formation.

Depending on the recipe of the finished product, 1 tonne of electrode production is consumed: baked coke (oil or pitch) – 700-840 kg, coal tar pitch – 180-310 kg. For baking 1 tonne of raw electrode coke 15-50 kg of fuel oil or 20-65 nm^3 of natural gas is consumed, and when coke is dried, the fuel consumption is reduced by 3-4 times. Pitch coke contains from 0.1 to 0.7% sulfur, and petroleum coke – 0.3% -1.6%. The sulfur content in fuel oil, depending on its grade, is from 0.5 to 3.5%, and in natural gas it is usually not great and can be neglected. To determine the amount of sulfur, the reporting indicators for the consumption and quality of raw materials in a particular production should be used. When raw coke is used, it is preliminarily subjected to baking at a temperature of 1100 - 1250 °C; if a pre-baked coke is used, it is dried at a temperature of 250 to 400 °C. As shown by numerous studies, baking of coke at a temperature up to 1250 °C reduces the sulfur content by no more than 5%.

During the coke baking the volatile compounds are combusted, but they practically do not contain sulfur compounds. The composition of the mass also includes pitch, the sulfur content in which, on average, does not exceed 0.2%. The sulfur contained in the pitch passes into the mass or into the burnt anodes. In the process of anodes baking, their desulfurization practically does not occur, and all sulfur contained in the coke and pitch completely passes into the fired block.

Thus, the estimation of sulfur dioxide emissions in the production of aluminum is necessary to find ways to improve its environmental safety.

2. Method for calculation of sulfur dioxide emissions

In the production of electrode mass, sulfur dioxide is formed due to the oxidation of sulfur contained in the fuel, and its quantity per 1 tonne of electrode mass Q can be determined according to the formula

$$Q = 2 \cdot \left(\frac{q_f \cdot C_f}{100}\right) \cdot \left(\frac{K_c}{1000}\right), \qquad (1)$$

where 2 – coefficient of conversion of sulfur content into sulfur dioxide;

q_f – fuel consumption for baking (drying) 1 tonne of coke;

C_f – sulfur content in fuel, %;

K_c – consumption of coke for production 1 tonne of electrode mass, kg.

If the sulfur content in the fuel oil is 0.5 – 3.5%, and its consumption during the baking 1 tonne of coke is in the range 15-50 kg, in this case the amount of the formed sulfur dioxide is 0.15-3.5 kg. When drying the baked coke, the amount of sulfur dioxide will be 3 to 4 times lower. To determine the actual amount of sulfur dioxide produced during the thermal treatment of coke, it is necessary to use the reporting data of the plants on the sulfur content in the coke, the fuel consumption per 1 tonne of coke and the consumption of baked coke for the production of 1 tonne of electrode mass.

During production of baked anodes, the additional amount of sulfur dioxide is produced due to combustion of fuel oil used for kilns, the amount of which, as a rule, is about 150 kg per 1 tonne of ready blocks. Therefore, the amount of sulfur dioxide in this case can increase by 1.5-10.5 kg/tonne.

The amount of sulfur in 1 tonne of electrode mass or baked anode Q_{cm} (kg) can be calculated by the formula

$$Q_{cm} = \left(\frac{q_c \cdot C_c}{100}\right) + \left(\frac{q_p \cdot C_p}{100}\right), \qquad (2)$$

where q_k – the content of baked coke per 1 tonne of mass, kg;

q_p – content of pitch per 1 tonne of mass, kg;

C_c – the sulfur content in coke, %;

C_p – the sulfur content in the pitch, %.

In the production of aluminum by electrolysis, sulfur enters the electrolyte with alumina, fluorides (in the form of sulfates) and with anode material (sulphides). Sulphates react with cryolite

$2\,Na_2SO_4 + 2Na_3AlF_6 = Al_2(SO_4)_3 + 12NaF.$

Sulfate dissolved in the electrolyte is reduced by aluminum

$3\,Na_2SO_4 + 8Al = 3Na_2S + 4Al_2O_3,$

and the resulting sulphide is transferred to the anode and oxidized to sulfur, which in turn reacts with carbon dioxide

$S + 2CO_2 = SO_2\uparrow + 2CO.$

The total amount of sulfur involved in the production of 1 tonne of aluminum depends on the consumption of each type of raw material and the sulfur content in them. The overwhelming amount of sulfur contained in all kinds of raw materials is oxidized during electrolysis to sulfur dioxide, the amount of which is equal to doubled sulfur content.

The sulfur content in alumina does not exceed 0.02% according to the data in [1], and in fluoride salts (in terms of sulfur) it is about 0.3%, but for calculations the actual sulfur content should be clarified according to the reporting data of the specific production.

3. Results of calculations of sulfur dioxide emissions and their discussion

Thus, to calculate the amount of sulfur dioxide produced during electrolysis, it is necessary to know the consumption factors for 1 tonne of aluminum of all types of raw materials and the sulfur content in them. Table 1 gives data on the consumption coefficients of the main types of raw materials for the production of 1 tonne of aluminum, the maximum permissible sulfur content in them and the amount of sulfur dioxide produced during electrolysis from each type of raw material. These data represent the maximum possible amount of the produced sulfur dioxide. If foam is removed from the electrolyte surface during the process, the amount of anode mass consumed for electrolysis should be reduced by the amount of carbon in the foam, which usually does not exceed 30% [2].

Table 1. The main types of raw materials for aluminum electrolysis and sulfur content in them.

Type of raw material	Consumption, kg	Sulfur content,%	Amount of SO_2, kg
Anode material	560	1.5	16.8
Alumina	1890	0.02	0.76
Fluorides	70	0.3	0.42
Total			18.0

From the data given in table 1 it can be seen that the main source for sulfur dioxide formation is the anode material, which accounts for more than 93% of the sulfur content in the raw material.

Therefore, in order to determine the amount of sulfur dioxide produced during electrolysis Q_{ce} (kg), the following expression can be used

$$Q_{ce} = 2 \cdot \frac{\left[\left(q_a - q_{cp}\right) \cdot C_a\right]}{100} + \left(\frac{q_{al} \cdot C_{al}}{100}\right) + \left(\frac{q_{fl} \cdot C_{fl}}{100}\right), \tag{3}$$

where q_a, q_{al}, q_{fl} – consumption, respectively, of anode material, alumina and fluorides per 1 tonne of aluminum, kg;

q_{cp} – carbon content in the removed pitch, calculated per 1 tonne of Al;

C_a, C_{al}, C_f – sulfur content, respectively, in the anode material, alumina and fluorides, %.

Thus, the maximum amount of sulfur dioxide produced during electrolysis does not exceed 18 kg/tonne of aluminum, and the amount of sulfur dioxide released into the environment depends on the availability and type of gas cleaning systems.

The maximum amount of sulfur dioxide emissions will be in the facilities that are not equipped with gas purification system, which, unfortunately, can be found at domestic aluminum plants (NkAZ, BAZ, KAZ, UAZ).

If there is a system for wet cleaning of waste gases, sulfur dioxide reacts with soda solution

$$Na_2CO_3 + SO_2 + 0.5O_2 = Na_2SO_4 + CO_2,$$

and the resulting sulphate is excreted in the process of production of the regenerated cryolite. The efficiency of wet gas cleaning for sulfur dioxide does not exceed 85%. Therefore, all gas which is not caught by the gas gathering system get into the surrounding atmosphere. The efficiency coefficient of gas gathering devices (curtains, gas gathering bells, shelters) is 60-85% and therefore from 30 to 50% of the formed sulfur dioxide gets into the environment.

The situation is further exacerbated by the use of a dry flue gas cleaning system, which is not capable of capturing sulfur dioxide, and therefore all the sulfuric anhydride formed gets into the

IOP Conf. Series: Materials Science and Engineering **411** (2018) 012046 doi:10.1088/1757-899X/411/1/012046

surrounding atmosphere. This type of gas cleaning is widely used in cells with baked anodes, and for reduction of emissions of sulfur dioxide wet-gas purification system can be installed after dry cleaning, but this should be justified for each particular production.

Analyzing the statistical reports, the data on emissions of sulfur dioxide at aluminum plants of Russia seem to be not well grounded. So, emissions of sulfur dioxide for 1994 at different plants are in the range from 1.51 (IrkAZ) to 28.64 (KAZ) kg/tonne Al. At Kandalaksha (KAZ) and Nadvoytsky (NAZ) plants they are about 25-29 kg/tonne, while at Bogoslovsk (BAZ) plant they do not exceed 6.5 kg/tonne. At the same time, KAZ and NAZ use the anode mass produced at BAZ. The reporting data on emissions of sulfur dioxide at IrkAZ, which are 4 to 5 times lower than at Bratsk and Krasnoyarsk plants, hardly correspond to reality. The production of anode mass, electrolysis and gas cleaning technology, and the quality of the raw materials used are practically the same in all three plants.

Such contradictory and inexplicable data on emissions of sulfur dioxide are, in our opinion, a consequence of the lack of calculation method, which would not be based on single measurements (as it is at present), but on actual data characterizing the quality of the used raw materials for the production of electrode mass and aluminum.

The proposed method, in our opinion, will allow the amount of sulfur dioxide formed and emitted at aluminum and electrode plants to be significantly clarified.

4. Conclusion

The calculation method is proposed for determining the amount of sulfur dioxide formed in the electrolysis process, taking into account the sulfur content in the anode, alumina, fluorides, carbon content in the coal foam, consumption and formation of these materials. It is shown that the maximum amount of sulfur dioxide formed in the process of electrolysis does not exceed 18 kg/tonne of aluminum, the amount of sulfur dioxide released into the environment depends on the availability and type of gas cleaning system of the pot line.

References

[1] *Handbook of a Metallurgist on Non-ferrous Metals. Production of Aluminum* 1971 (M: Metallurgy) p 559

[2] Galevsky G V, Kulagin N M and Mintsis M Ya 1997 *Ecology and Utilization of Waste in Aluminum Production* (Novosibirsk: Nauka) p 158

IOP Conf. Series: Materials Science and Engineering **411** (2018) 012047 doi:10.1088/1757-899X/411/1/012047

The emission of PAHs from self-baking anodes in aluminum production

M Ya Mintsis[1], G V Galevsky[1], V V Rudneva[1] and S G Galevsky[2]

[1]Siberian State Industrial University, 42 Kirova street, Novokuznetsk, 654007, Russia
[2]St. Petersburg Mining University, V.O., 21 Line, 2, St. Petersburg, 199106, Russia

E-mail: kafcmet@sibsiu.ru

Abstract. The sources of PAH emission from the Soderberg anode surface of aluminum electrolysis cells are analyzed and measures for their minimization due to evaporative cooling of the anode to temperature 100 °C are offered.

1. Introduction

In the aluminum industry, the main source of emission of resinous substances containing various PAHs is coal tar pitch, used as a binder in the production of anode mass, from which the Soderberg anode is formed. Coal pitch used in the production of anodic paste (GOST 10200-83) contains 53-62% of volatile substances that are a complex mixture of organic compounds, a significant proportion of which are PACs.

The US regulatory authorities identified 129 the most dangerous compounds [1], and 16 of them are the most toxic affecting the chromosome apparatus of the cell, as well as their genotoxicity index. There are several ways to determine the toxicity of PAHs, but in most methods the toxicity of a particular type of PAH is compared to the toxicity of benzo(a)pyrene (BaP), the value of which is assumed to be unity [1].

Because of the extremely low concentrations of PAHs in the ambient air, unique analytical equipment is required to determine the content of each PAH, which domestic aluminum plants do not have. Therefore, regular monitoring of these emissions is not carried out, and periodic monitoring is limited to sporadic measurements of the total amount of PAHs in the atmosphere that is performed out by specialized organizations.

A considerable part of the resinous compounds is released through the aeration lanterns of the electrolysis cells bodies and enters the surrounding environment without purification and, despite the low concentration of PAHs in the lantern emission, their contribution to the total emissions is dominant [1]. In this regard, the reduction in PAHs emissions is a prerequisite for improving the environmental safety of electrolytic aluminum production.

2. State of the problem

Reduction of PAHs emission can be ensured only if the reported data on emission, on the basis of which measures are developed to reduce PAHs emissions, adequately reflect the actual state of the matter. The analysis of the current methodology for determining emissions [3] has shown that it needs to be improved, since calculations based on it give underestimated results far from the actual state of the matter.

Accordance to this method, in electrolysis cell with self-baking anodes with an upper current lead during electrolysis, the resinous substances formed during the coking of the anode mass are released into the atmosphere of the electrolysis cell when the anode studs are moved and under the bell gas collector. Thus, the method does not take into account the emissions of tarry substances from the anode surface in the period between rearrangement of the studs, which can not be accepted for the following reasons.

The melting point of at least six of the 16 most toxic PAHs is below the surface temperature of the anode, which is about 150-160 °C, rising in the summer is much higher.

Back in 1959, M.I. Alivoyvodich [4] established that at a temperature of 100 °C the yield of resinous compounds from the BT anode surface reaches 10 $g/m^2 \cdot h$, and with an increase in temperature to 190 °C, their amount increases by 12 times.

The laboratory tests carried out by the authors [5] showed that with a binder content approximately 25% at a temperature 100 °C the volatile matter yield is 1.32-2.48 kg/tonne Al, and at a temperature 190 °C the emission rises to 14.7-45.7 kg/tonne Al.

M. Sorle [6], in particular, showed that with an increase in the temperature of the Soderberg anode surface from 140 to 185 °C, the emission rate of common PAHs, including the BaP, increases by 10 times. Thus, the total emission of resinous substances from the anode surface reaches up to 50% of their content in the anode mass, that is, 40-50 kg/tonne Al.

The work targeting the reduction of PAHs emissions into the environment by factories and specialized research institutes is constantly carried out. To date, several methods have been proposed [2] to reduce PAHs emissions: the use of aluminum ribs to increase heat removal from the anode, the use of a dry anode mass, the use of a pitch with a higher softening temperature, an increase in the height of the anode casing, the creation of a colloidal anode, etc. Each of the measures in one way or another can provide a reduction in PAHs emissions, and some of them are applied in production. However, there has been no significant reduction in PAHs emissions and the proposed measures can not ensure that.

Radical reduction in PAHs emission can be ensured only by replacing Soderberg anodes with baked anodes, but the costs of this modernization on the scale of the industry amount to tens of billions of dollars, and therefore such modernization becomes more than problematic.

3. Analysis procedure and results

The most effective way to reduce PAHs emissions is to reduce the surface temperature of the anode up to 100-110 °C, which will reduce the mass consumption and improve the quality of the anode.

In a number of branches of non-ferrous metallurgy, the evaporative cooling method is widely used, in which the heat exchange agent is water. Indeed, water has the greatest latent heat of evaporation - at atmospheric pressure and at a temperature of 100 °C it is 2256.25 kJ / kg. For all other substances and chemical elements, the latent heat of evaporation is practically an order of magnitude smaller.

Therefore, a method of cooling the surface layer of the Soderberg anode by applying fine water droplets (water fog) onto its surface, which, in order to evaporate, will remove heat from the anode and lower the temperature of its surface. To quantify the proposed cooling method, we determine the specific water demand.

The density of the anode mass at a temperature of 200 °C is equal to d_{am} = 1.5 g/cm_3 (1.5 t/m^3) [7]. The volume (V, m^3) of the mass layer of 1 m^2 and the thickness of 5 cm (0.05 m) is 0.05 m^3, and its mass (weight) (P_{am}) is equal to:

$$P_{am} = 1.5 \cdot 0.05 = 0.075 \text{ tonne} = 75 \text{ kg}.$$

At the temperature of the mass surface T = 150 °C and the specific heat capacity C_{am} = 1.8 kJ/kg·°C, the heat content of this quantity of mass Q_{am} (kJ) is:

$$Q_{am} = 75 \cdot 1.8 \cdot 150 = 20250 \text{ kJ}.$$

To reduce the temperature of this anode layer to 100 °C, one third of the accumulated heat, i.e., 6750 kJ, must subtracted from this volume. For heating to 100 °C and evaporation of 1 kg of water, it is necessary to spend about 2672 kJ. Therefore, to reduce the temperature of 75 kg of the anode surface by 50 °C, it is sufficient to evaporate only 6750/2672 = 2.53 kg of water. This amount of water is sufficient to reduce the temperature of 1 m^2 of the anode surface to 100 °C. However, when heat is removed, heat from the lower layers of the anode will flow to the top layer of the anode.

Due to the low value of the coefficient of thermal conductivity λ (1.5 W/m·°C = 5.4 kJ/m·h·°C) and the temperature difference between the cooled and adjacent lower layer t, which is assumed to be 50 °C, each hour he cooled layer of the mass receives about 3600 kJ of heat, for the removal of which it will be necessary to use water in the amount 3600/2672 = 1.34 l/m^2. With an anode area of a C8BM cell being equal to 24 m^2, the water flow will be approximately 24·1.34 = 32.3 l/h.

The given considerations are approximate, since they do not take into account the presence of heat-conducting studs in the anode, and also the fact that the heat capacity and thermal conductivity of the mass can vary within fairly wide limits. Of course, more accurate data can only be obtained after industrial tests.

If 90 electrolysis cells are installed in the pot room, which provide the total aluminum output of about 4750 kg/h, 32.3·90 = 2907 l (2.9 m^3) of water will have to be evaporated hourly, or about 612 l/tonne of aluminum. Evaporation, this amount of water forms about 3000 nm3 / h of steam. According to [8] the amount of ventilation air supplied to the pot room equipped with Soderbergh VT cells is (1.8-2.0)·10^6 nm^3/ton of aluminum.

Consequently, within an hour the volume of ventilation air will be about 9·10^6 nm^3. In this case the volume of water vapour formed during the anode cooling will amount to (3000/9·10^6)·100 = 0.03 % of the total volume of ventilation air, which, apparently, can not have a noticeable effect on the freezing of water vapour in winter season.

The development of the device and its control circuit can be started only after carrying out the industrial tests of the proposed method. However, the efficiency of this method give grounds to hope that the considerable difficulties that will inevitably arise when an industrial plant for anode cooling by vaporization is created will be successfully overcome.

4. Conclusion

The feasibility of reducing PAHs emissions by reducing the anode surface temperature by 100 °C through evaporative cooling is substantiated. With the area of electrolytic cell anode 24 m^2, the water flow rate is 32.3 l/h. The volume of water vapor is 0.03% of the total volume of the ventilation air of the pot room. The implementation of technological measures allows us to predict a 10-fold reduction in PAHs emissions.

References

[1] Kulikov B P and Istomin S P 2004 *Processing of Aluminum Waste Products* (Krasnoyarsk: Classic-Center) p 477
[2] *News bulletin* 2001 **(8)14** 29–43
[3] 2005 *Settlement instruction (methodology) for determining the composition and amount of harmful (polluting) substances emitted into the air during electrolytic aluminum production. Approved by the Order of the Federal Service for Environmental, Technological and Nuclear Supervision of March 31, 2005 No. 182*
[4] Alivoyvodich M I 1959 *Non-ferrous metals* **12** 42–47
[5] Mintsis M Ya 2006 *RUSAL Technical and Economic Bulletin* **16** 3–5
[6] Sorle M 2005 *RUSAL Technical and Economic Bulletin* **11** 33–37
[7] Korobov M A and Dmitriev A A 1972 *Self-baking Anodes of Aluminum Electrolysers* (Moscow: Metallurgy) p 207
[8] 2007 *Aluminum of Siberia-2007* (Krasnoyarsk) p 29

Modernization of the cathode assembly of aluminum electrolysis cells with Soderberg anode during the AAFS introduction

M I Mintsis and G V Galevsky

Siberian State Industrial University, 42 Kirova street, Novokuznetsk, 654007, Russia

E-mail: kafcmet@sibsiu.ru

Abstract. The state of the cathode assembly on Soderbergh electrolysis cells is analyzed, and measures for its improvement are proposed.

1. Introduction

Aluminum electrolysis with Soderberg anodes now constitute about 70% of all types of electrolysis cells in Russian plants. The dominant part of these cells was developed in the 50's of the last century and was designed for the technology in which the electrolyte was fed by alumina periodically (every 3 or 4 hours). In these conditions the electrolyte mass should absorb as much alumina as it is necessary for the cell to work for 3 to 4 hours. The labour costs for the delivery of alumina to the cell and its immersion into the electrolyte melt account for 40% of the total labour of technological personnel.

At the moment the work on the introduction of the automated alumina feeding system (AAFS) at Krasnoyarsk Aluminum Plant has been completed, and in a number of plants the work on the introduction of AAFSs is still being carried out. The introduction of AAFSs not only increases the productivity of technological personnel, but also completely changes the electrolysis technology – the frequency of anode effects is sharply reduced, the concentration of alumina in the electrolyte and the magnitude of the inter-pole distance, etc. are stabilized. However, this technology change is not accompanied by a corresponding change in the cell design.

With the introduction of AAFSs the duration of alumina supply to the melt is sharply reduced – from 3 to 4 hours to several minutes. In these conditions the required volume of electrolyte is reduced.

Unfortunately, in the educational and technical literature there is no method for determining the volume of electrolyte in the shaft, which complicates the justification of the minimum value of the technically necessary volume of electrolyte in the conditions of its feeding with alumina by AAFSs.

2. Calculation of the electrolyte volume in the cathode assembly

Below is a procedure for determining the volume of electrolyte in the cathode assembly, and its scheme, used to calculate the mass of electrolyte in the cathode assembly, is shown in figure 1.

The volume of electrolyte V_e (dm^3) in the cathode assembly is determined by the expression:

$$V_e = (S_m \cdot H_e) - S_a \cdot (H_e - L_{ipd}), \tag{1}$$

where S_m – the melt surface, dm^2;

H_e – the height of the electrolyte column above the metal, dm;

S_a – the anode area, dm^2;
L_{ipd} – the inter-pole distance, dm.

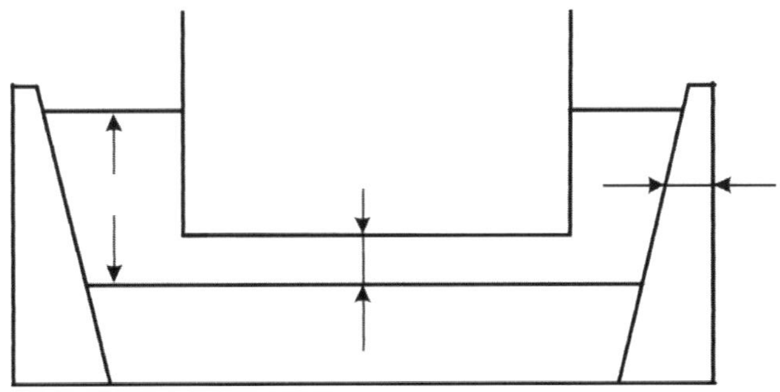

Figure 1. Calculation scheme.

The surface of the melt S_m, taking into account the average thickness of the side accretions n (dm), is determined by the following expression:

$$S_m = (L_{ca} - 2sa) \cdot (B_{ca} - 2sa), \quad (2)$$

where L_{ca} and B_{ca} – the length and width of the cathode assembly, dm.
The anode area S_a (dm^2) is defined by the expression:

for Soderbergh electrolysis cells $$S_a = L_a \cdot B_a, \quad (3)$$

and for BA cells $$S_a = n_b \cdot S_b = n_b \cdot L_b \cdot B_b, \quad (4)$$

where L_a, B_a – the length and width of the anode, respectively, dm;
n_b – number of anode blocks, pcs;
S_b – the area of the anode block dm^2;
L_b and B_b – the length and width of the anode block, respectively, dm.
The electrolyte mass M_e (kg) in the cathode assembly is equal to the product of its volume V_e (dm^3) by the density of the electrolyte d_e, which is about 2.08 kg/dm^3:

$$M_e = V_e \cdot d_e. \quad (5)$$

For calculations it is convenient to use the concept of the electrolyte specific mass, m_e (kg/kA), which is equal to the ratio of the electrolyte mass, M_e, to the current I_c (kA):

$$m_e = \frac{M_e}{I_c}. \quad (6)$$

3. Results and discussion

To determine the required amount of electrolyte in the cathode assembly, we determine the demand for alumina q_{al} (kg/kAh) per 1 kA for one hour. To do this, we determine the specific capacity of the cell q_a (kg/ kA·h) for a time t = 1 h at a current strength I_c = 1 kA and an output current η = 0.9:

$$q_a = 0.3356 \cdot I_c \cdot t \cdot \eta = 0.3356 \cdot 1 \cdot 1 \cdot 0.9 = 0.30,$$

and the specific demand for alumina q_{al} for the production of aluminum q_a will be

$$q_{al} = 1.889 \cdot q_a = 1.889 \cdot 0.30 = 0.567 \text{ kg/kA·h}.$$

The solubility of alumina in the electrolyte depends on many factors, which are discussed in detail in [1]. During in-line treatment of the crust, the maximum value of the alumina concentration in the electrolyte C_{max} occurs some time after it is fed into the electrolyte. In the process of electrolysis not all alumina dissolved in the electrolyte is consumed, since an anodic effect occurs when a certain critical value of its concentration C_{cr} is reached. In order to avoid the occurrence of anodic effect, it is necessary to introduce the next portion of alumina into the electrolyte at the moment when its concentration value will be the minimum allowable C_{min}, but above the value C_{cr}. Thus, the amount of alumina that can be produced from the electrolyte depends on the difference in values:

$$\Delta C = C_{max} - C_{min}. \tag{7}$$

Depending on the quality of alumina, electrolyte parameters, anode current density, the value of C_{max} with modern technology is about 6.0% [1], and C_{min} – about 2.5%. Then the value of ΔC is about 3.5% of the electrolyte mass.

The specific mass of the electrolyte q_e required to dissolve q_{al} of alumina in the agreed conditions will be:

$$q_e = \frac{q_{al}}{\Delta C} = \frac{0.567}{0.035} = 16.2 \text{ kg/kA·h}. \tag{8}$$

In the case of 8-fold treatment of the crust, alumina is fed into the melt approximately every 3 hours. Consequently, the required specific gravity of the electrolyte must be $16.2 \cdot 3 = 48.6$ kg/kA·h.

When the electrolysis cell is equipped with the AAFS system, the process is carried out with the alumina concentration about 2.5% in the electrolyte, since the electrolyte electrical resistance is minimal. Another portion of alumina is fed into the melt with a periodicity of several minutes. That is why the required amount of electrolyte should be much smaller.

For comparison we calculate the specific volume of electrolyte for the most widely used in the plants of Russia Soderberg cell of S8BM type and the modern electrolysis with burned anodes of RA-300 type installed in KhAZ, the parameters of which are presented in the table 1.

We calculate the volume V_e, its mass M_e and the specific mass m_e of the electrolyte for each type of cell.

Assuming the height of the electrolyte column H_e is 1.8 dm, the inter-pole distance L_{ipd} – 0.5 dm, and the thickness of side accretions in the electrolyte zone – 1.0 dm, we get:

- for electrolysis cell S8BM:

$$V_e = (S'_m \cdot H_e) - S_a \cdot (H_e - L_{ipd}) = V_e = (3606.4 \cdot 1.8) - 2394 \cdot (1.8 - 0.5) = 3379.3 \text{ dm}^3;$$

$$M_e = V_e \cdot d_e = 3379.3 \cdot 2.08 = 7029 \text{ kg}; \quad m_e = \frac{M_e}{I_c} = \frac{7029}{172} = 40.86 \text{ kg/kA}.$$

- for electrolysis RA-300:

$$V_e = (S'_m \cdot H_e) - S_a \cdot (H_e - L_{ipd}) = V_e = (5037.4 \cdot 1.8) - 3641 \cdot (1.8 - 0.5) = 4333.5 \text{ dm}^3;$$

$$M_e = V_e \cdot d_e = 4333.5 \cdot 2.08 = 9013.6 \text{ kg}; \quad m_e = \frac{M_e}{I_c} = \frac{9013.6}{320} = 28.16 \text{ kg/kA}.$$

Table 1. Parameters of electrolysis cell S8BM and RA-300.

Parameter	Measurement unit	S8BM	RA-300
Anode			
- length	dm	84	14.45

- width	dm	28.5	7
- number of blocks	pcs	1	36
- area	dm^2	2394	3641.4
- current density	A/cm^2	0.722	0.878
Bottom			
- length	dm	94	143.5
- width	dm	41.2	37.6
- area	dm^2	3872.8	5395.6
- current density	A/cm^2	0.446	0.593
Calculated parameters:			
- volume	dm^3	3397.3	4333.5
- масса weight	kg	7029	9013.6
- specific weight	kg/kA	40.86	28.16

Thus, at the electrolysis cell RA-300 equipped with an APG system, at a distance between the anode mass and the cathode assembly side of 350 mm, the specific weight of the electrolyte is 28.16 kg/kA, which is 1.45 times less than in the cell of the S8BM type.

In modern foreign cells with BA, the distance between the anode massif and the side of the cathode assembly is 250 mm, and the specific volume of the electrolyte does not exceed 25 kg/kA, which is 1.63 times less than in the S8BM cells.

Reducing the width of the cathode assembly will lead to a significant increase in the cathodic current density, by which is meant the ratio of the current to the mine area. Currently, in the electrolysis cells with burned anodes, the anode current density, according to Gibson Crest Pty Ltd, is 0.85-0.9 A / cm2. In this case, the ratio of the area of the anode to the area of the bottom is recommended to be within 0.85 - 0.9, that is, the cathode current density should be at least 0.72-0.80 A/cm^2.

The anodic current density at the Soderberg anodes is unattainable in the anode cells of Soderberg anodes. At the 173 kA electrolysis cells, the anode current density has reached 0.725 A/cm^2, but the cathode current density is only 0.447 A/cm^2. Such a low cathode current density leads to a decrease in the current output, subcooling of the bottom, the formation of floorboards under the anode, and increases the material consumption of the cathode device.

Apparently, the effective anodic current density without upgrading the Soderberg BT anodes will not exceed 0.75 A/cm^2, but it is advisable to increase the cathode current density, which will certainly reduce the metal losses due to the reverse reactions, or, in other words, increase the current output. There is reason to believe that the life of a narrow cathode device will increase.

4. Conclusion

The obtained results show that there is a possibility of reducing the melt volumes in the cathode assembly of the bath by reducing the distance between the anode and the side of the cathode assembly. This will reduce the material consumption of the cathode assembly by approximately 20 % and will adequately reduce the costs for general repairs. In this case, the surface area of the electrolyte will also decrease by approximately 20%, which will positively affect the emission of toxic gases into the atmosphere of the casing.

References

[1] Isayeva L A and Polyakov P V 2000 *Alumina in the Production of Aluminum by Electrolysis* (Krasnoturinsk: Publishing house "BAZ") p 198

The impact of basic boroaluminizing factors on diffusion layer thickness in low-carbon steels and its mathematical modeling

U L Mishigdorzhiyn and N S Ulakhanov

East Siberia State University of Technology and Management, 40V Kluchevskaya street, Ulan-Ude, 670013, Russia

E-mail: nulahanov@mail.ru

Abstract. The paper presents the study of pack boroaluminizing carried out on steel 20 using mathematical planning methods. A mathematical model and nomograms showing the dependence of diffusion layer thickness on basic parameters factors of thermochemical treatment (technological factors such as temperature, treatment time and powder mixture composition) were obtained by means of conducting a full factorial experiment. The degree of impact which technological factors make on the final result and the ranges of their regulation are revealed. The temperature of pack boroaluminizing varied from 900 to 1000 °C, the duration of the process was 2-4 hours, and the ratio of the treatment components of B_2O_3 / Al was 40/60 to 50/50 wt-%. The metallographic analysis has revealed that the depth of the diffusion layer is from 80 to 260 μm, depending on technological factors.

1. Introduction

Currently in mechanical engineering, to strengthen the surface layer of machine parts, it is possible to effectively use multicomponent thermochemical treatment (TCT) consisting in simultaneous or sequential diffusion saturation with several chemical elements [1,2]. Multicomponent methods of TCT, such as boroaluminizing, allow to increase wear resistance, improve heat resistance, oxidation resistance and a number of other properties of surface layers of machine parts [3-6]. Boroaluminizing is implemented in different ways: in treatment pastes, in powder mixtures of boron and aluminum-containing substances, in liquid and gaseous media.

For a broader practical application of this technology, mathematical models are needed. They generalize experimental data and allow to reliably control the technological process and predict the result of processing. Such models of TCT processes in the form of regression equations or in the form of power functions are presented in [7-10]. They are convenient for calculating output parameters, optimizing technological modes and can be quite accurate (~ 8-10%). To construct a mathematical model of the TCT process, in order to reduce the number of experimental studies, the method of mathematical planning of the experiment was used. The TCT process comprises a fairly large number of factors affecting the output properties of the surface layer. Full factorial experiment method was used to determine the optimal parameters of technological modes and to obtain the dependences of the examined properties from these parameters.

The purpose of this work is to investigate the dependence of the diffusion layer thickness on the temperature, treatment time and saturating mixture composition at pack boroaluminizing. On the basis of the dependence obtained, a mathematical model will be constructed that describes the influence of the main boroaluminizing factors on the diffusion layer thickness.

To achieve the proposed aim, it is necessary to solve the following objectives:

- to create and implement an experiment plan that provides an opportunity to obtain an adequate mathematical model of the boroaluminizing process;
- to carry out a metallographic analysis to estimate the diffusion layers thickness after boroaluminizing;
- to obtain a graphoanalytical description of the variation in the diffusion layer thickness, that will be convenient for selecting modes and predicting the result of the treatment.

2. Materials and methods

The steel 20 was chosen as the object of the research. Boroaluminizing of the steel samples was carried out in a powder mixture with composition of 70% Al2O3 + 30 (x% B2O3 + y% Al) + 0.5% NaF in a container with a fusible gate in isothermal exposure. The thickness of the boroaluminized layers was evaluated on optical microscope "Neophot-21". To obtain the relationship between treatment parameters and diffusion layer thickness, the method of mathematical planning of the experiment was applied. Data analysis of a priori information, as well as the results of experiments [2] allowed to distinguish three main factors (in code values): X1 - treatment time in hours; X2 – treatment temperature in degrees Celsius; X3 - composition of treatment mixture in percentage ratio of B2O3 and Al. The experiments were carried out according to the full factorial experiment algorithm of type 23, where the number of factors is k = 3, the number of levels is p = 2. To calculate the coefficients, it is necessary to perform N = 8 experiments, the number of repeated experiments is n = 5. The ranges of regulation of the main technological factors are given in Table 1.

Table 1. Technological factors and their variation levels.

Factors	Variation levels			Variation interval
	-1	0	+1	
x_1 – treatment time, h	2	3	4	1
x_2 – temperature, °C	900	950	1000	50
x_3 – B_2O_3/Al ratio, %	40/60	45/55	50/50	5/5

3. Experimental results

Creating a mathematical model comes down to obtaining an equation of the first or second order in the form [10]:

$$Y = b_0 + b_1x_1 + b_2x_2 + ... + b_nx_n + b_{11}x_1^2 + b_{22}x_2^2 + ... + b_{12}x_1x_2 + ... + b_{nm}x_nx_m \quad (1)$$

where Y – calculated boroaluminized layer thickness, μm;
x_1, x_2, x_n, x_m – variation parameters (Table 2);
b_0, b_n, b_{nm} – regression coefficients.
Table 2 shows the planning matrix and averaged results of the performed experiments.

Table 2. Planning matrix of the experiment.

Number of experiment, No	Planning matrix								Average value of boroalumi-nized layer thickness, Yi, μm
	x_0	x_1	x_2	x_3	x_1x_2	x_1x_3	x_2x_3	$x_1x_2x_3$	
1	+1	-1	-1	-1	+1	+1	+1	-1	110.6
2	+1	+1	-1	-1	-1	-1	+1	+1	180.8

3	+1	-1	+1	-1	-1	+1	-1	+1	160.2
4	+1	+1	+1	-1	+1	-1	-1	-1	259.8
5	+1	-1	-1	+1	+1	-1	-1	+1	53.6
6	+1	+1	-1	+1	-1	+1	-1	-1	80
7	+1	-1	+1	+1	-1	-1	+1	-1	209.8
8	+1	+1	+1	+1	+1	+1	+1	+1	244.4

Experiments and processing of results allowed to obtain an equation for calculating a boroaluminized layer thickness in the given interval of main factors variation: treatment temperature, treatment time and composition of saturating mixture:

$$Y = 162{,}4 + 28{,}85x_1 + 56{,}15x_2 - 15{,}45x_3 + 4{,}7x_1x_2 - 13{,}6x_1x_3 + 24x_2x_3 \quad (2)$$

The obtained mathematical model was tested for adequacy by Student's and Fisher's criteria. The procedures of evaluating the coefficients of mathematical model, verification of its adequacy and the statistical analysis of accuracy was calculated according to sources [8,9,10].

Coefficients significance examination is implemented by comparing the absolute coefficient value with the confidence interval. To do this, is to be calculate the variance of reproducibility by following formula:

$$S_{\{y\}}^2 = \frac{\sum_1^N S_i^2}{N} = 14{,}25 \quad (3)$$

Next, is to be find the confidence interval by formula:

$$\Delta b_i = \pm \frac{t \cdot S_{\{y\}}}{\sqrt{N}} = 3{,}746 \quad (4)$$

where t is the tabular value of criterion at the accepted significance level and the number of degrees of freedom f with which the variance S_y^2 was determined. The value of t for five repeated experiments and the confidence probability of 0.95 is 2.78 [10]. Coefficients significance evaluation of the model with the use of Student's t-test showed that the coefficient b_{123} equal to 2.65 can be considered not significant.

The adequacy dispersion, characterizing the empirical values scattering relative to the calculated values Y, is to be found by the formula:

$$S_{ad}^2 = \frac{\sum_1^N (Y_{iu} - Y_i)^2}{f} = \frac{\sum_1^N \Delta Y_i^2}{f} = 67{,}86 \quad (5)$$

where Y_{iu} is the parameter arithmetic mean in the i-th experiment; Yi - value of the investigated parameter, calculated by the model for the i-th experiment conditions; f is the number of degrees of freedom equal to N-(k+1); where k is the number of factors.

The adequacy of the mathematical model is to be determined by means of the Fisher's criteria:

$$F_{расч} = \frac{S_{ад}^2}{S_{\{y\}}^2} = 4,672 \qquad (6)$$

Mathematical model is considered adequate as long as $F_{calc} \leq F_{table}$.

The adequacy dispertion calculation is given in Table 3, where the number of degrees of freedom: $f_5 = f_7 = 5 - 1 = 4$. The results of the model validation on adequacy are shown in Table 4.

Table 3. Adequacy dispertion.

No. of experiment	Boroaluminized layer thickness, Y, µm		ΔY	ΔY^2
	Experimantal values	Calculated values		
1	110.6	107.95	2.65	7.0225
2	180.8	183.45	-2.65	7.0225
3	160.2	162.85	-2.65	7.0225
4	259.8	247.75	12.05	145.2025
5	53.6	56.25	-2.65	7.0225
6	80	86.75	-6.75	45.5625
7	209.8	216.55	-6.75	45.5625
8	244.4	247.05	-2.65	7.0225
	ΣΔY2			271.44

Table 4. Prove of model adequacy calculation.

$S^2_{\{y\}}$	$S^2_{ад}$	F_{calc}	F_{table}
14.525	67.86	4.6719	6.4
$F_{calc} \leq F_{table}$, the model is adequate			

4. Graphical processing of experimental results and discussion

Surfaces of the response and nomograms, depending on the technological factors of the boroaluminizing process, are shown in Figs. 1-3. As follows from the results of the investigation and modeling, the optimal values of the boroaluminized layer thickness correspond to the region of maximum values of the duration and temperature of the process. The most practical is a graphoanalytical representation that will allow make selection and predict the diffusion layer thickness based on two technological factors: the treatment temperature and time.

The composition of treatment mixture (the third technological factor) is recommended to be selected depending on the required surface properties. For higher oxidation resistance at high temperatures, a predominantly aluminized layer is required, and for enhanced wear resistance, a predominant boronized layer is needed [3]. The nomograms shown in Figs. 1b, 2b and 3b are the cross sections of the response surface at a given level of output variables. Such a graphoanalytical description allows one to predict the boroaluminized layer thickness under the combined influence of the temperature and the time of the TCT process for various composition the saturating mixture.

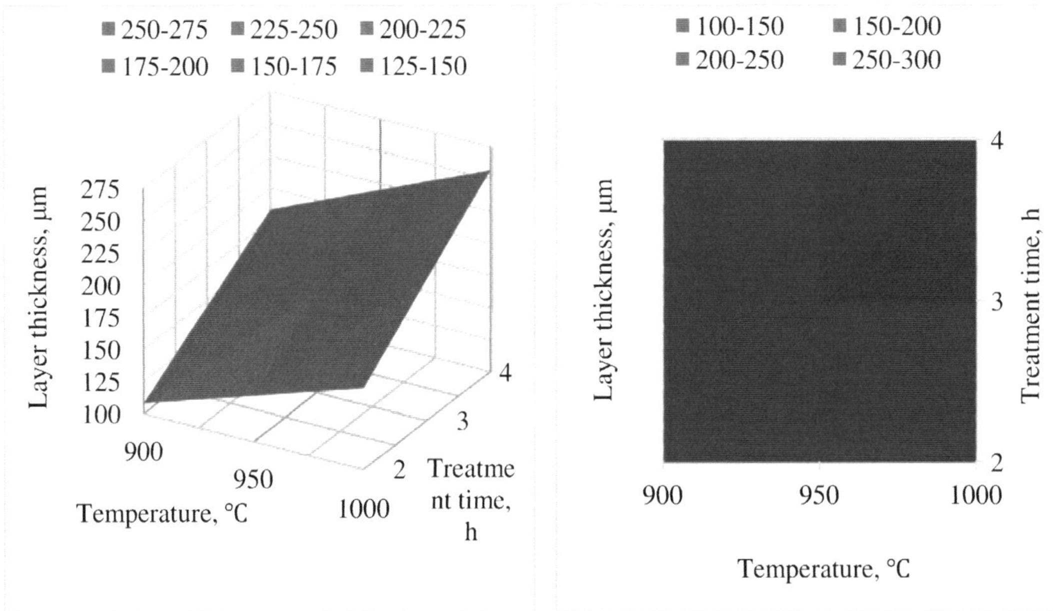

(a) (b)

Figure 1. Response surfaces (a) and nomographs (b) in the factor space of the boroaluminized layer thickness variation depending on the treatment temperature and time. The composition of the saturating mixture 40% B_2O_3 + 60% Al (lower level).

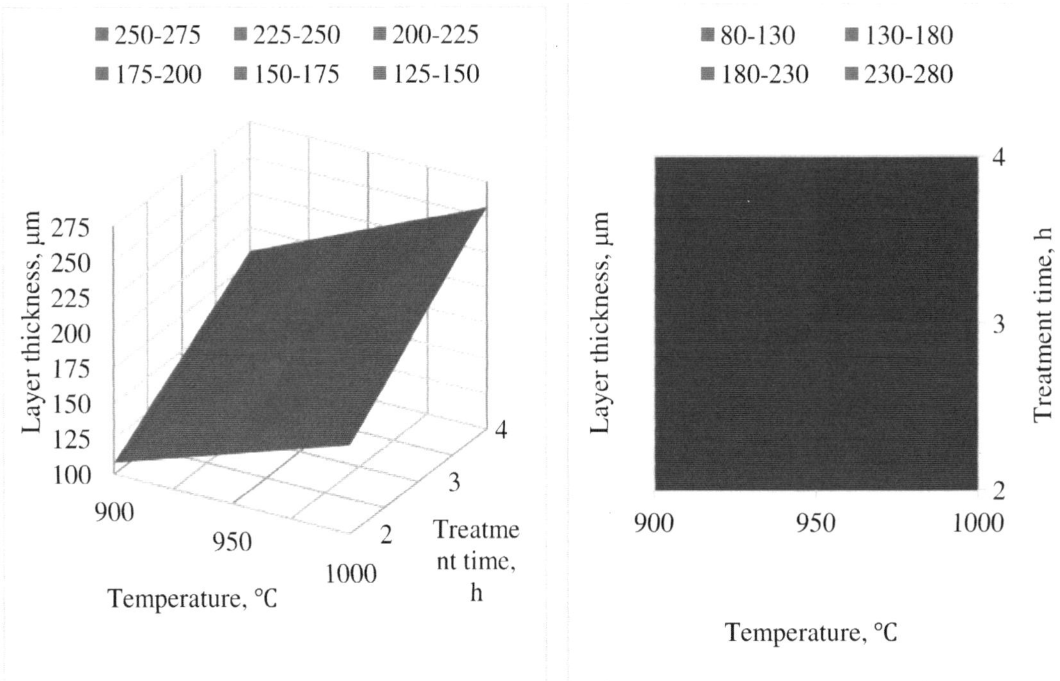

(a) (b)

Figure 2. Response surfaces (a) and nomographs (b) in the factor space of the boroaluminized layer thickness variation depending on the treatment temperature and time. The composition of the saturating mixture 45% B_2O_3 + 55% Al (zero level).

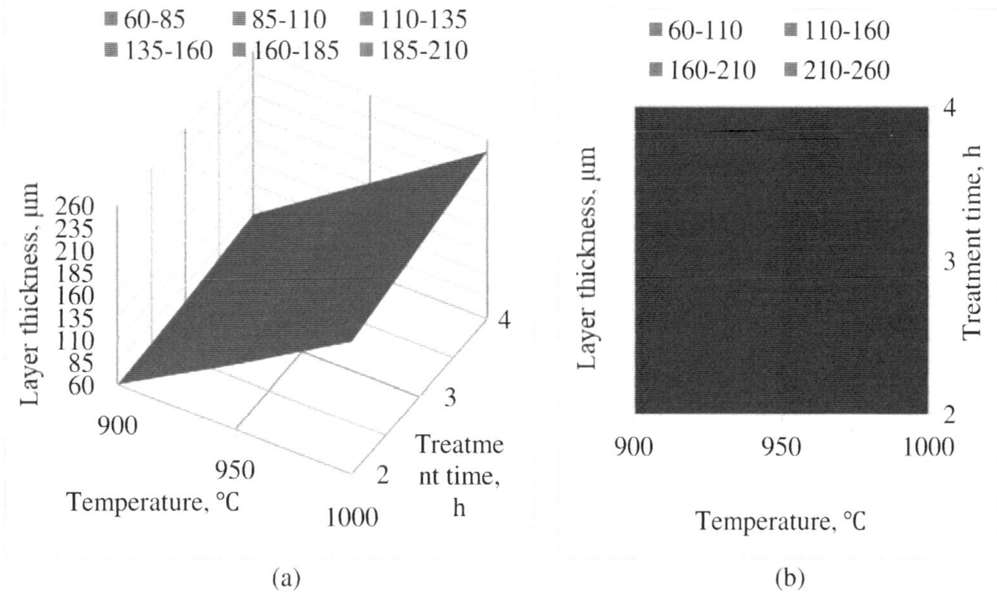

Figure 3. Response surfaces (a) and nomographs (b) in the factor space of the boroaluminized layer thickness variation depending on the treatment temperature and time. The composition of the saturating mixture 50% B_2O_3 + 50% Al (upper level).

Figure 4 shows the microstructures of boroaluminized layers.

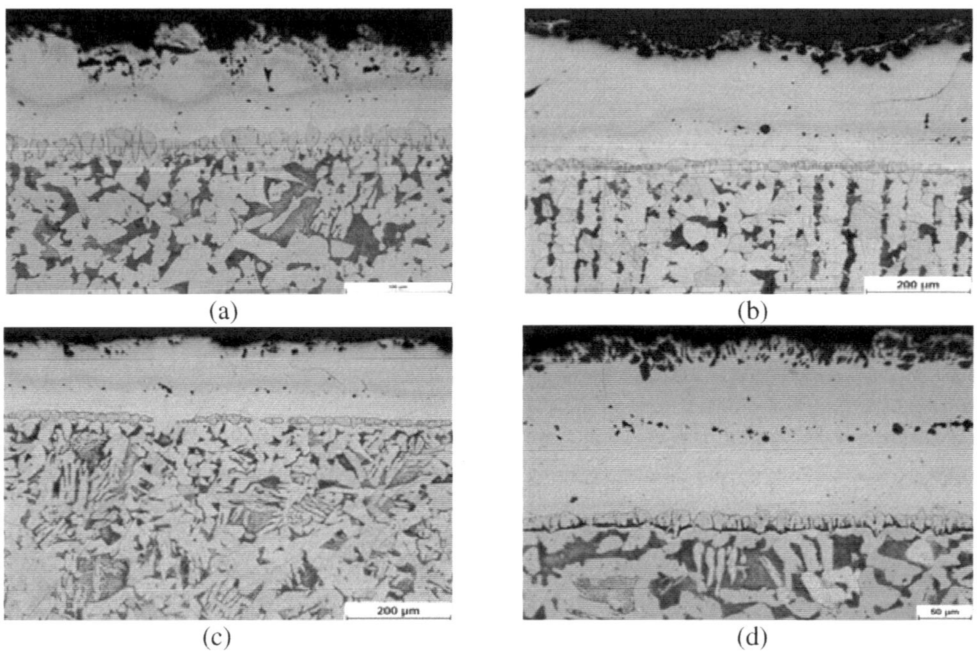

Figure 4. Microstructures of boroaluminized layers on steel 20 after treatment in saturating mixture with percentage ratio of 40%B_2O_3 + 60%Al: a) 900 °C, 2 hours, average layer thickness is 110 µm; b) 900 °C, 4 hours, average layer thickness is 160 µm; c) 1000 °C, 2 hours, average layer thickness is 180 µm; d) 1000 °C, 4 hours, average layer thickness is 260 µm.

5. Conclusions

- Implemented experiments and processing of the obtained results allowed to obtain an adequate mathematical model of the dependence of the boroaluminized layer thickness on the values of the temperature, the treatment time and the composition of the saturating mixture.
- The obtained graphoanalytical description in the form of response surfaces and nomograms is a convenient tool for predicting the diffusion layer thickness.

Acknowledgements
The work was supported by the grant of the President of the Russian Federation for the state support of young Russian scientists – PhD holders Grant number MK-2641.2018.8 and State support number 11.9999.2017/DAAD for scientific investigations.

References

[1] Voroshnin L G et al 2010 *Theory and Technology of Chemical and Thermal Processing* (Minsk: Novoje znanije) p 304
[2] Sizov I G et al 2012 *Metals Sci. and Heat Treatment* **53(11-12)** 592–597 doi:10.1007/s11041-012-9440-4
[3] Lakhtin Yu M and Arzamasov B N 1989 *Thermo-chemical Treatment of Metals* (M.: Metallurgiya) p 256
[4] Sizov I G et al 2014 *J. Surf. Eng.* **30(2)** 129–133 doi:10.1179/1743294413Y.0000000208
[5] Garmayeva I A et al 2008 *Uprochnyayushchiye tekhnologii i pokrytiya* **10** 30–32
[6] Gerasimov S A et al 2013 *Nauka i Obrazovaniye: Nauchnoye Izdaniye MGTU im. N.E. Baumana* **1** 313–332
[7] Krukovich M G 2004 *Metal Science and Heat Treatment* **46** 25–31 doi:10.1023/B:MSAT.0000029596.11102.d8
[8] Novik F S 1971 *Mathematical Methods for Planning Experiments in Metallurgy* (Moscow: MISiS) p 228
[9] Novik F S and Arsov Ya B 1980 *Optimization of Metal Technology Processes by Methods of Experiment Planning* (Moscow: Mashinostroyeniye; Sofiya: Tekhnika) p 304
[10] Adler Yu P et al 1976 *Planning an Experiment when Searching for Optimal Conditions* (Moscow: Nauka) p 280

IOP Conf. Series: Materials Science and Engineering **411** (2018) 012050 doi:10.1088/1757-899X/411/1/012050

Research, monitoring and diagnosis of continuous control systems IT capacity of information flows in of metallurgical enterprises

E N Bulakina, O N Nedzelskaya, V V Moiseev and A N Bikineeva

Siberian Federal University, 79 Svobodniy prospekt, Krasnoyarsk, 660041, Russia

E-mail: elenagb09@mail.ru

Abstract. The article analyzes the effectiveness of the development, implementation and operation of information systems, IT infrastructures the system of metallurgical enterprises and government agencies. Identifying existing problems of the life cycle of IT systems. Some aspects of the development of information technology and the prospects for their use. Prospects of virtualization technologies with the existing development pattern element base processors. Developed IT strategy, which is the link between the strategy of development of the enterprise and the IT department. A comparative analysis of international experience in the application of foreign methodologies, ITIL, COBIT, offering tools to build systems that can provide solutions to highly targeted IT issues. Defined the scope of application of these methodologies in the development of information systems of national importance. A model of the efficiency of public services through information technology. We consider methodologies to gather in a single system the various methodologies, guidance, and legal aspects. The methods for evaluating the effectiveness of information systems. The analysis of simulation systems of information technologies in which they are applied.

1. Introduction

The main purpose is: the study of technological virtualization solutions to improve the efficiency of information systems their interpenetration; Development of methodological approaches to the management of complex, centralized IT infrastructure; consideration of the use of adaptive management of IT facilities. To develop a plan for the development of IT infrastructure, you can use modeling to predict the trends of development. In the case of a simple analysis of the carrying capacity of the hardware and software system there is a representation only possible acceptable operating parameters. Mathematical modeling allows to obtain data on the system in a wide range of possible parameters [1]. During the simulation we can get accurate data about the performance of hardware and software system and use it to select the required technical platform [2]. You should also consider testing any available hardware to create a complex environment with real operating conditions in the data center. This environment allows you to experience the power limit values, work in emergency situations, and other options. Recommendations for technological solutions should contain information about the expected level of performance and the necessary software and hardware and the costs of implementation and support. When designing the technical platform for large enterprises with complex IT infrastructures requires in-depth analysis, taking into account all aspects of the integration into a single system. Technological solution should provide guaranteed performance in such clear terms, such as performance computing subsystem, the number of IOPS per unit time, the number of transactions per unit of time, and others.

2. Trends in the development of IT infrastructures, taking into account the performance of computer systems

Determine the performance of hardware and software system helps to calculate workload indicators, such as growth in the number of transactions by 50% at peak load. These data are used to predict the state of the system when the load increases, as well as for determining the parameters of the expansion system. There is an important smooth interaction with the hardware vendor in terms of harmonized test methods for correct comparison of performance information systems. Also worth noting is that the introduction of a complex distributed IT environments requires a comprehensive approach, as a large number of technological interfaces creates many interdependencies performance parameters [3].

By itself, the development of IT infrastructure is a tool for effective development of the company. In the development of IT, there are various empirical observations, in particular, of Moore's Law and its derivatives in various fields, Koomey, Amdahl, Grosch, Nelsen, Kreider and technologies shaping the development of IT, such as Web 2.0, BYOD (Bring Your Own Device), Tablet PC virtualization and others. Of course, between the laws and technologies have relationships to help you determine the horizons of development of a technology. For example, to implement virtualization technologies on the existing hardware components can be attributed Moore's law in respect of the element base, Amdahl's law regarding the effectiveness of parallelism computation Koomey's law with respect to energy efficiency, Grosch's law in respect of the cost, the Bekenstein limit in respect of the limited computing systems [4]. Thus, modeling the trend of development of information technology, or perhaps taking into account existing laws, figure 1.

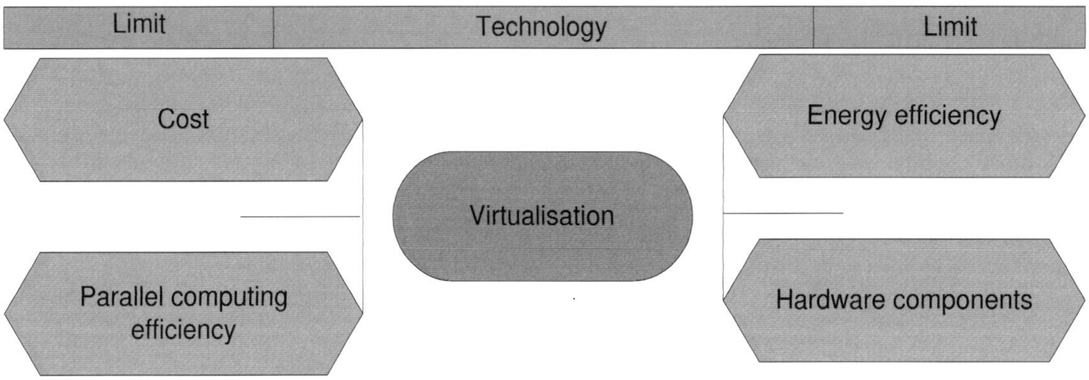

Figure 1. The limits of virtualization technologies.

3. Simulation and testing of the hardware complex to create an environment with real operating conditions.

Thus, Amdahl's Law shows growth restriction of the computing system computing increase parallelism "In the case when the task is divided into several parts, the total time of execution on a parallel system cannot be less than the execution time of the longest fragment":

$$S_p = \frac{1}{\alpha + \frac{1-\alpha}{\rho}} \tag{1}$$

Where α - the proportion of the total volume of calculations that can only be obtained by successive calculations, the share (α-1) - can be parallelized ideally, ρ - the number of computing nodes.

From this we see that the algorithm contains only parallel computing provides a linear increase in performance with increasing number of calculators. If the proportion of the algorithm in parallel computing is 90%, the increase in the number of times processor 10 provides the acceleration to 5.26 times, 100 times – a 9.17 fold increase to 1,000 and gives 9.91 times acceleration. Thus, the increase of efficiency of parallel computing depends on the algorithm and limited at $\alpha > 0$. In turn, this shows the limitations of concurrency computing for virtualization

Koomey law describes a multi-year trend in the development of hardware computing systems since 1946. Koomey observations show that the number of calculations relating to the joule energy dissipation increases approximately every 1.57 years.

Grosch's law states that the more expensive computer, the better balance of performance and price for it. In practice, this law corresponds to that of a large scalability and high performance easier to achieve using centralized data centers than on small computer systems [5].

Restriction Bekenstein an upper limit quantity of information which may be contained within a given volume using a predetermined amount of energy. That gives us an idea of the physical limitations of productivity computing systems.

Virtualization technologies allow you to create alternative solutions in the construction of information structure. Traditional solutions, in this case, include desktops. According to surveys, the main aspects of the classic PC environment, stimulating interest in alternative solutions are price, security, manageability, flexible remote access, reducing energy costs [6]. The most interesting for us is a flexible remote access.

For an employee working environment should be varied, accessible from anywhere, to support different loads and devices for flexible access. From the point of view of the person responsible for the management of the working environment, it should provide control over all user data, operating system images, settings and so on, and at the same time be accessible remotely, figure 2.

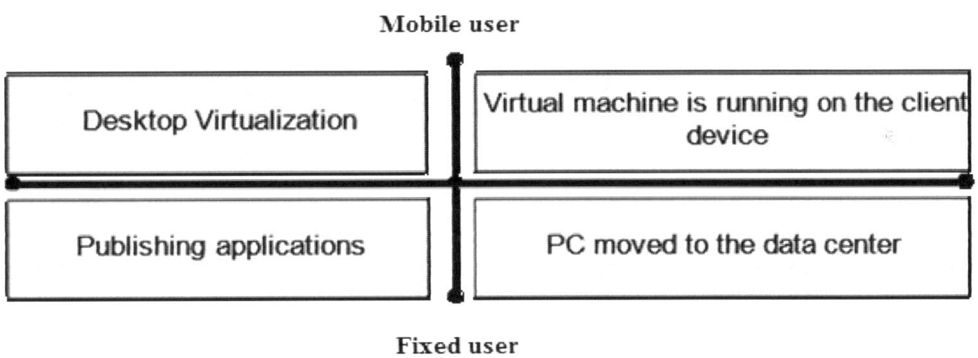

Figure 2. Approaches that provide flexibility in the production environment.

4. Implementation of virtualization technologies on the existing element base of the enterprise, providing flexibility in the work environment

In this regard, we can turn our attention to the fleet management devices, whose number is constantly growing to control the working environment. This will solve the problem of constant upgrades, updates, hardware and software parts. In fact the differences in remote access technologies consist in what part of the client environment, we move into the data center [7]. Consider the basic approaches (Figure 3).

The PC is loaded with the external storage system. In this case, the delivery will take an image of the operating system of the data center. Thus, in the data center, we move everything about the storage infrastructure. That is, in personal computers completely or partially missing hard drives, but all that relates to computational processes - is on the side of the client device.

1. Virtualization jobs. Using this approach, the data center shall be made not only all that relates to the storage and computing resources. In this case, all that relates to the processing of image data remains on the side of the client device.
2. Publication of the application is the most common approach. In this case, all applications run in the data center. The client device carried postage image, i.e. the device acts as a client terminal device.
3. PC is moved to the data center. In this case, the employee of the display remains and input-output device that is not a computer.
4. The virtual machine running on the client device. This approach allows us to empower approach desktop virtualization, providing the ability to perform tasks without the employee's connection to the data center on his personal computer in an isolated virtual environment.

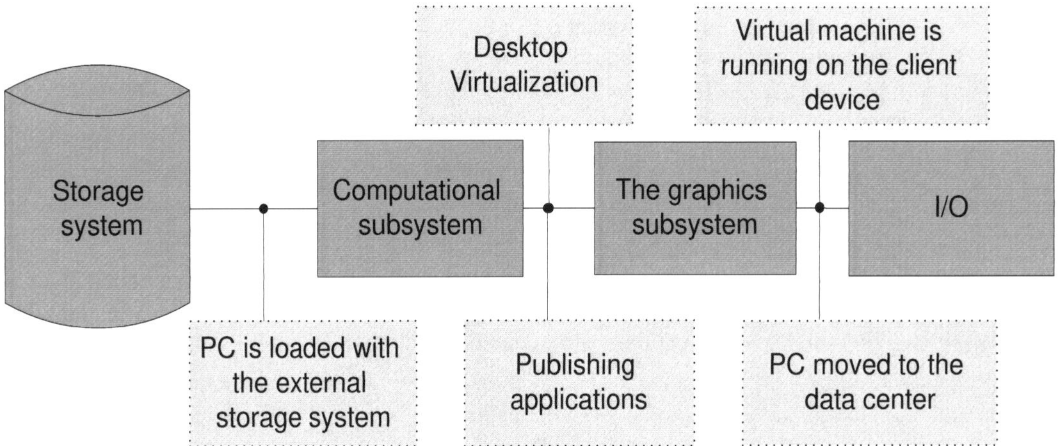

Figure 3. Different approaches centralized management

It should be noted that it is difficult to name some of these universal solutions [8]. Depending on the requirements of the infrastructure operating environment may use one or the other approach, or combinations thereof. It should be noted that it may remain a task that will remain outside of these approaches. Consider approaches that provide flexibility in the organization of the working environment.

For users who need to provide mobility, using the classical set of applications most appropriate solution is desktop virtualization. If you want more processing power, centralized data protection approach is used in which the PC is loaded with the external storage system. In those cases where it takes a lot of computing resources, such as graphic workstations, PCs displacement method is applicable to a data center. This will optimize the management of power and cooling, as well as improve the ergonomics of the workspace. In the case of performing specialized applications on mobile computers may be the best approach is the implementation of the virtual machine on a PC.

Thus, the conclusion that before using any technology necessary to analyze the needs of the enterprise. This will create an integrated system that links a variety of technologies, methodologies, and processes occurring in the enterprise. Currently developed a database that allows to evaluate IT infrastructure, based on the results to develop an action plan for the modernization and further development of the IT infrastructure to meet the requirements necessary enterprise methodologies.

5. Conclusions

It should be noted that it is difficult to name any of the above solutions universal [8]. Depending on the requirements for the infrastructure of the work environment, it is possible to apply this or that approach, or their combinations. Here it should be noted that there can be a task that will remain outside the considered.

For a user who needs to provide mobility, when using the classic set of applications, the most appropriate solution is the virtualization of the PC. In the event that it is necessary to provide more computing power, centralized data protection applies the approach in which the PC is booted from an external storage system. In those cases where a lot of computing resource is required, for example, graphic stations, we apply the way of moving the PC to the data center. This will optimize the management of power and cooling systems, as well as improve the ergonomics of the workspace. In the case of performing specialized applications on mobile computers, the best approach may be to run a virtual machine on a PC.

Thus, the conclusion suggests that before the application, this or that, the technology requires an analysis of the needs of the enterprise. This allows you to create an integrated system that connects various technologies, methodologies, and processes occurring in the enterprise. Currently, a database has been developed that allows to assess the IT infrastructure of the enterprise, based on the results obtained, an action plan is being developed to modernize and further develop the IT infrastructure in accordance with the requirements of the necessary enterprise.

When choosing technologies for the development of the company's IT infrastructure, one should pay attention to the prospects of using the chosen technology. So, according to the principles of building computing systems, virtualization technologies have a limited development limit, at the end of which it is possible to replace innovative technologies. But for today this technology is one of the most effective in application.

In the organization of flexible management of the working environment, a comprehensive assessment of the tasks facing us should be carried out. It is necessary to conduct a detailed collection of data on the interaction of the collective of the enterprise to create an optimal information structure. Only taking into account all requirements, it is possible to achieve results that meet the requirements of the enterprise and increase the efficiency of the work.

References

[1] Azaryeva V V and Stepanov S A 2013 *"Quality Management": Proc. of the Regional Sci. and Pract. Conf.* (Voronez: Voronezh Gos. Tekhn. Un-t) pp 101-104

[2] Dukhanov A V 2010 *Simulation Modeling of Complex Systems* (Vladimir: Vladimir State Un-ty) p 115

[3] Jan Van Bong 2012 *IT Service Management, Introduction* (SMF)

[4] Jacob D Bekenstein 2013 *Scholarpeda* http://www.scholarpedia.org/article/Bekenstein_bound

[5] Jerry Wu and Yin-Lin Shen 2013 *Computational Intelligence and Soft Computing* article ID 426962

[6] *A survey among **565** IT directors in the US and Europe* 2007 Source Desktop and Application Virtualization, Forester Research Inc.

[7] Moiseev V V, Bulakina O N and Bulakina E N 2013 *Fundamental Research* **6** (part 3) 709–14

[8] www.rae.ru/fs/?section=content&op=show_article&article_id=10000789

[9] Moiseev V V, Ketov A V and Lebedkin P V 2010 *Proc of the VII All-Russian Conf. of students, graduate students and young scientists in Microsoft Technologies in Programming Theory and Practice* (Tomsk: TPU) pp 126–128

IOP Conf. Series: Materials Science and Engineering **411** (2018) 012051 doi:10.1088/1757-899X/411/1/012051

Receiving of layered composite materials with shape memory effect of medical appointment

E O Nasakina, A S Baikin, S V Konushkin, K V Sergiyenko, M A Kaplan, I M Fedyuk, M A Sevost'yanov and A G Kolmakov

Baikov Institute of Metallurgy and Material Sciences, 49 Leninsky ave, Moscow, 119991, Russia

E-mail: nacakina@mail.ru

Abstract. Tantalum and titanium surface layers were created on nanostructural TiNi substrates when the conditions of magnetron sputtering are varied. The state of the TiNi surface after various treatments has been studied. Structure and composition of samples were defined by SEM, AES. With increase in time and output of sputtering and decrease in deposition distance surface layer thickness not linearly increases. The transitional layer provide high adhesion of a surface layer to a substrate.

1. Introduction

Materials with shape memory effect (SME) are the best candidates for the creation of medical implants that are plastically deformed in a cooled state to an extremely compact form (facilitating easier and less traumatic delivery to the necessary site of the organism without cavitary surgery) and independently assuming functional form in the specified operational conditions without additional impact [1-12]. The most famous medical material from this class is nitinol, endowed with mechanical characteristics similar to the behavior of living tissues, which helps it to adapt to physiological stresses, providing the necessary operating conditions. But in addition to positive mechanical characteristics, this alloy is endowed with a number of shortcomings: the difficulty of processing in the manufacture of products, high content of a toxic element, a disputable level of biocompatibility.

In the modern world, an effective way to improve operational characteristics and eliminate the drawbacks of classical materials is the formation of composite materials on their basis [13-14].

This work is devoted to the comparison of the surface state of nitnol depending on various treatments and of metal layered composites based on it with surface layers of tantalum or titanium of medical appointment.

2. Materials and methods

An object for investigations was wires with diameter 280 μm made of nanostructural nitinol (55.91 wt % Ni–44.03 wt % Ti) and composites based on it with tantalum or titanium surface layers, obtained by the magnetron sputtering method [14].

The wires in the initial state were sequentially polished by abrasive paper from 180 to 1000 grit and processed finally by GOI (State Optical Institute) paste to a mirror surface with the purpose to surface quality improve. The decrease in the diameter was to 10 μm in comparison with the original. For end stabilization of the B2 phase and shaping, wires in the initial state or after polishing were annealed at 450°C for 15 min in air. The wire diameter was invariable.

Composites were produced in argon at residual and working pressures of $4x10^{-4}$ and 0.4 Pa, respectively. The magnetron, with a reagent-grade target, was operated at a sputtering distance on the order of 150 mm, at direct current of 860 mA and voltage of 400 V. The sputtering time was 80 min on the main surface with rotation and 30 min on end faces, and the substrate rotation rate was 9 rpm. Preparatory ion etching of the substrate (cleaning, activating, and polishing the surface by bombarding it with argon ions) was carried out with the discharge parameters Ue = 900 V, Ie = 80 mA.

The surface morphology and the layer-by-layer composition were investigated on a TESCAN VEGA II SBU scanning electron microscope equipped with an INCA Energy energy dispersive spectrometer system and on a JAMP 9500F JEOL Auger spectrometer in combination with ion etching at argon bombardment under an angle of 30°.

3. Results and discussion

The initial TiNi surface is heterogeneous and covered by spots, dark areas alternate with bright ones, and the high roughness and defects because of drawing during production are clearly expressed (figure 1a). The compositions of bright and dark spots are different: a high content of titanium oxide is revealed in bright ones; and carbon, in dark ones. Both layers attain 4.5 μm in thickness and are not placed over each other. Such a thick surface layer, as was believed, was a result of the long intermediate thermal treatment during the wire production.

After annealing, the surface externally is similar to the initial one, except for small increase in a share of light oxide-containing sites (figure 1b).

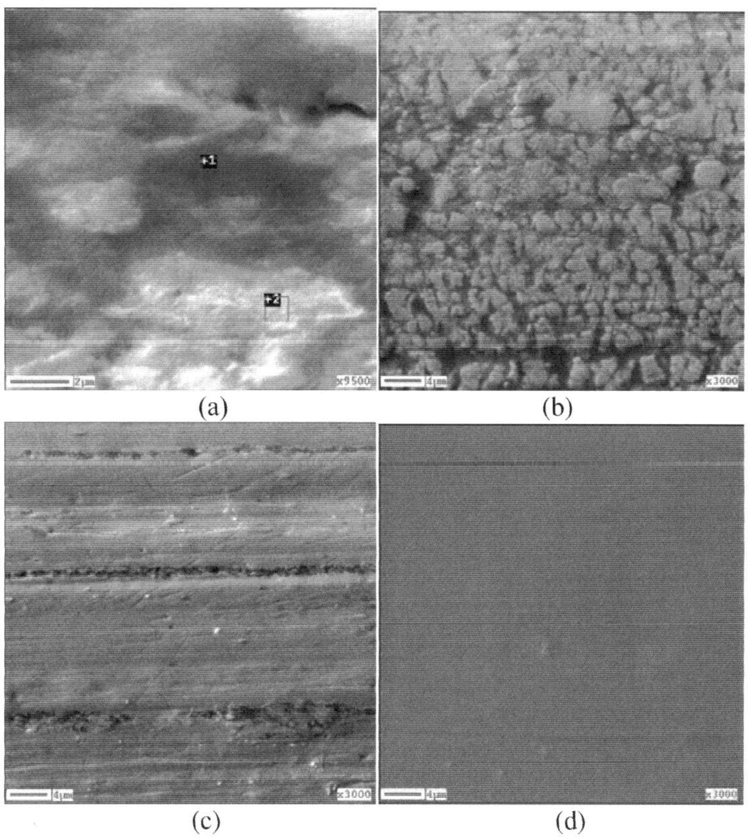

Figure 1. Morphology of the surface of nitinol: a) in the initial state, b) after annealing, c) after grinding, d) after grinding and annealing.

After polishing practically all defects and the roughness are smoothed, spots are absent, and only traces of treatment are visible (figure 1c). Surface microdefects are grooves with a depth and width of less than 1 μm. The composition of the polished surface is homogeneous: the entire wire is covered by an oxide layer less than 20 nm in thickness. The minus is the presence of nickel in the surface layer, though in an insignificant amount.

The subsequent annealing promotes formation of an oxide-nitride layer up to 80-150 nanometers in depth, to 50–60 nanometers free from nickel that could affect corrosion resistance of material positively. Surface defects were not observed at all, the entire roughness was smoothed, an extremely high surface uniformity was observed (figure 1d).

The received composite materials had layered structure "a surface layer from the deposited substance (thickness ~ 0.9 microns) – the transitional layer containing elements both of the surface layer and of a basis (thickness ~ 0.2 microns) – a basis" (figure 2a), the morphology of the surface corresponds to the morphology of the surface of the substrate after grinding and annealing (figure 2b).

With increasing time and power of sputtering and a decrease in the sputtering distance, the thickness of the surface layer increases nonlinearly. The regions of saturation of the surface and transition layers are marked, after which the thickness practically does not change when the conditions change. The transition layer provides high adhesion of the surface layer to the substrate.

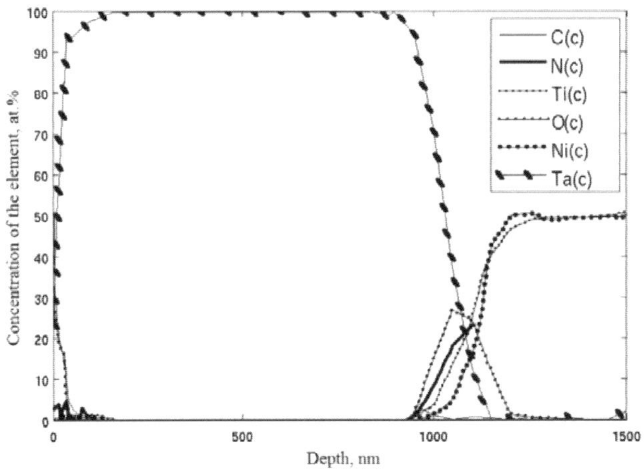

(a)

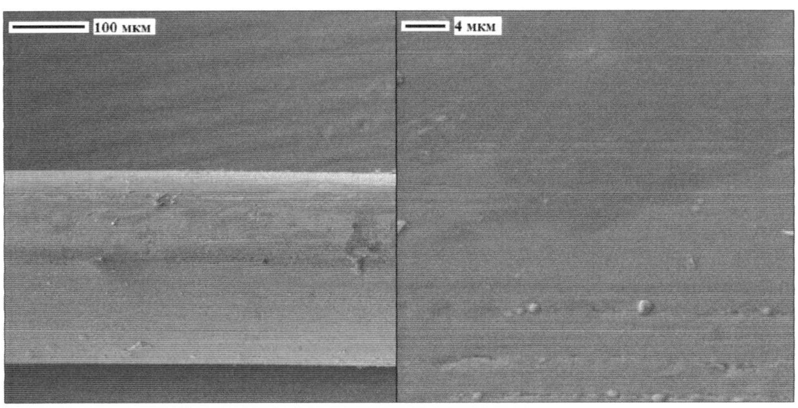

(b)

Figure 2. The layered composition (a) and morphology (b) of the surface of composites with a tantalum surface layer.

4. Conclusion

Layered composite materials with surface layers of tantalum and titanium and a base of nanostructured titanium nickelide, as well as a base material in a different state, were obtained and investigated.

Acknowledgments

The work was supported by the grant of the President of the Russian Federation for state support of young Russian scientists - candidates of sciences (MK-4521.2018.8).

References

[1] Petrini L and Migliavacca F 2011 *Journal of Metallurgy* **2011** 1–15
[2] Duerig T W, Melton K N, Wayman C M and Stockel D 1990 *Engineering Aspects of Shape–memory Alloys* (Oxford: Butterworth Heinemann Ltd) pp 181–194
[3] Shabalovskaya S 1996 *Bio Med Mater Eng* **6** 267–289
[4] Marjan Bahrami Nasab and Mohd Roshdi Hassan 2010 *Trends in Biomaterials and Artificial Organs* **24(1)** 69–82
[5] Surdell D, Shaibani A, Bendok B and Eskandari M K 2007 *J Vasc Interv Radiol* **18(10)** 1297–1299
[6] Bose A, Hartmann M and Henkes H A 2007 *Stroke* **38** 1531–1537
[7] Duerig T W, Tolomeo D E and Wholey M 2000 *Minim Invasive Ther Allied Tech* **9** 235–246
[8] Horowitz M B and Purdy P D 2005 *Progress in Neurological Surgery* **17** 18
[9] Rösch J, Keller F S and Kaufman J A 2003 *J Vasc Interv Radiol* **14** 841–853
[10] Balazic M and Kopac J 2007 *J. Achiev. Mater. Manuf. Eng* **25(2)** 31–34
[11] Meier B 2006 *Hospital Chronicles* Suppl. 99–103
[12] Stoeckel D, Pelton A and Duerig T 2004 *Eur Radiol* **14** 292–301
[13] Kim J H, Shin J H, Shin D H, Moon M W, Park K and Kim T H et al 2011 *Br J Radiol* **84** 210–215
[14] Sevostyanov M A, Fedotov A Yu, Kolmakov A G, Zabolotnyi V T, Barinov S M and Goncharenko B A et al 2014 *Inorg. Mater. Appl. Res.* **5(4)** 344–346

IOP Conf. Series: Materials Science and Engineering **411** (2018) 012052 doi:10.1088/1757-899X/411/1/012052

On the possibility of carbonaceous dust waste use of prebaked anode production in silicon metallurgy

N V Nemchinova, A A Tyutrin, N A Korepina and S S Belskii

Irkutsk National Research Technical University, 83 Lermontova street, Irkutsk, 664074, Russia

E-mail: ninavn@yandex.ru

Abstract. During the prebaked anodes production used in aluminium electrometallurgy a considerable amount of fine-dispersed waste with a carbon content of more than 90% is formed. Analytical studies of the waste properties such as chemical, particle size analysis, moisture content and some others have been conducted. Electro-paramagnetic resonance (EPR) method was used to study the structures of charcoal, oil coke and a sample of carbon-containing dust from mixing and pressing compartment of aspiration equipment. EPR results showed that the studied sample is similar in structure to natural traditional reductants (its EPR spectrum has a pronounced narrow Gaussian signal). This indirectly shows its satisfactory reactivity and can be recommended as an additive (in pelletized form) to the charge for the silicon production

1. Introduction

The process of silicon technology in electric furnaces and the process operation depend on the use of different carbonaceous reducing agents or carbonaceous reductants (CR). The dominant requirements applicable to them are low ash level, high reactivity (R), high specific electrical resistivity (SER) and low cost [1–3].

Nowadays at ZAO SILICON UC RUSAL charcoal, oilcoke, mineral carbon from different deposits (Kazakhstan, Colombia) and woodchips (as a desintegrator) have wide spread occurrence as CR.

Charcoal is one of the best carbonaceous materials complying with the requirements of crystalline silicon production technique. However, considering its insufficiency and high cost it is necessary to search for new carbonaceous reductants able to partially substitute some traditional reductants in order to cut down prices for the finished product.

JSC “RUSAL Sayanogorsk” is the third largest aluminium smelter in the UC RUSAL structure. The plant uses the technique of baths equipped with prebaked anodes (PA) [4] and has a workshop producing electrodes. Producing PA, a great amount of finely dispersed carbon-containing waste is formed, that, in turn, has a negative influence on the ecology near the enterprise.

The present research is aimed at the study of the properties of the given type of technogeneous raw stock.

2. Research materials

At present electrode production at aluminium smelters is an independent multistep works with a branched transport and technology structure. Thus, electrode production workshop of JSC “RUSAL

Sayanogorsk" consists of three departments: Green Anode Plant (GAP), green anode baking area (BA) and anode assembly area (AAA).

The multistep process of PA production includes some procedures like grinding, breakage, sizing, mixing, stub-end cleaning and etc. During the steps there is a large amount of dust partially captured by gas-cleaning plant. The dust partly precipitates in the production unit place:

1. GAP dust from dust exhausting plants;
2. GAP dust from plant cleaning;
3. BA dust from pot tending machines;
4. BA dust from coke filling unit cleaning;
5. BA dust from dust exhausting plants;
6. AAA dust from a barreling machine;
7. AAA dust from dust exhausting plants;
8. AAA dust from stubs and spoiled units fr. -0.5 mm breaking;
9. AAA dust from stubs shot blasting.

Practically, all waste, except the dust from dust exhausting plants of Green Anode Plant, is not used in the production again and stockpiled on the landfill.

3. Study results

The tests of the study specimens according to the size grade were carried out with the help of a lazer particle-size analyzer, Analysette 22 NanoTecplus (Fritsch, Germany). The device tests particle size in a liquid fluid. As test specimens have poor wettability, the suspension of dust and surface-active substances was prepared for measuring.

The results of the granulometric composition of the test specimens of the bake furnace dust are presented in the form of the figures with an integral curve and differential distribution as well as tables. The table of the size grade distribution consists of two main columns: size (µm) and size yield (%). The particle size is represented in the form of the size grade from 0 µm to the indicated number, i.e. if the size is 40 µm and the size yield is 99.3 %, it means that the particles of the size less than 40 µm make 99.3 %. From the table of the size grade distribution one can conclude what is prevailing in the size grade. For example, if the size yield is 90 % and the size is 26.4 µm, this means that 90 % of the probe has the size less than 26,4 µm.

The example of the granulometric composition study of GAP dust from dust exhausting plants is shown in Table 1 and 2 as well as Figure 1.

GAP dust from dust exhausting plants (Specimen 1) has size of -50 µm (figure 1, table 1), and 90 % is grade of -26.4 µm (table 2).

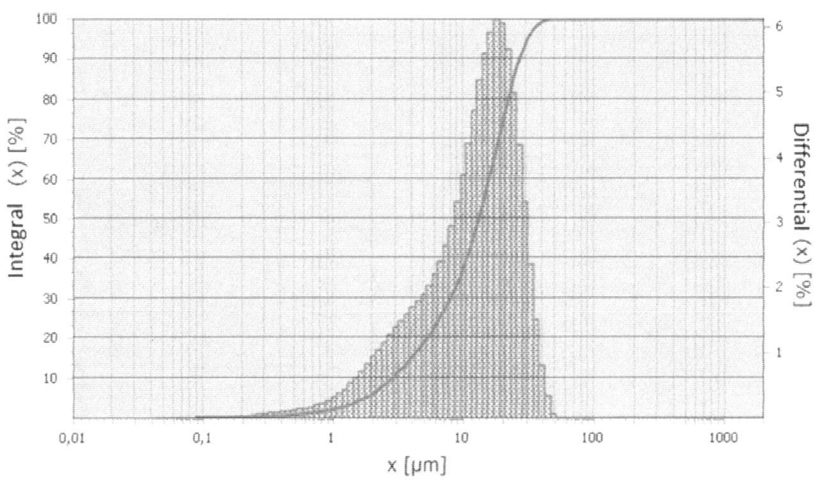

Figure 1. Distribution of dust particle specimen from dust exhausting plants GAP (Specimen 1) by size.

Table 1. Specimen 1 particle distribution by size grade.

Size, μm	Size yield, %					
	Average	Measuring number				
		1	2	3	4	5
0.3	0.2	0.2	0.2	0.2	0.2	0.2
0.5	0.6	0.5	0.6	0.6	0.6	0.6
1	1.8	1.6	1.9	1.9	1.9	1.9
10	40.1	44.4	41.3	39.2	38.2	37.4
20	76	78.6	77	75.5	74.8	74.1
30	94.4	95.1	94.7	94.3	94	93.7
40	99.3	99.4	99.4	99.3	99.3	99.2
50	100	100	100	100	100	100

Table 2. Specimen 1 particle distribution by size yield.

Size yield, %	Size, μm					
	Average	Measuring number				
		1	2	3	4	5
5	1.8	1.8	1.7	1.7	1.8	1.8
10	2.7	2.6	2.7	2.7	2.8	2.9
20	4.9	4.4	4.7	4.9	5.1	5.3
30	7.4	6.5	7.1	7.6	7.8	8
40	10	8.9	9.7	10.2	10.5	10.7
50	12.5	11.4	12.2	12.8	13	13.2
60	15.2	14.1	14.8	15.4	15.6	15.9
70	18	17	17.7	18.2	18.5	18.7
80	21.4	20.5	21.1	21.6	21.9	22.1
90	26.4	25.5	26.1	26.6	26.8	27.1
95	30.6	29.9	30.3	30.7	30.9	31.2
98	35.3	34.4	34.9	35.4	35.6	36
100	51.1	51.1	51.1	51.1	51.1	51.1

The most part of finely dispersed carbon-containing waste of the three departments of the electrode production workshop (Green Anode Plant, green anode baking area, anode assembly area) is represented by the fraction of -100 μm, that creates a problem of the further direct use of the materials [5]. However, their main value is high content of carbon (more than 90 %) and low ash level of some specimens (0,29-1,35 %):

No. 1 – 0.53 %; No. 2 – 0.45 %; No. 3 – 6.03 %;
No. 4 – 3.58 %; No. 5 – 0.29 %; No. 6 – 61.71 %;
No. 7 – 1.20 %; No. 8 – 1.35 %; No. 9 – 31.33 %.

The ash content of the specimens was rated according to GOST 22692-77. The essence of the method is in ashing of the specimen of the tested material in a muffle furnace and baking of the ash residue to the fixed mass at temperature of (850±20) °C.

The chemical composition of ash residue was analyzed with the help of a wavelength-dispersive spectrometer, S8 TIGER (Bruker, Germany), using X-ray fluorescence spectroscopy, table 3.

The carried out researches identifying the chemical composition of the ash residue of the carbonaceous dust specimens do not demonstrate considerable contrasts to the component analysis. Moreover, the largest amount of ferric oxide is registered in the specimen of GAP dust from the workshop cleaning, that can be indicative of the availability of some contaminating impurities of inorganic nature (metallic dust and etc.).

Table 3. Chemical composition of ash residue of tested specimens.

Constituent	Content in test specimens, % mass.								
	1	2	3	4	5	6	7	8	9
Fe_2O_3	19.85	62.71	11.69	7.88	16.85	61.44	28.44	34.69	48.1
SiO_2	29.96	13.74	38.27	41.23	20.07	5.32	6.13	10.03	1.71
Al_2O_3	19.13	6.94	28.71	33.83	29.94	14.3	33.2	24.4	23.91
CaO	8.66	2.73	2.94	2.55	6.17	0.77	3.48	3.36	1.74
V_2O_5	4.86	4.35	4.31	2.53	7.67	0.05	2.76	2.68	0.14
Na_2O	4.5	2.52	4.29	5.06	6.56	10.11	10.19	9.98	12.82
SO_3	3.98	1.21	0.07	0.12	3.05	6.68	11.38	8.65	8.87
NiO	3.27	2.46	2.01	0.91	4.58	0.06	1.53	1.43	0.22
MgO	3.25	1.68	1.25	1.04	1.31	0.17	0.82	0.69	0.17
P_2O_5	0.68	0.31	1.66	0.92	0.71	0.24	0.34	0.28	0.08
TiO_2	0.48	0.29	1.56	0.95	0.45	0.05	0.17	0.15	0.01
ZnO	0.28	0.17	0.18	0.07	0.74	0.0078	0.28	2.54	0.02
MnO	0.15	0.28	0.19	0.12	0.18	0.32	0.19	0.22	0.2
CuO	0.11	0.12	0.08	0.05	0.22	0.08	0.18	0.16	0.08
Other	0.84	0.49	2.79	2.74	1.5	0.4022	0.91	0.74	1.93

The moisture determination of the analysis sample of the dust specimens from different places is fulfilled according to GOST 11014-2001.

All 9 specimens meet the requirements of carbon content and moisture, imposed to CR in silicon production. Specimens No.1,2,4,5,7 and 8 have the best results of ash level.

The analysis of carbon-base material properties was realized by electro-paramagnetic resonance method (EPR) [6]. EPR-spectra were registered by a spectrometer, X-band Brüker ELEXSYSE-580 (X-band 9.7 GHz). The specimen spin concentration was estimated with a well-known computational method [7] using diphenyl picryl hydrazyl (DPPH) as a standard.

In EPR spectra of the specimens of charcoal, petcoke and GAP dust from dust exhausting plants a pronounced narrow Gaussian signal is observed. Similar signals are conditioned by free radicals (with g-factor, close to the value of a free electron of the standard of DPPH – g_e 2.0023), their electrons are strongly delocalized in a highly conjugated substance system [8].

In EPR spectra of the specimens of GAP dust from the workshop cleaning there are uniform broad anisotropic signals (signal width is ΔH = 1500-2000 G, g_u-factor ~3.5, line form – Dyson), which are defined by high density of delocalized π-electrons and characteristic of ferromagnetic materials. Such signals are similar to the broad lines for the composites based on the graphene matrices, graphene itself after high-temperature processing, carbon nanotubes [9]. In the spectra there is also a narrow signal in the range of g-factor 2.00, which is weakly intensive.

4. Conclusion

The carried out analyses by EPR method demonstrate that the specimens of carbonaceous GAP dust from dust exhausting plants are similar to the structure of natural CR, that indirectly indicates their satisfactory reactivity worth and they can be used as addition agents to the traditional reducing agents in silicon metallurgy. The specimen of GAP dust from the workshop cleaning has slightly worse results from the viewpoint of structure and activity. However, for more complete information it is necessary to conduct some supplementary investigations. Besides, to use the data of the carbonaceous dust waste in silicon production it is essential to apply the methods of their preliminary balling [10].

Acknowledgments
This work was performed in Irkutsk National Research Technical University. The authors like to acknowledge the financial support from the Ministry of Education and Science of the Russian Federation (Agreement No. 11.7210.2017/8.9).

References
[1] Gasik M I and Gasik M M 2011 *Electrothermics of Silicon* (Dnepropetrovsk: National Metallurgical Academy of Ukraine) p 487
[2] Katkov O M 1999 *Smelting of Technical Silicon* (Irkutsk: Irkutsk State Technical University) p 243
[3] Nemchinova N V 2017 *Physicochemistry and Carbothermy of Silicon* (Irkutsk: Irkutsk National Research Technical University) p 286
[4] Yanko E A 2001 *Anodes of Aluminium Electrolyzers* (Moskow: Ore and Metals) p 671
[5] Nemchinova N V, Tyutrin A A, Piavkina A A, Mikhalev V A and Fedotova E A 2016 *Proc. IX Int. Congress on Non-ferrous Metals and Minerals* (Krasnoyarsk: Science and Innovation Center) p 94
[6] Wertz J E and Bolton J R 1975 *Theory and Practical Applications of EPR method* (Moskow: Mir) p 550
[7] Poole C P 1997 *Electron Spin Resonance: A Comprehensive Treatise on Experimental Techniques/Second Edition* (Dover Publications) p 810
[8] Fedin M Gromov I and Schweiger A 2004 *Journal of Magnetic Resonance* **171** 80–9
[9] Bocheng Qiu et al 2015 *Scientific Reports* **5** 8591
[10] Nemchinova N V, Leonova M S and Tyutrin A A 2017 *Proceedings of Irkutsk State Technical University* **1** 209–17

IOP Conf. Series: Materials Science and Engineering **411** (2018) 012053 doi:10.1088/1757-899X/411/1/012053

Energy-saving method of cutting a pre-bent bar on the shear machine

A G Nikitin[1], E I Demina[1], E Ya Zhivago[1], L T Dvornikov[1] and L A Saruev[2]

[1]Siberian State Industrial University, 42 Kirova street, Novokuznetsk, 654007, Russia
[2]Tomsk Polytechnic University, 30 Lenina ave, Tomsk, 634050, Russia

E-mail: nikitin1601@yandex.ru

Abstract. For transverse cutting of rolled products on mills, a shear machine with parallel knives is used. The permissible cross-sectional dimensions of rolled products, cut on this type of shear machine, are determined by the maximum force needed for cutting the bar in the cold state. It is obvious, that the lower the cutting force required, the lower the energy consumption. In order to reduce energy consumption during the rolled metal cutting at Siberian State Industrial University a cutting device for cutting a pre-bent bar has been developed in which the bar in the cutting zone under the action of gravity is bent due to the fact that the level of the outgoing roller table is located below the level of the approach roller table. A technique is proposed for calculating the force of cutting a pre-bent bar on the shear machine. The conditions are determined (the ratio of the length of the bar overhang and the difference in the levels of the approach and outgoing roller tables) for obtaining a qualitative cut when cutting the pre-bent bar. It is established that when cutting a pre-bent bar, the maximum cutting force is less than when cutting a straight bar.

1. Introduction

For transverse cutting of the rolled products on mills, shear machine with parallel knives are used. The permissible cross-sectional dimensions of the bar cut on such shear machine of this type are determined by the cutting force of the bar in the cold state P, which can be determined [1] from the dependence

$$P = \tau \cdot F, \tag{1}$$

where: τ – tangential stresses arising during the cutting of a straight bar; F – the cross-sectional area of the bar. The cutting force reaches its maximum value at $\tau = \tau_u$, where τ_u – the ultimate strength of the cut material.

It is obvious that the value of cutting force P determines the energy consumption of the cutting process, and therefore it can be adopted as a criterion for the optimality of the process being studied.

2. Results and discussion

In order to reduce energy consumption during cutting of rolled metal at Siberian State Industrial University a cutting device for a pre-bent bar has been developed [2] with the level of the outgoing roller table (figure 1) located below the level of the approach roller table, which leads to spontaneous bending of the bar under its gravity and appearance of normal stresses in it before the beginning of the cutting process.

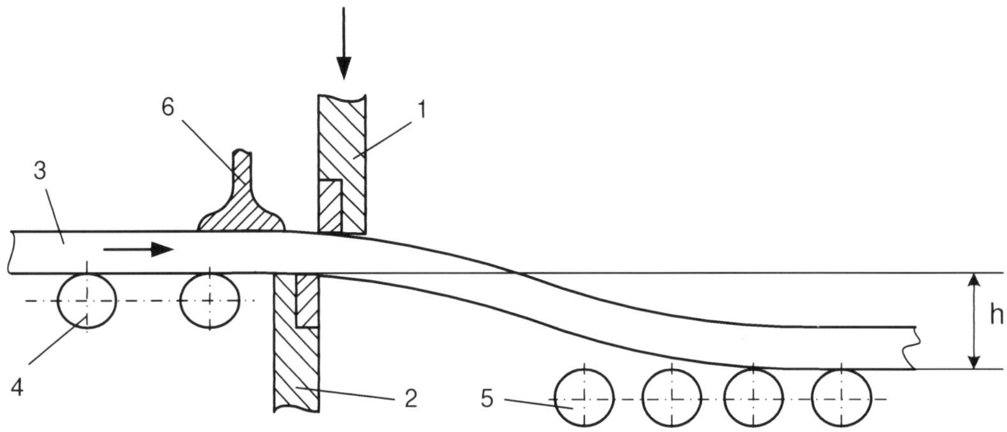

Figure 1. Scheme of cutting the pre-bent bar of rolled metal: 1 – upper knife; 2 – lower knife; 3 – bar; 4 – approach roller table; 5 – outgoing roller table; 6 - clamp.

The cutting process is as follows. Before the beginning of cutting, the bars of knives are diluted and the bar passes between them along the roller table; the upper knife is above the level of the upper surface of the bar and does not interfere with the movement of the bar. Then, by using the stop, the bar stops at a predetermined position along the length. Under the action of gravity the descending part of the bar, lying on the approach roller table, bends, and the part of the bar located on the feeding roller table is rigidly fixed with the help of a clamp.

In this case, normal stresses σ appear in the bent part of the bar even before the cutting process begins, which reach their maximum value in the marginal fibers along the height of the workpiece in the cutting zone. When the upper knife is lowered, the cutting process takes place, during which the shearing stresses arise in the cutting plane due to the force of the moving knife. Thus, in the cutting zone of the bar there is a complex stressed state, in which the magnitude of the stresses arising in the section is determined from the third strength hypothesis [3]:

$$\sigma_{eq} = \sqrt{\sigma^2 + 4\tau^2} \le \sigma_u \approx 2\tau_u, \tag{2}$$

where σ_u - the ultimate strength of the bar material under the action of normal stresses. Expressing the action of tangential stresses from (1) and substituting in (2) we obtain

$$\sqrt{\sigma^2 + 4\frac{P^2}{F^2}} \approx 2\tau_u, \tag{3}$$

where the cutting force P is defined as

$$P \approx \sqrt{\frac{4\tau_u^2 - \sigma^2}{4} \cdot F^2} \approx \sqrt{\tau_u^2 - \frac{\sigma^2}{4}} \cdot F. \tag{4}$$

Analysis of dependencies (1) and (4) shows that the required cutting force of the pre-bent bar is less than that required force for cutting a straight bar.

The straightness of the workpiece is a prerequisite for obtaining a quality product in the form of a measured product (cut billet), therefore, the deformation of the bending of the bar before the cut must not exceed the elastic deformations, that is, the normal stresses in the bar during its bent, should be less than the yield strength of the bar material:

$$\sigma = M_{bent} / W \leq \sigma_{\mathrm{t}}, \tag{5}$$

where M_{bent} – the internal bending moment arising in the bar under the action of gravity not supported by the rollers of the outgoing roller table; W – resistance moment of the section of the bar. The magnitude of the bending moment of the bar is determined by [3]

$$M_{bent} = ql^2/2, \tag{6}$$

where $q = F\gamma g$ – distributed load acting on the bent part of the bar under the action of gravity; γ – specific mass of the bar material; g – acceleration of gravity; l – length of the bar at which the normal stresses remain elastic in the part of the bar that does not rest on the rollers of the approach roller table.

The maximum allowable length of the bar, with allowance for (5), will be determined by the dependence $l_{\mathrm{ad}} = \sqrt{\dfrac{2W \cdot \sigma_{\mathrm{t}}}{\mathrm{F}\gamma \mathrm{g}}}$. If the length of the cut bar is less than l_{ad}, then the bar will be deformed elastically without additional support in the form of the upper surface of the roller of the outgoing roller table. If the length of the cut bar is greater than l_{ad}, then the residual deformations will appear in the bar without additional support, which is not permissible.

The distance from the top surface of the roller of the approach roller table to the upper surface of the roller of the outgoing roller table is determined from the condition that with the maximum allowable deflection, the normal stresses arising in the bar will be less than the yield strength of the bar material. The maximum allowable deflection given by the difference in the upper levels of the approach and outgoing roller tables is determined from the allowable length of the bar, under elastic deformation of the bar [4]: $h_{\max} = \dfrac{ql_{\mathrm{ad}}^4}{8EJ} = \dfrac{W^2\sigma_{\mathrm{t}}^2}{2\mathrm{F}\gamma \mathrm{gEJ}}$, where E is the modulus of elasticity of the first kind of bar material; J – the moment of inertia of the bar section.

Obviously, the larger the deflection, the larger the normal stresses appearing in the bent bar and the lesser force must be applied to perform the cutting operation. However, because of the dispersion of the mechanical characteristics of the material at the maximum calculated value of the deflection, a residual plastic deformation of the bar is possible [5], therefore the maximum permissible deflection should be taken less than the calculated deflection, that is, $h = 0.8\ h_{max}$.

To verify the correctness of the analytical comparison of the cutting forces of the straight bar and the pre-bent one, a cutting force of 40 x 40 mm square was made from steel 40, for which the tensile-compression strength σ_u = 568 MPa, the shear strength τ_u = 340 MPa, and the yield stress at tension-compression σ_t = 333 MPa.

According to formula (1), the cutting force of the straight bar is 544 kN, and the cutting force of the bar material pre-bent to the yield strength by formula (4) is 472 kN. As it follows from the calculation, the cutting force of the pre-bent bar is 14% less than the cutting force of the straight bar.

3. Conclusion

Conditions are determined (the ratio of the length of the bar overhang and the difference in the levels of the approach and outgoing roller tables) for obtaining a qualitative cut when cutting a pre-bent bar.

IOP Conf. Series: Materials Science and Engineering **411** (2018) 012053 doi:10.1088/1757-899X/411/1/012053

It is established that when cutting a pre-bent bar, the cutting force is less compared to cutting a straight bar, which makes the proposed cutting process more energy efficient.

References

[1] *Machines and Aggregates of Metallurgical Plants* 1988 ed A I Tselikov vol 3 (Moscow: Metallurgy) p 560
[2] Nikitin A G, Belov E G and Poltaratsky L M 2010 *Device for Cutting Rolled Products* patent of the Russian Federation No. 91911 Bul. 7
[3] Belyaev N M 1965 *Resistance of Materials* (Moscow: Nauka) p 312
[4] Stepin P A 2012 *Resistance of Materials* (St. Petersburg: Lan) p 320
[5] *Reference Book of a Metal Worker* 1986 ed P N Orlov and E A Skorokhodov (Moscow: Mechanical Engineering) p 960

IOP Conf. Series: Materials Science and Engineering **411** (2018) 012054 doi:10.1088/1757-899X/411/1/012054

Improvement of jaw crushers reliability using elastic pneumatic elements in the connection of kinematic pairs

A G Nikitin[1], A V Abramov[1], E Ya Zhivago[1], L T Dvornikov[1] and L A Saruev[2]

[1]Siberian State Industrial University, 42 Kirova street, Novokuznetsk, 654007, Russia
[2]Tomsk Polytechnic University, 30 Lenina ave, Tomsk, 634050, Russia

E-mail: nikitin1601@yandex.ru

Abstract. Downtimes associated with the replacement of the plain bearing liners because of their tear and wear are among the numerous technical reasons for relatively short-term but rather frequent downtimes in the conditions of operation of jaw crushers. The rapid wear of the liners is the result not only of contact friction between axle journals and liners, but also the action of dynamic forces. Dynamic forces that arise during the operation of jaw crushers are due to the presence of a gap in the kinematic pairing of the links (journals and bearing lining) and discrete speed values of the relative movement of the links inside the clearances of plain bearings. Thus, the reliable operation of the machine largely depends on the creation of conditions that ensure a gapless contact of the conjugated links. It was experimentally proved that the use of mechanisms for the selection of gaps with elastic elements, which during the whole cycle of operation of the crank and rocker mechanism select a gap in the joints of kinematic pairs, prevent the appearance of additional dynamic loads and, thereby, increase the reliability of the jaw crushers.

1. Introduction

In almost all metallurgical operations crushed material is required as a raw product obtained using crushing machines, as well as processing products of metallurgical industry: slags, ferroalloys and others. The main indicator of the quality of crushers, characterizing their technical level and competitiveness, are their reliability and durability under operating conditions, which can be quantitatively estimated by the number of machine failures in the process of obtaining finished products [1]. It is obvious that any reduction in machine downtime for elimination of failures increases the productivity of the machine as a whole.

Downtimes associated with the replacement of the plain bearing liners because of their tear and wear are among the numerous technical reasons for relatively short-term but rather frequent downtimes in the conditions of operation of jaw crushers.

The rapid wear of the liners is the result not only of contact friction between axle journals and liners, but also the actions of dynamic forces that are the source of machine vibrations, noise, premature breakdown of threaded joints and breakdowns of parts caused by an increase in the number of loading cycles of alternating loads, fatigue strength of its elements.

Dynamic forces arising during the operation of jaw crushers are due to the presence of gaps in the kinematic pairs of links connection (journals and bearing linings) and discrete speed values of the of the relative movement of the links inside the clearances of the plain bearings.

Nodes and parts of jaw crushers in the process of their operation are exposed the dynamic forces caused by shock loads that arise when the load is released, when the (technological) force is removed from the previously elastically deformed mechanical system when the crushing zone of the jaw crusher is discharged, since under the force of the technological resistance of the journal is pressed against the surface of the bearing lining located on the opposite side of the crushing zone, and after removing the force of the technological resistance the journals under the action of elastic forces moves to the opposite side by the size of the gap, bumping against the liner [2], with a runaway of the gap followed by bounces that decay in strength due to damping. Emerging dynamic (shock) effects lead to the excitation of elastic vibrations of the mechanical system: bearings, bearing housings, the frame as a whole.

Clearances in kinematic pairs, the presence of which are unavoidable in order to ensure relative mobility of links, with increasing duration of operation of jaw crushers, gradually increase, which leads to a decrease in the smooth operation of the machine and the quality of the product obtained (change in fractional composition), and to the increase in additional dynamic forces.

Thus, the reliable operation of the machine largely depends on the creation of conditions that ensure a gapless contact of the connected links. In practice, this problem is traditionally solved by the application of either conical conjugate surfaces [3], which cause difficulties in their manufacture or mechanical systems with spring compression of the half-sleeves of plain bearings [4]. If steel springs with a small damping capacity are used, their parameters change significantly over time, which does not allow realistic monitoring of their condition. In addition, it is impossible to create automatic or automated control systems for devices with the selection of gaps by means of spring elements. All of the foregoing stipulates the need for the development of constructive measures for the creation of a gapless connection of the elements of kinematic pairs.

2. Results and discussion

Studies showed that the necessary effect can be obtained by using small elastic pneumatic elements (figure 1), embedded in a kinematic pair [5]. Constantly acting on the movable body with the antifriction liner fixed on it, the elastic element chooses the gap between the journal and the liner. It is installed from the side opposite to the effect of the force of technological resistance on the bearing, which contributes not only to the selection of the gap in the connection, but also provides compensation for the liner wear. The magnitude of the overpressure inside the elastic element is set to such a value that, as a result of the action of the elastic forces acting on the journal after the reset of the technological resistance forces, there is no opening of the gap in the connection.

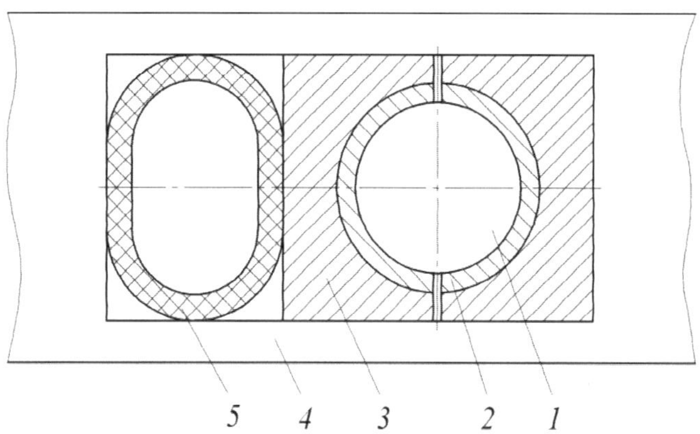

Figure 1. Scheme of support with an elastic pneumatic element: 1 – journal; 2 – liner; 3 – mobile semi-support; 4 – frame; 5 – elastic pneumatic element.

The stiffness of the elastic pneumatic element, made in the form of a cylinder with a limited axial deformation, is determined [6] by the relation

$$C = \frac{\pi \cdot l \cdot p_0}{2},$$

where p_0 – the value of excess pressure; l – the length of the cylinder.

The experiments were carried out on a research facility – jaw crusher. The level of vibration was estimated indirectly through the values of the frame acceleration, in which accelerometers are installed in the horizontal and vertical planes.

Experiments showed that in the presence of gaps in plain bearings, the level of acceleration in the horizontal plane (figure 2a) is 0.4 ÷ 0.5 m/s^2 at idle speeds. Under the influence of the technological resistance forces (in the process of single crushing) and during their relief, peaks of acceleration appear in magnitude up to 5 m/s^2, and in the vertical plane (figure 3a), respectively, 0.3 ÷ 0.4 m/s^2 and 1.5 ÷ 2 m/s^2.

If the gaps in the supports are chosen with the help of elastic pneumatic devices, then the level of accelerations decreases insignificantly in the horizontal plane at idle speeds and amounts to 0.3 ÷ 0.4 m/s^2, and in the process of crushing and when the load is dropped, the magnitude of the accelerations is much less than in operation of bearings with gaps 2 ÷ 2.5 m/s^2 (figure 2 b).

In the vertical plane, the selection of gaps practically does not affect the level of acceleration (figure 3 b). Obviously, the lower the level of acceleration that occurs when a jaw crusher works, the less is the vibration level of the machine.

It should be noted that attenuation of acceleration values in the presence of gaps in bearings during load relief occurs over 3 ÷ 4 periods of oscillation, while using devices for gaps selection damping occurs almost immediately, which indicates a high damping capacity of elastic pneumatic elements.

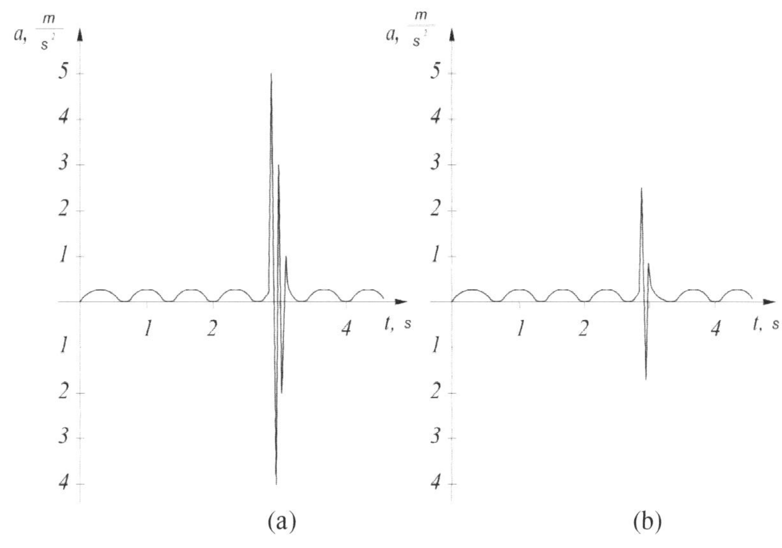

Figure 2. Oscillograms of the frame acceleration in the horizontal plane: a – in the presence of gaps in the support; b – for selected gaps.

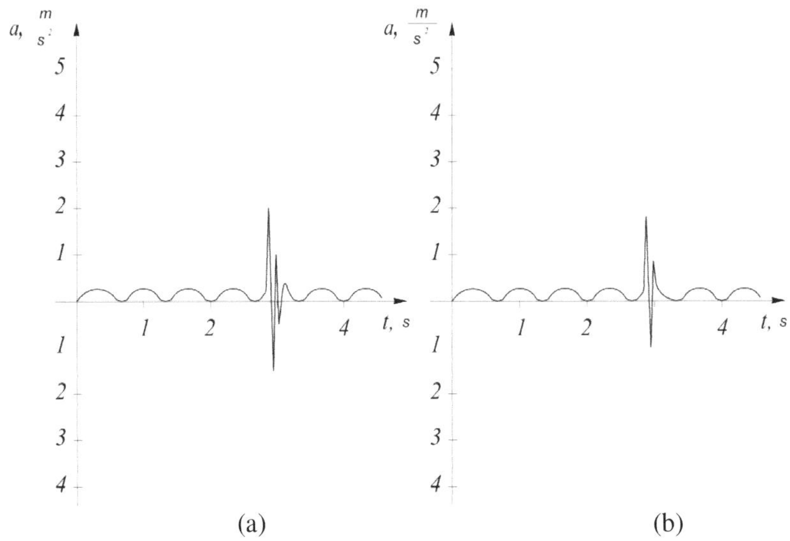

Figure 3. Oscillograms of the frame accelerations in the vertical plane: a – in the presence of gaps in the support; b – for selected gaps.

3. Conclusion

The influence of the vibrational action due to the presence of gaps in the plain bearings on the reliability of the jaw crushers is determined. The design of the elastic pneumatic device for the selection of gaps in plain bearings during the jaw crusher operation is described and the reduction of the vibration level by eliminating gaps is experimentally proved.

References

[1] Grebennik V M and Tsapko V K 1980 *Reliability of Metallurgical Equipment* (Moscow: Metallurgy) p 343
[2] Nikitin A G, Chaynikov K A and Zinovieva E V 2011 *Izv. Vuzov. Ferrous Metallurgy* **10** 46–7
[3] Orlov P I 1988 *Fundamentals of Design. Volume 1* (Moscow: Mechanical Engineering) p 560
[4] Zhivov L I, Kolesnik F I, Mishchanin V G and Bulat V I 1974 *Forging and Stamping Production* **5** 29–31
[5] Nikitin A G, Sakharov D F, Chaynikov K A and Prilukova N Z 2012 *Roller Crusher* patent of the Russian Federation No. 2453371 Bul. 17
[6] Nikitin A G et al 2012 *Izv. Vuzov. Ferrous Metallurgy* **4** 68–70

IOP Conf. Series: Materials Science and Engineering **411** (2018) 012055 doi:10.1088/1757-899X/411/1/012055

Direct iron production: the state of the problem, trends

O I Nokhrina[1], I D Rozhikhina[1] and I E Khodosov[2]

[1]Siberian State Industrial University, 42 Kirova street, Novokuznetsk, 654007, Russia
[2]JSC "NPO "Aconite", 34 Udarnikov street, Vologda, 160000, Russia

E-mail: Khodosov@mail.ru

Abstract. The current technologies for obtaining iron-containing materials without use of coke and blast-furnace processing are described. Brief description of mastered industrial technologies for direct iron production is given. Possible trends in the development of direct iron production technologies are indicated.

1. Introduction

The primary metal is produced by the reaction of iron reduction from the oxide-containing material. At present, the main unit for the reduction of iron is a blast furnace. The technology of blast furnace melting has been developing for 600 years and, according to a number of scientists working in this field, has reached its limit of efficiency [1]. The most obvious shortcomings of blast-furnace operations are high energy-consumtion, the need for agglomerate production, the use of expensive metallurgical coke and high carbon dioxide emissions into the atmosphere. At the same time, the development of the theory and practice of reduction processes, the search for new ways of obtaining metal led to the creation of technologies that allow high-quality iron-containing materials to be obtained without use of coke and blast furnaces. Such technologies were named: technologies of direct iron production, non-blast furnace n or non-coke metallurgy [2].

The main differences between the blast-furnace and the non-coke methods of obtaining iron are presented in table 1.

The reduction processes that take place in the blast furnace, and with the non- blast furnace production of iron proceed according to general thermodynamic laws. That is, the reduction of iron from stable oxides (Fe_3O_4, Fe_2O_3, FeO) formed by it is carried out with the participation of solid carbon, the source of which in the blast furnace is coke, and also with the participation of gases – CO, H_2 according to the reactions:

$$Fe_mO_{n(s)} + nC_{(s)} = mFe_{(s)} + nCO_{(gas)};$$
$$Fe_mO_{n(s)} + nCO_{(gas)} = mFe_{(s)} + nCO_{2(gas)};$$
$$Fe_mO_{n(s)} + nH_{2(gas)} = mFe_{(s)} + nH_2O_{(gas)}.$$

Note: reactions are presented in general form, for temperatures above 570 °C.

Reactions in the solid phase, that is, the solid-phase reduction of iron from oxides, formed the basis of most direct iron production technologies.

Further on, these processes will be called "metallization processes", and the resulting materials – "metallized materials".

Over the past century, more than 130 methods of non-blast furnace production of iron have been proposed and tested [3]. The common fact for most processes, is that the raw materials and the resulting materials are in the solid state, and the process temperatures do not exceed their softening temperatures. The quality of the materials obtained is generally assessed by the content of metallic iron, carbon, harmful impurities in the material and the degree of metallization, which shows the ratio of the content of metallic iron to the total [4].

Table 1. comparison of processes.

Characteristic features:	Blast furnace production	Non-blast furnace production
Temperature of processes, °C	~ 1550	~ 1200
Raw material preparation	Agglomerate production	Production of pellets or briquettes
Equipment	Blast furnace	Pipe furnace, shaft furnace, rotary hearth furnace
Reducing agent	Coke, pulverized-coal fuel	Natural gas, synthetic gas, coal

2. Development of processes

The development of metallization processes has led to the creation and industrial implementation of such technologies as SL/RN, Midrex, HyLIII, Corex, Fastmet, ITmk3, etc. The metallized materials obtained by these technologies were used as a charge for melting electric steel.

Metallization technology Midrex © Process (figure 1) is the most widespread in the industry. Currently, this technology is used for production of 70% of the total volume of metallized raw materials in the world. A vertical shaft furnace is used as a metallization unit. According to this technology, the pelletized iron ore in the form of pellets is fed into the furnace and when it comes into direct contact with the reducing counterflow gas gets to the bottom of the furnace. The temperature of the process does not exceed 900 °C, which is necessary to prevent the build-up of incrustation on the furnace lining. A mixture of natural and recycled gases is used as a reducing agent, which is passed through catalytic tubes to produce CO and H_2 [5].

Midrex Technologies, Inc. offered the technology – MXCOL©Technology that allows the necessary gas by gasification of coal to be received.

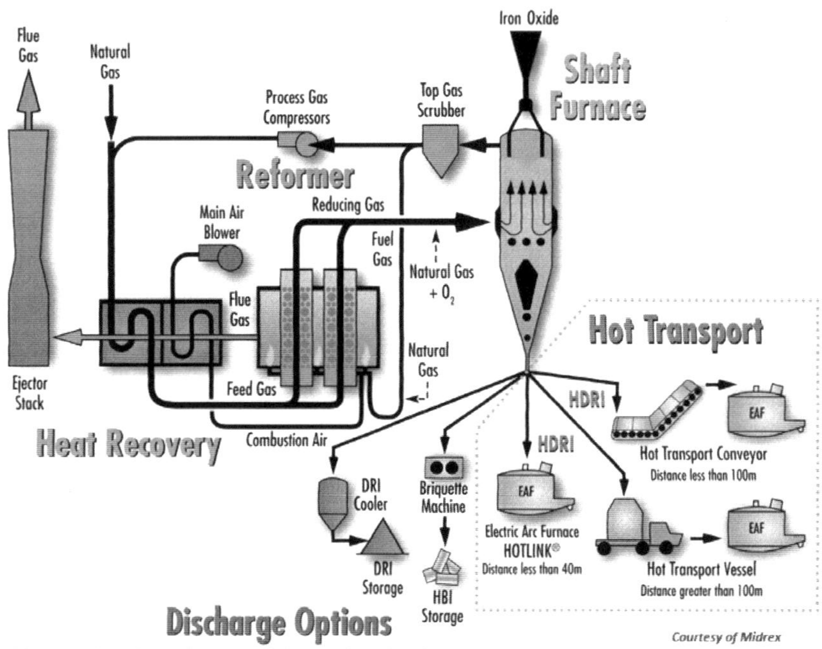

Figure 1. The scheme of the technological process Midrex Process (source: Midrex Technologies, Inc. [5]).

IOP Conf. Series: Materials Science and Engineering **411** (2018) 012055 doi:10.1088/1757-899X/411/1/012055

As a result of the process, metallized pellets with a spongy structure (international designation - DRI) are formed. The resulting pellets can be forwarded for hot briquetting for production of hot briquetted DRI (international designation – HBI). HBI allows safe storage and transportation of materials.

In Russia "Metalloinvest" (JSC "OEMK") in 2016 using Midrex technology produced 5.7 million tonnes of metallized pellets. According to the company's plans, it is expected to launch production of HDR with a capacity of 1.8 million tonnes per year at the site of Lebedinsky Mining and Processing Combine in Gubkin [6].

Another widespread technology of metallization is the technology HYL III © Process. Currently, according to this technology up to 15% of all metallized raw materials is produced in the world. This technology is implemented in a shaft furnace equipped with a lifter. The reduction processes take place at temperatures of 1000 °C and overpressure that can reach 6 atm. According to HYL III technology, the reduction gas is produced by steam conversion of natural gas and waste gases taken from the upper horizons of the shaft furnace.

Metallized raw materials, obtained by the technologies described above, are in high demand among enterprises smelting high-quality electric steels. This is due to the fact that the metallized materials do not contain impurities of non-ferrous metals, which in turn reduce the quality of steel and require additional methods of removing them. Metallized materials are also characterized by stable chemical composition and technical characteristics, which in turn positively affects the melting process as a whole. Of particular interest is the possibility of supplying hot pellets – HDRI directly to the EAF, i.e. until they are cooled. This technique allows the technical and economic indicators to be significantly increased due to the reduction of heat costs for melting the material. Thus, metallized materials are a direct alternative to scrap metal when melting steels in the EAF. A common disadvantage for this raw material is that there is a rock refuse in the material and, in the case of coal, the ash is the residue of the reducing agent. This fact limits the use of lean ore and iron-containing waste production [7].

3. Development of processes using solid fuel

A well-known metallization method is the SL/RN (Stelco-Lurgi Steel-National Lead) technology, which is based on the use of a rotary kiln (firing type). As a raw material, a pelletized mixture of iron ore and coal is used, it is also possible to feed coal directly to the furnace. The metallization process takes place at temperatures 1000-1200 °C, due to which it is possible to achieve a sufficiently high degree of metallization (· 98%) when using lean ores. The advantage of this technology is the technological simplicity and the possibility of using different solid carbonaceous reducing agents, including low-grade ones. As a fuel for burners circulating and synthetic gases can also be used. There are examples of the use of hot waste gases for power generation (figure 2).

It should be noted that in Russia at the plant "Sibellectrostal" the technology of metallization was developed and mastered, in which as a reducing agent brown coal was applied. Metallization was carried out on a complex installation "lattice-tube furnace". Metallized pellets were fed for melting into the furnace in the hot state at temperatures of 800-1000 °C. Such decision showed an increase in EAF productivity by 1.5 times as compared to work on cold fired pellets [8].

The further development of metallization technologies led to the creation of about 20 methods of obtaining metal in the form of a melt or granules in special metallurgical aggregates. Since the beginning of the 90's of the last century, the industrial development has received the Corex process (figure 3), according to which in the conjugated aggregates initially solid metallic material is obtained, which is then sent to the melting gasifier. The coal is loaded into the melting gasifier through the crown. Oxygen is injected in the middle zone of the gasifier, that promotes the formation of a fluidized (pseudo-liquefied) layer and hot reducing gases in the coal layer. As a result, liquid cast iron and slag are formed at the bottom of the melting gasifier. Thus, the Corex process was the first large-scale process of non-coke iron production, alternative to blast furnace.

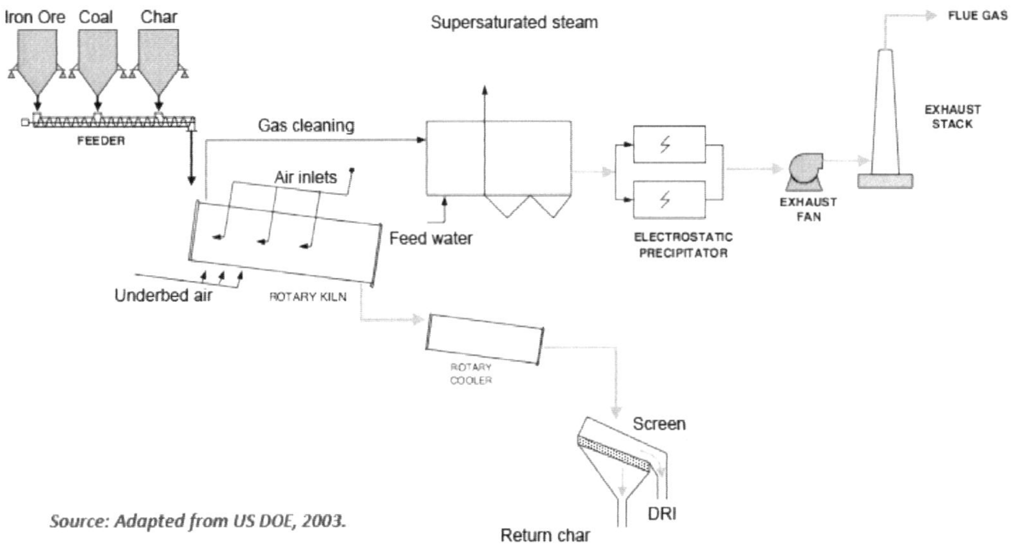

Figure 2. The scheme of the technological process SL/RN (source: IETD [7]).

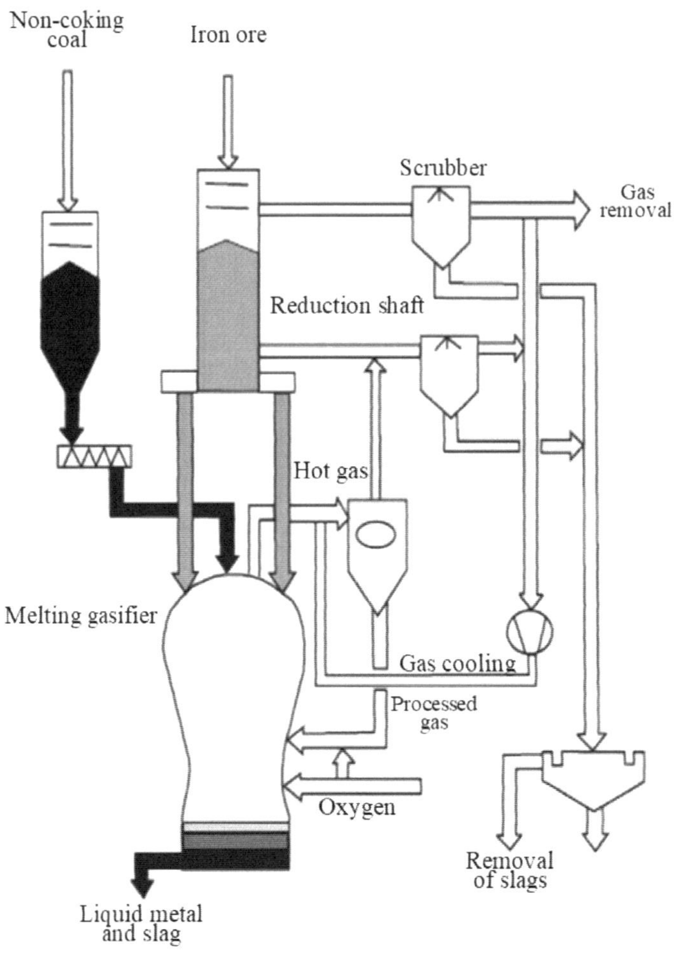

Figure 3. The scheme of Corex technological process.

Studies of metallization processes conducted by Kobe Steel ltd. and Midrex Technologies, Inc. led to the creation of technology Fastmet©. This technology is realized with the help of a rotary hearth furnace (RHF) and the output is DRI. At present, this technology has found its application in the metallization of steelmaking waste. As a solid carbonaceous component of the charge, coal, charcoal, and coke are used. Iron-containing materials and a reducing agent in the form of pellets or briquettes are fed to RHF in such a way that a layer of 2 to 5 cm is formed on the surface of the movable hearth. The metallization is carried out at a temperature of 1350 °C for 6-12 minutes. Such metallization modes make it possible to produce metallized materials with a metallization degree 95-98%. The resulting hot gases containing a significant amount of CO are re-directed to the furnace by burners located along the furnace contour. Later, this technology was supplemented with a melting plant according to the EAF type. This process was called Fastmelt. According to this technology it is possible to obtain pure non-ferrous metals and sulfur grades [9].

The development of this technology led to the creation of process ITmk3 – "production of pig iron of the highest quality" (figure 4). This technology is also based on the use of RHF. The difference is that in the process of metallization, the melting of materials takes place, which leads to the formation of granular cast iron. The processes are carried out at temperatures of 1350-1450 °C [10].

The use of annular furnaces for the production of metallized materials makes it possible to use efficiently the heat of the exhaust gases and more accurately control the temperature of the materials as they move through the furnace space. The rotary hearth furnace also allows materials to be heated to high temperatures (1400-1500 °C) during a short time period of time (5-15 minutes), which makes it possible to control the separation of metal and slag phases in the production of granulated iron [10]. Later, this technology was supplemented with a melting plant according to the EAF type. This process is called Fastmelt. According to this technology it is possible to obtain profiled steels with no sulfur and non-ferrous metals with no impurities [9].

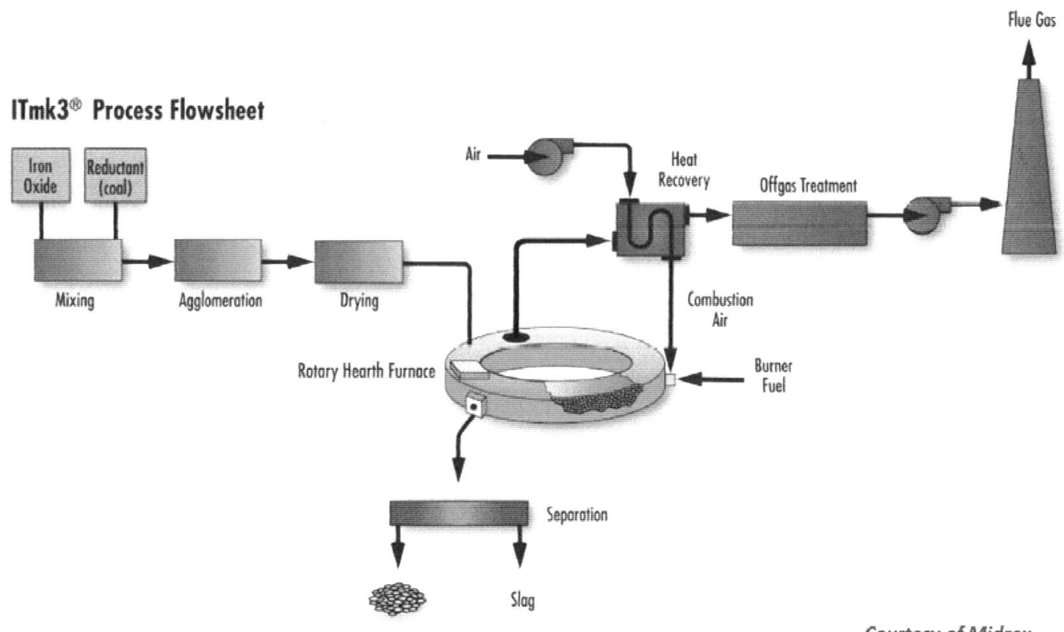

Figure 4. The scheme of ITmk3 process (source: [10]).

4. Conclusion

Metallization technologies during the last half century became a progressive trend in the iron and steel industry. Analysis of commercialized metallization technologies makes it possible to identify possible directions in the development of technologies for non-blast furnace production of iron:

IOP Conf. Series: Materials Science and Engineering **411** (2018) 012055 doi:10.1088/1757-899X/411/1/012055

- combination of metallization processes and remelting of metallized materials in AEF with the production of marketable products;
- use of alternative heat sources and reducing gases, including those obtained by gasification of non-deficient solid carbonaceous materials;
- improvement of hardware processes;
- use of lean ores, iron and carbon containing technogenic waste;
- improvement of technical and economic indicators of existing methods for the production of granular cast iron.

References

[1] Tovarovsky I G 2003 *Blast Furnace – Evolution, Process Flow, Problems and Perspectives* (Dnepropetrovsk: Porogi) p 597
[2] Nokhrina O I et al 2016 *Steel in Translation* **46(4)** pp 237–244
[3] Kurunov I F 2012 *Metallurg* **4** 40–44
[4] Yusfin Yu S and Pashkov N F 2007 *Metallurgy of Iron* (Moscow) p 320
[5] *Research & Development Department, Iron Unit Division, Natural Resources & Engineering Business, Midrex Technologies, Inc* http://www.midrex.com
[6] http://www.metalloinvest.com/
[7] *Organisation for Economic Co-operation and Development. Energy Technology Transitions for Industry: Strategies for the Next Industrial Revolution* 2009 (OECD Publishing)
[8] Tulin N A et al 1987 *The Development of Non-Coke Metallurgy*
[9] Ziebik A, Lampert K and Szega M 2008 *Energy* **33(2)** 199–205
[10] Kikuchi S et al 2010 *Kobelco Technology Review* **29** 77–84

AFM investigation of silumin structure modified by Al-Y_2O_3 coating using the method of electric explosive alloying

K A Osintsev[1], K A Butakova[1], S V Konovalov[2], D V Zagulyaev[1] and V E Gromov[1]
[1]Siberian State Industrial University, 42 Kirova street, Novokuznetsk, 654007, Russia
[2]Samara National Research University, 34 Moskovskoye shosse, Samara, 443086, Russia

E-mail: zagulyaev_dv@physics.sibsiu.ru

Abstract The paper reports on coating of silumin samples with Al-Y_2O_3 by electric explosive alloying. When testing the micro-hardness, two appropriate processing modes of samples were revealed; the porosity of their surface profile was examined using atomic-force microscopy. A multi-layer structure was disclosed by metallographic microscopy, the thickness of each layer in the coating was determined as well. The mechanism of this structure formation was suggested.

1. Introduction
To date, silumins are broadly used in aircraft and ship construction, airspace industry and mechanical engineering but their insufficient strength properties and crack-resistance make them less utilized in comparison to more durable materials [1, 2].

The researchers [3-5] emphasize that the coating formation by electric explosive alloying makes it possible to enhance strength, durometric and tribological properties of the modified material. The strengthening effect is achieved due to the formation of coatings with fine-dispersed phases in the viscous metallic matrix [3].

The objective of this study is to investigate the effect of electric explosive alloying with a powder portion of Al-Y_2O_3 on the strength properties of silumin AK10M2H, as well as on the structure of its sub-surface layers.

2. Materials and methods
For the purpose of research the silumin samples AK10M2H were used; they were cut from one-piece bar, had a form of a parallelepiped with dimensions 20×20×10 mm^3. Using electric explosive alloying the silumin substrate was coated with composites of the Al-Y_2O_3 system by the electric explosive unity EVU 60/10 (Siberian State Industrial University, Novokuznetsk). The electric explosive unit has the following parameters of alloying by plasma, which is formed in the electric explosion of aluminum foil with a powder portion of yttrium oxide: exposure time to plasma ~100 μs, absorbed power density on the jet axle ~8.2 GW/m^2, pressure in the shocked layer close to the surface ~18.8 MPa.

Front explosion was used [3-5] to intensify the heat impact on the material surface and further the melting process, providing conditions for alloying. The aluminium foil (0.0589 g) was hold between coaxial electrodes, which were supplied with adjustable voltage through a vacuum arrester. The coatings were formed in thermal conditions, which lead to heating-up of the substrate surface up to the melting temperature at two different voltages: U_1 = 2.6 kV (samples 1, 3, 5) and U_2 = 2.8 kV (samples 2, 4, 6). Three samples with different weight of Y_2O_3 powder portion were used in each voltage mode. A powder portion of 0.0589 g was used for samples 1, 2; 0.02945 g – for samples 3, 4, and 0.0883 g – for samples 5, 6, respectively.

When discharging the capacitive energy storage, a periphery of the foil close to the outer electrode-nozzle is a source of a condensed phase of explosion products; a foil above the central electrode, where the powder of yttrium oxide is supplied, is a source of ionized vapour [4].

Completing the electric explosive alloying, Vickers micro-hardness tests were carried out (micro-hardness measuring device HVS-1000A). A load of 0.05 HV was constant for all six samples. The micro-hardness was measured in grains and eutectic.

The porosity of the coating and the heat-affected zone was investigated using an atomic-force microscope NT-MDT Solver "NEXT", since it has high-performance and helps to carry out up to date procedures of research into the morphology and local properties of the solid surface with a high space resolution [6-8]. The topography of the surface profile was imaged.

The structure of the material was explored using a metallographic microscope Olympus GX-51. The images for measuring surface layer thicknesses were taken; a form and sizes of grains were analyzed, and the conclusion on uniformity of the formed coating was drawn.

3. Experiment results

In accordance with the experimental measurements of micro-hardness (table 1) the correlation of micro-hardness to the distance of the coated layer (coating included) was plotted (figure 1). From the statistical data of the experiment the maximal micro-hardness of samples 2 and 5 is apparent. A significant increase in the micro-hardness is disclosed in these samples close to the surface, both in grains and in eutectic. The micro-hardness of the surface in samples 2 and 5 is higher than samples, in total, by 50 % and 65 %, respectively.

Table 1. Micro-hardness test results of silumin samples subjected to electric explosive alloying with Al-Y_2O_3.

Sample No.	Phase	520 μm from the edge	90 μm from the edge	70 μm from the edge	50 μm from the edge	Composite surface layer
1	Grain	54	56.03	51.16	78.48	78.55
	Eutectic	67.04	66.15	72.73	94.17	
2	Grain	63	59.36	63.92	87	130.51
	Eutectic	73.99	89.5	89.22	87.92	
3	Grain	58.67	58.55	57.53	61.64	57.12
	Eutectic	79.96	76.18	74.78	84.15	
4	Grain	59.53	58	67.88	73.34	77.34
	Eutectic	80.37	72.76	80.62	85.65	
5	Grain	63.09	65.61	67.58	66.58	161.2
	Eutectic	108.7	105.17	122.34	111.57	
6	Grain	86.56	66.84	77	81.31	101.56
	Eutectic	88.31	94.56	86.49	103.51	

From the curves (figure 1) it can be seen that the micro-hardness both in grains and eutectic of the modified samples increases close to the coated layer.

The metallographic study on the cross-section structure of the samples with the maximal micro-hardness (samples 2 and 5) revealed a multi-layered structure consisting of a highly-porous coating with non-uniform thickness, a layer of liquid-phase alloying, and a heat-affected layer c (figure 2 a, b). The thickness of the modified layer ranges from 17 to 117 μm for sample 2 and from 33 to 60 μm for sample 5 owing to the precisely determined weight of the coated powder – less Y_2O_3 was used for sample 2 than for sample 5. Therefore, the coating of sample 5 was more uniform in the width than of sample 2. The thickness of the heat-affected layer is 150–260 μm in sample 2 and 53-80 μm in sample 5, respectively. The heat-affected layer in sample 2 is significantly thicker than in sample 5 due to higher energy mode of coating for sample 2.

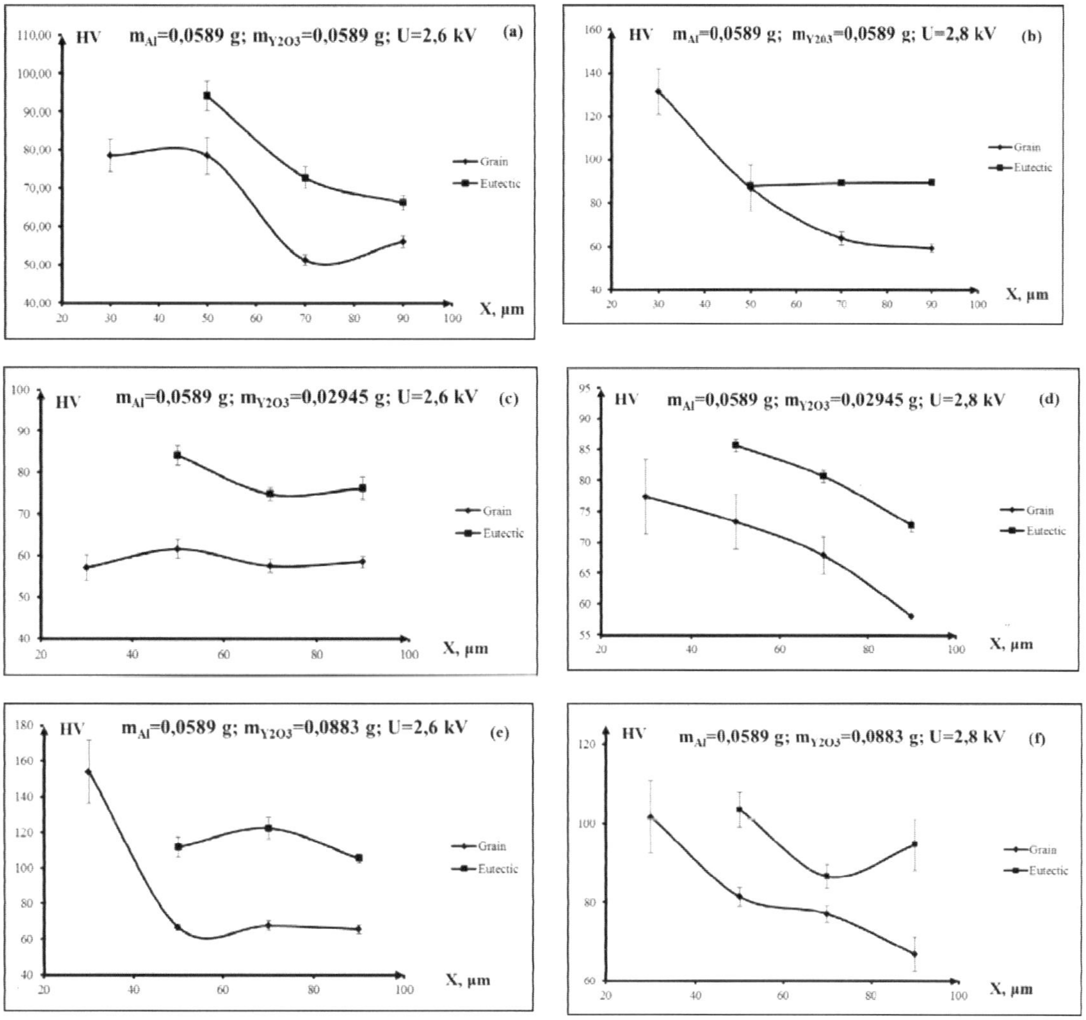

Figure 1. Micro-hardness in grains and eutectic of silumin samples in different processing modes vs. distance to the sample surface.

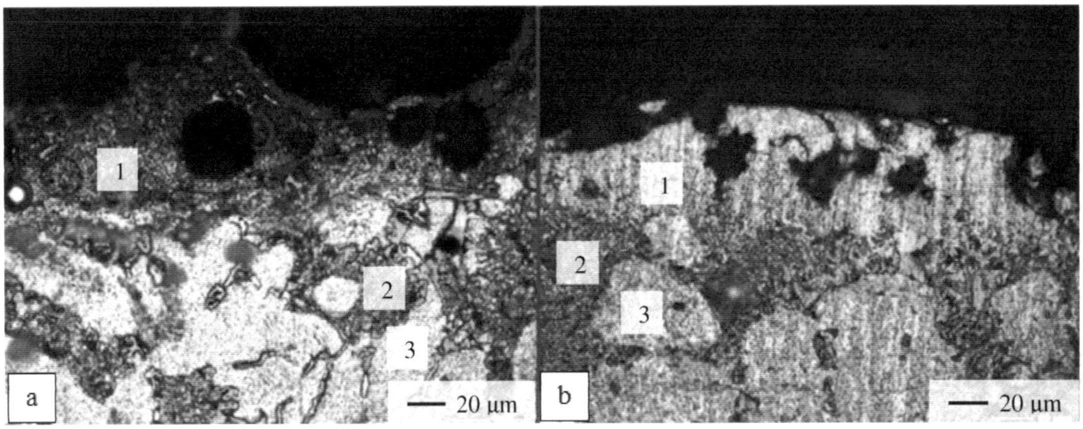

Figure 2. Optical microscopy of the surface profile in silumin samples processed in modes 2 (a) and 5 (b): 1 – modified layers, 2 – eutectic, 3 – grains.

The topography of the surface profiles imaged by atomic-force microscopy is shown in figures 3 – 6. 3D and 2D images, as well as asperity profiles along the base line in the sample (coating excluded) and on the coatings are presented.

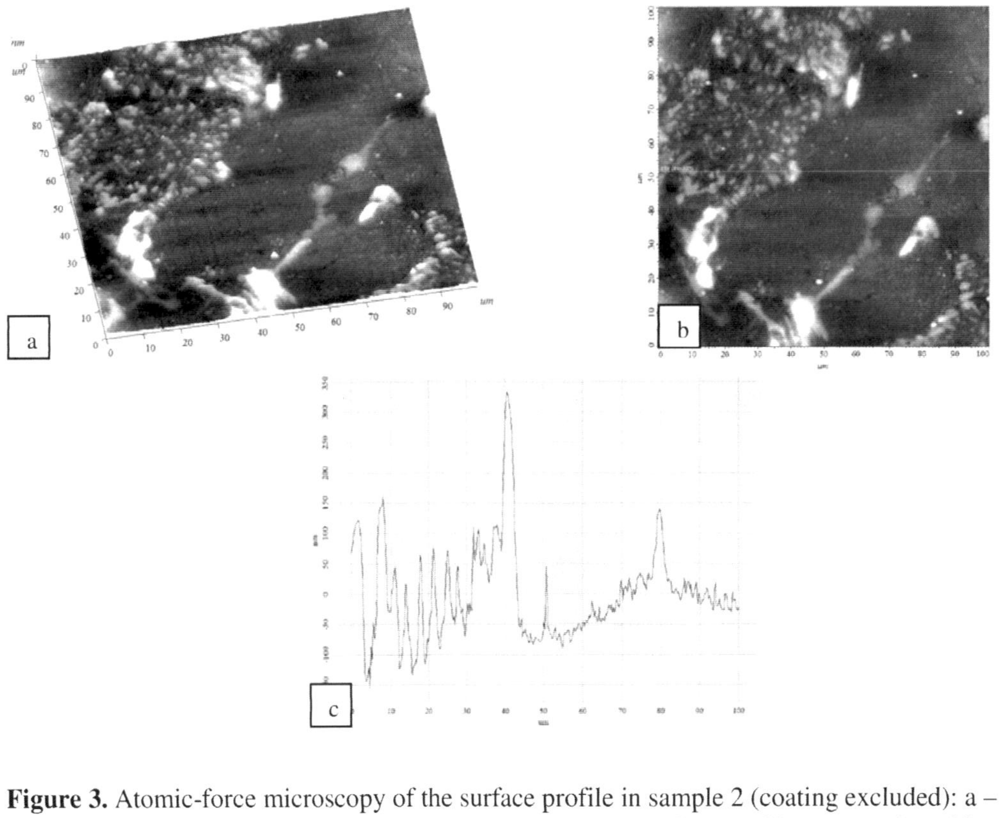

Figure 3. Atomic-force microscopy of the surface profile in sample 2 (coating excluded): a – profile asperity according to the height in 3D, b – 2D image of the surface profile topography with a drawn section, c – asperity distribution along the base length.

The structure of the modified silumin in the sample (coating excluded) comprises grains with a height of 350 nm and relatively smooth eutectic.

IOP Conf. Series: Materials Science and Engineering **411** (2018) 012056 doi:10.1088/1757-899X/411/1/012056

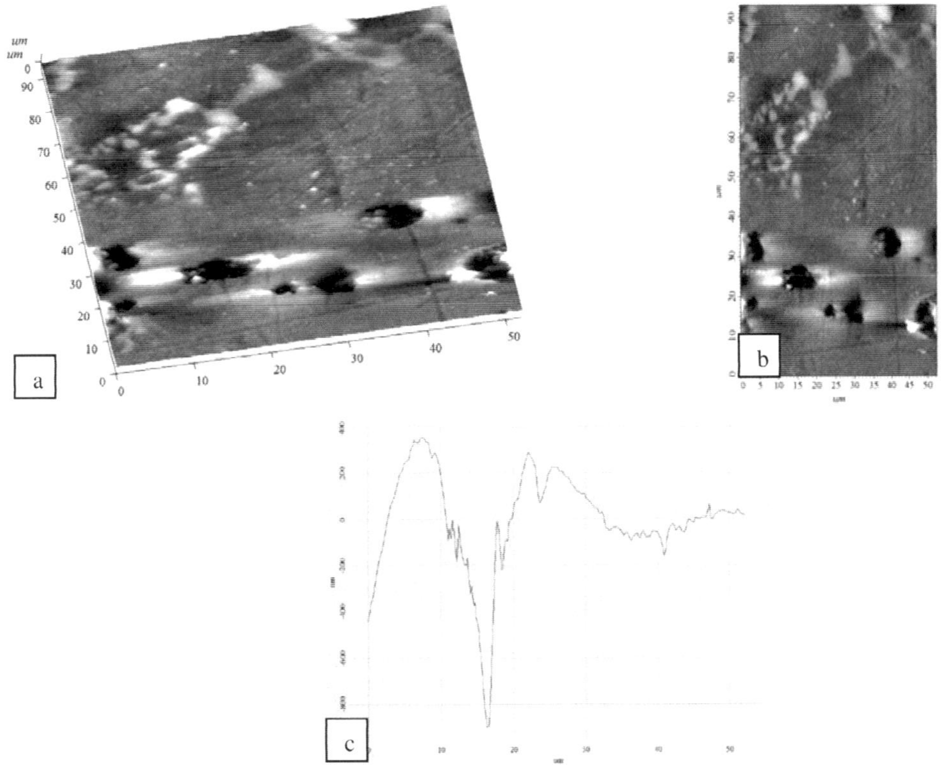

Figure 4. Atomic-force microscopy of the surface profile in the coated layer of sample 2: a – profile asperity according to the height in 3D, b – 2D image of the surface profile topography with a drawn section, c – asperity distribution along the base length.

As seen in figure 4, the coating in sample 2 is a highly-porous one. Statistical processing of images revealed a maximum depth of pores (1500 nm). The diameter of pores is up to 10 μm.

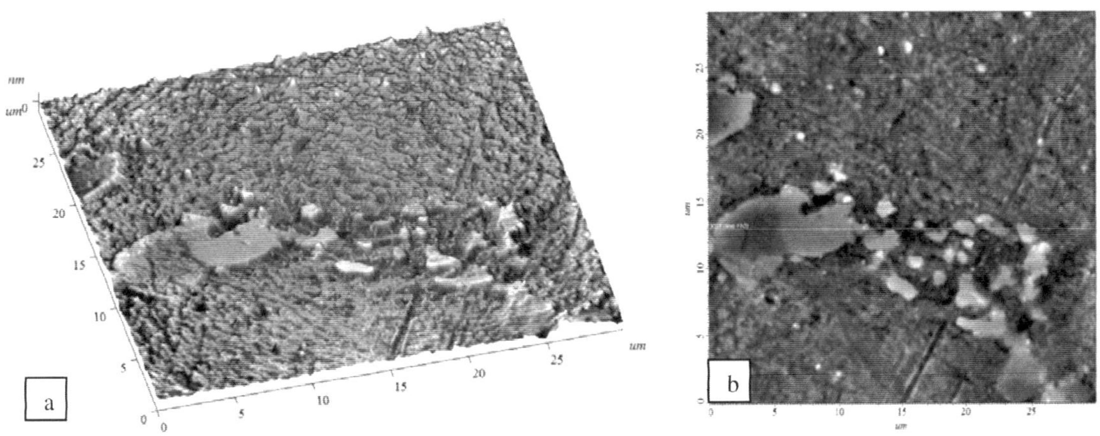

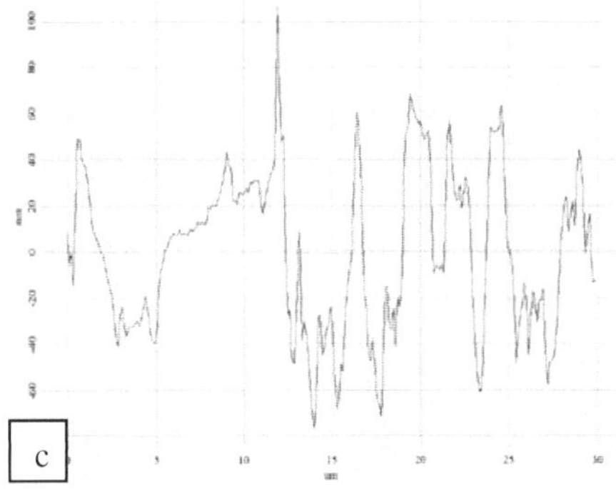

Figure 5. Atomic-force microscopy of the surface profile in sample 2 (coating excluded): a – profile asperity according to the height in 3D, b – 2D image of the surface profile topography with a drawn section, c – asperity distribution along the base length.

Sample 5 (coating excluded) similar to sample 2 (coating excluded) has a structure comprising eutectic and grains.

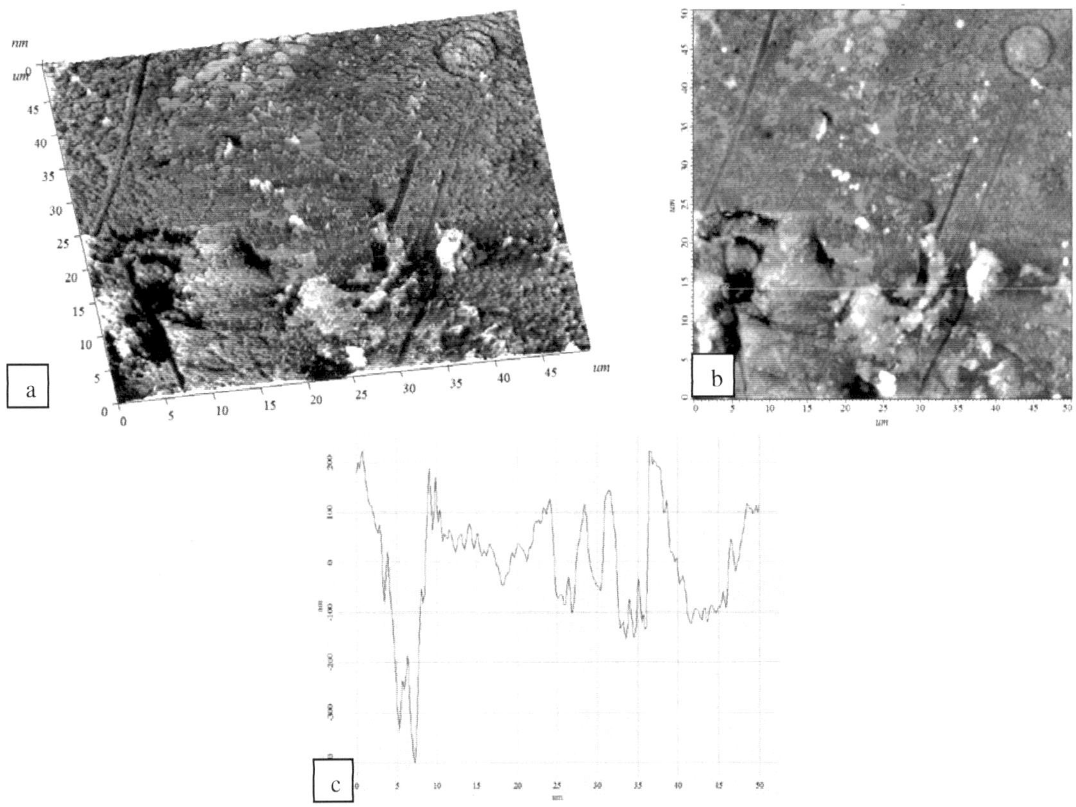

Figure 6. Atomic-force microscopy of the coated layer surface profile of sample 5: a – profile asperity according to the height in 3D, b – 2D image of the surface profile topography with a drawn section, c – asperity distribution along the base length.

From figure 6 it is apparent that the coating in sample 5 is porous. The number and depth of pores are less significant than in the coated layer in sample 2. The maximal depth of pores is 500 nm. The explanation is that owing to higher weight of a portion of the sprayed powder and lower energy impact, particles of a plasma jet with lower energy are distributed more uniformly. Kelvin–Helmholtz instability is possible as well, i.e. remixing of coated and melted layers; as a consequence, there are fewer pores than in sample 2.

4. Conclusion

In the course of the work, it is found that electric explosive alloying leads to the increase in microhardness in the silumin surface layer and optimal treatment regimes are determined, in which the microhardness value increases by 50-65%.

Study of the structure of silumin surface profile by metallographic and atomic force microscopy shows that thanks to electric explosive alloying a multilayer structure is formed, which consists of a highly porous coating, non-uniform in thickness, a layer of liquid-phase alloying and a layer of thermal influence.

An assumption is put forward that the formation mechanism of the resulting structure is the flow of Kelvin-Helmholtz instability, i.e. the mixing of the sprayed and melted layers, resulting in a reduced number of pores.

Acknowledgements
This work was supported by the State Order 3.1283.2017/4.6.

References

[1] Martyushev N V, Bashev V S and Zykova A P 2017 *IOP Conf. Ser.: Mater. Sci. Eng.* **177** 012118
[2] Konovalov S V, Alsaraeva K V, Gromov V E and Ivanov Yu F 2015 *IOP Conf. Ser.: Mater. Sci. Eng.* **91** 012029
[3] Vashchuk E S, Romanov D A, Budovskikh E A and Ivanov Y F 2011 *Steel in Transl.* **41** (6) 469–474
[4] Romanov D A, Budovskikh E A and Gromov V E 2011 *J. Of Surf. Invest.: X-Ray, Synch. And Neutr. Tech.* **5** (6) 1112–17
[5] Ivanov Y F, Budovskikh E A, Gromov V E, Bashchenko L P, Soskova N A and Raikov S V 2012 *Steel in Transl* **42** (6) 499–501
[6] Haviland D B 2017 *Current Op. in Coll. & Int. Sci.* **27** 74–81
[7] Yuan Li and Cheng Y F 2016 *App. Surf. Sci.* **366** 96–103
[8] Lee S W 2016 *Synt. Met.* **216** 88–92
[9] Sarychev V D, Vashchuk E S Budovskikh E A and Gromov V E 2010 *Tech. Phys.* **36(7)** 656–659
[10] Granovskii A Y, Sarychev V D and Gromov V E 2013 Tech. Phys. **58(10)** 1544–47

IOP Conf. Series: Materials Science and Engineering **411** (2018) 012057 doi:10.1088/1757-899X/411/1/012057

The analysis of the effect of iron-ore raw material composition on blast-furnace smelting parameters by the mathematical model method

A V Pavlov[1], O P Onorin[2], N A Spirin[2], A A Polinov[1], V V Lavrov[2] and I A Gurin[2]

[1]Magnitogorsk Iron and Steel Works, 93 Kirov street, Magnitogorsk, 455000, Russia
[2]Ural Federal University n.a. the first President of Russia B.N. Yeltsin, 51 Lenina ave, Ekaterinburg, 620075, Russia

E-mail: olnasta@inbox.ru

Abstract. The paper considers capabilities of the developed model-based decision support system (URFU-MMK model of the blast-furnace process) for diagnostics of operation and prediction of production situations at blast furnaces. It presents some results of practical application of the developed system and it also gives recommendations for certain aspects of process task solving on the basis of the model-based decision support system. It presents calculation results of blast furnace operation with a high portion of non-fluxed pellets from Sokolovsk-Sarbay Ore Mining and Processing Group in burden of MMK's blast furnaces with different content of local sinter. It shows that with a higher portion of non-fluxed pellets (up to 60%) in the charged burden it is necessary to charge limestone in the furnace to achieve the required basicity of blast-furnace slag. In order to reduce the amount of crude limestone to be charged, it is necessary to produce sinter of high basicity. As far as the portion of pellets in the charged burden increases, concentration of magnesia in blast-furnace slag decreases. To increase concentration of magnesium oxides in slag, it is efficient to use combined basic flux consisting of 70% normal limestone and 30% dolomite limestone.

1. Introduction

Due to a great energy intensity and complexity of blast-furnace production, the role of scientifically proven solutions aimed at solving a complex of process tasks relating to definition of the optimum composition of blast-furnace and sinter burdens, selection of efficient slag, blast, gas dynamic and heat conditions of blast-furnace smelting is getting more and more important [1–10].

2. Model for selecting a composition of iron-ore raw materials and fluxes

The generalized structure of the model for selecting a composition of iron-ore raw materials and fluxes is given in figure 1. The analysis of input and output parameters makes it possible to state that the mathematical model shall include the following interrelated calculation blocks: sinter composition, heat conditions, slag conditions, gas dynamic conditions.

The model is divided into two parts, i.e. basic state model and predictive model. The basic (reference) state model allows assessment of the process state by averaged parameters during the base (reference) period of blast furnace operation. In this case, the actually available information on blast furnace operation is used. The predictive model allows assessing parameters of the blast-furnace process on the basis of the results obtained with the use of the base state model in case of changes in smelting conditions. The works [11-21] describe in detail the main design ratios and example implementation of this model.

Modelling of final slag properties includes the following blocks:

- definition of final slag output, composition and viscosity polytherm (slag viscosity dependence on slag temperature);
- calculation of sulphur-carrying power of slag and sulphur content in iron;
- diagnostics of slag conditions.

The calculation of *the slag output and composition* is based on the material balance equation for main chemical elements and entities. The slag output is calculated by the balance of slag-forming elements. The analytical *calculation of slag viscosity* is made by mathematical treatment of the threefold slag system diagram CaO–Al2O3–SiO2 at 1400 and 1500°C in the range of the real viscosity values of blast-furnace slags and use of known relationship between homogeneous melted slag viscosity and temperature. A provision is made for introduction of correction for content of other oxides (MgO, TiO_2) in slag, etc.

The parameters of the obtained viscosity polytherm are given below:

- slag viscosity at the given temperature: at the output – η_{sl}, 1400°C, 1450°C, 1500°C;
- slag viscosity gradients $\Delta\eta_{1400}^{1500}$ show slag stability in the range of final (operating) temperatures numerically equal to a change in slag viscosity in case of a change in the slag temperature by 1°C in the range from 1400°C to 1500°C, poise/°C.

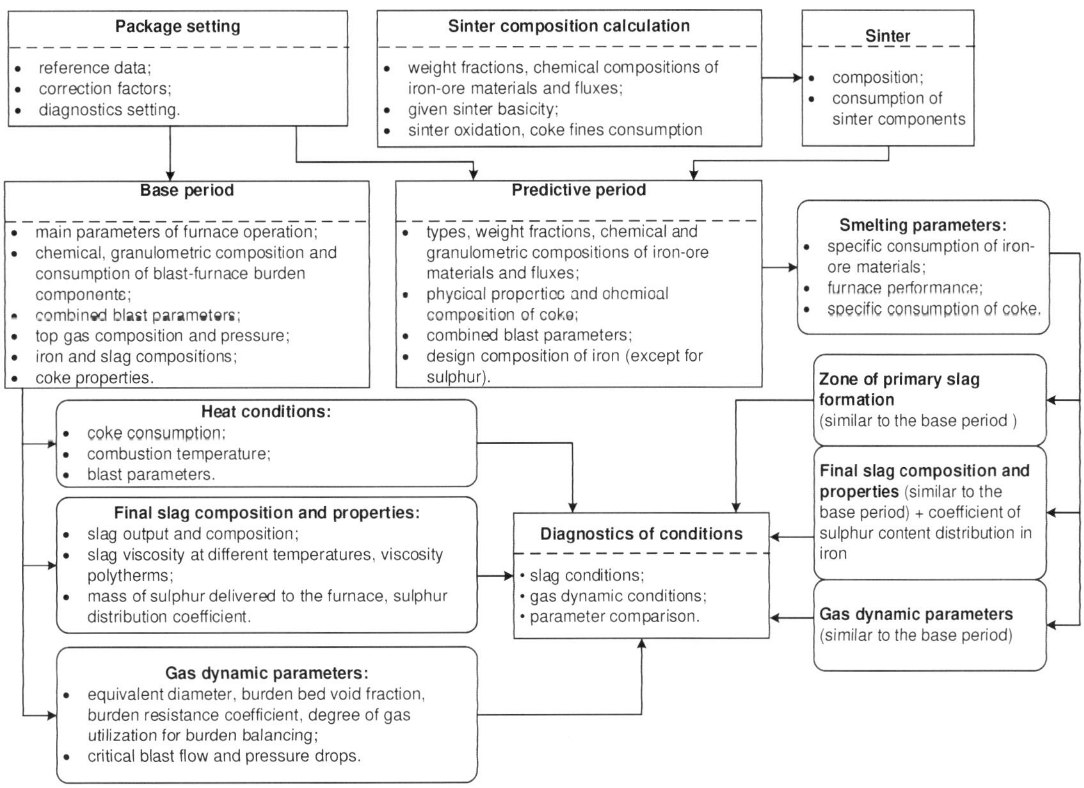

Figure 1. Structure of the model for selecting a composition of iron-ore raw materials and fluxes.

3. Model calculation variants

A provision is made for multivariance of predictive period modelling. The variants of problem solving are given below.

3.1. Project 1

Prediction of performance, coke consumption, consumption of iron-ore materials, properties of preliminary and final slag, gas dynamic parameters of smelting and thermal state is performed with given weight fractions and compositions of all iron-ore materials and fluxes charged in the furnace.

3.2. Project 2

Correction of blast furnace operation in case if changes in properties of iron-ore raw materials and fluxes is made by composition of one component of iron-ore part of blast-furnace burden, e.g. local sinter. Unlike the previous variant, this variant calculation is made with known compositions of all materials charged in the blast furnace, except for local sinter. Additionally, in a result of the calculation the required composition (consumption of iron-ore components, coke fines and fluxes) of sinter burden is determined.

In general, the rational conditions of blast-furnace smelting with changes in supplies of iron-raw raw materials for blast furnaces, iron-ore components of sinter burden as well as with changes in coke properties are provided by correcting:

- weight fractions and chemical composition of iron-ore components of sinter burden;
- flux consumption for sinter burden;
- flux consumption for blast-furnace burden;
- weight fractions and chemical composition of iron-ore components of blast-furnace burden;
- blast parameters and parameters of combined blast;
- any combination of the above options.

4. Process task solving

Below capabilities of process task solving are illustrated through the example of blast-furnace burden composition selection.

Due to the planned reconstruction of the sintering plant at Magnitogorsk Iron and Steel Works (MMK), the ratio of sinter and imported non-fluxed iron-ore pellets of Sokolovsk-Sarbay Ore Mining and Processing Group (SSGPO) in blast furnaces will change, the portion of pellets in burden can increase from 35 % (at present) to 60 %. With a bigger amount of non-fluxed pellets, a number of problems related to different chemical compositions of pellets and sinter as well as to different physical properties of these iron-ore materials will appear. With an increased consumption of non-fluxed pellets in burden, basicity of charged iron-ore materials decreases and charging of basic flux, i.e. limestone, into the furnace is necessary to get the required slag basicity.

To reduce the amount of crude limestone to be charged in the furnace, it is practical to be aimed at production of high-basicity sinter. The maximum basicity of sinter produced in MMK's sintering machines reached the value CaO/SiO_2 = 2.10. The tentative calculations show that sinter basicity (when the portion of pellets in burden is 60%) shall be CaO/SiO_2 = 2.7 to completely exclude charging of crude limestone into the furnace. Production of sinter with this basicity at MMK's sintering plant is problematic and, therefore, the basicity of produced sinter is taken CaO/SiO_2 = 2.10 for further calculations.

The next problem related to a larger amount of SSGPO's pellets in the iron-ore part of burden is a lower MgO concentration in final blast-furnace slag. This will lead to a larger slag viscosity and worse slag filtration through the coke packing as well as loss in sulphur-carrying power of slag and deterioration of iron quality because of larger sulphur content.

Due to a wider melting temperature range of pellets in comparison with sinter – by 30-50 °C (according to studies [6, 7]), an increase in the melting temperature range of the iron-ore mixture and increase in the lower gas pressure drop should be expected. Therefore, when making predictive calculations of blast-furnace smelting with a high portion of pellets in burden, a provision is made for addition in sinter burden of components increasing MgO content in sinter, i.e. dolomite limestone and Bakal siderite ore.

Calculations of blast-furnace burden with a high portion of pellets in the mixture with sinter for MMK's blast furnaces were made in the following sequence. In accordance with the data of sinter production, the averaged composition of sinter burden was determined. It shall be noted that at present the iron-ore part of sinter burden includes about 30 components. In one of the predictive calculation variants, Bakal siderite ore was added to the components of sinter burden. Normal limestone and limestone/dolomite limestone mixture were used as fluxing agents.

When calculating the composition of sinter with basicity CaO/SiO_2 = 2.10, we considered that sinter burden included blast-furnace dust (1.7%). As a solid fuel, coke fines are used and we were taking its consumption equal to 5.0% in all calculation variants. The calculation of sinter composition was made in accordance with the method of Prof. V.I. Korotich [8]. FeO content in sinter was taken 10.0 % for all calculation variants.

Out of many calculation variants, we have chosen four variants of the sinter composition calculation. In the *first* variant, normal limestone is used as a flux for sintering: in the *second* variant – a mixture of basic fluxes, i.e. 72.5 % of normal limestone and 27.5 % of dolomite limestone; in the *third* variant, sinter burden included Bakal siderite ore (10%); in the *fourth* variant, a combined flux (70% of normal limestone and 30 % of dolomite limestone) was used for MMK's sintering process. The purpose of calculating flux added for sintering is to obtain MgO concentration in final blast-furnace slag within 9-10 %. The chemical composition of sinter mixtures is given in table 1 in four variants. 1.

Table 1. Calculated compositions of MMK's sinter with basicity CaO/SiO_2 = 2.10 with addition in sinter burden of different fluxes and Bakal siderite ore

Type of produced sinter	Content of elements and oxides in sinter, mass % %											
	Fe	Mn	S	P	Fe_2O_3	FeO	CaO	SiO_2	Al_2O_3	MgO	TiO_2	MnO
Sinter-1 (flux – limestone)	53.47	0.21	0.01	0.02	65.27	10.00	13.88	6.61	1.86	2.01	0.27	0.27
Sinter-2 (flux – 72.5% limestone + 27.5% dolomite limestone)	53.06	0.21	0.01	0.02	64.69	10.00	13.64	6.49	1.83	2.97	0.27	0.27
Sinter-3 (addition of Bakal siderite ore 10.0% in sinter)	52.95	0.28	0.01	0.02	64.53	10.00	13.86	6.60	1.89	2.65	0.26	0.36
Sinter-4 (flux – 70% limestone + 30% dolomite limestone)	53.00	0.21	0.01	0.02	64.61	10.00	13.61	6.48	1.83	3.10	0.27	0.27

On the basis of these data we made calculations of blast-furnace burdens with determination of the total consumption of iron-ore materials (MMK's sinter and non-fluxed pellets), coke consumption and required consumption of crude limestone, softening start temperature and melting temperature of the iron-ore mixture, melting temperature range, slag composition and output as well as slag properties, i.e. viscosity at 1400, 1450 and 1500°C and viscosity gradient determining slag stability in the temperature range from 1400 to 1500°C [6].

The composition of blast-furnace burden was as follows: for calculation variants 1-3 – 40 % of MMK's sinter and 60 % of SSGPO's pellets and for calculation variant 4 – 40 % of MMK's sinter, 30 % of SSGPO's pellets and 30 % of Mikhailovsky GOK' pellets. All the calculations were made on condition that Si content in iron should be 0.70 % with final slag basicity CaO/SiO_2 = 1.05 for the following input data: blast temperature is 1178 °C; oxygen content in blast is 25.88 %; natural gas flow rate is 103 m^3/t of iron; ash content in coke is 13.08 %; sulphur content in coke is 0.50 %; coke strength M_{25} = 87.99 %. When solving the problem, we assumed that production technology and chemical composition of SSGPO's iron-ore pellets remain unchanged. The coke consumption was calculated by the method of Prof. A.N. Ramm [9].

5. Model calculation results

The calculation results are given in table 2.

Table 2. Some parameters of blast-furnace smelting variants in MMK's furnaces.

Parameter	Blast-furnace smelting variants			
	I[a] Sinter-1	II[a] Sinter-2	III[a] Sinter-3	IV[b] Sinter-4
Total consumption of iron-ore materials, kg/t of hot metal	1603.4	1607.8	1608.5	1604.0
Specific consumption, kg/t of hot metal:				
coke	443.8	445.6	445.1	449.4
crude limestone	31.8	32.4	28.4	56.4
slag	314	321	320	328
Slag composition, %:				
(CaO)	38.47	37.53	37.94	39.64
(SiO_2)	36.64	35.74	36.13	37.75
(Al_2O_3)	13.88	13.56	13.75	10.88
(MgO)	7.65	9.86	8.83	8.63
(TiO_2)	1.59	1.56	1.55	1.47
Slag viscosity, poise/(Pa·s) at temperature, °C:				
1400	17.3/(1.73)	16.7/(1.67)	17.0/(1.70)	13.9/(1.39)
1450	8.2/(0.82)	7.5/(0.75)	7.8/(0.78)	5.8/(0.58)
1500	4.7/(0.47)	4.2/(0.42)	4.4/(0.44)	3.2/(0.32)
Slag viscosity gradient, poise/°C	0.22	0.24	0.23	0.29
Viscoplastic characteristics of iron-ore mixture, °C:				
softening start temperature	1157	1167	1163	1162
melting temperature	1453	1444	1447	1436
melting temperature range	295	278	284	274

Note. Slag basicity CaO/SiO_2 = 1.05.
[a]Blast-furnace slag composition: 40% MMK's sinter with basicity CaO/SiO_2 = 2.1 + 60% SSGPO's pellets.
[b]Blast-furnace slag composition: 40% MMK's sinter +30% SSGPO's pellets + 30% Mikhailovsky GOK's pellets.

The calculations show that when charging the burden of Variant 1 into the furnace the composition of the resulting slag is extremely unfavorable - Al2O3 concentration = 13.88% and MgO = 7.65 % (refer to table 2). At the temperature of 1450°C slag has an unacceptably high viscosity – 8.2 poise (0.82 Pa·s) [6]. According to the previous studies of slag conditions of MMK's blast-furnace smelting, it was determined that an increase in slag viscosity η_{1450} > 6.5 poise (0.65 Pa·s) is accompanied by a degraded slag flow ability and reduced efficiency of hot metal desulphurization. A probability of an increased lower gas pressure drop and limited smelting rate also grows up.

In order to increase magnesia concentration in slag for the purpose of increasing slag sulphur-absorbing ability and improving temperature characteristics of iron-ore material melting, Calculation

Variant II was performed (refer to table 2): use of combined flux for sintering. Blast-furnace smelting of Sinter-2 together with SSGPO's pellets in the ratio of 40:60 % leads to an increased magnesia concentration in slag up to MgO = 9,86 % (refer to table 2) and improved viscoplastic characteristics of the iron-ore mixture charged in the blast furnace: Δt_{melt} decreases from 295°C (in Variant I) to 278 °C. Slag viscosity at 1450°C also decreases to η_{1450} = 7.5 poise (0.75 Pa·s). However, this parameter is also too high and does not conform to normal progress of blast-furnace smelting in slag conditions.

With the same ratio of sinter and SSPGO's pellets (40:60%) in blast-furnace burden, Calculation Varian III was performed to check if it is efficient to add Bakal siderite ore in MMK's sinter burden to obtain blast-furnace slags with in an increased MgO concentration. In fact, 10% addition of Bakal siderite ore leads to final slags with content of MgO = 8.83 %. However, characteristics of slag conditions are worse in efficiency than in Variant II – MgO content in slag is less (8.83 instead of 9.86) and Al_2O_3 concentration in slag is even slightly higher. The temperature characteristics of iron-ore material melting are worse – the melting range Δt_{melt} increases to 284 °C (in Variant II Δt_{melt}=278 °C). In Variant III slag viscosity at 1450 °C is not acceptable – η_{1450} = 7.8 poise (0.78 Pa·s). Thus, addition of Bakal siderite ore into MMK's sinter burden does not solve the problem of decreasing in slag of Al_2O_3 = 13.75%.

To get the required values of Al_2O_3 and MgO concentration in slag, Calculation Variant IV was performed: with burden consisting of 40% sinter and different pellets (30% SSGPO's pellets and 30% Mikhailovsky GOK's pellets). The calculations show that in this variant the characteristics of slag viscosity in the operating temperature range conform to the required characteristics at the temperature of 1450°C – η_{1450} = 5.8 poise (0.58 Pa·s) and at the temperature of 1500 °C – 3.2 poise (0.32 Pa·s). The melting temperature range determining the viscoplastic zone thickness is minimum for all the above variants – Δt_{melt} = 274°C. However, in order to get the assigned slag basicity CaO/SiO_2 = 1.05 with the chosen ratio of iron-ore materials, it is required to charge crude limestone in the quantity of 56.4 kg/t of hot metal and in a result the specific coke consumption is somewhat higher – 449.4 kg/t of hot metal.

6. Conclusions

1. We developed a model-based decision support system which allows diagnostics of blast-furnace operation on the basis of actual and design data as well as solving a set of process task in the predictive mode.
2. With a higher portion of non fluxed pellets (up to 60%) in the charged burden, it is necessary to charge limestone in the furnace to achieve the required basicity of blast-furnace slag.
3. In order to reduce the amount of crude limestone to be charged in the furnace, high-basicity sinter is required. MMK's sintering plant uses sinter with basicity CaO/SiO_2 = 2.1 for the sintering process and this value is accepted as the maximum value for predictive calculations.
4. As far as the portion of SSPGO's pellets in the charged burden increases, MgO concentration in blast-furnace slag decreases. To increase MgO concentration in slag, it is efficient to use combined basic flux consisting of 70% normal limestone and 30% dolomite limestone for sinter production.
5. The calculation analysis shows that a higher MgO concentration in blast-furnace slag achieved by addition of Bakal siderite ore into sinter burden is less preferable as compared to use of a mixture of normal and dolomite limestone for sinter production.
6. Additional calculations show that Al_2O_3 content in slag grows up when the portion of SSGPO's pellets in burden is increased. To reduce its content to the level acceptable for normal slag conditions (Al_2O_3 ≈ 10–12%), it is efficient to replace partially SSGPO's pellets with Mikhailovsky GOK's pellets. It is established by calculations that Al_2O_3 concentration in blast-furnace slag reaches ≈10.9 % when blast-furnace burden consists of 40 % MMK's sinter with basicity CaO/SiO_2 = 2.1, 30 % SSGPO's pellets and 30% Mikhailovsky GOK's pellets.

References

[1] Novikov V S et al 1972 *Ferrous Metallurgy: NTIEI News* **3** 1–14

[2] Riznitsky I G et al *1984 Metallurgical and Ore Mining Industry* **1** 4–6
[3] Nekrasov Z I 1978 *Steel* **1** 11–18
[4] Yakovlev Yu V et al *Steel* **1** 17–25
[5] Spirin N A et al 2011 *Model-Based Decision Support Systems in ICS of Blast-Furnace Smelting* (Ekaterinburg: UrFU) p 462
[6] Onorin O P et al 2005 *Computer Methods of Blast-furnace Process Modelling* (Ekaterinburg: UGTU-UPI p 301
[7] Spirin N A et al 2014 *Mathematical Modelling of Metallurgical Processes in ICS: Study Guide* (Ekaterinburg: UrFU) p 558
[8] Korotich V I et al 2009 *The Basic Theory of Pelletizing Technology for Metallurgical Raw Materials* (Ekaterinburg: UGTU-UPI) p 417
[9] Ramm A N 1980 *Modern Blast-furnace Process* (M.: Metallurgy) p 304
[10] Bolshakov V I 2007 *Technology of High-efficient Energy-saving Blast-furnace Smelting* (K.: Naukova Dumka) p 411
[11] Spirin N A et al 2016 *Metallurgist* **60(5-6)** 471–477
[12] Rybolovlev V Y et al 2015 *Metallurgist* **59(7)** 653–658
[13] Spirin N A et al 2011 *Metallurgist* vol **54(9-10)** 566–569
[14] Lavrov V V and Spirin N A 2016 *IOP Conf. Series: Materials Sci. and Eng.* **15**0 012010
[15] Spirin N A et al *2016 IOP Conf. Series: Materials Sci. and Eng.* **150** 12011
[16] Spirin N A et al 2015 *AISTech 2015 Iron and Steel Technology Conference and 7th International Conference on the Science and Technology of Ironmaking, ICSTI 2015* (USA: Cleveland Convention Center Cleveland) 1 article No. 113707 pp 1225–1232
[17] Pavlov A V et al 2017 *Use of Model Systems for Solving New Technological Problems in Blast-Furnace Production* **61(5-6)** 448–454
[18] Onorin O P et al 2017 AISTech 2017 *Iron and Steel Technology Conference Proceedings* (USA:Nashville) vol 1 article No. 128474 pp 771–778
[19] Pavlov A V et al 2016 *Metallurgist* **60(5-6)** pp 581–588
[20] Pavlov A V et al 2016 *Metallurgist* **60(7-8)** pp 653–657
[21] Spirin N A et al 2015 *Metallurgist* **59(1)** pp 104–112

Alternate drawing as the way for improving mechanical properties of medium carbon steel wire

M A Polyakova, A E Gulin and E M Golubchik

Nosov Magnitogorsk state technical university, 38 Lenin ave, Magnitogorsk, 455000, Russia

E-mail: m.polyakova-64@mail.ru

Abstract. Drawing in monolithic dies is the basic operation of wire manufacturing. Despite the obvious advantages the drawing in monolithic dies is characterized by relatively tough stress-strain scheme of the processed wire which results in decreasing of its plasticity. During alternate drawing the wire goes through the conical die after which the drawing direction is changed. The results of investigation of alternate drawing process of medium carbon steel wire with 0.5 %C are presented in the paper. After simulation in software Deform-3D it was stated that strain intensity distributes uniformly across the wire transversal cross section. Using coordinate grid method the data about the metal flow both at traditional and alternate drawing was obtained. Results of carbon steel wire mechanical properties investigation proved the advantages of alternate drawing process. After alternate drawing the medium carbon steel wire possesses high plasticity at the same level of strength properties as compared with traditional drawing process. At the same time there were no considerable changings of medium carbon steel microstructure after alternate drawing.

1. Introduction

Wire manufacturing takes the first place in the whole volume of metal ware production. Steel wire can be used both as the final consumer product (for example, string wire, needle wire, polygraphic wire etc.) and as the semiproduct for mesh, electrodes, cables and ropes production etc. Technological process of wire manufacturing consists of several traditional operations: descaling and surface preparation before drawing, drawing to the definite diameter, thermal treatment and final processing operations. Basics tendencies in industrial wire manufacturing are the following: increasing the capacity of equipment with maximum saving of technological process current, decreasing of both basic and nonproduction material and energy sources, increasing strength and plastic wire properties, enhancement of manufacturing of new wire types with improved customer properties.

Drawing in monolithic dies is the basic operation of wire manufacturing. Despite the obvious advantages the drawing in monolithic dies is characterized by relatively tough stress-strain scheme of the processed wire which results in decreasing of its plasticity especially at multiple drawing. Frequently high-strength wire is rejected exactly because of untimely loss of plasticity and exfoliation. One of the main reasons of these effects is the high level of residual stresses in wire which formed after drawing. Especially frequently it is observed on the wire with large diameter [1]. From this point of view one of the main tendency both in theoretical and practical investigations is the choice of new approaches for drawing process development which allow to preserve the plasticity reserve in the processed wire.

Combined and integrated processes which are designed in the basic operation of drawing make it possible to solve this problem [2-4]. Advantages of these technologies consist not only in the reduction of technological operations in the manufacturing process and, hence, energy- and material saving, but also they allow to achieve the product with new complex of properties [5, 6]. But in most cases designing of such processes demands the construction of special instruments and tools, using special approaches for concordance of the processing speed on different parts of the wire for ensuring the process continuity which in total to high extent limits the implementation of such processes in the industrial wire manufacturing technologies. From this point of view finding of new methods for traditional wire drawing which make it possible to preserve its plasticity reserves is the actual task.

It was shown in [7-13] that after alternate drawing the wire have high level of strength and plastic properties. The essence of this method consists in the drawing the wire through the conical die after which the drawing direction is changed. Such method of processing is characterized by low level of share deformation near the wire surface which saves its high plasticity. It was proved experimentally on the wire from nonferrous metals and alloys and high carbon steel. In order to get knowledge about this type of drawing it is reasonable to study the peculiarities of alternate drawing of wire made from steel with another carbon content.

2. Methodology

The aim of this investigation is to test the method of alternate drawing (figure 1) on medium carbon steel wire grade 50 (0.5 % C) with 4.0 mm in diameter, which chemical composition is presented in table 1. For the experiments medium carbon steel wire was used in as received state.

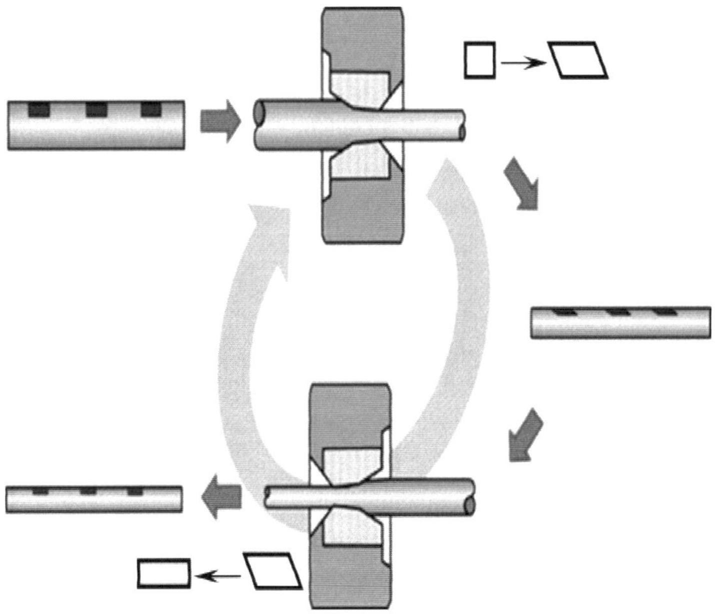

Figure 1. Alternate drawing process [10].

Table 1. Chemical composition of steel grade 50.

Steel grade	Element content, %									
	C	Si	Mn	Ni	S	P	Cr	Cu	As	Fe
Grade 50	0.47 - 0.55	0.17 - 0.37	0.5 - 0.8	till 0.25	till 0.04	till 0.035	till 0.25	till 0.25	till 0.08	~97

In table 2 experimental technological parameters of wire processing by alternate drawing are presented.

Table 2. Parameters of alternate drawing of medium carbon steel wire grade 50.

Die diameter *D*, mm	3.71	3.37	3.05	2.82	2.59	2.3	2.00
Reduction degree in die ε. %	13.97	17.48	18.08	14.51	15.64	21.14	19.56
Total deformation degree. %	13.97	29.01	41.85	50.29	58.07	66.93	75

Alternate drawing was carried out on the laboratory single-stage drawing mill. Mineral soap was used as the lubricant. Tensile test was carried out in accordance with ISO 6892-1:2009(en) "Metallic materials – Tensile testing – Part 1: Method of test at room temperature".

3. Results and discussion

Finite-element simulation in the software Deform-3D was carried out for investigation the regularities of metal flow at alternate drawing. Strain intensity distributions at different kinds of drawing of medium carbon steel wire grade 50 with 4.0 mm in diameter are presented in figure 2.

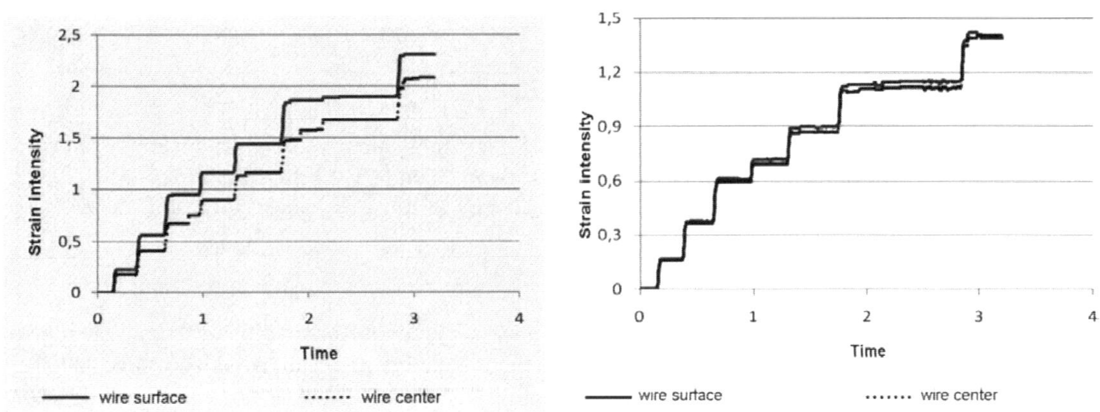

(a) (b)

Figure 2. Strain intensity of wire from steel grade 50 at traditional drawing (a) and alternate drawing (b).

Results of simulation show that at alternate drawing because of changing drawing direction strain intensity distributes uniformly across the whole wire. At the same time strain intensity has different quantitative meanings: it is lower at alternate drawing. This is one of the essential factors which ensures the preservation of wire plasticity.

Using grid method the data about metal flow at traditional and alternate drawing was obtained (figure 3).

As one can see, because of applying pulling force to the wire front end (drawing force) the coordinate grid lines elongate along the drawing direction resulting in formation of typical drawing texture, i.e. grains elongate along the drawing axis. In table 3 coordinate grid geometrical parameters obtained by metal flow simulation of different types of drawing of wire from steel grade 50 with 4.0 mm in diameter are presented.

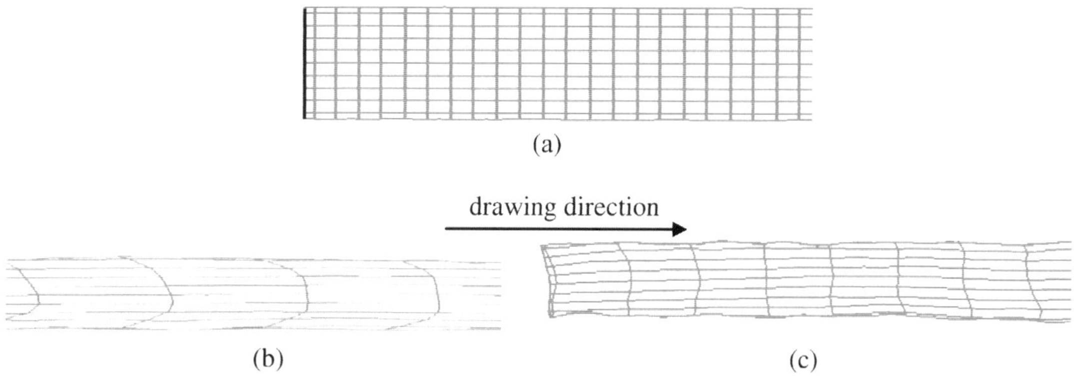

Figure 3. Coordinate grid changing on the longitudinal wire cross section: (a) original grid; (b) coordinate grid after traditional drawing; (c) coordinate grid after alternate drawing.

Table 3. Coordinate grid geometrical parameters changing obtained by simulation of metal flow at traditional and alternate drawing of wire from steel grade 50 with 4.0 mm in diameter.

Wire diameter, mm		4.0	3.71	3.37	3.05	2.82	2.59	2.3	2.0
Cell length *L*, mm	traditional drawing	0.18	0.18	0.3	0.77	0.81	1.16	1.5	1.92
	alternate drawing	0.18	0.18	0.36	0.63	0.35	0.45	0.38	0.20
Angle of coordinate mesh vertical line α, grad.	traditional drawing	0	3.54	6.8	8.5	13.5	19.94	33.6	39.9
	traditional drawing	0	6.91	7.11	7.84	5.88	5.30	4.15	2.60

After tensile test of medium carbon steel wire grade 50 after different types of drawing it was found that after alternate drawing the processed wire possessed high plasticity at the same level of strength as after traditional drawing (figure 4). In table 4 mechanical properties of medium carbon steel wire grade 50 after different types of drawing are presented.

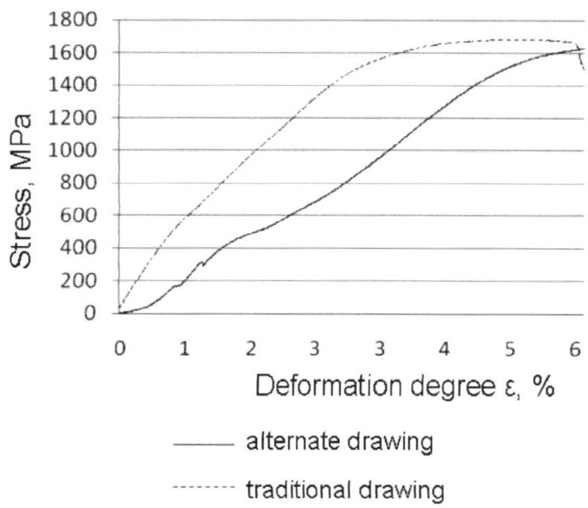

Figure 4. Load-extension diagram of medium carbon steel wire grade 50 after traditional and alternative drawing.

Table 4. Mechanical properties of medium carbon steel wire grade 50 after traditional and alternative drawing.

Method of deformational processing	Ultimate tensile strength σ_S, MPa	Yield strength $\sigma_{0.2}$, MPa	Relative elongation δ, %
Traditional drawing	1677	1142	1.55
Alternate drawing	1660	1259	1.77

Results of mechanical tests of medium carbon steel wire grade 50 proved the advantages of alternate drawing: yield strength increases to 9,3 %; relative elongation increases to 12,4 %. At the same time considerable changes in microstructure of medium carbon steel wire after traditional and alternate drawing are not observed (figure 5).

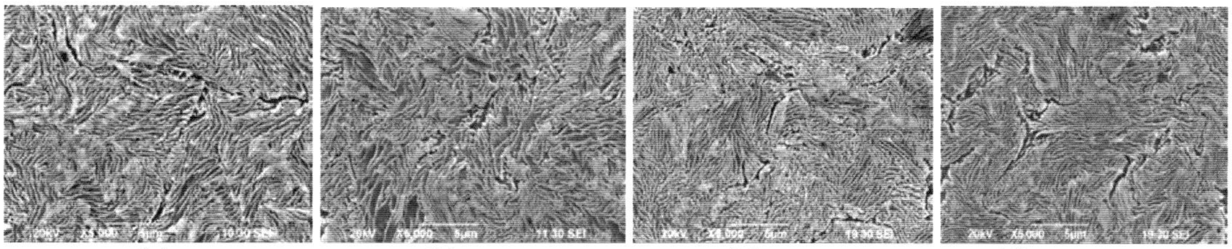

(a) (b) (c) (d)

Figure 5. Microstructure of medium carbon steel wire grade 50 after traditional (a, b) and alternate (c, d) drawing: surface area (a, c), central area (b, d).

4. Conclusion

For preservation of plasticity reserve of carbon steel wire at drawing it is expediently to use such methods which make it possible to reduce the strain unevenness. During alternate drawing the drawing direction is changed resulting in decreasing strain unevenness in the processed wire. On medium carbon steel wire it was proved experimentally that after alternate drawing plastic properties increase with saving the same level of strength as after traditional drawing. The obtained results are promising for further implication of alternative drawing to the industrial technologies of wire manufacturing for obtaining the metal product with improved level of customer properties.

Acknowledgments
The study was financially supported by Ministry of Education and Science of the Russian Federation (Contract 02.G25.31.0178 from 01.12.2015; № MK204895 from 27.07.2015).

References

[1] Korchunov A, Polyakova M, Koptseva N, Brunelli K, Breda M and Dabalà M 2016 *La Metallurgia Italiana* February 5–11
[2] Sidelnikov S, Dovzhenko N and Zagirov N 2005 *Integrated and Combined Methods of Non-ferrous Metals and Alloys Processing* (Moscow: MAKS Press) p 344
[3] Kharitonov V and Gallyamov D 2016 *Ferrous Metals* **8** 71–7
[4] Chukin M, Polyakova M, Gulin A and Nikitenko O 2015 *Key Engineering Materials* **685** 487–91
[5] Korchunov A, Chukin M, Polyakova M and Emaleeva D 2011 *Vestnik of Nosov Magnitogorsk State Technical University* **1** 43–6
[6] Chukin M V, Polyakova M A and Gulin A E 2016 *Steel in Translation* **46(8)** 548–51
[7] Stefanov Hristov V and Yoshida K 2016 *Key Engineering Materials* **716** 13–21
[8] Yoshida K 2004 *Steel Grips 2 Suppl. Metal Forming* 199–201

[9] Yoshida K and Doi K 2013, 2014 *Interwire* 77–9
[10] Nagashima H and Yoshida K *Key Engineering Materials* **716** 22–31
[11] Yoshida K and Doi K 2014 *Proceedia Engineering* 706–11
[12] Kikuchi S 2011 *Wire Journal International* 76–80
[13] Nagashima H and Yoshida K May 2015 *Journal of Achievements in Materials and Manufacturing Engineering* **70(1)** 29–35

IOP Conf. Series: Materials Science and Engineering **411** (2018) 012059 doi:10.1088/1757-899X/411/1/012059

Comparative analysis of modern trends in the use of industrial wastes of coke chemistry

O A Polyakh, N S Ponomarev and A D Zhuravlev

Siberian State Industrial University, 42 Kirova street, Novokuznetsk, 654007, Russia

E-mail: kafcmet@sibsiu.ru

Abstract. Modern methods of using industrial wastes of by-product-coking industry are analyzed. The paper contains the review of the main types of coke production waste. The environmental aspects of waste disposal and utilization are considered. The analysis of the use of liquid and solid wastes in technological processes is carried out.

1. Introduction

The use of wastes from by-product-coking industry is an important issue widely discussed in the literature [1 - 23]. When coking coal a significant amount of waste is formed, which until now do not find any effective use. In addition, the toughening of environmental requirements for the production of coke-chemical products dictates the need to find solutions for the recycling and use of various types of waste. One of the pressing problems of the coke-chemical industry is the problem of expanding the raw material base of coking by using low-quality coal in the coke to replace the better quality one.

The purpose of this work is a comparative analysis of current trends in the use of industrial wastes of coke chemistry.

Among the wastes of by product coke production are liquid wastes (heavy coal-tar products, acid tar, polymers, etc.) and solid wastes (coke dust).

2. Resinous wastes

Tar sludges – heavy residues of coal tar, containing 40-50% of coal and coke dust, carried by coke oven gas. They settle on the bottom of mechanical sedimentation tanks and clarifiers and represent a heavy viscous mass [1].

Tar sludges are used as a fuel or as part of a charge for coking and gasification. They are mixed and pelleted with basic components and other types of combustible waste. In many factories, due to the lack of equipment, a significant mass of tar sludges is not used and is sent to storage tanks.

The second area of application of tar sludges is the construction industry. On the basis of tar sludges, materials are made for protective coatings of concrete, reinforced concrete and metal products. Such compositions are obtained by dissolving tar sludges in white alcohol and other solvents with the addition of polyvinyl chloride resin and subsequent settling. Coatings have a stable hydrophobicity, high strength and water resistance.

Acid tar – residues of condensation of light tar from coke oven gas and polymerization products of unsaturated compounds present in coke oven gas under the influence of sulfuric acid during the process of gas purification from ammonia.

Acid tar is added to the coke during coking and used in the production of bitumen of different grades, for the production of sulfur dioxide followed by processing it into sulfuric acid [2].

The tar is also used as an additive to cement clinker for intensification of grinding and activation of cement hardening, as swelling additives to the charge in the production of expanded clay. After neutralization, it can be used for the production of road tar oils. Neutralization is carried out with alkaline waste and reagents. It is possible to use sour resin instead of joiner's glue.

Thus, the resinous waste of by-product-coking industry is used in the construction industry and can be returned to the main cycle as part of the charge for coking.

3. Coke dust

Coke dust is trapped from dry quenching plants when blowing with inert gases of unloaded coke, as well as from installations for dustless coke discharge from coke-oven battery. It is almost not used due to the complexity of unloading and transportation, therefore it is usually returned to the charge of coking in the amount of 1% of the charge mass or processed "in situ" using various methods of compaction and packing, or packed into containers (sacks). In general, due to the finely dispersed state and high ash content, the coke dust is not suitable for direct use [11, 12].

The volumes of coke dust are very high, on average, about 18-20 thousand tonnes of coke dust per year is generated at one coke plant. Traditionally, dust accumulated in storages (which are already filled) or added to the charge without prior technological preparation. Other ways of wastes disposal (burial, burning, biodegradation) were also ineffective.

At present, the raw material base for coking and the replacement of coking coal grades with low-caking grades have changed significantly, and there are changes in the technology of charge preparation, including with the use of coke dust.

The modern coal raw material base is very unstable in terms of its grades composition and technological properties, coal is unevenly supplied to the plants, and the charge for coking is multicomponent. Fluctuations in the quality of the charge can not be eliminated only through organizational measures in the coal industry. It seems that the problem of obtaining blast-furnace coke from the charge of modern grades composition can be solved by introducing new efficient technological processes for the preparation of this charge with an increase in the loading density [11, 12].

The effectiveness of all new methods of preparation of coal charge before coking, including briquetting, is largely due to the increase in the density of coal charge in the furnace chambers. Moreover, the increase in density not only serves for increase in the productivity of the furnaces, but also improves the conditions for sintering the products of the destruction of coal grains of various sizes and degrees of metamorphism, and, consequently, the mechanical strength of coke.

To expand the coking feedstock base, different methods are used to prepare the coal charge for coking, in particular, compaction of coal charge with the addition of fluid waste of by-product-coking production. A number of authors propose to use such wastes as a binder for partial briquetting of coal charge for coking, which will allow not only the amount of low-caking coal grades to be increased, but also the production wastes to be disposed. The increase in the charge density is 800-870 kg/m^3, which leads to an increase in the strength of coke [13, 14].

Briquetting of coke dust, including mixing of the crushed solid fuel with a binder component, is considered in the sources [13, 14]. As a pulverized solid fuel, pre-enriched coke dust is used for the oil agglomeration, and the preheated carbamide is used as a binder component. It is also possible to use tar as additives (after neutralization) to the charge for coking (gasification), formed as a waste during interaction of tar products of coke oven gas with sulfuric acid in the process of ammonium sulfate production.

Thus, the possibility of using coke dust together with other production wastes in the charge for coking coal is shown in figure 1.

The use of highly dispersed materials of technogenic origin, that have no value and have large amounts of accumulation, as raw materials is a relatively new, promising and energy-saving direction in electrothermal processes, in particular, silicon carbide technology [15, 16].

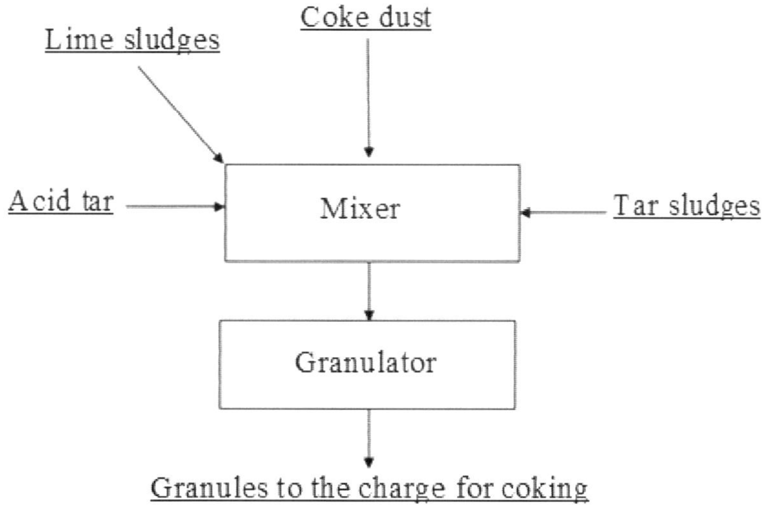

Figure 1. Schematic diagram of charge preparation from wastes of by-product-coking production.

According to the existing thermodynamic concepts, the process of carbon thermal reduction of silica proceeds with the active participation of gaseous silicon oxides. The size effect, arising in the one-component system "gas-dispersed crystalline phase", consists in changing the saturated vapour pressure above the surface of crystalline particles, depending on the degree of dispersion of the solid phase. It is supposed that the interaction rate can be increased by increasing the evaporation surface of SiO_2 and using a carbonaceous reductant with a high adsorption capacity and a developed surface [17]. Coke dust may be proposed as such reducing agent.

To realize the electrothermal process on the basis of microsilica and coke dust, the technology of furnace synthesis of highly dispersed silicon carbide has been developed, which makes it possible to obtain a product with a specific surface of 3000-4000 m/kg and a carbide content of 90-92 wt% [16].

Thus, the coke dust together with other wastes of by-product-coking production can be successfully applied in the composition of coke charge and in some electrothermal processes.

4. Wastewaters of by-product-coking production

By-product-coking production is the source of a significant number of highly toxic secondary products (wastes). The greatest part of them (about 99%) is phenolic wastewater generated in the process of coking coal charges; the share of other wastes (acid tar, tar sludges, oil, etc.), formed in the shops for the trapping and processing of coking products, is about 1%. Wastewater contain various oils, suspended solids and other impurities of organic and inorganic origin, most of which are harmful, making it difficult to use wastewater in production.

The amount of wastewater and the concentration of impurities in them depend on the quality of the coked coals, operating conditions and the state of the chemical equipment and is usually 0.25-0.3 m^3/h per 1 ton of coke charge, the absolute amount is up to 150-170 m^3/h [18-21] .

Phenolic waters disrupt the natural processes of self-purification. Surface water reservoirs, especially small ones, are converted into sewage ditches without animal and plant world. They can not be used for cultural, recreational and household needs. In addition, surface water reservoirs contaminated with phenolic wastewater contribute to the deterioration of the quality of underground water sources.

Currently, there is a number of environmentally-friendly technical solutions that can significantly reduce the harmful impact of wastes from by-product-coking production on the environment, taking into account the specificity of their physical and chemical properties. Reduction is carried out by means of mechanical, extraction and biochemical methods [22].

Purified and neutralized wastewater from the by-product-coking production is used in its circulating cycles and in related industries. Scientists [3] write about the possibility of using supramolecular water as an inhibitor of carbonate salt deposition, while observing optimal concentrations.

5. Coke oven gas

Coke oven gas is formed during the coking of coal as a result of thermal decomposition and is a by-product in the production of coke. Approximate composition of coke oven gas, % vol.: CO_2 1.6 – 3%, O_2 0.4 – 0.8%, N_2 2 – 3.5%, C_mH_n 2 – 2.5%, CO 5 – 6.5%, CH_4 24.5 – 26.5%, H_2 58 – 62% [3].

Direct coke oven gas undergoes compulsory treatment, during which vapours of resins and water condense, and ammonia and benzene hydrocarbons are trapped. After such treatment the gas is called the inverse and it is used mainly for combustion as a fuel alone or in the mixture with a blast furnace one.

6. Conclusion

The review of main types of waste from by-product-coking production on the basis of modern literature is presented. It is established that one of the most urgent environmental problems is the need to recycle industrial waste. Modern ways of using by-product-coking productio wastes, including as charge materials of electrothermal processes (silicon carbide production) are analyzed.

References

[1] *Heavy Coal-tar Products, processing* http://ru-ecology.info/term/43173
[2] *Ecology of the city. Waste of by-product coke production* http://portaleco.ru/
[3] Volynkina E P 2008 *Proc. of Int. Sci. Conf in Waste Management – the bBasis for Restoring Ecological Balance in Kuzbass* (Novokuznetsk: SibSIU) p 313
[4] *Phenols* http://www.eurolab.ua/encyclopedia/3863/34232/
[5] Polyakh O A, Anikin A E, Yakushevich N F and Galevsky G V 2014 *Proc. of All-Russian Sci. Conf in Modern Problems of Coke Production and Processing of Coking Products* (Kemerovo) pp 10–15
[6] Ponomarev N S and Polyakh O A 2015 *All-Russian Scientific Conference of Students, Post-Graduates and Young Scientists "Science and Youth: Problems, Searches, Solutions"* (Novokuznetsk: SibSIU) pp 246–248
[7] Terentieva I M and Polyakh O A 2012 *All-Russian Scientific Conference of Students, Post-Graduates and Young Scientists "Science and Youth: Problems, Searches, Solutions"* (Novokuznetsk: SibSIU) pp 26–29
[8] Konovalova Kh A and Polyakh O A 2016 *All-Russian Scientific Conference of Students, Post-Graduates and Young Scientists "Science and Youth: Problems, Searches, Solutions"* (Novokuznetsk: SibSIU) (Novokuznetsk: SibSIU) pp 330–334
[9] Ponomarev N S and Polyakh O A 2016 *All-Russian Scientific Conference of Students, Post-Graduates and Young Scientists "Science and Youth: Problems, Searches, Solutions"* (Novokuznetsk: SibSIU) pp 339–342
[10] Brown N V and Glushchenko I M 1989 *Promising Trends in the Development of Coke Production* (Moscow: Metallurgy) p 271
[11] Ukhmylov G S 1987 *General Information of the Institute "Chermetinformation". Series Coke Production* **1** 45
[12] Polyakh O A, Rudneva V V, Yakushevich N F and Galevsky G V 2013 *Bulletin of the Mining and Metallurgical Section of the Russian Academy of Natural Sciences. Department of Metallurgy* **32** 112–131
[13] Polyakh O A and Rudneva V V 2007 *Nanomaterials and Nanotechnologies in the Production of*

Silicon Carbide. Vol 1 Microsilica in the Production of Silicon Carbide ed G V Galievsky (Moscow: Flint: Science) p 248
[14] Yakushevich N F and Galevsky G V 1999 *Interaction of Carbon with Calcium, Silicon, Aluminum Oxide*s (Novokuznetsk: SibSIU) p 250
[15] Zhukov A I 2008 Methods for Purification of Industrial Wastewater (Moscow: Stroyizdat) p 114
[16] Komarova L F and Poletaeva M A 2010 Water Use in Enterprises and Wastewater Treatment in Various Industries (Barnaul: AltSTU) p 174
Rodionov A I, Kuznetsov Yu P and Soloviev G S 2005 Biosphere Protection from Industrial Emissions. Basics of Designing Technological Processes (Moscow: Chemistry, Kolos) p. 392
[17] Sabirova T M and Zaidenberg M A 1999 *Coke and Chemistry* **10** 35–38
[18] Zhischenko V V and Margins O A 2016 *All-Russian Scientific Conference of Students, Post-Graduates and Young Scientists "Science and Youth: Problems, Searches, Solutions"* (Novokuznetsk: SibSIU) 321–324
[19] Sokur A A, Tretyakov P V and Kutnyashenko A I 2010 *III Int. Conf. of Students, Postgraduates and Young Scientists in Chemistry and Chem. Tech.* http://masters.donntu.org/2010/fimm/kutnyashenko/library/0kiev_2/kiev_2 .html
[20] *Coke Production* http://metalspace.ru/education-career/osnovy-metallurgii/koks.html
[21] Kulkova T N et al 2007 *Metallurgist* **4** 17–20
[22] Zaitsev V A 2015 *Industrial Ecology* (Moscow: BINOM, Laboratory of Knowledge) p 385

IOP Conf. Series: Materials Science and Engineering **411** (2018) 012060 doi:10.1088/1757-899X/411/1/012060

The aging mode for reducing the thermal expansion of the piston from alloy AK21

A N Prudnikov, V A Prudnikov, M V Popova, V K Afanasyev and V Ya Zellermayer

Siberian State Industrial University, 42 Kirova street, Novokuznetsk, 654007, Russia

E-mail: a.prudnikov@mail.ru

Abstract. The effect of temperature and aging time (1÷10 h) at 100-300 °C on the microstructure and thermal expansion of the engine piston YaMZ 5363 from the hypereutectic silumin of AK21 type was studied. It is established that low-temperature aging at 100 °C for 1 hour, allows the temperature coefficient of linear expansion (TCLE) to be reduced in the operating temperature range of heavy-loaded engines (250-350 °C) by 3-6%. The increase in the aging time of pistons up to 3, 5 and 10 h does not lead to the further decrease in TCLE, as well as the increase in the aging temperature to 150, 200, 250 and 300 °C. Moreover, low-temperature aging does not lead to significant changes in the structure of the pistons of the engines in the delivery state, with the exception of coagulation of some of the volume fraction of the eutectic component of Al-Si or Al-Si-Cu.

1. Introduction

At present the need for powerful and multifunctional cars has increased, which necessitated a balance of fuel economy, durability and reliability of the engine. The automakers always faced the problem of not only constructive changes of parts and the engine itself, but also the development of light materials with improved properties, including mechanical, physical and operational ones. One of the most important parts of the engine is the piston. It withstands mechanical loads from pressure forces of gases and inertia forces, as well as thermal loads from contact of the bottom with hot gases and friction of its lateral surface against the wall of the cylinder. Therefore, high requirements are imposed on piston alloys, such as high thermal conductivity and heat resistance, good antifriction properties, low specific gravity, low temperature coefficient of linear expansion (TCLE). The most widespread among piston alloys for heavily loaded engines were the hypereutectic alloyed siluminins of the Al-Si-Cu-Mg-Ni system cast by various casting methods on chill casting machines. The wide application of these alloys is associated with the complex of their properties.

However, up to now, the full service life of piston alloys used for heavily loaded internal combustion engines has not been developed, especially in the field of physical properties (thermal expansion, thermal conductivity, specific gravity). To improve the structure and characteristics of the ICE pistons uses a variety of processing methods at all stages of their manufacture, from preparation of the charge and ending with the production of the finished product.

In particular, thermal or deformation preparation of charge [1-3]; metallurgical – refining and modifying with the use of reagents and physical methods of influencing the melt in the production of castings and ingots (ultrasonic and electromagnetic processing) [4-12]; special methods of processing thermal, thermocyclic (TP, TCP) [13-16] and deformation-thermocycling treatment (DTCO) [3,7,17-

20]. It is more attractive to use the most economically efficient and effective ones for improving the properties of pistons, and also implemented at the stage of the finished product.

2. Methods and materials

The piston of YaMZ 5363 engine manufactured by Yaroslavl Motor Plant of MAZ 5550 series in the delivery state was selected as a study object. The chemical composition of the piston was determined on the emission spectrometer ARL 3460 and is given in table 1.

Table 1. Chemical composition of the studied piston.

Material	Chemical composition, % (aluminum remaining)								
	Si	Cu	Mg	Ni	Mn	Fe	Cr	Ti	Zn
Piston alloy	22.1	2.91	0.64	1.23	0.08	0.49	0.26	0.08	0.01

The composition of the studied alloy corresponds to the composition of the piston hypereutectic silumin AK21M2.5H2.5 (GOST 1583-93) and differs somewhat less in nickel, manganese and slightly magnesium, which is due to the requirements of the company specification. Samples cut from the piston skirt were subjected to thermal treatment (aging) in SNOL-1,6.2,5.1/9-I3 electric furnace in the following modes: heating and holding for 1÷10 hours at 100 and further through 50 °C to 300 °C with cooling in the air.

Metallographic studies were carried out on the optical microscope LaboMet-I1. For dilatometric analysis, a high-temperature optical differential dilatometer of DIL type was used. When determining the true value of TCLE, the error of the method was $1 \cdot 10^{-6}$ K^{-1}.

3. Results and discussion

The results of the study of the effect of the aging temperature in the interval 100-300 °C after 50 °C with the holding time of 1 hr for the TCLE of the piston of the YaMZ engine from the AK21 alloy are shown in figure 1.

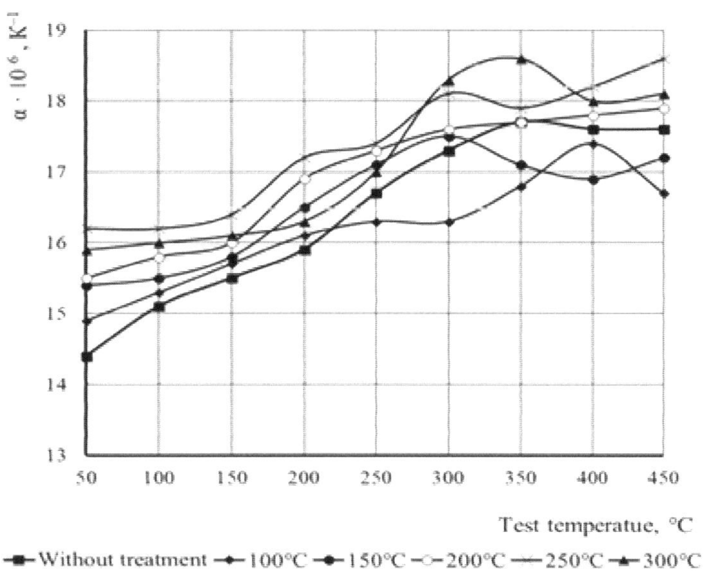

Figure 1. Influence of the aging temperature with a holding time of 1 h on the linear expansion of the piston from AK21 alloy.

Analysis of the α-T curves showed that aging at temperatures above 100 °C with a holding time of 1 h increases the value of the TCLE of the piston alloy as compared with the alloy without thermal treatment throughout the temperature range of the tests. The magnitude of this increase for each aging temperature is different. Thus, for aging temperatures of 150 and 200 °C it is on average 2 ÷ 7%, for temperatures 250 and 300 °C – 1.1 ÷ 12 and 1.8 ÷ 10% respectively. For low-temperature aging (100 °C) of samples from the piston, a different change in the magnitude of the thermal expansion is noted. In the low-temperature range of the piston operation (up to 200 °C) the TCLE slightly increases by an average of 1÷3% relative to the initial state of the alloy. At high engine operating temperatures (above 200 °C) the capacity of piston alloy to expand is reduced by an average of 3÷6% in the range 250÷350 °C compared to non-heat-treated alloys.

The consequent increase in the holding time of the samples from the piston at the most promising temperature of 100 °C (see figure 1) to 3, 5 and 10 h does not lead to a further decrease in TCLE over the whole temperature range of the test (figure 2). The tendency to increase the value of thermal expansion with increasing aging time in the aging process is almost the same as the temperature factor. Moreover, the maximum increase in TCLE of the piston alloy corresponds to the greatest aging time (10 h) at 100 °C and is from 3 to 7.5% for operating temperatures of 20-300 °C compared to samples aged for 1 h. And only at the test temperature 350 °C increase in TCLE becomes insignificant (about 1%).

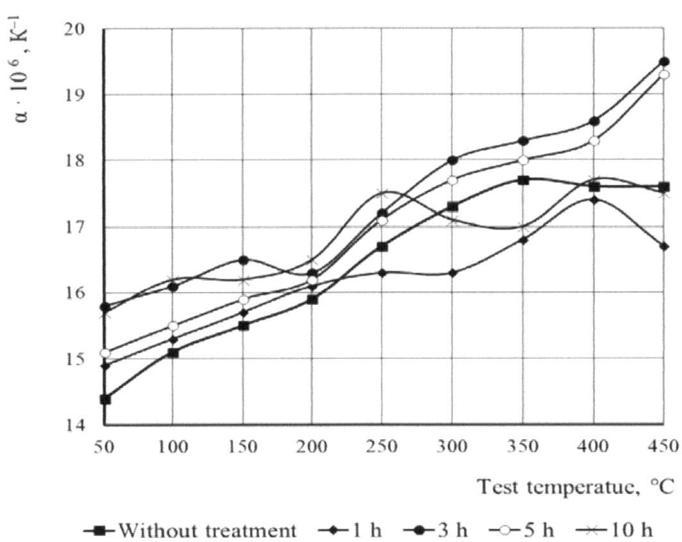

Figure 2. Influence of the aging time at the aging temperature of 100 °C on the linear expansion of piston from AK21 alloy.

The piston alloy of YaMZ engine has a rather complex phase composition, which is characteristic for casting piston hypereutectic siluminins such as AK21. The main structural components of the piston silumin are α-solid aluminum solution (Al_α), primary silicon crystals Si_β, aluminum-silicon eutectic (Al + Si). In addition, copper-containing phases are present in the structure, first of all, the strengthening phase of W-$CuSi_4Mg_5Al_4$ and a certain amount of $CuAl_2$. Phases containing Ni and Mg and more complex eutectics based on Al, Ni, Fe, Si are also detected. In the structure of the investigated piston, the CPSs have an average size of 30-64 μm and are fairly evenly distributed in the piston structure, and the size of the CPSs is larger in the bottom of the piston than in crystals in the skirt. This is primarily due to the greater thickness of the bottom and, correspondingly, the lower cooling rate during crystallization of the billet in the chill mold. The microstructure of the piston alloy after aging at 100-300 °C for 1 hour is shown in figure 3, and the effect of holding time during aging from 1 to 10 hours is shown in figure 4.

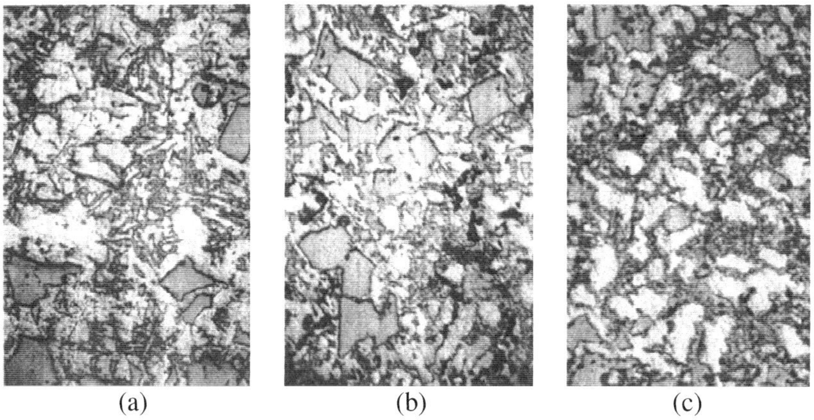

Figure 3. Effect of aging temperature on the piston microstructure of AK21 alloy (holding time 1 h); a – without treatment, b – 100 °C, c – 300 °C, × 300.

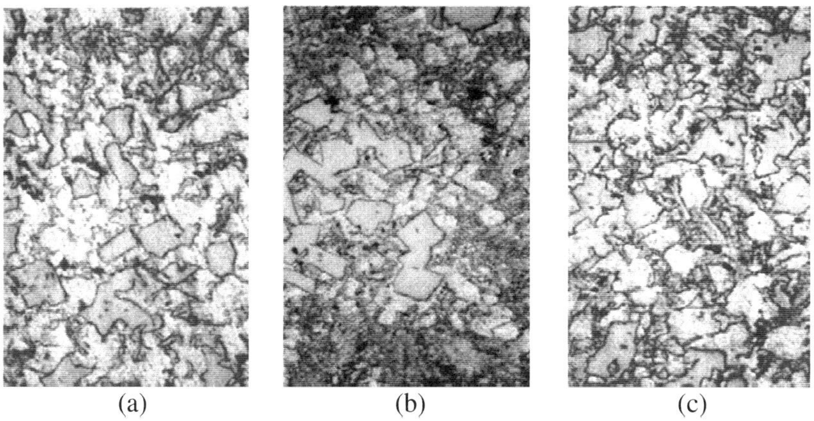

Figure 4. Effect of aging time at the aging temperature 100 °C on the piston microstructure made of AK21 alloy; a – 1 h, b – 5 h, c – 10 h, × 300.

It can be noted that as the aging time increases, a certain fragmentation and spheroidization of the eutectic silicon takes place. Moreover, the higher the aging temperature, the earlier this process begins in time and proceeds more intensively. However, holding for 1 hour at the aging temperature 100 °C is not sufficient for a significant change in the structure of the piston silumin. Thus, it has been established that practically neither time nor temperature has a significant effect on the size and distribution of CPSs, the size of which remains within 30-64 μm, which is apparently associated with a low aging temperature 100-300 °C for hypereutectic silumin.

4. Conclusion

As a mode for reducing the linear expansion of the piston from AK21 alloy, it is possible to recommend additional low temperature aging at 100 °C for 1 hour, which makes it possible to reduce the TCLE in the operating temperature range of heavy-loaded engines (250-350 °C) by 3-6%. And low-temperature aging does not lead to significant changes in the structure and properties of engine pistons.

References

[1] Prudnikov A N 2014 *Technology of Metals* **3** 16–22

[2] Puga H, Barbosa J, Soares D, Silva F and Ribeiro S 2009 *J. of Mat. Proc. Tech.* **209(11)** 5195–203
[3] Prudnikov A N 2013 *Structural and Technological Basis for the Development of Precision Silumin with Regulated Hydrogen Content* (Novosibirsk: NSTU) 40
[4] Deyev V B et al 2008 *Polzunovsky Almanac* **3** 77–81
[5] Prudnikov A N 2009 *Foundry Production* **2** 2–5
[6] Friedrich B and Kräutlein C 2006 *Metalurgija - Journal of Metallurgy* **12(4)** 252–266
[7] Prudnikov A N 2014 *Deformation and Destruction of Materials* **2** 14–20
[8] Naidek V L, Narivsky A V 2008 *Improving the Quality of Aluminum and Copper Alloys Castings by Means of Plasmareagent Treatment of Melts* (Kiev: Naukova Dumka) p 183
[9] Prudnikov A N 2009 *Metallurgy of Machine-building* **3** 28–31
[10] Deyev V B et al 2012 *Foundry* **5** 16–18
[11] Prudnikov A N 2009 *Metal Processing* (Technology, Equipment, Tools) **2** 26–31
[12] Grachev V A and Turakhodjaev N D 2017 *Archives of Foundry Engineering* **17(4)** 61–66
[13] Taskin M, Orhan M and Ozan S 2006 *J. of Engineering Mater. Sci.* **12** 362-7
[14] Prudnikov A N 2004 *Izv. Vuzov. Ferrous Metallurgy* **4** 40–2
[15] Song D, Kang G and Kan Q 2014 *Smart Materials and Structures* **23(1)** 1–7
[16] Prudnikov A N 2009 *Metal Processing* (Technology, Equipment, Tools) **1** 8–11
[17] Furuya Y and Park Y 1992 *Nondestructive Testing and Evaluation* **8(1)** 541–54
[18] Prudnikov A N and Prudnikov V A 2015 *Metallurgy: Technologies, Innovations, Quality* vol 2 (Novokuznetsk: SibSIU) pp 15-18
[19] Fedyukin V K and Smagorinsky M E 1989 *Thermocyclic Processing of Metals and Machine Parts* (Leningrad: Mechanical Engineering) p 255
[20] Yu T, Wang L and Zhao Y Q 2012 *International Journal of Fatigue* **35(1)** 31–36

IOP Conf. Series: Materials Science and Engineering **411** (2018) 012061 doi:10.1088/1757-899X/411/1/012061

Dimensions for reducing the consumption of fluoride salts in the production of aluminum

E P Rzechitskiy[1], A A Petrovskiy[1], N V Nemchinova[1], I A Sysoev[1] and Yu V Bogdanov[2]

[1]Irkutsk National Research Technical University, 83 Lermontov street, Irkutsk, 664074, Russia
[2]UC RUSAL, RUSAL Engineering Technology Centre, 40 Pogranichnikov street, Krasnoyarsk, 660111, Russia

E-mail: begimotus@mail.ru

Abstract. This article is devoted to the dimensions for reducing the consumption of fluoride salts in the production of aluminum. The main dimensions are: production and use of alumina with a lower sodium content; processing of a part of the excess electrolyte using a new technology to produce aluminum fluoride; processing of the heat insulating part of the spent lining with the production of regenerative cryolite, which is processed into a low-modulus cryolite with the help of a new technology. With the use of new technologies, secondary wastes are used in the production of cement, and aluminum smelters can become completely non-waste ones.

1. Introduction

The current consumption of aluminum fluoride in the use of electrolyzers with baked anodes and "dry" gas scrubbing is in the range of 16-22 kg per ton of aluminum, which, in terms of fluorine, is about 10-14 kg [1,2]. The balance of the distribution of fluorine losses in electrolytic aluminum production is given in table 1.

Table 1. Balance of the distribution of fluorine flowrate.

S No	Reason of losses	Kg/tn Al	%
1	Losses with gases	1.0	8.3
2	With excess electrolyte introduced to maintain the cryolite ratio (with Na2O content in alumina 0.35-0.45%)	5.0	41.7
3	With spent lining including:	5.5	45.8
	Carbon part	3.0	25
	Heat insulating part	2.5	20.8
4	Coal foam and mechanical losses	0.5	4.2
5	Losses of everything	12	100

Losses with gases do not exceed 1 kg/t of aluminum, further reduction of fluorine losses is constrained by existing designes of sealing shelters.The formation of excess electrolyte is associated with the need to maintain the cryolite ratio specified by the technology in the electrolysis process. Cryolite ratio is the molar ratio of the content of sodium fluoride to the content of aluminum fluoride. Cryolite ratio, with modern technologies of aluminum production, is maintained at 2.4-2.6. The main raw material for the production of aluminum - alumina - contains tenths of a percent of sodium oxide, which leads to a shift of the cryolite ratio to the alkaline side. To maintain the cryolite ratio at a given level, aluminum fluoride is introduced into the electrolysis cell, and excess electrolyte is removed from it.

2. Calculation of excess electrolyte formation is given below

Average content of Na_2O in alumina is ~ 0.35% wt. With a specific consumption of alumina of 1920 kg/t of produced aluminum, the sodium entrance into the electrolysis cell is:

$$1920 * \frac{0.35}{100} * \frac{46}{62} = 4.986\, kg \text{ Na}/t \text{ Al}$$

To impregnate the lining materials of the cathode shell, ~ 30% is spent:

$$4.986 * 0.3 = 1.496\, kg \text{ Na}/t \text{ Al}$$

The loss of electrolyte during the processing of turnover is ~ 5%:

$$4.986 * 0.05 = 0.249\ \ kg \text{ Na}/t \text{ Al}$$

Thus, the accumulation of sodium in the electrolyte is:

$$4.986 - 1.496 - 0.249 = 3.241\, kg\ /t \text{ Al}$$

With a sodium content in the electrolyte of ~ 22% wt. The accumulation of electrolyte is:

$$3.241/0.22 = 14.73\, kg\ /t \text{ Al}$$

This correlates with the data obtained during the operation of the pilot section "Electrolysis 300" of the Ural Aluminum Plant, where the accumulation of electrolyte was 10-15 kg/t of produced metal [3].

Abroad, heat insulating part of the spent lining is neutralized and stored in dumps, or used in the production of cement and construction materials [4]. In Russia, the heat insulating part without neutralization is stored in specially equipped storage tanks. We carried out research to study the composition of the heat insulating part of the spent lining and the methods of its processing [5, 6]. The average chemical composition by the results of the analysis of samples from five electrolysers of the С8БМ type wt.%: C – 1-5; F – 7-10; Al – 12-16; Na – 7-10; Ca – 0.5-1.0; Si – 16-20; Mg – 0.5-1.0; Fe – 1.0-2.0; others – 35-45 (oxygen in the form of aluminum oxides and silicon).

The following substances were detected as the main phases in the heat insulating part of spent lining by X-ray diffraction analysis: mullite $3Al_2O_3.2SiO_2$, sodium fluoride NaF, cryolite Na_3AlF_6, kyolite $Na_5Al_3F_{14}$, silicon oxide SiO_2 (in the form of cristobalite, quartz, tridymite), calcium fluoride CaF_2, aluminum oxide Al_2O_3 , nepheline $NaAlSiO_4$, albite $NaAlSi_3O_8$.

3. Water leaching

As a result of water leaching of the heat insulating part of the spent lining, a solution of the following composition is formed: g/dm^3: NaF-15-25; Al_2O_3 - 0-0.1; Na_2SO_4 - 1-3; SiO_2 - 0.05-0.7. From the solutions, obtained after leaching of the heat insulating part of the spent lining, cryolite can be precipitated with the help of one of the known methods [7,8]. This solution is similar to solutions of wet gas purification of aluminum electrolysers and can be processed with the production of cryolite according to a conventional sodium bicarbonate scheme [7,8]. For the precipitation of cryolite, additional addition of reagents in the form of an aluminate solution and sodium bicarbonate is necessary.

Aluminate solution is prepared from aluminum hydroxide and caustic alkali. The process of cryolite formation is carried out according to the reaction:

$$6NaF + 4\ NaHCO_3 + NaAlO_2 = Na_3AlF_6 + 4\ Na_2CO_3 + 2H_2O \quad (1)$$

Reduction of the silicon content in solutions from the leaching of spent lining can be achieved using various ways as described below.

In the technology of hydrochemical processing of the carbon part of the spent lining, desiliconization is achieved by contacting the solution after leaching with its own slime for 6-8 hours due to the formation of sodium aluminosilicates [9]. In case of the heat insulating part of the spent lining, this process is accompanied by difficulties associated with a low content of aluminum compounds in the slime.

It is also possible to regulate the leaching process at a level of pH 6-9 by introducing of acidic additives (acid salts of aluminum, acids), however this complicates and increases the cost of the process.

In addition, the use of the cryolite produced with the help of this technology is currently limited in the production of aluminum due to the transition to acid electrolytes in the production of aluminum. A radical solution to the problem is the process of obtaining a low-modulus cryolite (a mixture of cryolite and chiolite), which will be considered below. The low-modulus cryolite obtained by us contains 0.2-0.3% of silicon.

An increase in the consumer properties of cryolite can be achieved by processing cryolite with a high sodium content of 30-32% per cholite ($Na_5Al_3F_{14}$). The technology of processing cryolite with aluminum sulfate solution was carried out at the Irkutsk Aluminum Smelter on a pilot scale [10].

The composition of the initial regeneration cryolite, cryolite treated with aluminum sulfate solution and the precipitate after leaching is given in table 2.

Table 2. The composition of regenerative cryolite, cryolite treated with aluminum sulfate solution.

Name of product	Content, wt. %							
	F	Na	Al	SO_4	C	Ca	Fe	Si
Initial regenerative cryolite	45.47	30.27	12.34	6.18	0.74	0.61	0.11	0.60
Cryolite treated with aluminum sulfate	47.74	22.56	17.32	1.35	0.62	0.72	0.23	0.3

The main phases of the initial regeneration cryolite are cryolite and sodium sulfate, cryolite treated with sulfate aluminum - cryolite and chiolite.

4. Conclusion

The main directions of reducing the consumption of aluminum fluoride are:

- production and use of alumina with a lower sodium content;
- processing of a part of the excess electrolyte using the new technology to produce aluminum fluoride;
- processing of the heat insulating part of the spent lining with the production of regenerative cryolite processed with the help of a new technology into a low-modulus cryolite.

With the use of new technologies, secondary wastes are used in the production of cement, and aluminum plants are approaching to become completely non-waste ones.

Acknowledgements
This article was prepared using the results of work carried out in the framework of the federal target program "Research and development on priority directions of scientific-technological complex of Russia for 2014 - 2020 years". Theme: "Development and testing of effective pyrolytic method of processing spent lining of aluminum electrolyzers" The agreement with the Ministry of Education for a grant No. 14.577.21.0190. Unique identifier PNIER RFMEFI57715X0190.

References
[1] Kondratev V V, Rzhechitsky E P, Gorovoi V O, Shakhrai S G and Karlina A I 2016 *International Journal of Applied Engineering Research* vol 11 **23** 11374–81
[2] Kondratev V V, Rzhechitsky E P, Bogdanov Yu V, Zakharov S V and Karlina A I 2016 *International Journal of Applied Engineering Research* vol 11 **23** 11369–73
[3] Rzhechitsky E P, Kondratiev V V, Karlina A I and Shakhrai S G 2016 *Non-ferrous Metals* **4(880)** 23–26
[4] Holywell G and Breault R 2013 *JOM* **65** 1441–51
[5] Kondrat'ev V V, Rzhechitskii E P, Karlina A I, Sysoev I A and Shakhrai S G 2016 *Metallurgist* vol 60 **5-6** 571–5
[6] Kondrat'ev V V, Ershov V A, Shakhrai S G, Ivanov N A and Karlina A I 2016 *Metallurgist* vol 60 **7-8** 877–882
[7] Shakhrai S G, Budnik E V and Kondratiev V V 2013 *Metallurgist* vol 56 **9-10** 700–4
[8] Sysoev I A, Ershov V A and Kondrat'ev V V 2015 *Metallurgist* vol 59 **5-6** 518–525
[9] Ershov V A, Sysoev I A and Kondrat'Ev V V 2013 *Metallurgist* vol 57 **3-4** 346–351
[10] Kondrat'Ev V V, Nemchinova N V, Ivanov N A, Ershov V A and Sysoev I A 2013 *Metallurgist* vol 57 **5-6** 455–9

Composition of the heat insulating part of the spent lining – waste of overhaul of aluminum electrolysers

E P Rzechitskiy[1], A A Petrovskiy[1], N V Nemchinova[1], A A Ivanov[1] and Yu V Bogdanov[2]

[1]Irkutsk National Research State Technical University, 83 Lermontov street, Irkutsk, 664074, Russia
[2]UC RUSAL, RUSAL Engineering Technology Centre, Krasnoyarsk, Russia,

E-mail: begimotus@mail.ru

Abstract. The main part of losses of fluorine during the electrolytic production of aluminum accounts for the wastes of the overhaul of electrolysers - spent lining. To assess the possibility of creating a technology for the regeneration of fluorine from the heat insulating part of the spent lining of aluminum electrolysers, its chemical and phase composition was studied. Presence of water-soluble fluorides in the heat insulating part of the spent lining makes it possible to extract from it up to 70% of fluorine by leaching.

1. Introduction

At the present time, the main type of solid wastes in the production of aluminum are materials for the overhaul of electrolysers, including spent lining. Annually, in Russia, it is formed and stored in storage tanks, or exported to the dump about 130-150 thousand tons of lining of dismantled electrolysers [1]. The spent lining during being stored in landfills can interact with water and air, forming toxic compounds and alkaline fluoride-containing solutions. In this regard, spent lining is an environmentally hazardous waste.

2. Study of the material

To study the material composition and properties of the heat insulating part of spent lining of aluminum electrolysers, dismantling products of the cathode device of five C8БМ with different operating times were used at the Krasnoyarsk and Irkutsk alluminium smelters.

Samples, with the exception of chamotte backfill, which was taken from different locations along the perimeter of the pool, were taken from each product by a point method in terms of area and cross-section. Samples were crushed with a hydraulic press, then crushed into powder with a jaw crusher and rubbed in a disk attritor since the fraction is of -0.2 mm. Then, by quartizing several samples in certain ratios, an average representative sample was obtained. Heat insulating part of the spent lining has a complex phase and material composition. That's why, various modern analysis methods were used to estimate the material and phase composition of the averaged samples: micro-X-ray one, X-ray phase one and X-ray fluorescence one.

Chemical composition according to the results of the analysis of samples performed on a S8 TIGER X-ray fluorescence spectrometer from five C8БМ electrolysers of various plants is given in table 1.

Table 1. Results of X-ray fluorescence analysis of samples from five C8БМ electrolysers of various plants, % wt.

Electrolyser	C	F	Na	Mg	Al	Si	Ca	Fe	K	Others
1	1.60	13.30	10.30	0.17	16.20	21.40	0.60	1.20	0.90	34.33
2		33.90	24.52	0.21	8.20	10.00	3.10	2.60	0.60	17.08
3		1.80	3.90	0.16	16.20	26.10	0.27	0.78	0.57	50.22
4	0.4	9.42	18.36	0.21	9.72	22.42	0.36	1.32	0.61	37.18
5		2.1	4.1	0.17	16.12	26.3	1.11	1.349	0.53	48.22
At average	0.40	12.10	12.24	0.18	13.29	21.24	1.09	1.45	0.64	39.31

As follows from the data given, composition of the heat insulating part of the spent lining for each electrolyser is individual and can vary significantly. Average values were used in the study.

Elemental composition of constituents of the heat insulating part of the spent lining of the electrolyser No. 1 is shown in table. 2

Table 2. Elemental composition of constituents of the heat insulating part of spent lining,% wt.

Sample	C	F	Na	Mg	Al	Si	Ca	Fe	K	Others
Dissolved chamotte	3.20	20.60	17.20	0.12	16.30	16.80	0.70	0.70	0.90	23.48
First layer of chamotte	2.80	12.80	19.10	0.14	16.00	19.50	0.90	1.50	0.70	26.56
Second layer of chamotte	-	1.30	3.60	0.23	16.50	26.10	0.27	0.78	0.44	50.78
Third layer of chamotte	-	0.20	2.20	0.30	16.50	26.90	0.68	1.90	0.44	50.88
Diatomite	-	6.90	3.00	8.40	4.50	22.40	0.47	7.20	3.00	44.13

The others are mainly represented by oxygen in the form of Al_2O_3, SiO_2.The diffractograms, obtained with the help of Shimadzu XRD-7000S powder diffractometer, of the averaged sample of the heat insulating part of the spent lining of electrolysers 2 and 5 of the Irkutsk and Krasnoyarsk aluminum smelters of OK "RUSAL" are shown in figure 1 and 2, respectively.

In the first diffractogram, the peaks which correspond to sodium fluoride are most intensively isolated, since the electrolyser was emergency and the fluorine content in this sample was very high. As seen in the second diffractogram, peaks which correspond to nepheline, cristobalite and mullite are more pronounced.

In the first diffractogram, the peaks which correspond to sodium fluoride are most intensively isolated, since the electrolyser was emergency and the fluorine content in this sample was very high. As seen in the second diffractogram, peaks which correspond to nepheline, cristobalite and mullite are more pronounced.

As main phases, with the help of the X-ray phase analysis, the following substances were found in the heat insulating part of the spent lining were found: mullite $3Al_2O^3.2SiO_2$, sodium fluoride NaF, cryolite Na_3AlF_6, kyolite $Na_5Al_3F_{14}$, silicon oxide SiO_2 in the form of cristobalite, quartz, tridymite, calcium fluoride CaF_2, aluminum oxide Al_2O_3, nepheline $NaAlSiO_4$, albite $NaAlSi_3O_8$.

Below, we consider the main properties of the phases which appear in the heat insulating part of the spent lining af the aluminium electrolyser.

The solubility data are presented in figure 3 [3].

Cryolite Na3AlF6 and chiolite $Na_5Al_3F_{14}$ are slightly soluble in water 0.41 g/dm^3 (at 200 °C). They react with strong acids to form salts of aluminum and sodium and hydrogen fluoride. Cryolite and chiolite are decomposed by strong alkalis with the formation of metal tetrahydroxoaluminates of the corresponding alkalis and fluoride. Under normal conditions, silicone dioxide SiO2 is most often found in the polymorphic modification of α-quartz, which at a temperature above +573 °C reversibly passes into β-quartz.

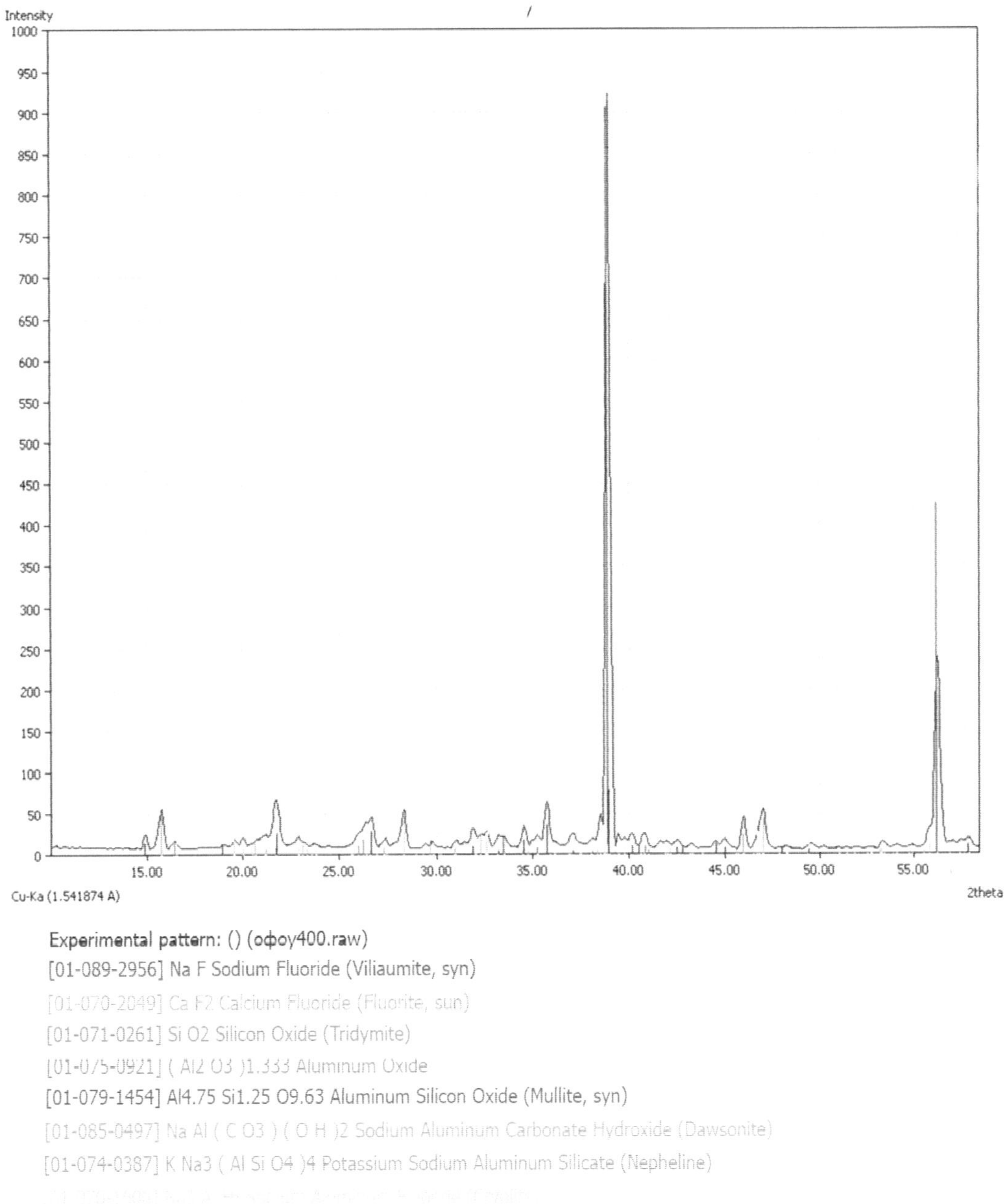

Figure 1. Diffractogram of the sample of the heat insulating part of the spent lining (electrolyser No. 2).

With a further increase in temperature, quartz becomes tridymite and cristobalite. These polymorphic modifications are stable at high temperatures and low pressures. Cristobalite is more stable than tridymite at temperatures below 1770 °C [8]. Silicon dioxide practically does not dissolve in water, solubility is, mg/dm^3: 7 for quartz, 12 for cristobalite, 16 for tridymite, 120 for amorphous silica [9]. However, as the pH increases, solubility of silicon dioxide in water sharply increases.

IOP Conf. Series: Materials Science and Engineering **411** (2018) 012062 doi:10.1088/1757-899X/411/1/012062

Calcium fluoride CaF_2 is practically insoluble in water 0.017 g/dm^3 (at 200 °C) Calcium fluoride is chemically relatively passive. It reacts with sulfuric acid to form calcium sulphate and hydrogen fluoride, which is used in industry to produce hydrofluoric acid. At high temperatures, it undergoes hydrolysis.

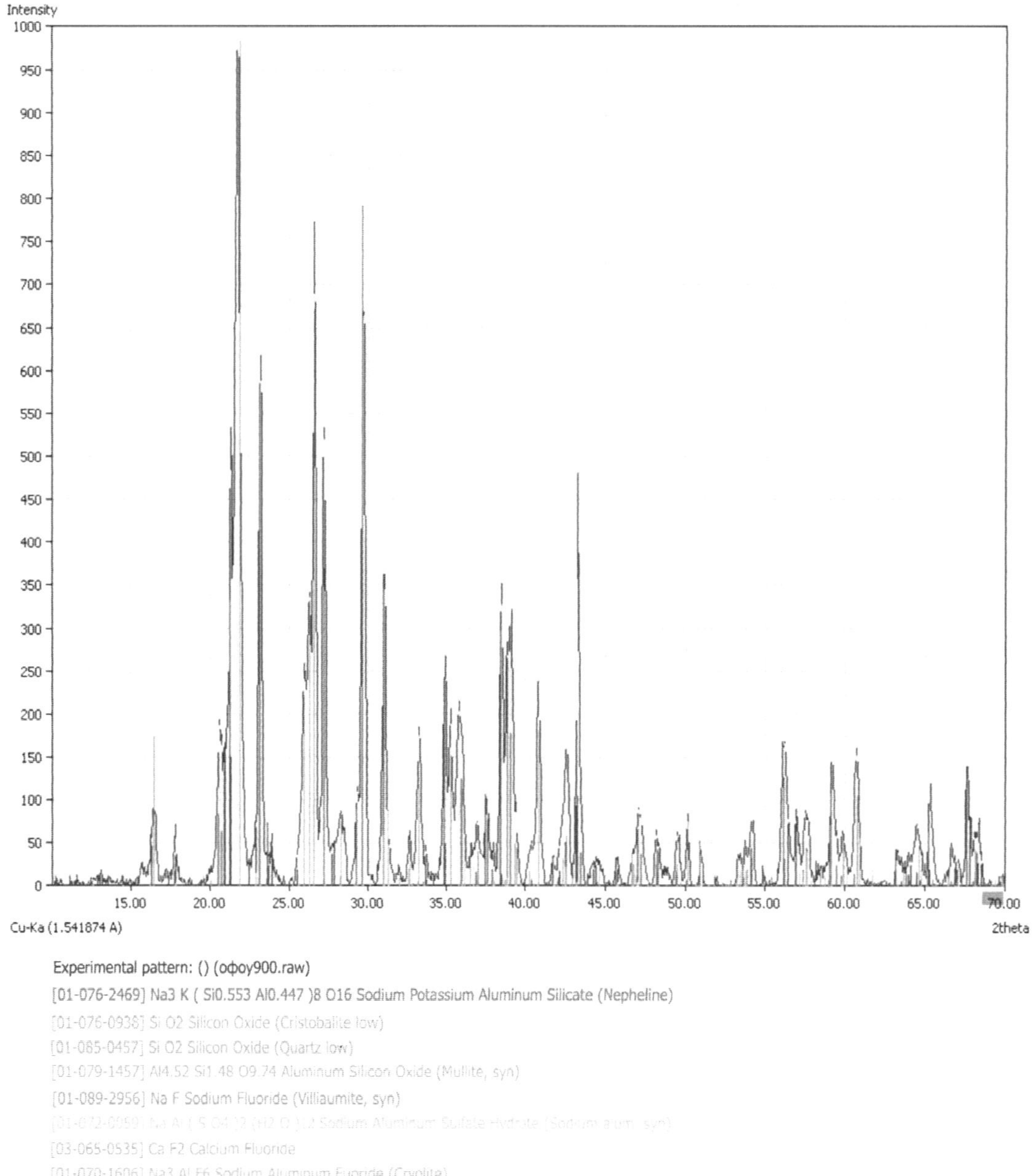

Figure 2. Diffractogram of the sample of the heat insulating part of the spent lining (electrolyser No. 5).

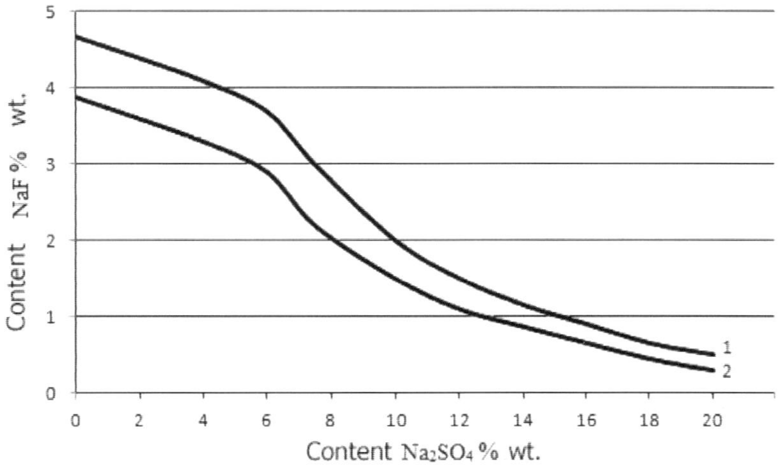

Figure 3. Solubility in the NaF-Na_2SO_4-Na_2CO_3-$NaHCO_3$-H_2O system: 1 – isotherm of the NaF-Na_2SO_4-H_2O system; 2 – the same in the presence of a constant sum of salts (about 4%) of Na_2CO_3 and $NaHCO_3$.

3. Conclusion

Aluminum oxide Al_2O_3 in the heat-insulating part of the spent lining is represented by α and β phases (corundum and alumina, respectively). Both phases are insoluble in water, but interact with acids to form aluminum salts and water. Reacting with water solutions of alkali, alumina forms tetrahydroxoaluminate, which decomposes in air to aluminum hydroxide and bicarbonate.

Nepheline $NaAlSiO_4$ is easily decomposed by mineral and organic acids with the release of gelatinous silica. In case of prolonged boiling in water, the solution acquires an alkaline reaction. Nepheline does not interact with alkalis.

Albit $NaAlSi_3O_8$ is insoluble in water, slightly soluble in acids, decomposed by hydrofluoric acid with the release of gelatinous silica.

Acknowledgements
The article was prepared with the financial support of the Ministry of Education and Science of the Russian Federation with the use of the results of a project on the topic "Development and testing of an effective pyrolytic method for processing spent lining of aluminum elecrolysers" No. 14.577.21.0190, unique identifier RFMEFI57715X0190.

References
[1] Kovács V and Kiss L 2015 *Light Metals* 529–34
[2] Isler R 1982 *Chemistry of Silica* vol 1 (M.: Mir) p 416
[3] Kondrat'ev V V et al 2016 *Metallurgist* vol 7-8 877–82
[4] Shakhrai S G et al 2013 *Metallurgist* vol 56 **9-10** 700–4
[5] Sysoev I A et al 2015 *Metallurgist* vol 59 **5-6** 518–25
[6] Ershov V A et al 2013 *Metallurgist* vol 57 **3-4** 346–51
[7] Holywell G and Breault R 2013 *JOM* **65** 1441–51
[8] Kondratev V V et al 2016 *Int. J. of Applied Eng. Research* vol 11 **23** 11374–81
[9] Kondrat'ev V V et al 2013 *Metallurgist* vol 57 **5-6** 455–9
[10] Kondrat'ev V V et al 2016 *Metallurgist* vol 60 **5-6** 571–5

IOP Conf. Series: Materials Science and Engineering **411** (2018) 012063 doi:10.1088/1757-899X/411/1/012063

Optimization of technological parameters of manganese metal aluminothermic smelting

I D Rozhikhina and O I Nokhrina

Siberian State Industrial University, 42 Kirova street, Novokuznetsk, 654007, Russia

E-mail: rogihina_id@mail.ru

Abstract. In the paper using the mathematical model, the technology parameters of manganese metal smelting by aluminothermic method from manganese concentrate of chemical enrichment with the use of synthesized monophase material CaMnO3 are predicted. Based on thermal analysis and mathematical modeling data, the technology of manganese metal smelting was developed, including the preliminary preparation of charge materials to produce a synthetic material that contains the highest manganese oxide MnO_2 and at the same time is an effective flux. With optimum proportions in the charge of monophasic material $CaMnO_3$ and concentrate of chemical enrichment, it is possible to completely eliminate the use of fluorspar. The developed optimal technological regimes for out-of-furnace aluminothermic smelting of metallic manganese using monophase material $CaMnO_3$ and concentrate of chemical enrichment made it possible to obtain a through-rate manganese recovery factor at the level 75-80%, which is almost twice as high as the through-extraction coefficient of manganese by the known technologies of manganese metal smelting.

1. Introduction

For the smelting of manganese metal by aluminothermic method, it is necessary to use high-quality manganese raw materials [1]. To ensure the required GOST 6008-80 chemical composition of manganese metal, such raw materials are peroxide manganese ores [2] or low-phosphorous manganese slags [3].

Concentrate of chemical enrichment, having in its composition the minimum content of impurities (Fe_2O_3 up to 0.1%, P up to 0.009%, 0.5-1.2% SiO_2, S – traces, less than 2% $CaCl_2$), can be an appropriate base material for the production of metallic manganese by an aluminothermic process [4,5].

When manganese metal is smelted by aluminothermic metallurgy, the problem of the most efficient use of scarce and expensive high-quality manganese ores and secondary aluminum powder appears. The main characteristics of the efficiency of the out-of-furnace process are undoubtedly the degree of extraction of components into the alloy, the chemical composition of the alloy and the associated slag, the multiplicity of slag. In the present work, an attempt was made to analytically determine the slag multiplicity and the optimal composition of the slag phase in the aluminothermic smelting of a manganese alloy of the given composition. The remaining characteristics of the process can be determined on the basis of these quantities.

2. Calculation procedure

To solve the problem, we used equations obtained by processing empirical data on out-of-furnace aluminothermic melting of complex alloys with manganese and titanium [6]:

$$c\left[Me^{/}\right]=\frac{c\left[Me^{//}\right]^{\frac{n}{x}}\cdot c\left(Me_x^{/}O_y\right)^{\frac{1}{x}}\cdot\gamma\left[Me^{//}\right]^{\frac{2n}{x}}\cdot\gamma\left(Me_x^{/}O_y\right)^{\frac{2}{x}}}{c\left(Me_n^{//}O_y\right)^{\frac{1}{x}}\cdot\gamma\left[Me^{/}\right]^{2}\cdot\gamma\left(Me_n^{//}O_y\right)^{\frac{2}{x}}\cdot\exp\left(\frac{\Delta G_T^0}{x\cdot R\cdot T}\right)}, \quad (1)$$

where c[Me$^{/}$] and c[Me$^{//}$] – the molar fraction of the metal to be reduced and the reducing metal in the metallic melt, respectively;
c[$Me^{/}_{x}O_y$] and c[$Me^{//}_{n}O_y$] – the molar fraction of oxides of the metal to be reduced and the reducing metal in the slag melt, respectively;
γ[Me$^{/}$] and γ[Me$^{//}$] – the activity coefficients of the metal to be reduced and the reducing metal in the metallic melt, respectively;
γ($Me^{/}_{x}O_y$), γ($Me^{//}_{n}O_y$) – the activity coefficients of oxides of the metal to be reduced and the reducing metal in the slag melt, respectively;
ΔG_T^0 – standard change in the Gibbs energy of the metal-thermal reduction reaction, J;
R – the universal gas constant, J/(mol · K);
T – the process temperature, K.

For the reduction reaction of manganese monoxide with aluminum, the coefficients x and n in equation (1) will be respectively equal to 1 and 2. The process temperature, measured directly in the laboratory conditions during reduction of 1 kg of manganese ore concentrate, was approximately 2300 K. The standard change in Gibbs energy of the aluminothermic reduction of monoxide manganese at the given temperature was calculated from the data in [6].

Thus, equation (1) takes the following form:

$$c\left[Mn\right]=\frac{c\left[Al\right]^{2}\cdot c\left(MnO\right)\cdot\gamma\left[Al\right]^{4}\cdot\gamma\left(MnO\right)^{2}\cdot 10^{6}}{1.016\cdot c\left(Al_2O_3\right)\cdot\gamma\left[Mn\right]^{2}\cdot\gamma\left(Al_2O_3\right)^{2}}. \quad (2)$$

Similarly, equations for the reactions of aluminothermic reduction of iron monoxide and silicon dioxide were obtained. They have the following form:

$$c\left[Fe\right]=\frac{c\left[Al\right]^{2}\cdot c\left(FeO\right)\cdot\gamma\left[Al\right]^{4}\cdot\gamma\left(FeO\right)^{2}\cdot 10^{17}}{4.265\cdot c\left(Al_2O_3\right)\cdot\gamma\left[Fe\right]^{2}\cdot\gamma\left(Al_2O_3\right)^{2}}, \quad (3)$$

$$c\left[Si\right]=\frac{c\left[Al\right]^{4}\cdot c\left(SiO_2\right)\cdot\gamma\left[Al\right]^{8}\cdot\gamma\left(SiO_2\right)^{2}\cdot 10^{9}}{3\cdot c\left(Al_2O_3\right)\cdot\gamma\left[Si\right]^{2}\cdot\gamma\left(Al_2O_3\right)^{2}}. \quad (4)$$

The high-percent manganese alloy obtained by the aluminothermic method can be described in the first approximation by the Mn-Fe-Al-Si system. The slag phase is a system of MnO-Al_2O_3-SiO_2-FeO-CaO-MgO-Na_2O-K_2O-BaO. Th coefficients in the equations were obtained as a result of statistical processing of experimental melting data of metallic manganese:

$$\ln\gamma\left[Mn\right]=-0.033\cdot\frac{c\left[Fe\right]}{c\left[Mn\right]}+1.058\cdot\frac{c\left[Al\right]}{c\left[Mn\right]}+0.171\cdot\frac{c\left[Si\right]}{c\left[Mn\right]}, \quad (5)$$

$$\ln\gamma\left[Fe\right]=-0.424\cdot\frac{c\left[Mn\right]}{c\left[Fe\right]}+5.252\cdot\frac{c\left[Al\right]}{c\left[Fe\right]}-2.278\cdot\frac{c\left[Si\right]}{c\left[Fe\right]}, \quad (6)$$

$$\ln\gamma[Al] = 0.11\cdot\frac{c[Fe]}{c[Al]} - 0.112\cdot\frac{c[Mn]}{c[Al]} + 0.105\cdot\frac{c[Si]}{c[Al]}, \tag{7}$$

$$\ln\gamma[Si] = -2.845\cdot\frac{c[Fe]}{c[Si]} + 1.362\cdot\frac{c[Mn]}{c[Si]} + 28.135\cdot\frac{c[Al]}{c[Si]}, \tag{8}$$

$$\ln\gamma(MnO) = 0.104\cdot\frac{c(Al_2O_3)}{c(MnO)} - 0.292\cdot\frac{c(SiO_2)}{c(MnO)} + 0.176\cdot\frac{c(FeO)}{c(MnO)} + 0.176\cdot\frac{\Sigma(c)}{c(MnO)}, \tag{9}$$

$$\ln\gamma(SiO_2) = -0.802\cdot\frac{c(Al_2O_3)}{c(SiO_2)} + 0.465\cdot\frac{c(FeO)}{c(SiO_2)} + 0.465\cdot\frac{c(MnO)}{c(SiO_2)} + 0.465\frac{\Sigma(c)}{c(SiO_2)}, \tag{10}$$

$$\ln\gamma(FeO) = -1.733\cdot\frac{c(Al_2O_3)}{c(FeO)} + 3.262\cdot\frac{c(SiO_2)}{c(FeO)} - 0.668\cdot\frac{c(MnO)}{c(FeO)} - 0.668\cdot\frac{\Sigma(c)}{c(FeO)}, \tag{11}$$

$$\ln\gamma(Al_2O_3) = -0.071\cdot\frac{c(SiO_2)}{c(Al_2O_3)} + 0.052\cdot\frac{c(FeO)}{c(Al_2O_3)} + 0.052\cdot\frac{c(MnO)}{c(Al_2O_3)} + 0.052\cdot\frac{\Sigma(c)}{c(Al_2O_3)}, \tag{12}$$

where $\Sigma(c)$ – is the total molar concentration of CaO, MgO, BaO, Na_2O and K_2O in the slag.

Since the molar composition of the alloy is usually known or can be calculated, equations (2) - (12) can be reduced to a system containing thirteen unknowns. Such a system can not be uniquely solved. To obtain a single-valued solution, the following additional equations are proposed:

$$\sum c \approx 1 - c(Al_2O_3) - c(SiO_2) - c(MnO) - c(FeO), \tag{13}$$

$$K_{sl} = \frac{N_{sl}\cdot[102\cdot c(Al_2O_3) + 60\cdot c(SiO_2) + 71\cdot c(MnO) + 72\cdot c(FeO) + 81\cdot\sum(c)]}{55\cdot c[Mn] + 56\cdot c[Fe] + 27\cdot c[Al] + 28\cdot c[Si]}, \tag{14}$$

where K_{SL} – slag multiplicity;
N_{SL} – amount of slag, mole.

$$1 = c[Al] + c[Si] + c[Mn] + c[Fe]. \tag{15}$$

Thus, a system of fourteen equations (including nonlinear ones) with fifteen unknowns can be obtained. The missing equation can be obtained either on the basis of any additional technological requirements, or similarly to equations (2) - (4) for any of the following reactions:

$$2FeO + Si = 2Fe + SiO_2, \tag{16}$$

$$2MnO + Si = 2Mn + SiO_2, \tag{17}$$

$$FeO + Mn = Fe + MnO. \tag{18}$$

This system of equations was solved using the MathCAD program under the following conditions:

- composition of CEC,%: 87 Mn_3O_4, 7 CaO, 4.0 $CaCl_2$, 0.3 SiO_2, 0.1 Fe_2O_3, S – traces, P_2O_5 – traces;

- monophasic material $CaMnO_3$: ~ 40% CaO, ~ 60% MnO_2;
- aluminum powder;
- composition of manganese metal.

As a result of the calculation, the optimum slag phase composition was obtained, wt%: 6.0 – 8.2 MnO, 35.5 – 40.0 CaO, 50.0 – 58.0 Al_2O_3, 0.5 – 1.5 SiO_2, 2.8 – 4.8 $CaCl_2$, 0.08 – 0.20 FeO, traces P, 0.015 – 0.05 S, which corresponds to the optimum ratio of chemical enrichment concentrate (CEC) and monophase material ($CaMnO_3$) – $\frac{CEC}{CaMnO_3} = \frac{6.5-7.5}{2.5-3.5}$. The slag multiplicity is 2.2 – 2.7.

3. Results and discussion

Based on thermal analysis and mathematical modeling data, the technology of manganese metal smelting was developed, including the preliminary preparation of charge materials to produce a synthetic material that contains the highest manganese oxide MnO_2 and which is also an effective flux.

Previously, optimal conditions for the preparation of solid solutions of calcium and manganese oxides, including $CaMnO_3$ [7], were established.

The use of monophase material $CaMnO_3$ as a charge component for aluminumothermal smelting of manganese metal makes it possible to increase the thermal character of the process, and, consequently, to increase the manganese extraction from the concentrate to an average of 90%.

The charge contained CEC, a synthesis product and aluminum powder. The smelting was conducted in the hearth with an upper ignition. As a result of experimental meltings, a metal was obtained, the chemical composition of which is given in table 1.

From the results obtained it is clear that the chemical composition of the alloy meets the requirements

It should be noted that manganese metal has a low content of harmful impurities – phosphorus and sulfur, and the iron content does not exceed 1% [8]. The extraction of manganese from the concentrate amounted to an average of 90%.

Table 1. Chemical composition of the test metal.

Melting No.	Composition of metal, wt%						Content of MnO in the slag, %	Extraction of Mn,%
	Mn	Al	Si	Fe	S	P		
1	96.89	0.81	0.45	0.88	0.004	0.006	7.24	90.70
2	97.00	0.75	0.38	0.83	0.002	0.006	6.15	90.11
3	97.12	0.73	0.40	0.74	0.003	0.005	6.01	91.16
4	96.78	0.68	0.58	0.97	0.004	0.004	8.15	87.68
5	96.84	0.83	0.64	0.91	0.004	0.006	7.35	89.89

Melt temperature, measured during the experiment, was from 2300 to 2373 K at the optimum ratio of CEC to monophase material equal to (6.5 – 7.5) to (2.5 – 3.5).

End-to-end manganese recovery in the smelting of manganese metal from the concentrate of chemical enrichment, obtained by the developed technology of enrichment of carbonate ores by calcium-chloride method was 85.3 – 89.3%, which significantly exceeds the extraction of manganese in the smelting of manganese by the aluminothermic out-of-furnace process from peroxide manganese ores, which is 69 – 72%, and is at the level of manganese extraction with the use of CEC in the production of manganese metal using electric heating [2. 3].

As a result of experimental meltings, the slag of the composition was obtained, wt%: 6.1-8.15 MnO, 36.0-38.1 CaO, 51.2-56.8 Al_2O_3, 0.6-1.3 SiO_2, 3.2-6.4 $CaCl_2$, 0.1-0.4 FeO, 0.02-0.03 S, traces of P.

This slag practically does not contain harmful impurities. oxides of iron and silicon and can be used as a synthetic slag for processing steel in a ladle. and also this slag can be used in the construction industry to produce cements. The useful use of aluminum is 94 – 96%.

4. Conclusion

The obtained results of laboratory aluminothermic meltings of manganese metal are consistent with the predicted data of technological parameters of aluminothermic melting of manganese metal, which allows us to conclude about the adequacy of the method used for parameters calculation.

References

[1] Lyakishev N P, Pliner Yu L et al 1978 *Alumotermia* (Moscow: Metallurgy) p 424
[2] Elyutin V P, Levin Yu A et al 1957 *Production of Ferroalloys* (Moscow: Metallurgizdat) p 256
[3] Korshunov E A et al 2002 *Method of Production of Metallic Manganese* patent of the Russian Federation No. 2002124665/02
[4] Tolstoguzov N V and Selivanov I A 1986 *Method of Metallic Manganese Smelting* certificate of authorship of USSR No. 3784537 / 22-02
[5] Tolstoguzov N V and Selivanov I A *Method of Metallic Manganese Smelting* certificate of authorship of USSR No. 3812806 / 22-02
[6] Safonov A V et al 2006 *Proc. of the Int. Sci. and Practical Conf.* (Novokuznetsk, SibSIU) vol 1 p 39-43
[7] Rozhikhina I D et al 2008 *Izv. Vuzov. Ferrous Metallurgy* **10** 27–31
[8] Rozhikhina I D and Nokhrina O I 2008 *Steel* **7** 58–60

IOP Conf. Series: Materials Science and Engineering **411** (2018) 012064 doi:10.1088/1757-899X/411/1/012064

Assessment of the nanocrystalline silicon carbide effectiveness in electroplating, ceramics, surface modification

V V Rudneva[1], G V Galevsky[1], S G Galevsky[2] and G N Chernovsky[1]

[1]Siberian State Industrial University, 42 Kirova street, Novokuznetsk, 654007, Russia
[2]St. Petersburg Mining University, V.O., 21 Line, 2, St. Petersburg, 199106, Russia

E-mail: kafcmet@sibsiu.ru

Abstract. Technological advantages and conditions for ensuring a new quality of coatings and products achieved with the use of silicon nanocarbide are established in the processes of composite electrodeposition of coatings, formation of structural ceramics, surface hardening of steels with electric explosive alloying. Nanocarbide of silicon is recommended for use in wear-resistant and corrosion-resistant chromium-carbide electrodepositable coatings operating at elevated temperatures, for hardening tools and rigging, including with especially difficult microrelief of working surfaces. Nanocarbide of silicon in the compositions "silicon carbide – boron – carbon" can be used for the production of ceramic billets by solid-phase sintering in argon with pressure 0.1 MPa and temperature 2273 K. The use of silicon nanocarbide in the technology of surface hardening of tool steels by electric explosive alloying provides a protective layer of high microhardness, wear resistance and heat resistance with depth about 20 μm.

1. Introduction

The production of silicon carbide is the most important in the structure of modern metallurgy. The combination of exceptional thermal and physical, mechanical and physical-chemical properties of silicon carbide makes it a unique material for many branches of technology and predetermines such basic applications as bonded materials, ceramics, composite materials and coatings.

The purpose of this work was to evaluate the effectiveness of silicon nanocarbide in composite electrodeposited coating (EC) technology, structural ceramics, electric explosive alloying (EEA) of steels. In carrying out experimental studies silicon nanocarbide and its compositions were used, obtained on the basis of the innovative technology proposed by the authors [1].

2. Application of silicon nanocarbide in composite electrochemical coating technology

Nanocarbide of silicon was used as a strengthening phase in the processes of electrochemical deposition of coatings from chromium electrolytes. To solve the problems of composite hardening, nanocarbide obtained by the carbidization of silicon micropowder with methane – SiC (1) and plasma modification of micropowder – SiC (2) was used. The characteristics of nanopowders (NP) of silicon carbide SiC (1) and SiC (2) and their physical and chemical properties are described in detail in [1 - 3].

For the electrodeposition of EC chrome – nanocarbide, a chrome solution containing silicon carbide SiC (1) or SiC (2) NP was used of the following composition, kg/m^3: chrome anhydride 250, sulfuric acid 3, chromium trivalent 8. To prepare the electrolyte – suspension a small amount of pure electrolyte was added the the weight powder, the components were mixed until the powder was completely wetted, transferred to the cell, and diluted to the desired concentration. PH adjustment was performed using solutions of NaOH or H_2SO_4.

Analysis of the research results and industrial application of silicon nanocarbide SiC (1) and SiC (2) confirmed the parametric, concentration, structural and technological effects established earlier for nanocarbides and nanoborides and described in [4 - 6]. Thus, the chromium electrolyte used was characterized, depending on the dispersity of the filler, by the following change in the cathode current density (MP – micropowder of silicon carbide with particle size less than 5 μm):

Without filler (5 kA/m^2) → MP SiC (5.5-6.0) → NP SiC (2), SiC (4) (8-10).

The increase in the permissible cathode current density is apparently due to the stabilization of the pH values in the near-cathode layer when introduced into the electrolyte of the second phase. The saturation of the coating with carbide nanoparticles occurs with a smaller mass content in the coating and concentration in the electrolyte (table 1), which is caused by an increase in the number of nanoparticles per unit volume, contributing to the dispersion hardening of the metal matrix. The microhardness of composite coatings with chromium-based nanocomponents is 1.80-1.90 times higher than in chromium, and in 1.10-1.20 higher in comparison with composite coatings with micropowders (table 1), in all cases the microhardness value increases with the content of particles in the composite material. The increase in hardness is due to the high efficiency of nanoparticles as barriers to the propagation of dislocations. The inclusion of nanoparticles in the chromium matrix reduces the intensity of its wear when the content of nanosized powder varies from 0.15 to 0.45 wt% by 1.25-2.5 times. A significant increase in the wear resistance of Cr – NP SiC material is associated with the elimination of causes of local stress concentration – microcracks in the structure of the layer present in significant amounts in chromium layers. Determination of the corrosion resistance of EC Cr – NP SiC in liquid media shows that the ECs exceed the conventional chromium-based coatings by 1.3-1.5 times (corrosion frequency rate 0.3%) and can be classified as resistant.

Table 1. Dependence of the content of the hardening phase (a) and microhardness (H) of EC[a] on the concentration of NP and MP in the electrolyte.

Concentration of NP, MP in electrolyte, kg/m^3	a, wt%. / H, GPa		
	Cr – NP SiC (1)	Cr – NP SiC (2)	Cr – MP SiC
1.0	0.20/9.0	0.24/9.1	0.47/8.4
3.0	0.41/9.3	0.43/9.2	0.72/8.6
10.0	0.59/9.7	0.58/9.4	0.89/8.9
20.0	0.82/10.1	0.84/10.1	1.1/9.0

[a]Conditions of electrodeposition: EC Cr – NP, MP SiC – cathode current density 8.0 kA/m^2, pH 2.0, T = 323 K.

It can be seen that chromium-carbide coatings have comparable to chromium-diamond wear resistance, microhardness, corrosion resistance, higher service life in operation at temperatures above 473-573 K, achieved at lower concentrations of carbide nanopowder in the electrolyte and a significant reduction in the cost of 1 m^3 of electrolyte – suspension.

Table 2. Comparative characteristics of chromium-based coatings with nanopowders of silicon carbide SiC (1) and diamond.

Conditions of electrodeposition and the results achieved		Technological options					
		1	2	3	4	5	6
Electrolyte composition, kg/m^3	Chromic anhydride	250	250	250	250	250	260
	Sulfuric acid	3	3	3	3	3	0.34
	Trivalent chromium	8	8	8	8	8	–
	Potassium silicofluoride	–	–	–	–	–	19

	Barium sulphate	–	–	–	–	–	7
	NP of silicon carbide	4	6	7	8	10	–
	NP of diamond	–	–	–	–	–	20
Mode of deposition	Temperature, K	328	328	328	328	328	331
	Current density, A/dm^2	55	55	55	55	55	55
Results	Wear resistance [a]	0.96	1.0	1.05	1.07	1.08	1.0
	Microhardness, GPa	8.9	9.0	9.1	9.15	9.15	9.0
	Corrosion resistance[a]	0.96	1.05	1.07	1.10	1.10	1.0
	Service life at temperature above 473-573 K[a]	1.1	1.50	1.70	2.00	2.06	1.0
	Cost of 1 m^3 of electrolyte suspension	0.13	0.16	0.18	0.20	0.23	1.0

[a]Durability, corrosion resistance, service life at a temperature above 473-573 K, cost of 1 m^3 of electrolyte suspension are given in relation to option 6, for which the values of these indicators are taken as 1.0.

According to the research results, the method for obtaining composite electrochemical coatings based on chromium and technological processes of composite hardening with silicon nanocarbide have been developed. EC based on Cr – NP SiC are recommended for the protection of parts and equipment that are subject to wear (for example, threadlike sensor of torsion machines, shock absorber rods, etc.), simultaneous wear and corrosion (for example, molds for molding products from thermoplastics, parts of oil and gas field equipment, etc.). But economically and technologically it is most expedient to use them instead of chrome-diamond ECs for surface hardening of tools, rigging, parts of machines and mechanisms operating at elevated temperatures (above 473-573 K): press tools for powder metallurgy, tool production details - punches, stamps, molds, punch holders, etc., parts of the piston group of internal combustion engines, etc.

3. Application of silicon nanocarbide in the technology of structural ceramics

For the technology of structural ceramics, silicon nanocarbide was used obtained by the carbidization of silicon micropowder with methane-SiC (1) and modification of microcarbon powder in the plasma nitrogen flow – SiC (2) followed by refining.

Silicon nanocarbide according to the main characteristics (level of dispersion, phase and chemical compositions) is not inferior to silicon carbide powders of constructional designation produced by foreign firms. The technological advantage of the material is the possibility of its provision the form of a composition containing additives (reactive boron and carbon) that activate the sintering process.

In order to substitute the import of a non-grinding powder of silicon carbide of furnace synthesis by company "Hermann Starck Co." grades A-10 and B-10, the testing of silicon nanocarbide was carried out in the solid-phase sintering processes of blanks for the manufacturing ceramic sealing rings series 156.017 and 156.073 (outer diameter from 16 to 30 mm, height from 18 to 30 mm) for pumps and submersible motors that pump oil, oil products, liquefied hydrocarbon gases. The technological advantages of silicon nanocarbide are confirmed and the following results are obtained. The use of silicon nanocarbide makes it possible to exclude the introduction of organic compounds into it during preparation of the charge and their carbonization to obtain uniformly distributed reactive carbon and to restore the silica film on the carbide nanoparticles, stirring on rolls, which simplifies and reduces the cost of the technology and allows it to be realized in the following industrially acceptable variant: heat treatment of the powder in vacuum at 1073 K, the introduction of a plasticizer, mixing and pressing at a pressure of 50 MPa, firing for 2 hours in argon at a pressure of 0.1 MPa at a temperature of 2273 K. Compliance technology provides a relative density after molding 0.62-0.63, 0.95-0.96 after sintering and stable shrinkage 26-29%.

4. Application of silicon nanocarbide in the technology of surface hardening of steels by EEA

In the technology of surface hardening of steels by EEA, silicon nanocarbide SiC (1) containing 99.21 wt% β-SiC with a specific surface area of 38000 m^2/kg was tested for the first time as a powder additive. The formation of composite surface layers on samples of tool steel Kh12 during electric explosion of aluminum foil with introduction of silicon nanocarbide into the explosion zone was investigated using the technology described in [8].

The choice of this steel is due to the fact that it is widely used for the manufacture of tools designed for the processing of materials by pressure (dies, punches, etc.), thus, the increased resistance against abrasion is important for it. This circumstance made the researchers to focus on silicon nanocarbide. The phase composition of the surface layers of steel after processing was determined, as well as their properties – microhardness, wear and heat resistance.

As an exploding conductor, an aluminum foil with the mass weighing 30 mg was used, onto which a sample of silicon nanocarbide with a mass of 7.5 mg was placed. The steel samples with dimensions 25 x 20 x 5 mm were subjected to treatment. The time of surface exposure was 100 μs, the effective value of the absorbed power density was 6.0 GW/m^2, the dynamic pressure of the jet on the surface reached 14.2 MPa. In this mode of action the depth of the aloying zone was about 20 μm.

EEA in a high-intensity regime of plasma impact on the surface led, along with fusion, to saturation of the steel surface layers with the products of aluminum foil explosion and silicon carbide nanoparticles followed by self-heating and formation of austenitic structure. The microhardness of the samples surface without treatment proved to be equal to 2513±205 MPa, and after treatment it increased by 2.8 times in average. The loss of mass of the steel samples in the state of delivery during the wear tests was 27.4 ± 7.3 mg, and after electric explosion treatment – 3.3±2.5 mg, i.e. wear resistance increased by 8 times. Heat resistance tests showed that the values of the mass corrosion index of samples in the initial state were 2.7; 5.2 and 10.6 $g/(m^2 \cdot h)$ at temperatures 800, 850 and 900 °C, respectively. After the protective treatment, the values of 0.3; 1.5 and 5.3 $g / (m^2 \cdot h)$ were obtained. The increase in the heat resistance at the indicated test temperatures was by 9; 3.5 and 2 times, which on average is 2 times greater than the corresponding heat resistance of this steel after a two-component EEA, for example, boron and gadolinium.

Thus, the alloying has a high microhardness and resistance against abrasive wear and high-temperature oxidation in the air.

5. Conclusion

The high efficiency of silicon nanocarbide application in the technology of composite chrome plating, structural ceramics, and surface modification of steels by electric explosive alloying was confirmed. Chrome-carbide coatings have comparable to chromium-diamond wear resistance, microhardness, corrosion resistance, higher service life. Sintered silicon carbide ceramic billets are characterized by high density and stable shrinkage. The alloying zone containing silicon nanocarbide possesses high microhardness, resistance against abrasive wear, thermal and oxidative stability.

References

[1] Rudneva V V 2008 *Nanomaterials and Nanotechnologies in the Production of Silicon Carbide* ed G V Galevsky (Moscow: Flint: Science) p 387

[2] Rudneva V V and Galievsky G V 2007 Izv. Vuzov. Ferrous Metallurgy **4** 20–24

[3] Galevsky G V, Rudneva V V and Galievsky S G 2003 *The Bulletin of GMS RANS. Section of Metallurgy* **12** 78–86

[4] Nozdrin I V, Galevsky G V and Rudneva V V 2012 *Harvesting Production in Machine Building* **2** 46–48

[5] Shiryaeva L S, Nozdrin I V, Galevsky G V and Rudneva V V 2014 *Electroplating and Surface Treatment* vol XXII **2** 51–57

[6] Shiryaeva L S, Garbuzova A K and Galevsky GV 2014 *Scientific and Technical Bulletin of SPbSPU* **2(195)** 100–108

[7] Glevsky G V, Rudneva V V and Holyakh O A 2008 *Method for Producing Composite Electrochemical Coatings Based on Chromium* patent of the Russian Federation No. 2006129821/02 Bul. **6** p 5

[8] Tsvirkun O A, Budovskikh E A, Rudneva V V, Goryushkin V F and Gromov V E 2007 *Journal of Functional Materials* vol 1 **3** 117–119

Study on the conditions and composition of the plasma treatment products of silicon carbide micropowder

V V Rudneva[1], G V Galevsky[1], G N Chernovsky[1] and S G Galevsky[2]

[1]Siberian State Industrial University, 42 Kirova street, Novokuznetsk, 654007, Russia
[2]St. Petersburg Mining University, V.O., 21 Line, 2, St. Petersburg, 199106, Russia

E-mail: kafcmet@sibsiu.ru

Abstract. The conditions of plasma processing of silicon carbide microgrit, phase and chemical compositions, nanolevel and the mechanism of nanocrystalline silicon carbide formation are investigated.

1. Introduction

The modern production of silicon carbide is almost entirely based on Acheson's method. The process of carbon-thermal synthesis of silicon carbide is carried out by a periodic block process in electric resistance furnaces with a capacity 4.5-5.5 MVA at temperatures 2500-2900 K [1, 2]. A coke core serves as a heater, laid in the center of the furnace around which the charge of quartz sand and oil coke is placed.

The use of small, sufficiently pure and thoroughly mixed materials, slow heating (6-7 hours) and long aging (20-22 hours) cause the reactions to proceed, including solid-phase type, in regimes close to equilibrium. The process develops in a practically immobile mass of solid reagents with a slow movement of the reaction front from the core to the outer surface of the working space of the furnace. During the electrothermal process the crystalline silicon carbide is formed as a result of chemical reactions, as well as products of incomplete reduction: amorph, intergrowths and siloxicon. A significant part of the charge plays the role of thermal insulation and does not undergo a noticeable change in the composition. The chemical composition of the grain of technical silicon carbide is given in table 1 [3].

Table 1. Chemical composition of technical silicon carbide in the grain.

Type of silicon carbide	Content of components, % mass								
	SiC	Si	SiO_2	C	Fe	Al	Ca	Mg	Ti
Green	98.0-99.5	0.1-1.0	0.1-0.2	0.1-0.3	0.05-0.2	0.01-0.08	0.01-0.1	0.01-0.05	0.01-0.05
Black	98.0-99.0	0.1-1.5	0.1-0.5	0.1-0.3	0.1-0.3	0.1-0.2	0.05-0.2	0.01-0.1	0.01-0.1

The sorted piece of silicon carbide with a size less than 400 μm is forwarded to the production of grinding materials (grinded grain and microgrit), representing a multi-stage redistribution [2]. The grain fineness of very fine microgrits is denoted by the upper limit of the grain size of the main fraction in accordance with table 2 [4]. The chemical composition of ultra-fine microgrits of silicon carbide must meet the requirements of table 3.

2. Methodology of research

The nonopowders of silicon carbide can be obtained by plasma processing of microgronding powders.

Table 2. Grain fineness of ultra-fine micro microgrits.

Grain fineness	M3	M2	M1
Grain size of the main fraction, µm	3.0 – 2.0	2.0 – 1.0	1.0 – 0.5

Table 3. Chemical composition of ultra-fine microgrits from green silicon carbide.

Make	Grain fineness	Chemical composition, % mass		
		SiC	Fe	C_{free}
		not less	not more	
64C	M2	92.0	0.4	0.4
63C		90.0	0.6	0.7

For plasma processing, a microgrit K3 64S M5 with a specific surface 4500 m^2/kg, containing impurities of nitrogen, oxygen and iron in the amount of 0.7; 3.3 and 0.5 wt%, respectively. A microphotograph of the microgrit of silicon carbide is shown in figure 1, and the X-ray diffraction pattern is shown in figure 2. The X-ray diffraction pattern contains all the main reflexes of α-SiC_{II} and β-SiC, which agrees well with the data on the polytype composition of technical silicon carbide given in [5].

For the research nitrogen of technical purity was used as a coolant and a carrier gas. The initial temperature of the nitrogen plasma flow, depending on the power supplied to the plasmatrons, was 3300, 5000, 5400 and 5600 K. The quenching of the condensed products of processing was carried out at a temperature of 2800 K by adiabatic expansion of the gas stream at the reactor outlet.

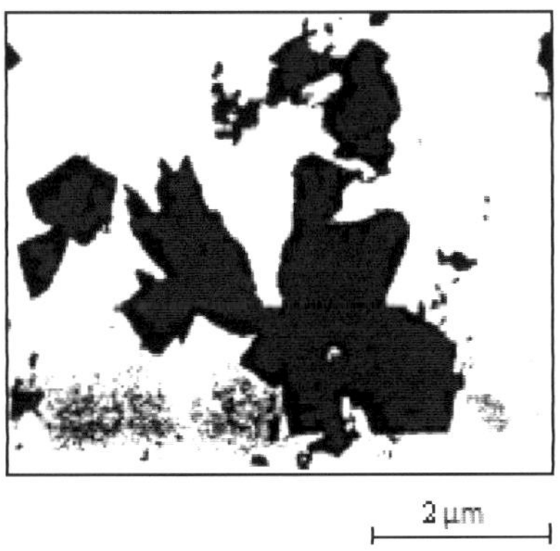

Figure 1. Micrograph of microgrit of silicon carbide K3 64S M5.

3. Research of conditions and results of plasma processing

The influence of the plasma flow and the mass flow rate of silicon carbide on the initial temperature process was studied. The physical and chemical evaluation of plasma treatment products was carried out using the methods of investigation of nanocrystalline materials described in [6 - 11]. The results are given in table 4 and in figure 3.

At a mass flow rate of silicon carbide 0.06 kg SiC/nm^3 N_2, the main product is β-SiC, the higher the initial temperature of the plasma flow, the greater the specific surface area, and hence the dispersion of the powder particles. At a higher mass flow rate of silicon carbide the structural conversion of α-SiC_{II}

→ β-SiC does not occur. In this case the dispersion of the plasma treatment products varies insignificantly with the increase in the initial temperature of the flow.

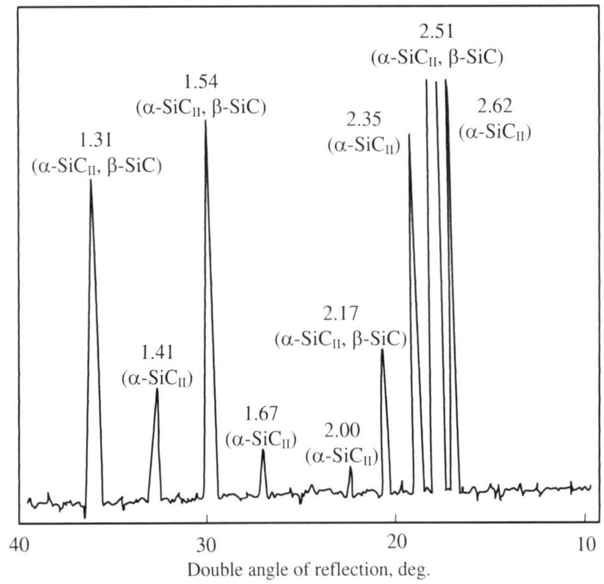

Figure 2. X-ray diffraction pattern of microgrit of silicon carbide K3 64S M5.

Table 4. Influence of the initial temperature of the plasma flow on the phase composition and dispersity of silicon carbide powders.

Initial temperature of the plasma flow, K	Mass flow rate, кг SiC/нм^3 N_2	Phase composition, wt%		Specific surface, m^2/kg
		β-SiC	α-SiC$_{II}$	
5600	0.06	95.0	5.0	30000
5400	the same	95.0	5.0	28000
5000	– " –	89.0	11.0	24000
3300	– " –	19.0	81.0	12000
5600	0.15	25.0	79.0	19000
5400	the same	23.0	77.0	18000
5000	– " –	22.0	78.0	17000
3300	– " –	15.0	85.0	13000

The influence of the mass flow of silicon carbide on the crystal structure, the specific surface and the chemical composition of the plasma treatment products was investigated at the initial flow temperature 5600 K and a hardening temperature 2800 K. As it can be seen from figure 3, the specific surface area of the resulting silicon carbide decreases with the increase in the mass flow of the initial microgrit. The oxygen content of the powder varies from 4.9 to 3.4 wt% (in all cases more than in the initial one, amounting to 3.0 wt%). The nitrogen content of the plasma treatment products with a minimum mass flow of silicon carbide reaches 5.2 wt%. However, X-ray phase analysis of the nitride and oxynitride phases did not detect silicon. This suggests that the nitrogen in the carbide-containing products of plasma treatment is in the form of a solid solution of substitution. The structure of the formed silicon carbide depends significantly on the mass flow of the initial microgrit and, consequently, the capacity of the plasma flow to homogenize the process, i.e. to transfer the initial powdered raw material into the vapour phase.

At a low mass flow rate (0.06-0.09 kg SiC/nm^3 N_2), during the heating silicon carbide is disproportionated, silicon is evaporated followed by carbidization, which is confirmed by the obtained specific surface values typical for powders of plasma-metallurgical synthesis. The natural result of the processes taking place in these conditions in the plasma flow is the formation of silicon carbide in the form of β-modification according to the mechanism "vapour-crystal". A X-ray diffraction pattern of the nanodispersed silicon carbide powder obtained by plasma processing at a mass flow rate 0.06 kg of SiC/nm^3 N_2 is shown in figure 4. It can be seen that, as a result of plasma processing, α-SiC_{II} reflections (2.62, 2.35 Å) decreased significantly; other reflexes α-SiC_{II} (2.00, 1.67, 1.41 Å) disappeared. At the same time, the height of joint reflections of α-SiC_{II} and β-SiC (2.51, 2.17, 1.54, 1.31 Å) remained almost the same. A significant decrease and disappearance of α-SiC_{II} reflections (2.62, 2.35, 2.00, 1.67, 1.41 Å) indicates a decrease in the content of this polytype in the nanopowder, which is confirmed by a quantitative determination of the phase content. Therefore, the nanopowder is basically β-SiC with an insignificant α-SiC_{II} content.

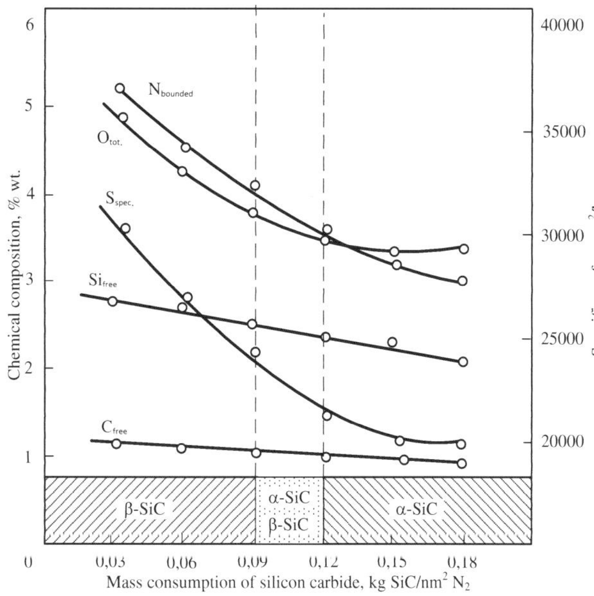

Figure 3. Dependence of phase and chemical compositions, the specific surface area of plasma processing products on the mass flow of silicon carbide.

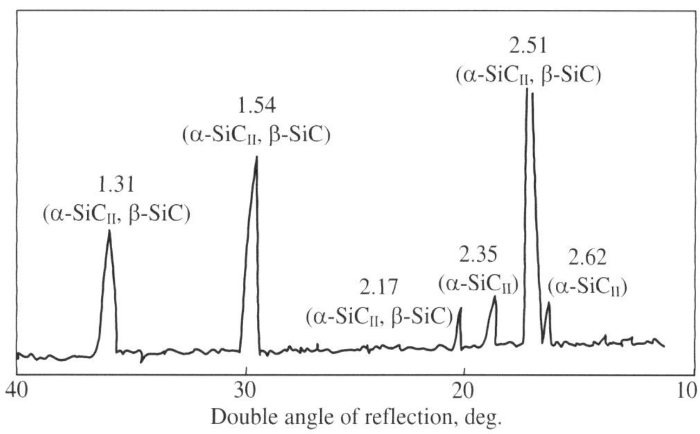

Figure 4. X-ray diffraction pattern of plasma-processed silicon carbide.

With a higher reactor load (0.15-0.18 kg SiC/nm^3 N_2) the heat content of the plasma flow is insufficient to vaporize the microgrit of silicon carbide and, consequently, to perform the structural transformation of α-SiC_{II} → β-SiC. In this case, the most likely mechanism for dispersing is the thermal cracking and destruction of particles of silicon carbide microgrit.

Thus, it is established that the products of plasma treatment are characterized by high dispersity (20000-30000 m^2/kg) corresponding to the nanolevel. However, the silicon carbide obtained by this method contains a significant (5-10 wt%) amount of SiO_2 due to the oxidation of products by oxygen and water vapour introduced by the plasma-forming gas. In connection with this, further experimental studies of plasma processing of silicon carbide microgrit were carried out in the following two directions:

1) Plasma processing in the drained and purified process gases.

2) Plasma processing of microgrit in process gases containing hydrocarbon additives to prevent oxidation of initial materials and synthesis products.

In experimental studies the initial temperature of plasma flow was 5600 K. Quenching of the synthesis products was carried out at a temperature 2800 K. In all experiments the mass flow of microgrit was 0.12 kg of SiC per 1 nm^3 N_2. Drying and cleaning of the plasma-forming gas was carried out in a column-type unit with a chromium-nickel filler and electric heating to a temperature of 750-850 K.

Additives of hydrocarbon – technical propane were introduced together with the initial powder into the zone of impact of plasma jets in amounts varying from half of the required by stoichiometry to bind oxygen in CO up to a fourfold excess. The results of the studies are given in table 5. For comparison, the table also includes the data obtained by plasma processing of microgrit in the untreated process gases.

Table 5. Influence of the composition of process gases on the characteristics of the products of plasma processing of microgrit.

No.	Composition of process gases	Phase composition, wt%		Chemical composition, wt%				S_{sp}, m^2/kg
		β-SiC	α-SiC_{II}	Si_{free}	C_{free}	SiO_2	Fe	
1	N_2+0.5%vol. O_2	57.0	43.0	2.25	1.07	6.6	0.32	21500
2	N_2+0.007% vol. O_2	56.0	44.0	3.00	0.88	2.7	0.37	26000
3	N_2+0.5% vol. O_2+ 0.1% vol. C_3H_8	55.0	45.0	2.05	0.82	3.8	0.29	23000
4	N_2+0.5% vol. O_2+ 0.44% vol. C_3H_8	52.0	48.0	1.05	0.78	2.5	0.31	21000
5	N_2+0.5% vol. O_2+ 0.84% vol. C_3H_8	59.0	41.0	1.01	0.91	2.5	0.27	19000
6	N_2+0.5% vol. O_2+ 1.4% vol. C_3H_8	56.0	44.0	1.12	1.1	2.3	0.29	22000

As it can be seen from table 5 the purification of process gases from oxygen and water vapour and the introduction of hydrocarbon in an amount from a 2-fold excess by stoichiometry and more leads to a significant decrease in the oxidation of products (the content 2.7 and 2.5 wt% of SiO_2 in experiments No. 2 and No. 4-5, respectively) compared to silicon carbide treated in the unprepared process gases (6.6 wt% SiO_2 in experiment No. 1) with comparable phase composition and dispersity. A microphotograph of the silicon carbide NDP obtained in the conditions of test No. 6 is shown in figure 5.

It can be seen that the powder contains particles of fragmented and faceted forms, which indirectly confirms the two-channel mechanism for particles dispersing of the initial microgrit: thermal cracking with retention of the α-SiC_{II} phase and evaporation followed by "vapour – crystal" condensation with formation of β-SiC.

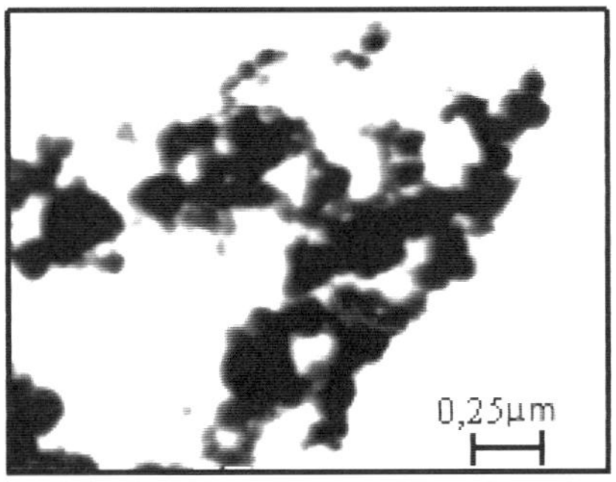

Figure 5. Micrograph of plasma-processed silicon carbide.

The optimum values of technological factors and the main characteristics of the plasma processing product, obtained with a fivefold duplication of experiments, are given in table 6.

Table 6. Parameters of the process of microgrit plasma processing and the main characteristics of the obtained nanodispersed silicon carbide (reactor power 80 kW).

Parameters of plasma processing and characteristics	Values
Composition of the gas – coolant,% vol. - nitrogen	 100
Raw material size, µm - microgrit - boron	 4-5 -1[a]
Productivity by raw materials, kg/h	1.60
The amount of reducing agent,% vol. of the volume of the gas – coolant	0.44-0.84
The content of boron in the charge,% wt.	3.00[a]
The initial temperature of the plasma flow, K	5600
Tempering temperature, K	2800
Phase composition	β-SiC, α-SiC_{II}, SiO_2 ам.
Chemical composition, % mass - main phase	 90.00-92.00
- associated impurities silicon oxides free carbon common boron volatile	 1.00-1.15 3.00-3.50 0.90-1.0 2.45-2.63[a] 2.12-2.72
Specific surface, m^2/kg	19000-21000
Average particle size, nm	100-110
Shape of particles	faceted, fragmentated
Productivity by target product, kg/h	1.45

[a]In the production of silicon carbide alloyed with boron

4. Conclusion

The process of plasma processing of silicon carbide, including the introduction of its microgrit K3 64 M5 in the plasma nitrogen flow with the initial temperature of 5600 K, plasma treatment and

IOP Conf. Series: Materials Science and Engineering **411** (2018) 012065 doi:10.1088/1757-899X/411/1/012065

quenching at a temperature of 2800 K. It is established that the product of treatment is silicon carbide in the nanodispersed state, the phase composition of which depends on the mass flow of the initial micropowder: β-SiC at 0.03-0.12 kg/nm^3 N_2, β-SiC + α-SiC_{II} at 0.12-0.15 kg/nm^3 N_2, α-SiC_{II} at 0.15 and above kg/nm^3 N_2. A two-channel mechanism for the dispersion of the initial carbide particles was revealed: thermal cracking with retention of the α-SiC_{II} phase and evaporation followed by "vapour – crystal" condensation with formation β-SiC. The possibility of increasing the oxidative stability of nanocrystalline silicon carbide with the addition of hydrocarbons (propane) in the amount of 0.44-0.84% vol. from the volume of the gas-coolant. The optimal values of technological factors and the main characteristics of nanocrystalline silicon carbide are determined: the content of the main phase is up to 92.00 wt%, silicon and free carbon is up to 1.00 wt%, oxides (calculated as SiO_2) – up to 3.50 wt%; specific surface area 19000-21000 m^2/kg; the average particle size is 100-110 nm.

References

[1] Dantsis Ya B et al 1984 *Electrothermal Processes of Chemical Technology* (L: Chemistry) p 464
[2] Porada A N and Gasik M I 1990 *Electrothermy of Inorganic Materials* (Moscow: Metallurgy) p 232
[3] *NeHudLit.Ru. Non-fiction library* http://nehudlit.ru
[4] 1983 OST 2 MT 71-1-82. *Grinding Materials. Particularly Fine Micro-grinding powders. Technical Conditions* introd. 01/01/83 (Moscow: Minstankoprom)
[5] Kosolapova T Ya et al 1985 *Non-metallic Refractory Compounds* (Moscow: Metallurgy) p 224
[6] Shiryaeva L S, Nozdrin I V, Galevsky G V and Rudneva V V 2014 *Electroplating and Surface Treatment* 2 vol. XXII 51–57
[7] Rudneva V V and Galevsky G V 2007 *Izv. Vuzov. Ferrous Metallurgy* **6** 6–8
[8] Galevsky G V, Rudneva V V and Galievsky S G 2003 *Bulletin of the Mining and Metallurgical Section of the Russian Academy of Natural Sciences. Department of Metallurgy* **12** 78–86
[9] Galevsky G V, Kulagin N M and Mintsis M Ya 2004 *Metallurgy of Aluminum. World and Domestic Production: Evaluation, Trends, Forecasts* (Moscow: Flint: Science) p 278
[10] Rudneva V V and Galevsky G V 2007 *Izv. Vuzov. Non-ferrous Metallurgy* **2** 59–63
[11] Rudneva V V and Galevsky GV 2007 *Izv. Vuzov. Ferrous Metallurgy* **3** 39–43

IOP Conf. Series: Materials Science and Engineering **411** (2018) 012066 doi:10.1088/1757-899X/411/1/012066

Instrumented system "Engineering-Metallurgy" for solving a wide class of engineering tasks

I A Rybenko

Siberian State Industrial University, 42 Kirova street, Novokuznetsk, 654007, Russia

E-mail: rubenkoi@mail.ru

Abstract. The instrumented system "Engineering-Metallurgy" is presented in the form of a set of programs and databases adapted for a number of metallurgical technologies that helps to perform multivariate calculations, conduct research, solve various optimization problems and determine the best conditions for implementation of metallurgical processes.

1. Introduction

One of the most important problems that ferrous and non-ferrous metallurgy industries are currently facing is the reduction in the costs per unit of raw materials and energy per unit of output, that is, the problem of creating energy and resource-saving technologies [1]. The development of new resource-saving technologies is impossible without computational experiments that allow the states and processes to be analyzed and conclusions about the behavior of the objects under investigation to be drawn on the basis of model representations. Therefore, at the present time it is actual to create instrumental systems that implement mathematical models and allow calculations, studies to be efficiently performed and various optimization problems to be solved. To solve these problems with the use of MS Excel and Delphi, the instrumented system "Engineering-Metallurgy" was created, which performs the tasks determining the optimal conditions for the implementation of metallurgical technologies, in which a mathematical description of the interconnection of flows and process parameters is implemented using a set of mathematical deterministic models, based on the first and second laws of thermodynamics [2 - 4].

2. Structure, functions and purpose of the instrumented system

The instrumented system, the structural diagram of which is shown in figure 1, is a set of programs adapted for a number of metallurgical technologies. The programs include a set of standard blocks: "Input data", "Balances", "Enthalpies", "Activities", "Reactions", "Technical and economic indicators", "Optimization", "Graphic", "Research", in which the following databases are used: "Chemical composition of materials", "Thermal properties of individual substances", "Parameters of interaction of the first order", "Atomic parameters", "Thermodynamic characteristics of reactions of phase transitions and dissolution of elements".

The algorithm for calculating the optimal modes of metallurgical processes in the "Engineering-Metallurgy" system is shown in figure 2.

Data of thermodynamic functions of substances are used for the calculation using the database "Thermodynamic properties of individual substances".

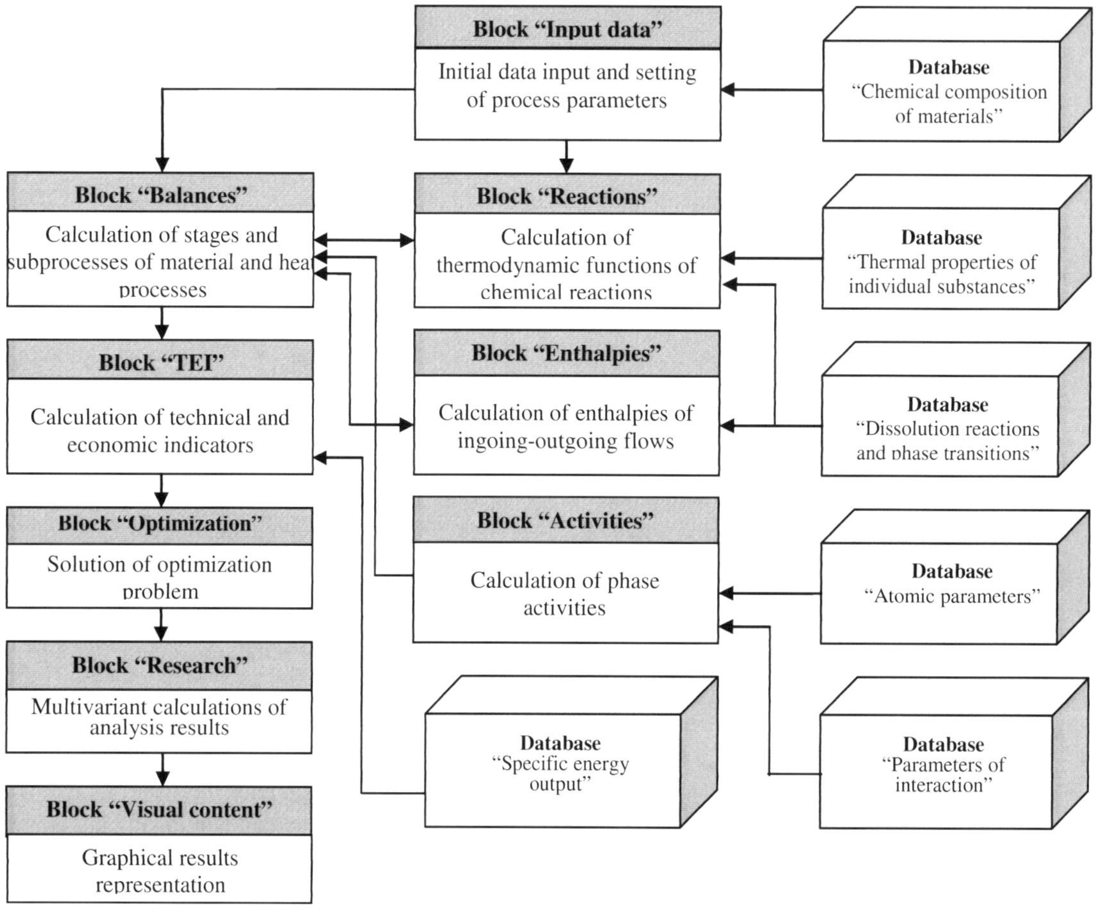

Figure 1. Scheme of the instrumented system "Engineering-Metallurgy".

In the "Activities" block, the activities of metal components are calculated using the apparatus of interaction parameters, slag – according to the theory of collectivized electrons and partial pressures of the gas phase. When calculating activities, the databases "Parameters of interaction of the first order" and "Atomic parameters" are used.Thermodynamic analysis of chemical reactions is carried out by the constant method in the "Reactions" block. The list of independent reactions is formed using the methods of thermodynamic modeling in the result of analysis of all possible reactions of the system transition from the initial state to the final one. The scheme for calculating thermodynamic characteristics of the main types of metallurgical reactions is shown in figure 3. For reactions of combustion, reduction, oxidative refining and reactions between individual substances, a change in the isobaric-isothermal potential and the equilibrium constant are calculated. For this purpose, we use data on the calculation of thermodynamic functions of individual substances using a directory on the thermal properties of substances and databases on the thermodynamic characteristics of reactions of phase transitions and the reactions of dissolution of elements in liquid iron.

An estimate is made of the degree of reactions deviation from thermodynamic equilibrium by analyzing the ratio of the product of the activities of the reacting substances to the equilibrium constant. For this purpose, data on the activity of metal and slag components are used.

The block "Technical and economic indicators" is presented by the table of calculation of the unit's productivity, unit costs of materials, cost and energy intensity of the product. In this case, a database of values of specific energy intensities of materials is used.

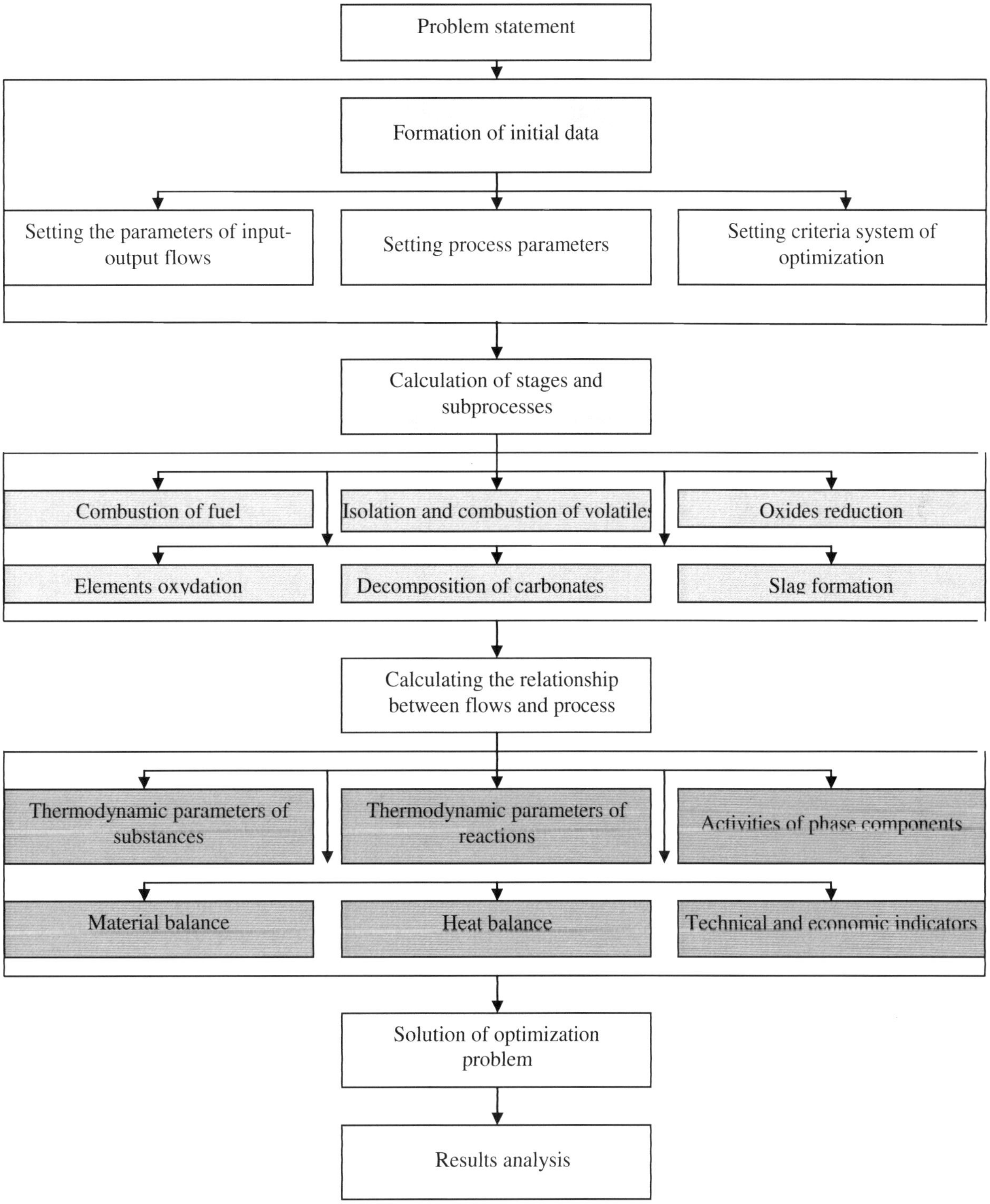

Figure 2. Algorithm for calculating the optimal modes of metallurgical process in the instrumented system "Engineering-Metallurgy".

IOP Conf. Series: Materials Science and Engineering **411** (2018) 012066 doi:10.1088/1757-899X/411/1/012066

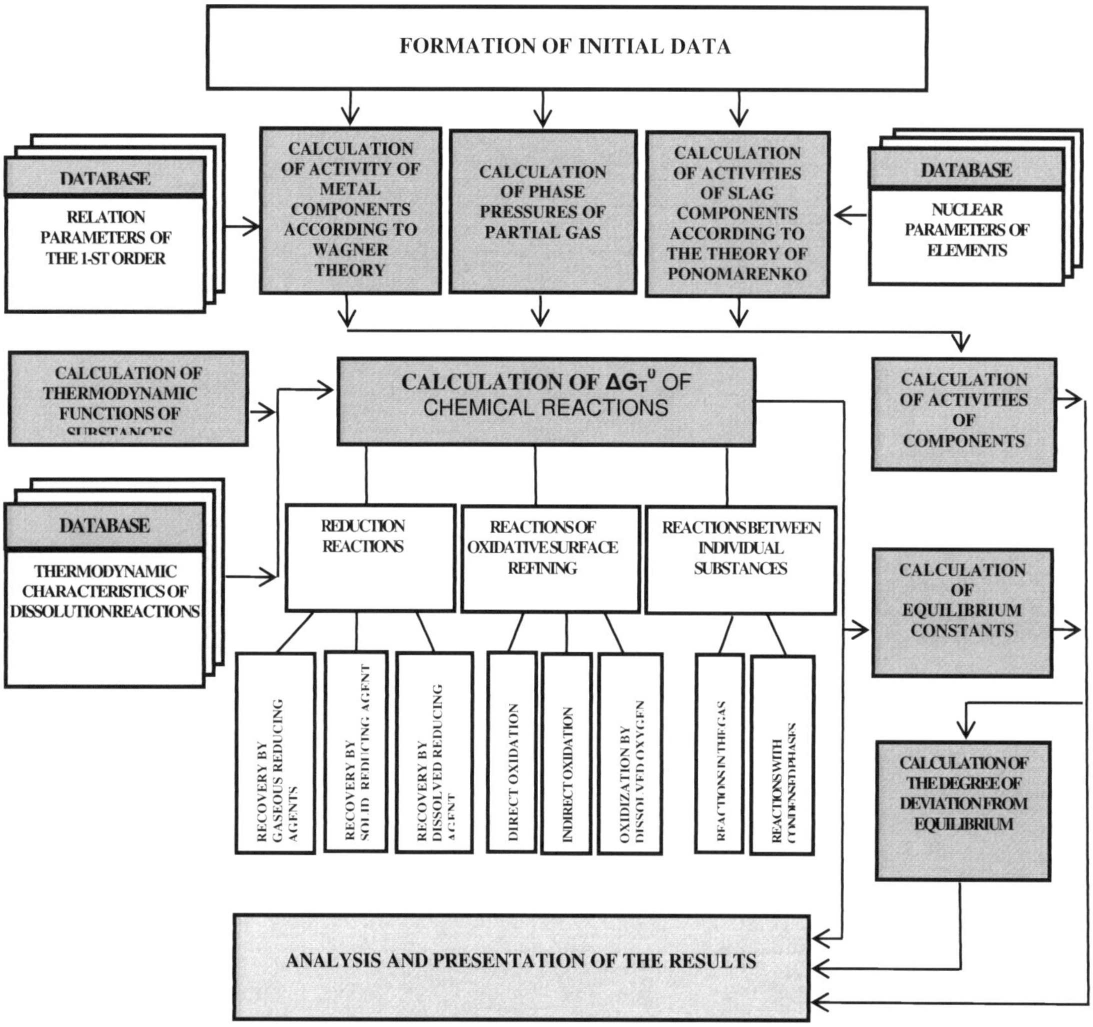

Figure 3. Calculation scheme of thermodynamic characteristics of chemical reactions.

In the "Optimization" block the problem of determining the optimal technological modes of metallurgical processes is solved, which consists in calculating control actions for obtaining a product with specified properties while optimizing one of the criteria. The following indicators can be chosen as criteria: total consumption of charge materials; the cost per unit of output, the cost of production, the energy intensity of the process or the capacity of the unit.

The solution of the optimization problem consists in determination of the extremum of one of the above criteria when performing limitations on the ranges of change in the flow components of the input streams; observance of the law of mass conservation at the level of flows, substances and elements; observance of the law of conservation of energy; compliance with restrictions on the parameters of the finished product; fulfillment of the normalizing relations and fulfillment of the target conditions by the remaining criteria.

The scheme for setting and solving the optimization problem is shown in figure 4. The problem is solved by the method of nonlinear programming – the generalized reduced gradient.

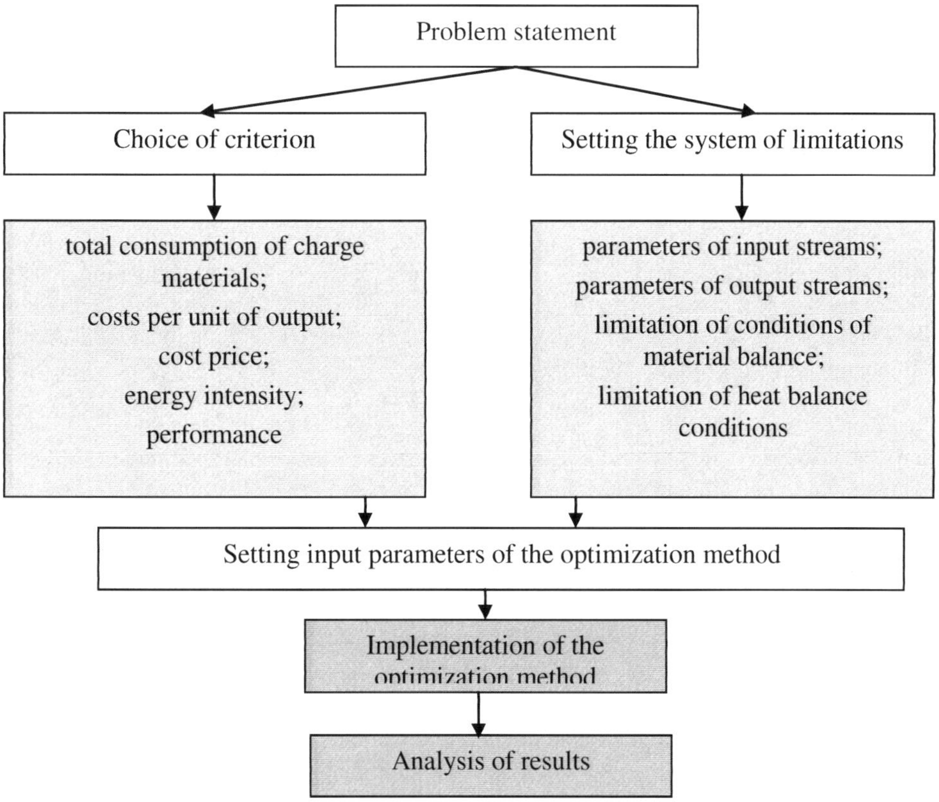

Figure 4. The scheme for solving the optimization problem.

In the "Optimization" block the summary results of the calculation are also presented: material balance by flows, total heat balance, technical and economic indicators and compositions of metal, slag and gas. In the "Visual content" block the graphical representation of the results in the form of histograms, pie charts and graphs is realized provides a full analysis of the results.

In the "Research" block a summary table with the results of multivariate studies is formed, followed by the possibility of plotting, including the values of unit costs of materials, technical and economic indicators, parameters of metal, slag and gas, and other necessary parameters for the chosen technology option.

The modules "Enthalpies", "Activities", "Reactions", "Optimization", "Graphics" and "Research", as well as the databases used are standard and can be used to calculate all types of processes by coordinating data with the "Input data" and "Balances" blocks, which, like the module "TEI", are adapted to the specific technology option.

In the software complex, various directories are implemented in the form of databases, the main of which is a directory on the thermodynamic properties of individual substances. It includes the following information: the chemical formula of the substance with the indication of the phase state, molar mass, enthalpy of formation and entropy of the substance at standard temperature, the temperature and the change in the enthalpy of phase transitions, and the coefficients of the approximation equation for the reduced Gibbs energy. The database of the software complex "Terra" [5] and the handbook on thermodynamic properties of individual substances "TSIV" [6] are taken as a basis.

The instrumented system includes a set of programs for the following technology options: oxygen-converter process, electric steelmaking process, steel alloying processes in the ladle, direct metal production from iron-containing materials, processing of titanium-magnesium concentrates and manganese ores, direct reduction of iron with co-production of synthesis gas in the unit of a jet-emulsion type, combustion processes of various types of fuel in the vortex adiabatic furnace, the process of obtaining copper in the anode furnace, process for the production of silicomanganese and a manganese production process in the plasma furnace. The developed software package was registered by the Federal Service for Intellectual Property of the Russian Federation (No. 2017617445).

3. Conclusion

Thus, the created instrumented system "Engineering-metallurgy" makes it possible to carry out multivariate calculations, conduct research and solve a complex of interrelated optimization problems for determining the best conditions for metallurgical processes [7 - 10]. The use of this software tool allows the labour productivity of researchers to be significantly increased and conditions for objective assessment of the obtained results reliability to be created.

References

[1] Lisin V S and Yusfin Yu S 1998 *Resource-ecological Problems of the 21st Century and Metallurgy* (Moscow: Higher School) p 447
[2] Rybenko I A and Mochalov S P 2005 *Izv. vuzov. Ferrous Metallurgy* **2** 55-8
[3] Tsymbal V P, Mochalov S P, Rybenko I A et al 2005 *Steel in Translation* **35(6)** 79–85
[4] Tsymbal V P, Mochalov S P, Rybenko I A et al 1996 *Steel in Translation* **26(12)** 44–8
[5] Trusov B G 2005 *III Intl. Symposium "Combustion and Plasma Chemistry"* (Almaty: Cazak University) pp 52–7
[6] Glushko V P 1965-1982 *Thermal Constants of Substances* (Moscow: VINITI)
[7] Rybenko I A 2015 *Proc. of Conf. in Modeling and Optimization of Stationary Modes of Metallurgical Processes* (Novokuznetsk: SibSIU) p 168
[8] Rybenko I A 2016 *Application of the Methodology and Instrumented system for Calculating Metallurgical Processes for the Development of Theoretical Foundations of Resource-saving Technologies* (Novokuznetsk: SibSIU) p 187
[9] Rybenko I A 2017 *Bul. Ferrous Metallurgy* **2** 37–43
[10] Rybenko I A 2017 *Bul. Ferrous Metallurgy* **3** 60–5

IOP Conf. Series: Materials Science and Engineering **411** (2018) 012067 doi:10.1088/1757-899X/411/1/012067

To the question of improving steel processing technology by converter vanadium slag

I A Rybenko, M A Golodova, O I Nokhrina and I D Rozhikhina

Siberian State Industrial University, 42 Kirova street, Novokuznetsk, 654007, Russia

E-mail: golodova_ma@mail.ru

Abstract. The results of thermodynamic modeling and laboratory study of the process of steel microalloying by vanadium from converter vanadium slag using coke carbon and silicon ferrosilicon are presented. These experimental studies on a laboratory arc furnace confirmed the results of thermodynamic modeling.

1. Introduction

At present, the technology of steel, based on the introduction of oxide materials and the provision of conditions for the reduction of alloying elements from their oxides, becomes a promising one in connection with the need to reduce the energy- and material intensity of production, solution of environmental problems. One of the most effective and therefore demanded alloying elements for the production of steels with high performance properties is vanadium. The source of vanadium oxides can be vanadium-containing metallized pellets, technical pentoxide vanadium and converter vanadium slag. It is more economically efficient to use converter vanadium slag, which is an intermediate product of the vanadium conversion, because its cost is much lower than the cost of technical pentoxide vanadium, and the vanadium content is higher than in metallized pellets. The actual task is to develop a technology for processing steel with converter vanadium slag using silicon and carbon as reducing agents, since at the present time high-priced aluminum, silicon, and calcium are used as reducing agents.

2. Methods of research

To solve the above stated problem using the previously developed thermodynamic simulation methods [1, 2], on the basis of the finished software product – "Terra" software package [3], calculations were carried out to determine the optimal conditions for the reduction of vanadium from vanadium converter slag, the laboratory investigations were carried out modelling the process of vanadium reduction from converter vanadium slag during out-of-furnace processing of steel (alloying the steel in the ladle at the outlet and processing at the unit furnace-ladle).

Calculations were carried out at a temperature of 1873 K, the amount of furnace slag getting into the ladle is 1 kg/t of metal, the additives of lime and ferrosilicon FS 75 are taken equal to 2 kg/t, respectively. The consumption of converter vanadium slag was 4, 6, 8, 10 and 12 kg/t of metal with a change in the consumption of coke in the range from 0 to 2 kg/t of metal.

During the calculations the following compositions of the initial materials were adopted:

- converter vanadium slag: 16 % V_2O_5, 25 % Fe_{tot}, 8 % MnO, 15 % SiO_2, 4 % TiO_2;
- lime: 85% CaO, 2% SiO_2, 1% Al_2O_3, 2% MgO, 9% $CaCO_3$;
- electric furnace slag: 25% FeO, 8% MnO, 45% CaO, 15% SiO_2, 2% MgO, 2% Al_2O_3, 2% P_2O_5;
- metal released from the electric arc furnace: 0.2% Mn, 0.02% S, 0.02% P, 0.1% to 0.7% C.

In order to confirm the results of the thermodynamic modeling of steel alloying by vanadium from converter vanadium slag, a series of experimental meltings was carried out on a laboratory arc furnace with a capacity of 2 kg with magnesite lining at a temperature of 1873 K. The base materials included carbon steel with the following composition 0.27% C, 0.49% Mn, 0.02% Si, 0.02% V, vanadium converter slag, reducing agents and lime. Crystalline silicon and graphite were used as reducing agents.

From pre-crushed and carefully mixed base materials, briquettes with different ratio of reducing agents were made. Liquid glass was used as a binder. The prepared briquettes were dried in the air. The composition and quantity of materials are given in table 1.

Table 1. The number of raw materials and additives, g.

No.	Mass of the briquette components			Briquette mass	Metal mass	Lime mass	Silicon mass (deoxidizer)
	slag	graphite	silicon				
1	100	17	0	25	1840	25	10
2	100	0	15	25	1695	25	10
3	100	15	3	15	1650	15	15
4	100	12	6	20	1675	20	10
5	100	7.6	9	25	1945	25	15
6	100	5	12	25	1845	25	25

The furnace was charged with carbon steel and slag-forming materials. After melting the materials and slag formation, the metal was deoxidized with silicon and briquettes were introduced. The effect of the steel processing with converter vanadium slag was achieved due to the mechanical mixing of the metal in the furnace.

The temperature of the experiments was controlled by a tungsten-rhenium thermocouple BP5/20. The time of the experiment after stabilizing the temperature was 10 minutes. After the end of the process, the metal was poured into a mold and cooled in the air.

3. Results and discussion

The obtained in the calculations dependences of vanadium content in the metal on the carbon content in the steel discharge and the consumption of the converter vanadium slag are shown in figure 1, and they have extrema. The calculation data showed that at a consumption of converter slag up to 8 kg/t of steel, the vanadium content in the metal is not less than 0.04% for both high carbon, medium and low carbon steels. For medium-carbon steel, it is possible to obtain a content of vanadium in the metal not more than 0.09% at a consumption of converter vanadium slag up to 20 kg/t of steel.

The completed thermodynamic calculations made it possible to determine the theoretical coefficient of vanadium extraction from the converter vanadium slag during the steel tapping and formation of ladle slag (figure 2).

From the results obtained, it follows that at the tapping time the coefficient of vanadium extraction is 0.5 with the consumption of converter vanadium slag 10 kg/t of steel, which makes it possible to obtain the vanadium content in the steel 0.04% (figure 2). For medium- and high-carbon steel with this slag consumption, the extraction coefficient is more than 0.9.

To evaluate the final result of the out-of-furnace treatment of steel, a thermodynamic simulation of the of vanadium reduction process from converter vanadium slag taking into account the introduced reducing agents (coke carbon and silicon ferrosilicon) was performed.

The calculations were carried out for the following additional conditions: medium-carbon steel, coke consumption – from 0 to 2 kg/t of steel, the consumption of vanadium slag from 4 to 16 kg/t of steel, the amount of furnace slag – 10 kg/t of steel. The results showed that in the indicated range of slag consumption with a coke consumption more than 0.6 kg/t of steel, the vanadium content in the metal does not change (figure 3) and the vanadium extraction coefficient exceeds 0.9 (figure 4) at any slag consumption.

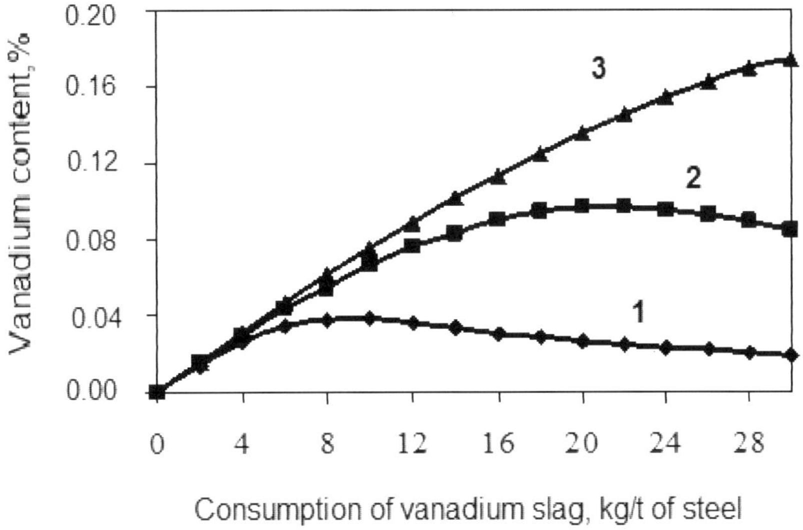

1 – 0.1 % [*C*], 2 – 0.3 % [*C*], 3 – 0.7 % [*C*]

Figure 1. Dependence of the vanadium content in steel on the carbon content of in metal and the consumption of converter vanadium slag.

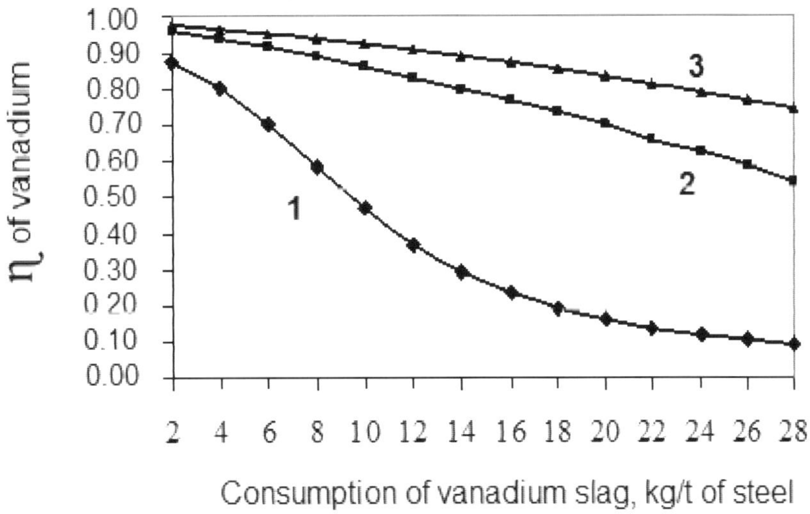

1 – 0.1 % [*C*], 2 – 0.3 % [*C*], 3 – 0.7 % [*C*]

Figure 2. Dependence of vanadium extraction coefficient on steel carbon and the amount of converter vanadium slag.

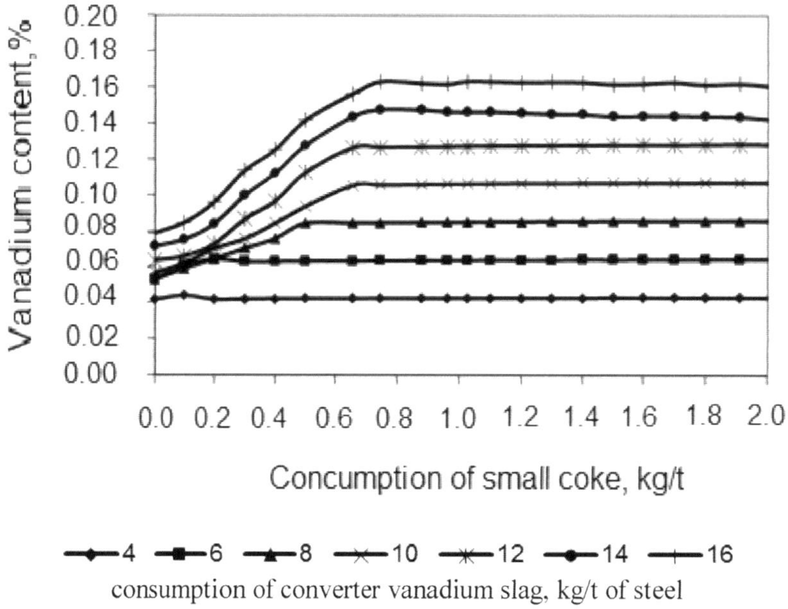

Figure 3. Dependence of the content of vanadium in steel on the costs of coke and converter vanadium slag.

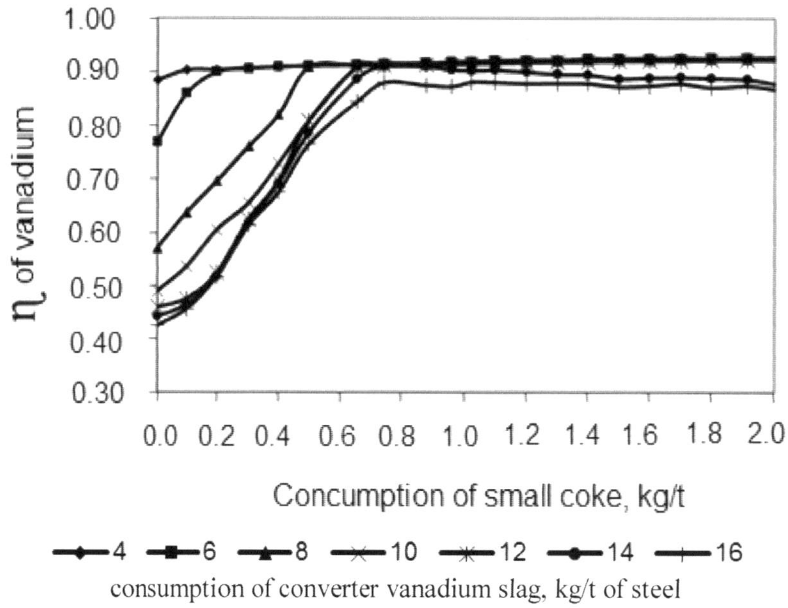

Figure 4. Dependence of vanadium extraction coefficient on coke consumption and converter vanadium slag.

The performed calculations also indicate that:

- vanadium oxides disappear at coke consumption from 0.5 to 0.6 kg/t of steel in the range of variation of the consumption of converter vanadium slag from 4 to 12 kg/t;
- the carbon content in the steel, depending on the consumption of coke, varies from 0.01 to 0.80% irrespective of the consumption of converter vanadium slag;

- manganese is reduced completely, its concentration in the metal increases in proportion to the consumption of slag;
- titanium is present in the metal insignificantly in the form of titanium carbides at high coke consumption and in the slag as $CaTiO_3$;
- slag is represented by oxides of calcium, magnesium, aluminum and silicon.

The metal, obtained after experiments, was subjected to spectral analysis, the results of which are given in table 2.

Table 2. Results of spectral analysis of the obtained metal.

No.	Mass fraction of the element, %								
	C	*Ni*	*Mn*	*Ti*	*Mo*	*Si*	*W*	*V*	*Al*
1	0.412	0.098	0.456	not det.	0.014	0.353	0.004	0.126	0.002
2	0.166	0.072	0.461	0.054	0.013	1.051	0.004	0.108	0.002
3	0.309	0.094	0.515	0.06	0.015	1.938	0.006	0.1	0.002
4	0.254	0.078	0.388	not det.	0.013	0.333	0.005	0.114	0.001
5	0.337	0.086	0.424	not det.	0.013	0.332	0.004	0.111	0.002
6	0.242	0.092	0.421	not det.	0.014	0.508	0.005	0.118	0.002
The error of the analysis results in accordance with GOST (State Standards) 18895 – 97, %									
	± 0. 003		**± 0. 012**	**± 0. 008**		**± 0. 015**		**± 0. 008**	

The analysis of the resulting metal showed a stable content of manganese and vanadium in the samples. Titanium was present only in two samples: when using silicon as a reducing agent (sample 2) and with a low consumption of lime (sample 3). Samples 2 and 3 also showed an increased silicon content in the metal, which is associated with a higher consumption of silicon in the charge materials.

According to the analysis, vanadium extraction coefficients η_V were determined taking into account the vanadium content in the base metal. The calculation results of the ratios are given in table 3.

Table 3. Ratio values of η_V.

No.	Briquette mass, g	Metal mass, g	Vanadium mass in the base metal, g	Vanadium mass in the briquette, g	Vanadium mass in the metal, g	η_V
1	25	1840	0.37	1.91	2.32	0.99
2	25	1695	0.34	1.91	1.83	0.96
3	15	1650	0.33	1.14	1.65	0.99
4	20	1675	0.34	1.51	1.91	0.99
5	25	1945	0.39	1.81	2.15	0.98
6	25	1845	0.37	1.91	2.17	0.95

4. Conclusion

The results of thermodynamic modeling show the possibility of steel microalloying by vanadium using converter vanadium slag in the ladle with coke and ferrosilicon.

The experimental studies on the laboratory arc furnace confirmed the results of thermodynamic modeling and showed the possibility of steel processing with converter vanadium slag using carbon and silicon as reducing agents.

References

[1] Rybenko I A 2017 *Bul. Ferrous Metallurgy* **2** 37–43
[2] Rybenko I A 2016 *Application of the Methodology and Development system for Calculating Metallurgical Processes for the Development of Theoretical Foundations of Resource-saving Technologies* (Novokuznetsk: SibSIU) p 187

[3] Trusov B G 2005 *III Intl. Symposium "Combustion and Plasma Chemistry"* (Almaty: Cazak University) pp 52–7
[4] Golodova M A et al 2011 *Izv. Vuzov. Ferrous Metallurgy* **4** 3–5
[5] Golodova M A and Rozhikhina I D 2016 *Bul. Ferrous Metallurgy* **10** 27–34
[6] Golodova M A, Rozhihina I D et al 2016 IOP *Conf. Series: Materials Sci. and Eng.* **150** 012016
[7] O I Nokhrina et al 2014 *Izv. vuzov. Ferrous Metallurgy* **2** 27–30
[8] Golodova M A et al 2016 *Modeling and Knowledge-intensive Information Technologies in Technical and socio-economic Systems* (Novokuznetsk: SibSIU) pp 112–117

Experimental verification of mathematical models adequacy of the new continuous metallurgical process JEP

I A Rybenko and V P Tsymbal

Siberian State Industrial University, 42 Kirova street, Novokuznetsk, 654007, Russia

E-mail: rubenkoi@mail.ru

Abstract. A comparative analysis of the experimental data obtained at a large-scale experimental installation and the results of modeling and optimization in the "Engineering-Metallurgy" system for various technological options in the unit of spray-emulsion type is carried out in the paper.

1. Introduction

Metallurgy is one of the most resource-intensive industries. Currently, the multistage scheme for metals production is the most common one, which includes agglomerate and steelmaking operations. The need to reduce the energy and materials consumption in metallurgical industry requires the improvement of traditional schemes of iron and steel production and the creation of new metallurgical processes and units for direct production of metals. Over the past 10 years, more than 100 plants of direct reduction have been built and some are still being constructed in the world implementing a variety of principles. Promising in this direction is the development of spray-emulsion processes and units, for which large phase-separation surfaces and high rates of physico-chemical interactions are typical. Such processes include the continuous metallurgical process of the spray-emulsion reactor (JEP) developed by the teams of researchers from Department of Information Technologies in Metallurgy (currently the Department of Applied Information Technologies and Programming) and West Siberian Metallurgical Combine (EVRAZ ZSMK) under the supervision of Doctor of Technical Sciences , Professor V.P. Tsymbal [1-3].

The process is based on the idea of implementing the principles of synergetics and nonequilibrium thermodynamics, which allowed theoretical foundations and a universal design of the unit to be created, which makes it possible to implement various technologies [4-5].

2. Structure, functions and purpose of the development system

The experimental JEP unit was constructed at KKTs-2 ZSMK and was intended for the adjustmen of developed technologies. Its main technological equipment is a technological unit, including reactor of a spray type and an emulsion-type refining clarifier, a feed system, a process gas supply system, a control system, an automated experiment system, and auxiliary systems consisting of a cooling system for the unit body and a product outlet. The general view of the pilot plant is shown in figure 1.

On the pilot plant 40 series of experiments were carried out, the results of which confirmed the advantages of the process and the spray-emulsion unit: high rates of heat and mass transfer processes (100 ÷ 200 times higher than in the converter); small sizes and low material consumption (10 ÷ 20 times less than in traditional metallurgy); low through energy consumption (15 ÷ 18 GJ/t);

environment friendly and non-waste technologies; controllability, universality and mobility, the possibility of implementing various technology options [6].

Figure 1. General view of the pilot installation of JEP.

To develop, research and optimize these technologies using the MS Excel and Delphi, the development system "Engineering-Metallurgy" was created, in which a mathematical description of the interconnection of flows and processes parameters is implemented using a complex of mathematical deterministic models based on the first and second laws of thermodynamics [7-8].

The instrumented system "Engineering-Metallurgy" made it possible to implement multivariate calculations, to carry out research and solve interrelated optimization problems in determining the best conditions for the processes during the adjustment of technologies on a pilot plant.

The control actions for each experiment were calculated with the help of the system, which resulted in the implementation of the following technology options: metal production from cast iron and scale, direct metal production from pulverized ores and wastes, processing of titanium-magnetite concentrate to produce the conditioned titanium slag, production of manganese alloys from poor dust-like manganese ores [9 -10].

The first 5 series of automated experiments were aimed at obtaining steel from cast iron using gaseous oxygen. Subsequent experiments after structural changes in the feed system of the charge to the first reactor were carried out without cast iron. Six series of experiments were conducted to develop the technology of direct reduction of pulverized materials, such as manganese concentrate, manganese concentrate – scale, scale.

In the next series of experiments the main attention was paid to the development of technology of iron direct reduction from a mixture of sludge and scale. At the same time, as the main task, the calculated and experimental choice of rational ratios of ore and reducing agents in the charge was made. These experiments helped to obtain a certain experience in controlling the process and determine the most rational ratio of 1:1 converter slag and scale for the ore part with the supply of plasticized charge, which confirmed the correctness of calculation results using the development system.

Also, experiments on processing titanium-magnetite concentrates for production of a conditioned titanium slag were successfully carried out.

One of the important tasks of conducting experiments was checking the adequacy of software-implemented mathematical models. A comparative analysis of the most representative experimental and calculated data is given in table 1 and in figure 2.

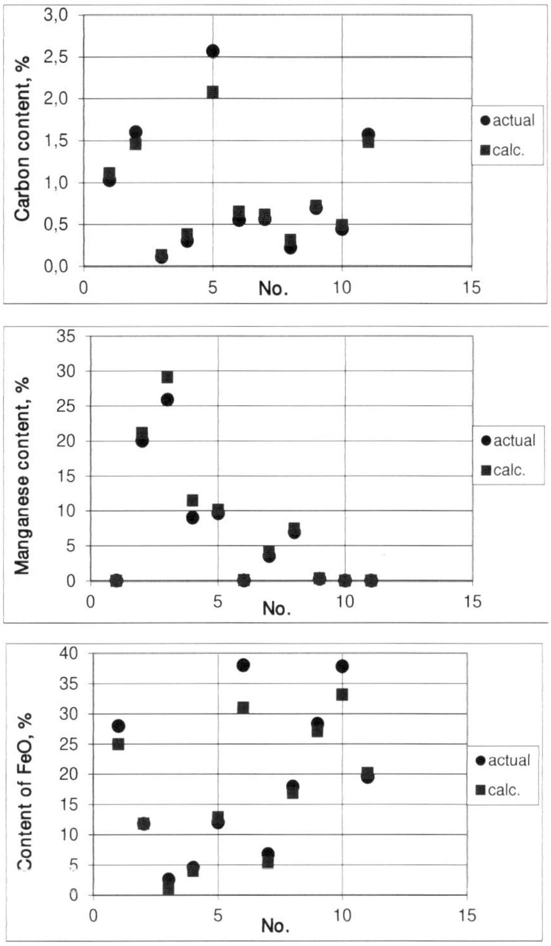

Figure 2. Comparative analysis of calculated and experimental data.

The adequacy of the models was estimated using a relative error, which was defined as the ratio of the root-mean-square deviation of the actual carbon and manganese content in the metal and FeO in the slag, calculated from the model to the mean value of the carbon, manganese and FeO content. The results of the calculations showed that the relative errors in the content of carbon, manganese in the metal and FeO in the slag are 15.5, 21.1 and 18.4%, which corresponds to the deviation from actual values for the carbon concentration 0.13%, manganese – 0.14%, FeO – 3.44%.

3. Conclusions

Experiments on adjustment of the technology of the spray-emulsion process at the pilot installation confirmed the adequacy of the models. Relative errors in the content of carbon, manganese in the metal and FeO in the slag can be considered quite acceptable in the conditions of adjustment of technology options in the presence of a large number of interferences, technical and technological limitations.

Table 1. Comparative analysis of calculated and experimental data.

No.	Exp. No.	Techno-logy	Consumption, kg/s (m^3/s)			Data	Composition of metal, %						*FeO*
			Cast iron	Charge	Oxygen		*C*	*Mn*	*Si*	*P*	*S*	*Al*	
1	3	cast iron + scale	20	4	1.78	act.	1.03	0.03	0.03	0.13	0.045	-	28.0
						calc.	1.15	0.05	0.02	0.023	0.018	-	25.00
2	7	*Mn* conc.	-	6.5	1.06	act.	1.6	20.0	20.4	-	-	3.5	11.8
						calc.	1.43	21.13	22.2	0.021	0.02	-	11.85
3	8	*Mn* conc.	-	5.06	0.67	act.	0.11	25.9	33.5	-	-	0.94	2.6
						calc.	0.51	29.1	34.5	0.018	0.019	-	1.19
4	10	*Mn* conc.+ scale	-	5.9	0.92	act.	0.3	9.0	6.5	-	-	8.0	4.5
						calc.	0.55	11.44	7.98	0.019	0.021	-	3.98
5	11	*Mn* conc.	-	3.4	0.78	act.	2.57	9.6	5.8	-	-	2.7	12.0
						calc.	2.21	10.11	6.19	0.023	0.021	-	12.8
6	12	scale	-	4.5	0.97	act.	0.55	0.03	0.01	0.013	0.01	-	38.0
						calc.	1.00	0.1	0.06	0.015	0.013	-	31.0
7	13	*Mn* charge	-	4.5	0.89	act.	0.56	3.53	7.5	0.070	0.023	-	6.76
						calc.	0.40	4.19	8.3	0.029	0.021	-	5.34
8	16	scale	-	5.45	0.72	act.	0.22	6.91	15.37	0.085	-	-	17.9
						calc.	0.58	7.44	18.53	0.025	0.024	-	16.89
9	18	scale	-	6.06	0.75	act.	0.69	0.26	0.3	0.1	-	-	28.35
						calc.	1.03	0.31	0.34	0.026	0.022	-	27.13
10	20	scale	-	5.4	0.78	act.	0.44	0.02	0.06	0.03	-	-	37.84
						calc.	0.77	0.031	0.076	0.023	0.022	-	33.15
11	24	scale + charge	-	4.8	0.81	act.	1.57	0.025	0.12	-	-	0.01	19.46
						calc.	1.19	0.031	0.14	0.02	0.021	-	20.12

References

[1] Tsymbal V P, Kozhemyachenko V I, et al 2016 IOP *Conf. Series: Materials Sci. and Eng.* **150** 012018

[2] Tsymbal V P, Mochalov S P, Rybenko I A et al 2014 *The JEP process – Metallurgical Spray-emulsion Reactor* (Moscow: Metallurgizdat) p 488

[3] Tsymbal V P, Mochalov S P Rybenko I A et al 2015 *Steel in Translation* **45(8)** 550–4

[4] Tsymbal V P, Kozhemyachenko V I, Rybenko I A et al 2015 *Izv. vuzov. Ferrous Metallurgy* **58(8)** 572–7

[5] Tsymbal V P, Kozhemyachenko V I and Rybenko A I 2016 *Proc. of Conf. in Modeling and High-Tech Information Technologies in Technical and Socio-Economic Systems* (Novokuznetsk: SibSIU) 27–36

[6] Aizatulov R S, Tsymbal V P, Mochalov S P et al 1997 *Proc. of Conf. in Contemporary Problems and Ways of Metallurgy Development* (Novokuznetsk: SibGIU Publishing Center) 107–8

[7] Mochalov S P, Tsymbal V P, Rybenko I A 1996 *Steel in Translation* **26 (12)** 44–8

[8] Rybenko I A 2015 *Modeling and Optimization of Stationary Modes of Metallurgical Processes* (Novokuznetsk: SibSIU) p 168

[9] Rybenko I A 2017 *Bul. Ferrous Metallurgy* **3** 60–5

[10] Rybenko I A 2016 *Application of the Methodology and Development system for Calculating Metallurgical Processes for the Development of Theoretical Foundations of Resource-saving Technologies* (Novokuznetsk: SibSIU) p 187

IOP Conf. Series: Materials Science and Engineering **411** (2018) 012069 doi:10.1088/1757-899X/411/1/012069

Plasma hardening hypereutectoid steel

E N Safonov and M V Mironova

Nizhny Tagil Technological Institute (branch) of Ural Federal University, 59 Krasnogvardeyskaya street, Nizhny Tagil, 622031, Russia

E-mail: e.n.safonov@urfu.ru

Abstract. There has been studied the effect of heat input plasma hardening steels process 0.9 C-Cr-Si, 1 C and 1.7 C-Cr-Ni-Mo on the structure, the depth of hardening and the hardness of the local area. It is shown that a graded structure is formed along the depth of the zone with regularly changing dispersion and microhardness. There have been specified the dependences, allowing to operate the structural state and performance properties of the hardening zone, ensuring their optimum ratios for different wear conditions.

1. Introduction

Hypereutectoid steels potentially have increased hardness and wear resistance under contact loading. They are usually applied in a thermally strengthened state (hardening with tempering) to provide the necessary strength. Such heat treatment does not allow the maximum hardness of the working surface of the parts to be realized. It is accompanied by a corresponding decrease in their service life. The solution of this problem is actual for massive products such as rolling rolls. It is associated with the use of thermal hardening of working surfaces, e.g., plasma hardening [1]. At the same time, the physico-chemical state and the properties of the material in the inner layers of the product do not change. They provide the predetermined structural strength. Surface hardening of such products increases their wear resistance and service life due to a favourable combination of high hardness of the surface working layer with a sufficiently strong core.

When the surface of a part is heated by a plasma arc of direct action, the main characteristics of the thermal cycle are the maximum temperature, the residence time of this point above the temperature of the point Ac1. The interdependent heating and cooling rates are determined by a combination of processing mode parameters. For example, with other things being equal, the maximum cycle temperature and residence time of the heated volume in the austenite region decrease with increasing arc travel speed. The cooling rate increases.

On the contrary, an increase in the maximum temperature, the depth of heating, the residence time of the heated volume in the temperature range of the existence of austenite increase the arc power. An increase in the grain size of the austenite and a corresponding increase in the M_n point accompany a possible decrease in the cooling rate.

The analysis of the results of experimental researches will reveal the patterns of size formation, structural-phase composition and properties of a layer heat-treated by a plasma arc. In the course of experiments, it is possible to determine the influence of technological process regimes on these indicators.

The target of the research is to determine the rational parameters of the surface plasma quenching of steels 9XC (0.88 % C), У10 (1.03 % C) and 170XHM (1.65 % C). The research is based on the study of the structure and properties of the local hardening zone.

2. Methods of research

The steels are preliminarily annealed. Such samples have perlite-cementite structure. Further, such samples are processed with an arc of direct polarity in argon.

Range of mode parameters change: arc current 130…300 A, speed of its movement 1…5 cm/s. The voltage on the arc was 21...23 V. The combinations of the mode parameters take into account the maximum process performance. The depth and width of the hardening zone are important in the absence of macro-melting of the surface.

Geometrical dimensions (depth and width), structure and microhardness of the local hardening zone were studied on cross-section microsections. Optical microscopes were used. These were Neophot-2, Zeiss Observer D1m with an increase 50-1000×. The software Thixomet PRO was used. Measurement of the microhardness of the Vickers pyramid under a load of 1.962 N (200 g) was carried out on a Future-Tech 300 and PMT-3 hardener under a load of 0.784 N (80 g). The microhardness was measured in sections with the maximum depth of the hardening zone. Phase analysis of the surface layer was carried out using X-ray diffractometers DRON-1, DRON-3 in Fe and Co Kα-radiations.

3. Results

As a result of the investigations carried out, it has been established that the depth, width and hardness of the local hardening zone increase with increasing current of the plasma arc.

An increase in the speed of arc displacement with a fixed value of the current is accompanied by a decrease in these parameters. This is explained by a decrease in the running energy of the heat input process. Running energy is the value of the specific heat flux through the surface. Running energy can be represented as the ratio of the effective thermal power of the arc to the speed of its translational motion. Thus, the running energy determines the geometrical dimensions of the local zone of plasma quenching, the characteristics of the temperature field and the thermal cycle of the heat treatment process. Dependence of the depth of the local zone of plasma quenching is presented in figure 1.

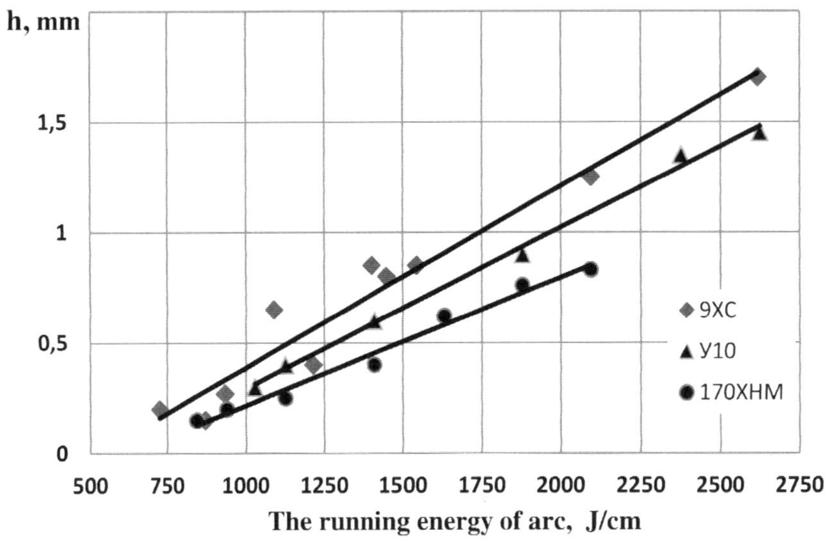

Figure 1. Influence of the running energy of arc on the depth (h) of the zone of plasma quenching of hypereutectoid steels.

In figure 2 shows the dependence of the change in microhardness on the depth of the quenched zone of the steels studied at different values of the running energy. It can be noted that the phase transformations occurring in the surface layer during plasma treatment make it possible to effectively increase the microhardness of all the investigated steels in the thermal-effect zone.

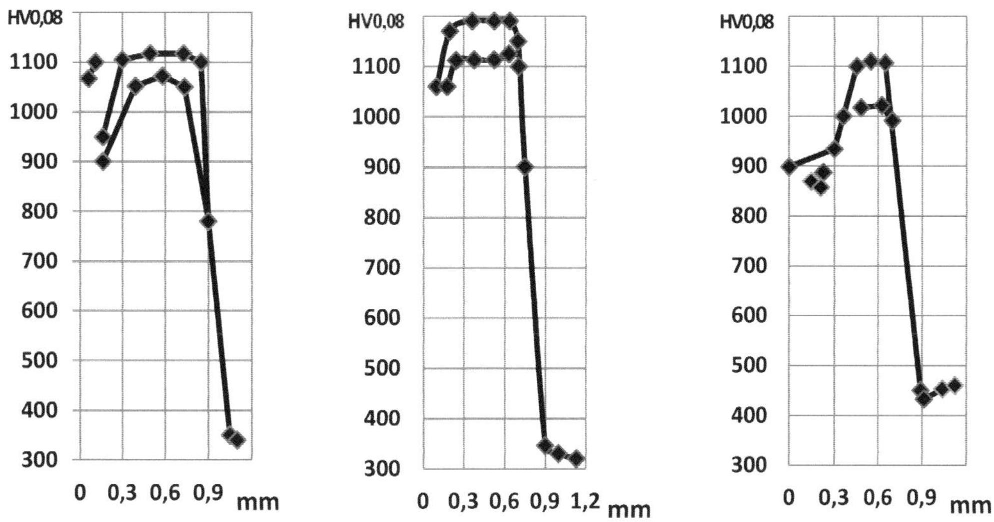

(2094) (2375) (2094)
(a) (b) (c)

Figure 2. A change in the microhardness (HV 0.08) over the depth of the plasma quenching zone of steels: 9XC (a), У10 (b), 170XHM (c) for different values of running energy of arc (J/cm).

For example, for an arc quenched arc with a running energy of 2618 J/cm, the surface layer of 9XC steel is characterized by the presence of residual austenite (up to 43 %), needle martensite of lamellar morphology (up to 51 %) with a dispersion of 5…15 μm and up to 6 % of the carbide phase. Starting from a depth of ~ 0.5 mm (for the central section), the residual austenite content is reduced. The carbide phase increases. Martensite and troostite appear in the structure with an interplanar distance of 0.1...0.15 μm. The microhardness of this zone decreases in accordance with the volume ratio of the phases present. At a depth of 1.3...1.5 mm troostite is gradually replaced by sorbitol with an interplanar distance of 0.3...0.5 μm and plate-like perlite. At a depth of 1.7 mm and further the effect of surface heating is not traced. The structure is perlite and secondary cementite with a characteristic microhardness (figure 3).

For other processing regimes, the general laws governing the formation of the structural-phase state of the hardening zone are preserved with a decrease in the length of sections of a homogeneous structure and a change in the ratio of structural components.

Figure 4 illustrates the effect of the running energy of the plasma arc on the ratio of the structural components in the surface layer of the local hardening zone of У10 steel. Phase analysis of the surface layer showed that asymmetry is observed on the γ-phase lines from the side of large reflection angles. This indicates the incompleteness of the processes of homogenization of austenite upon heating.

The value of the true austenite lattice parameter for all regimes was determined by extrapolation. This value was for all regimes higher than that corresponding to the total carbon content in the steel (~ 1 %). The amount of carbon in austenite, calculated by these parameters, depends on the running energy of heat treatment and varies within the range 1.3...1.6 %. This indicates that complete homogenization of austenite over carbon even in the surface layer, which is most heated by the arc, does not occur in the range of the investigated parameters of the regime.

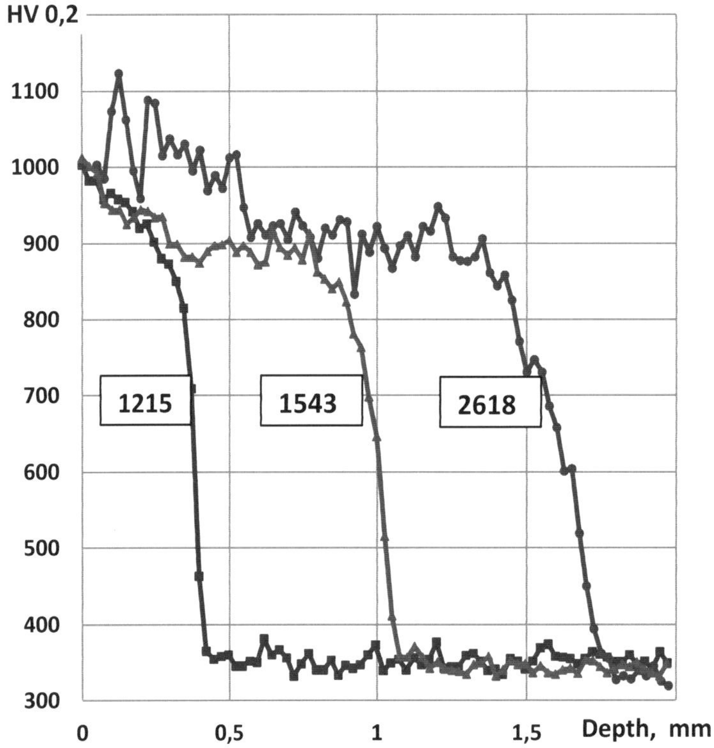

Figure 3. A change in the microhardness along the depth of the local zone of plasma quenching of 9XC steel at different running energy of arc values (J/cm).

Reflections from cementite on diffractograms have not been observed. It is logical to assume that local enrichment of austenite with carbon occurs due to its dissolution. The temperature-time diffusion conditions are not sufficient to completely equalize the carbon concentration. Carbon-enriched austenite has an improved resistance against decomposition upon cooling. Therefore, a high (up to 78 %) concentration of residual austenite is fixed in the surface layer of the samples. In the range of small values of running energy of arc, the residual austenite content does not exceed 30 %.

In the course of the X-ray structural analysis of У10 steel, the shift of the diffraction lines of the α-phase toward the smaller angles of reflection. This indicates the presence of an α-solid solution enriched in carbon. It has been experimentally established that the plasma treatment of У10 steel at elevated running energy of arc values allows to fix martensite with a carbon content of more than 1 %. The maximum amount of high-carbon martensite calculated at the graphical separation of the martensitic multiplet was 85 % of the total α-phase. At the initial values of the investigated range of running energy, the carbon concentration in the martensite does not exceed 0.6...0.7 %. The amount of high-carbon martensite in the α-phase decreases to ~ 40 % (figure 4). Thus, with an increase in the running energy of plasma quenching in hypereutectoid steels, the degree of dissolution of excess cementite. Austenite is saturated with carbon. This leads to the formation of a significant fraction of the residual austenite together with the carbon tetragonal martensite of increased hardness upon cooling.

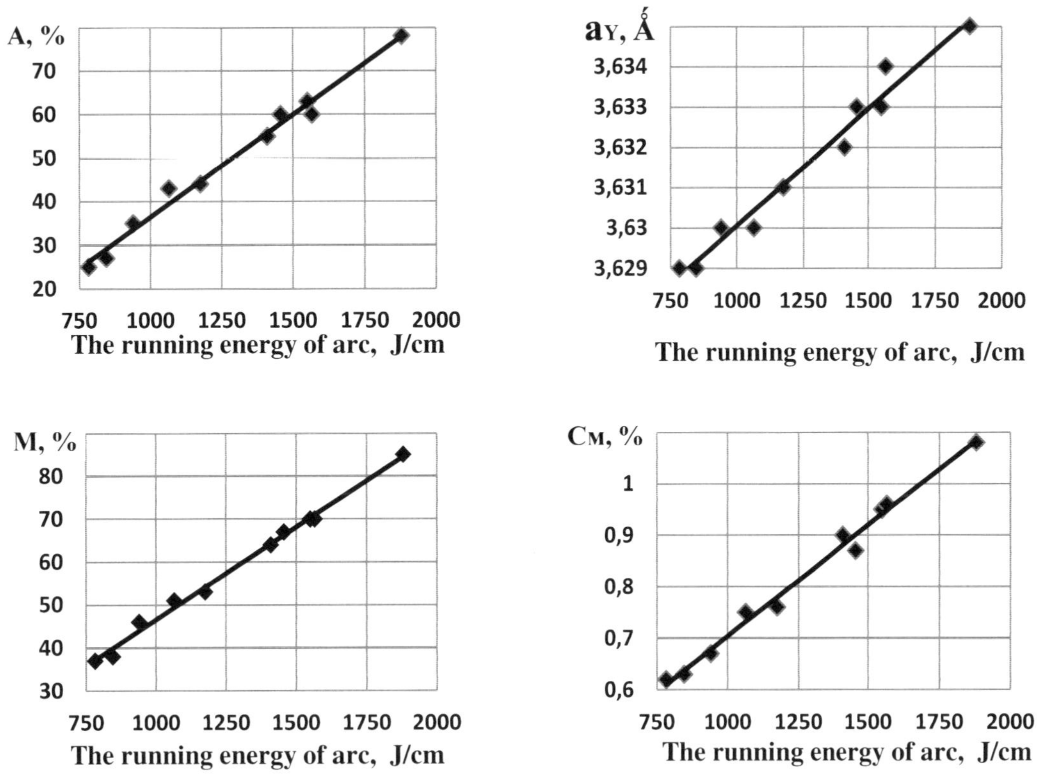

Figure 4. The effect of the running energy of arc in the plasma hardening of steel У10 on the volume fraction of the residual austenite in the surface layer (**A**, %), the austenite lattice parameter (**a**γ, Å), the volume fraction of the martensite of quenching in the composition of the α phase (**M**, %), the carbon concentration in martensite quenching (**Cм**, %).

4. Discussion of the results

The increase in heat input with increasing running energy during plasma hardening promotes the achievement of higher surface heating temperatures and an increase in the residence time in the temperature region above Ac_1. At the same time, the softer effect of treatment parameters on the increase in carbon concentration in hypereutectoid steels is explained by the presence of structurally-free cementite (for example, the amount of carbide phase in steel 170XHM reaches 15...20 %). Carbides have a lower thermal conductivity than steel. Energy costs to form a heat-treated layer of a certain depth increase with increasing content of the carbide phase.

Let us consider the influence of the running energy of the arc on the structural-phase composition and hardness of the hardening zone. When the hypereutectoid steels are heated above the phase transition temperature, the average carbon concentration in austenite increases from 0.8 % by dissolving the carbide phases. The stability of austenite increases.

It should be noted that the mechanism and kinetics of austenitic transformation during rapid heating, characteristic for plasma hardening, have not been studied in detail until now. Several hypotheses of the austenitization mechanism, analyzed in [2], are known, but for any mechanism an austenite is formed that is inhomogeneous in carbon [3, 4]. This heterogeneity in the case of rapid cooling is inherited by the products of its transformation. As a result, the high-carbon austenitic-martensitic structure is fixed after cooling in a thin surface layer (up to ~ 0.3 mm). Its microhardness varies with the ratio of the phases present and the martensite morphology (figure 2). In both cases, the degree of completeness of carbon redistribution is controlled by the temperature-time parameters of the thermal processing cycle. It depends on the initial structure of the steel. It can be different. This, in turn, determines a significant variety of structural states of the heat affected zone after treatment. For example, it is known that the hardness of martensite effectively increases as the carbon concentration in it increases to 0.6 %. Further, the increase in hardness slows down. Therefore, the treatment regime of hypereutectoid steel should provide a degree of dissolution of carbides sufficient to produce martensite with this concentration. Further growth of heat input and dissolution of the carbide phase increase the proportion of residual austenite.

Therefore, for hypereutectoid steels, when processing for the maximum hardness of the surface, heat regimes are preferred, which will provide the necessary degree of austenite saturation with carbon from the dissolving carbide phase.

A predominantly martensitic structure in the surface layer of hypereutectoid steels is formed in regimes with minimal heat input. It contains up to 30 % residual austenite. The carbon concentration in the martensite is ~ 0.65 %. The share of high-carbon martensite in the α-phase does not exceed 30 %. The hardness of the surface with such a structure reaches ~ 1000 HV 0.2 and gradually decreases in depth. The increase in heat input is accompanied by the formation in the surface layer of the investigated steels of a high (up to 78 %) fraction of high-carbon (~ 1.5 % C) residual austenite. In addition, we can observe an increase in the carbon concentration in martensite (over 1 %) and an increase in the volume fraction of high-carbon martensite in the α-phase composition.

Such a structure is a composition of solid constituents-martensite and carbides in less solid residual austenite. It is characterized by a considerable spread of microhardness (see figure 3). It is potentially unstable after cooling and has an increased margin of durability.

The martensitic transformation of the residual austenite proceeds in the course of contact-shock or abrasive interaction with the wear medium. It provides an increase in hardness and partial dissipation of fracture energy in combination with stress relaxation during the formation of martensite crystals of deformation. The wear resistance of the working surfaces of parts in the process of such a transformation of the structure is repeatedly increased.

Thus, the surface layer can be formed by varying the arc energy with the parameters of the plasma hardening mode of hypereutectoid steel. It will possess a dispersed martensitic-austenitic-carbide structure with a variable content of the constituent phases. With an increased content of carbon martensite and carbides, the hardness of such a layer will be the maximum for a given steel composition. The surface has increased wear resistance under conditions of contact friction of rolling and sliding. When the predominant phase in the composition of this layer is a high-carbon residual austenite, the hardness of the surface will be lower. But under appropriate operating conditions, it is possible to realize the mechanism of transformation of austenite into martensite deformation with increasing wear resistance.

5. Conclusion

The influence of the running energy of the plasma hardening process of hypereutectoid steels on the structure and properties (depth and hardness) of the local hardening zone was investigated. It is shown that a gradient structure is formed along the depth of the zone. It regularly changes the dispersity and microhardness of the components. The structural-phase state of this zone provides a smooth transition of mechanical properties from the quenched layer to the base metal.

It is shown that a dispersed martensitic-austenite-carbide structure is formed in the surface layer of the hardening zone. It has a variable content of components depending on the processing mode. The concentration of residual austenite and martensite decreases with depth. In the structure appears troostite, which is gradually replaced by sorbitol. The perlite-cementite structure of steels is located deeper. It has not undergone phase transformations as a result of thermal effects.

Based on the results of the research, regularities have been established that make it possible to purposefully manage the structural state and technological properties of the hardening zone of the investigated steels. You can achieve their optimal ratio for different wear conditions.

IOP Conf. Series: Materials Science and Engineering **411** (2018) 012069 doi:10.1088/1757-899X/411/1/012069

References

[1] Balanovsky A E 2015 *Strengthening Technologies and Coatings* **12** 18–30
[2] Dyachenko S S 1982 *The Formation of Austenite in Iron-carbon Alloys* (Moscow: Metallurgy) p 128
[3] Christian J W 2002 *The Theory of Transformations in Metals and Alloys* (Oxford: Pergamon Press) p 617
[4] Shtansky D V, Naka K and Onomori Y 1999 *Acta Mater.* vol 47 **9** 2619–32

Elaboration of steelmaking slag processing scheme for resource-saving

T V Sviridova, O B Bobrova, O Yu Ilina, A Yu Peryatinsky and N N Starostina

Nosov Magnitogorsk State Technical University, 38 Lenina ave, Magnitogorsk, 455000, Russia

E-mail: obproshkina@mail.ru

Abstract. The paper proposes a method for processing steelmaking slag for the purpose of resource saving. The authors propose to include partial granulation of steelmaking slag in the processing scheme. This embodiment includes draining of molten slag from the steelmaking unit in the bucket and separation of the molten slag into two parts, one of which is enriched in iron. With the use of statistical processing dependence for determination of content of FeO content in the converter slag was received. The resulting granulated product can be recommended for use in construction, in agriculture as a fertilizer, as well as a slag-forming flux in steelmaking.

1. Introduction

Now in the Russian Federation in dumps and tailings there are about 80 billion tons of waste from production and processing of mineral resources. In [1] it is noted that in 12 billion tons of accumulated solid waste content of valuable components is often higher than in natural ore deposits. In areas of placement of the metallurgical enterprises huge amounts of waste from various engineering procedures of a metallurgical complex (mining and enrichment of raw materials, production of cast iron, steel and, rolled products, coal combustion) which can be referred to technogenic deposits accumulated. The most significant number of technogenic fields was created in the Urals. In [2] it is noted that for the 300-year period of intensive mining and metallurgical production in the Urals billions of cubic meters of waste are saved up. In the late eighties PGO “Uralgeologiya” performed work on compiling geological and economic overview with assessment of possible use of waste of mining, metallurgical and fuel industry of the Ural economic region for receiving building materials. According to data of the overview, amounts of an annual output of waste in the Ural economic region as of 1986 were as follows: the overburden and host rock – 159.7 million m^3, coal mining waste – 73.2 million m^3, waste coal – 2.8 million m^3, tailings of dry magnetic separation – 7.6 million m^3, tailings of wet magnetic separation – 20.8 million m^3; metallurgical slag – 21.9 million tons, ash and slag wastes from TPPs– 17.9 million tons [3].

2. Theoretical part

With transition to the market relations in many factory cities of the Urals primary conversion of slag dumps started. Generally it is sorting and magnetic separation for the purpose of metal extraction (content of iron in slag of about 1.5 %) and sieving on screens for receiving slag crushed stone. Cost of slag crushed stone is 1.5-3.0 times lower than that of crushed stone from natural raw materials.

In large enterprises the current slag blast-furnace production at present is almost entirely processed, technologies of deep processing applied [4]. After granulation of fiery molten slag it is used as raw material for the production of slag Portland cement (SPC) and calcareous binders, can be fine aggregate of concrete and mortars based on these binders. The granulation step is required since the rapid freezing of the melt is formed the desired mineralogical composition of the material comprising the maximum number of active glassy phase. Shpts is made by joint grinding of PC clinker, granulated slag and gypsum or by carefully mixing these materials, crushed separately. The cost of shpts when introduced into a mixture of 30 to 60 % of granulated slag is 1.5-2.0 times lower than the SPC from mineral raw materials.

Slag of the current steel production is processed generally into crushed stone. More advanced processing, it isn't exposed. This is due to the high melting point of steelmaking slag and its rapid solidification after the flush from the kiln. Therefore, it isn't possible to granulate it similarly as blast furnace slag.

Development of technologies for deep processing of dump and steelmaking slag will create the conditions for full processing slag current production and reduction of the area of slag dumps, which will improve the environment. On the other hand it will solve the problem of cementing materials in places where there are slag heaps, but no natural raw materials.

Now in our country and abroad works on research of rational ways of processing of steelmaking slag in liquid state by granulation are carried out. For this purpose the most various designs of installations, using water, steam and air to cool the melt and as the energy source are applied [5-7].

It is experimentally proved that the steelmaking slag which passed a granulation stage sharply reduces hydraulic activity and tendency to self-disintegration. On chemical and mineralogical composition, the steelmaking slag which underwent thermocrushing is closer to the Portland cement than blast furnace slag, and can serve as raw material for production of cementing materials.

The main difference of steelmaking slag from Portland cement is the content of iron and manganese oxides, and also sulfur availability. Thus, in case of decrease of oxidation and reduction of the sulfur content in steelmaking slag possibilities of their application extend: as bases for slag-forming mixtures in ferrous metallurgy; for production of Portland cement and other cementing materials as well as fertilizers or active filters of sewage.

Various ways of granulation of steelmaking slag have been analyzed.

Granulation of steelmaking slag in the kiln hasn't found application due to the high concentration of the metal in slag flushed from a furnace. It led to unstable granulation process. claps and sharp emissions of steam gas mixture in case of metal contact with water. The second negative factor was the constant metalling of the oscillating screen grid. The third and main negative factor is the considerable difficulties of placement of the pool with water in direct proximity to the steelmaking unit.

Granulation on the special installations remote from steelmaking units at a distance from 1.5 to 5 km hasn't found application because of a low overheat of slag fusion over temperature of the beginning of solidification. Temperatures of final slag flush amounted to 1530-1550 °C during that period, and the beginning of solidification of slag -1350-1280 °C [8].

Currently, the process of tapping steel and slag from the melting units is carried out at temperatures of 1600-1650 °C. Slag is discharged into slag ladles at temperatures of 1600-1650 °C, and modern steel-making process is characterized by high intensity: the discharge of metal and slag is made practically at once after the oxygen lancing of the melt. The separation of the molten metal and slag is not perfect - slag-metal emulsion gets into the slag bowl. Further sludge is separated into hot metal that settles on the bottom of the bowl, and the molten slag. The practice of processing steelmaking slag of Magnitogorsk Iron & Steel Works shows that the slag in the slag bowls, received for processing in 60-80 minutes. after the flushing from the furnace keeps the ability to provide a flush.

3. Experimental part

After the conducted analysis of variants of granulation of steelmaking slag, it was decided to carry out the granulation in the laboratory. The study was carried out in the laboratory of the Department of metallurgy of ferrous metals of the Nosov Magnitogorsk State Technical University in the furnace of resistance of Tamman.

The steelmaking slag sulfur content is from 0.7 to 1 %. In case of water granulation of slag irrespective of a design of aggregates process happens by sharp chilling of the melted slag with water. Upon contact of the melt with water there is its intensive evaporation.

The smell of hydrogen sulfide during the granulation was felt. But the water the slag was poured into, did not change an indicator pH. This is due to the fact that parallel with the formation of acidic products of H_2S and SO_2 it occurs the hydration of CAO, followed by neutralization of acidic compounds with alkaline.

As a result of the experiment the product in black was received, which crumbled under slight mechanical action. The resulting product can be recommended for use in construction, as a result of hydration it is more resistant to degradation, in agriculture it can be used as a fertilizer and as a slag-forming flux in steelmaking.

Thus, it is proposed to include partial granulation of steelmaking slag in the scheme of processing.

Fluid steel slag in the ladles is delivered to installation of granulation in 40-60 minutes after discharge from the melting unit. This time is sufficient to separate slag-metal emulsion, but at the same time, the melt fluidity is maintained. After installing it's time to turn slag ladles over, and it is necessary to discharge only 30-40 % of the fusion which was in the ladle with melt to the pool. This will avoid discharging the metal into the pool that will affect the quality of the received granulated product. The remainder of the slag melt and deposited metal is sent for further processing. The iron concentration in the residue of the slag melt is increased, that will improve the recoverability of the metal in the processing.

The product (the granulated steelmaking slag) after a granulation and drying contains the minimum quantity of sulfur, iron and manganese, besides, it turns out without crystal structure therefore it can be used as lime containing material to slag in metallurgical process and for production of slag Portland cement and other cementing building materials.

For catching steam-gas emissions we suggest to establish the shelter with a general steam-gas withdrawal and to guide it to system of organizational cleaning in oxygen and converter shops. The shelter needs to be equipped with slide gates with electric drive [9].

Thus, we propose a method for processing steelmaking slag for the purpose of resource saving. This version includes discharge of slag melt from steelmaking unit into ladles and separation of the slag melt into two parts, one of which is enriched in iron The discharge is performed. at a temperature of over 1600 °C, which produces blowing of the jet of slag melt by air with an air flow rate 4.5 m / min, after discharge ladles are carried to the granulation installation where the mentioned separation of the slag melt into two parts by discharging 30-40 % of slag melt from the ladle into the pool to obtain a granulated product, and then the remaining part of the ladle enriched with iron, is carried to further processing to extract iron, while the slag melt in the ladle after discharge from the steelmaking unit before separating it into pieces is allowed to settle with in 40-60 min.

The results of experiments of converter slags does not allow to make an unambiguous conclusion about the decrease influence of the delay time of the melt slag on the recovery percentage of magnetic product. The experimental results are processed using the statistical method of analysis "steepest ascent" [8].

The following relationships were obtained to determine the content of FeO for converter slag:

$$\eta_{FeO} = 25,25 - 0,9 \cdot \tau - 0,015 \cdot T + 0,001 \cdot \tau \cdot T,$$

τ – is exposure time, [min]; T – is melt temperature, [°C].

The proposed scheme for processing and use of metallurgical slag in the conditions of the existing production is shown in figure 1.

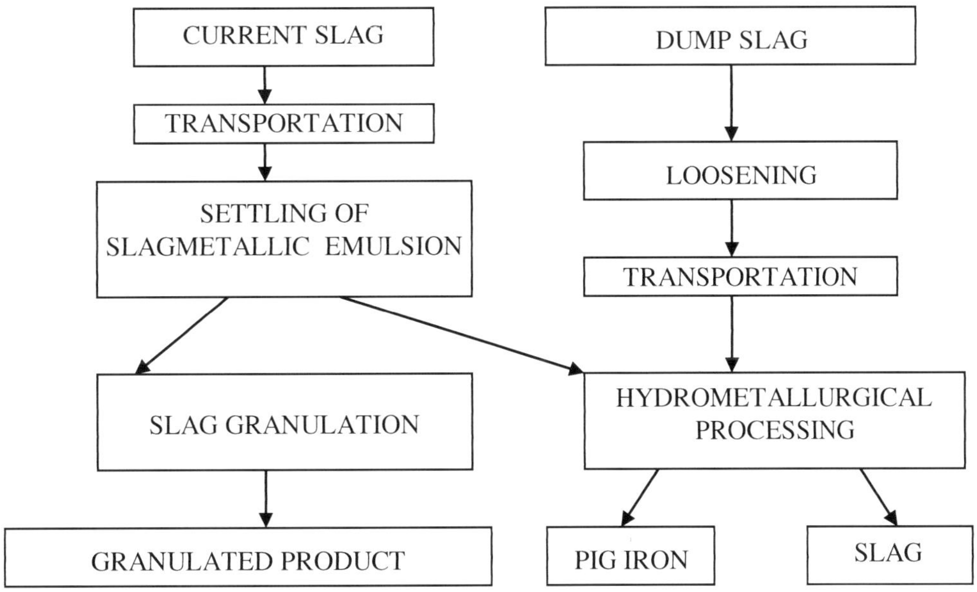

Figure 1. The proposed scheme for processing of slag.

The type of granulation installation has to provide utilization of the water vapor containing large amounts of sulfide compounds. A drum granulation method conforms to this requirement.

Dump slag after loosening by special crawler type machines is routed to the pyrometallurgical processing after which a melt and restored slag is obtained.

Thus, the inclusion of steel making slag in the scheme of partial granulation processing will produce more integrated processing of steelmaking slag, significantly reducing negative impacts on the environment and involvement of more man-made materials in the processing. Furthermore, the crushing stage is abandoned, that simplifies the procedure of steelmaking slag processing and reduces economic cost.

References

[1] Kozlovski I A 2003 *Mining Journal* **10** 4–9

[2] Zorya V N 2015 *Research of Technogenic Waste of Ferrous Metallurgy, Including Waste from Enrichment and Combustion of Coal, and Development of Technologies of their Processing* thesis for a degree of Candidate of Technical Sciences (Novokuznetsk) p 18

[3] Cherchintsev V D et al O B 2010 *Ecology and Industry of Russia* **2** 52–4

[4] Shatokhin I M et al 2016 *Int Conf-School on Chem. Tech.* (Volgograd: VSTU) pp 351–352

[5] Volkova E A, Sviridova T V and Bobrova O B 2015 *Integrated Use of Metallurgical Production Waste in Metallurgy: Technology, Innovation, Quality* ed E V Protopopov (Novokuznetsk) pp 361–364

[6] Drobniy O F, Harlov A A and Proshkina O B 2009 *Mineralogy Tehnogeneza* **10** 228–34

[7] Senik A I, Miliukov S V and Proshkina O B 2008 *Ecology and Industry of Russia* **9** 22–3

[8] Sviridova T V, Bobrova O B and Volkova E A 2016 *Materials Science Forum* **870** 466–70

[9] Akhmedjanova Z I and Bobrova O B 2016 *Actual Problems of Modern Science, Engineering and Education* (Magnitogorsk: Nosov MSTU) pp 228–230

IOP Conf. Series: Materials Science and Engineering **411** (2018) 012071 doi:10.1088/1757-899X/411/1/012071

Development of high-speed rolling modes of trim rails in the continuous reverse train

S V Smetanin[1], V N Peretyatko[2], V V Dorofeev[1], A G Nikitin[2] and E Ya Zhivago[2]

[1]EVRAZ Consolidated West Siberian Metallurgical Plant, 16 Kosmicheskoye shosse, 654043, Russia

[2]Siberian State Industrial University, 42 Kirova street, Novokuznetsk, 654007, Russia

E-mail: Sergey.Smetanin@evraz.com

Abstract. One of the main characteristics that makes it possible to use effectively a rolling mill is high-speed rolling mode. This paper presents the results of theoretical and practical investigations of tram rails rolling technology on the modern universal rail-and-structural steel mill of EVRAZ ZSMK. A new scheme for rolling tram rails in the finishing continuous reverse train was developed. The diagrams of the rolls revolution speed in the stand with the condition of the most effective use of technical characteristics of the main drives are developed. A new approach is offered for high-speed rolling modes at the stages of piece capture, steady-state process and rolled piece release during rolling in the continuous universal train including three stands. Based on the existing equipment of the rolling mill the recommendations are given concerning the frequency of rolls revolution separately for each pass and for each stand during continuous rolling in the finishing train. A rolling method is proposed in which the capture of the tolled metal in all passes is carried out at a constant speed of rolls revolution.

1. Introduction

Tramways around the world have a significant length, therefore both in Russia and abroad the tram rails need to meet rigid requirements. Currently in Russia construction projects of high-speed tram are at various stages of development and implementation, so tram rails of good quality are in high demand. In recent years, in the world practice of rolling production many metallurgical plants have undergone in-depth modernization and have constructed modern rail-and-structural steel mills. The fundamental difference between the modern rail-and-structural steel mill and the mill of the previous generation is that it is equipped with a high degree of mechanization and automation, while the modern rail-mill has a continuous finishing reversing train consisting of universal and two-roll stands. Modern rail-and-structural steel mills have significant advantages in comparison with the mills of the previous generation and make it possible to produce high-quality products.

2. Methods of research

The physical modeling was carried out on the existing rail-and-structural steel mill of "EVRAZ ZSMK" JSC, which is equipped with a powerful automation system and modern control and measuring system that allow different deformation-speed conditions for rolling to be set, and also a large amount of data to be recorded in real time.

One of the most important components of effective use of the rail-and-structural steel mill is the application of rational high-speed capabilities of the main drive.

The main goal of this work was to develop rational high-speed modes of rolling tram rails in the continuous reverse train based on the effective use of the technical characteristics of the modern rail-and-structural steel mill at "EVRAZ ZSMK" JSC.

In the literature there are methods for determining high-speed rolling modes in single-stand mills, as well as reverse stand with continuously connected stands [1, 2]. However, due to the fact that modern rail-and-structural steel mills have significantly better technical characteristics than the mills of previous generation, explanations on the choice of high-speed rolling modes are necessary [3-5].

The aim of the research is to study the features of rolling tram grooved rails at maximum speeds, provided by the main drive of the rolling mill, taking into account the emerging moments and rolling forces in the continuous reverse train in order to increase the productivity of the rolling mill.

The investigation of the speed modes was carried out during the rolling of tram grooved rails, which are one of the most complex profiles. The rail-and-structural steel mill has a continuous finishing reverse train consisting of two universal stands UR and UF, which have drive horizontal rolls and non-driven vertical rolls. Between the universal stands there is a drive stand duo ER. The sequence of the stands location is as follows: UR-ER-UF. The rolling scheme used on the rail-and-structural steel mill for the production of tram rails is as follows (figure 1): the first pass is made in two stands UR1 and ER1; the second pass after reverse also in two stands ER2-UR2 and the third finishing pass in three stands UR3-ER3-UF [6-8]. Thus, in three passes, the workpiece is processed in seven gauges. The characteristics of the rolling mill, the initial scheme for changing the rolls speed in the continuous reverse train and the features of the equipment and technological rolling modes are discussed in detail in [9-10]. In the technology under consideration, after each pass the gauges of the involved stands are tightened, and in the final third pass, in addition, the axial movement of the work stand ER to combine its finishing gauge and the rolling axis.

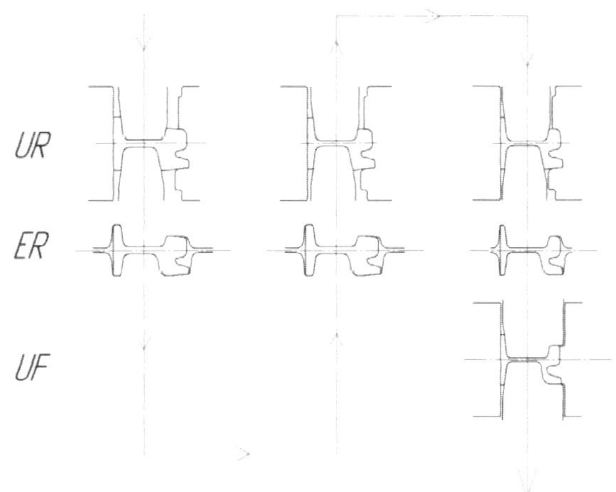

Figure 1. Scheme of tram rails rolling in the continuous train.

Let us consider high-speed rolling modes of tramway grooved rails, which were developed using the experimental-theoretical approach.

The first pass. In the first pass he rail metal is rolled in two working stands – a universal UR and a two-roll, only with horizontal rolls, Edger stand (ER). The speed diagram for the first pass is shown in figure 2. The UF universal stand rolls are parted for the free passage of the workpiece.

When the metal approaches the continuous train, the main engines of the universal stand UR and the auxiliary ER stand are automatically started. The acceleration of the rolls of the UR stand is from 0 to 152 rpm with an acceleration of 74.15 rpm*s. The acceleration of ER rolls is from 0 to 175 rpm with an acceleration of 85.37 rpm*sec. After that the UR and ER stands are automatically converted to roll revolution with a constant speed. The metal capture is carried out at a constant roll speed for the UR stand at 152 rpm, for the ER stand at 175 rpm. The number of revolutions between the stands is different due to the fact that the rolls of stands are of different diameters and it is necessary in each subsequent stand for continuous rolling to take into account the draft from the breakdown in the previous gauge. After the metal is captured, in two stands the UR and ER the metal is rolled for a while at a constant speed. Then simultaneously in two stands there is an acceleration of the revolution of the rolls with the metal. The acceleration is continued to the maximum rolling speed of UR – 297.55 rpm and ER – 338.65 rpm. Next, the metal is rolled at the maximum speed simultaneously in two stands.

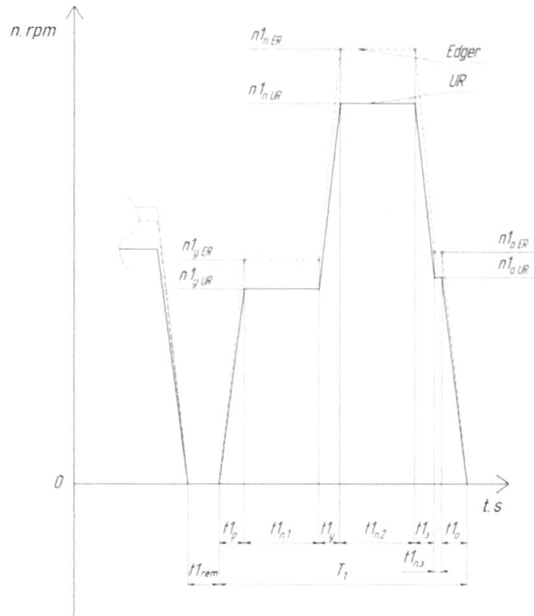

Figure 2. Diagram of speeds of the first pass, where: t_p – time of the stand acceleration without metal; t_{n1} – time of rolls revolution with constant speed; t_y – the acceleration time of the mill with the metal; t_{n2} – the time of metal rolling with a constant maximum speed; t_3 – time of the first stage of braking; t_{n3} – time of rolls revolution with constant speed; t_o – time of the second stage of braking. The number 1 after the letter t means the first passage of metal.

The rolled metal release from UR and ER stand takes place at the maximum speed and for a while the rolls rotate without metal. The braking of the rolls of UR and ER stands without metal is performed in two stages. For UR stand in the first stage of braking, the speed is reduced from 297.55 to 161.49 rpm after which the rolls rotate at a constant speed. At the second stage of braking, the revolution speed of the rolls decreases from 161.49 to 0 rpm. For ER stand in the first stage, braking is carried out and the speed of the rolls is reduced from 338.65 rpm to 181.52 rpm. After that, the rolls of ER stand rotate for a while at a constant speed. At the second stage of braking, the ER rolls revolution of decreases from 181.52 to 0 rpm (table 1).

Table 1. The work of the train "Tandem" in the first pass.

Operation	Stand UR, sec	Stand ER, sec
Acceleration time of rolls without a workpiece	2.05	2.05
Revolution speed of rolls at a constant speed	3.2	3.2
Time acceleration of rolls with a workpiece	1.75	1.75
Time of metal rolling at a constant maximum speed	4.85	4.85
Braking time at the first stage	1.65	1.65
Time of rolls revolution with a constant speed	0.65	0.65
Braking time of the second staged	2.05	2.05

The second pass. After the rolls are stopped and the gauges are tightened, the second pass starts in the reverse mode (figure 3). In the second pass the rolling is carried out in the continuous mode in two stands ER and UR. The rolls of the universal UF stand are parted for a free passage of the workpiece.

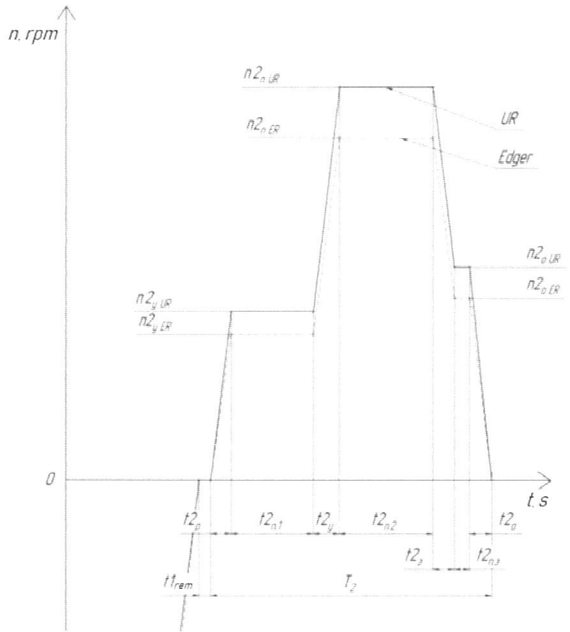

Figure 3. Speed diagram of the second pass.

The workpiece is first directed to the stand ER and then to the stand UR. Simultaneously, in automatic mode the two stands start to accelerate the rolls. In ER stand the rollers are accelerated from 0 to 114.97 rpm with an acceleration of 69.68 rpm*s. The acceleration of the rolls of UR stand is from 0 to 132.82 rpm with an acceleration of 80.5 rpm*s. After that ER and UR stands automatically transit roll revolution to a constant speed. The capture of metal is carried out at a constant revolution speed of the rolls, for ER stand at 114.28 rpm, for the UR stand at 132.82 rpm. After the metal capture in two stands ER and UR, the metal is rolled at a constant speed, after which the revolution of the rolls with the metal is simultaneously accelerated in two stands. Acceleration continues to the maximum rolling speed, which is in ER equal to 269.67 rpm, and in UR stand – 309.92 rpm. Then, for some time metal rolling occurs at the maximum speed simultaneously in two stands.

The rolled metal release from UR and ER stand takes place at the reduced speed. For ER stand after the start of braking of the rolls at 262.5 rpm the metal release from the stand takes place. Further, the rolls of ER stand rotate at a constant speed without metal. After the metal release from ER stand, braking continues at UR stand with metal and then the metal is released from it. Then the rolls of UR stand rotate without metal. The further braking of the rolling rolls in ER and UR stands is carried out in two stages. At the first stage of braking for ER stand, the speed is reduced from 269.67 rpm to 144.11 rpm, while for UR stand, the revolution speed is reduced from 309.92 rpm to 168.22 rpm. After that the rolls rotate for a while at a constant speed. Then the second stage of braking of the rolls takes place in ER stand from 144.11 rpm to 0 rpm and in UR stand from 168.22 rpm to 0 rpm (table 2).

Table 2. The work of the train "Tandem" in the second pass.

Operation	Stand UR, sec	Stand ER, sec
Acceleration time of rolls without a workpiece	1.65	1.65
Revolution speed of rolls at a constant speed	1.0	1.0
Time acceleration of rolls with a workpiece	2.1	2.1
Time of metal rolling at a constant maximum speed	7.55	7.55
Braking time at the first stage	1.75	1.75
Time of rolls revolution with a constant speed	1.25	1.25
Braking time of the second staged	1.8	1.8

The third pass. After the rolls are stopped, for the third pass ER stand moves axially for ER3 gauge setting in the rolling line of the finishing pass. The gauges dimensions of UR stands are set, and in UF the third finishing pass begins, which is the final one. In the third pass the rolling is carried out in a continuous mode in three stands UR-ER-UF (figure 4).

IOP Conf. Series: Materials Science and Engineering **411** (2018) 012071 doi:10.1088/1757-899X/411/1/012071

The rolls in all three stands are automatically accelerated by the main drives of the stands. In all stands acceleration starts simultaneously. In UR stand the acceleration of the rolls is from 0 to 183.44 rpm with acceleration of 73.38 rpm*sec. In ER stand acceleration of the rolls is from 0 to 205.35 rpm with acceleration of 82.14 rpm*sec. In UF stand acceleration of the rolls is from 0 to 216.77 rpm with acceleration 86.71 rpm*s. Then all three stands rotate at a constant speed.

The metal is captured at a constant speed of rolls revolution. In UR stand the workpiece is captured at a constant speed 183.44 rpm*s. In ER stand the workpiece is captured at a roll revolution speed 204.02 rpm and in UF stand at a roll revolution speed 216.77 rpm. Further the rolls of UR-ER-UF stands for a definite period of time roll the metal at a constant speed.

The rolled metal release takes place at a constant speed of rolls revolution in all stands – from UR stand at a constant roll speed 183.44 rpm, from ER at a roll speed 204.02 rpm and UF at a roll revolution speed 216.77 rpm. After release of the metal, the rolls of UR, ER and UF stands rotate for a while at a constant speed. Then, there is simultaneous braking of the rolls in all three stands from 183.44 rpm to 0 for UR; from 204.02 rpm to 0 for ER and from 216.77 rpm to 0 for UF (table 3).

After the rolls are stopped in all three stands, the gauge in UR1 stand is set, the ER stand is shifted to the initial position of ER1 gauge, and UF stand rolls are parted, after which the continuous reverse train is ready for rolling the next workpiece.

Table 3. The work of the train "Tandem" in the third pass.

Operation	Stand UR, sec	Stand ER, sec	Stand UF, sec
Acceleration time of rolls without a workpiece	2.5	2.5	2.5
Time before the metal is captured by rolls at a constant speed	1.6	3.3	4.95
Time of metal rolling at a constant maximum speed	18.44	18.44	18.44
Time of rolls revolution with a constant speed after workpiece release	5.95	4.25	2.6
Time of deceleration of the mill without a workpiece	2.5	2.5	2.5

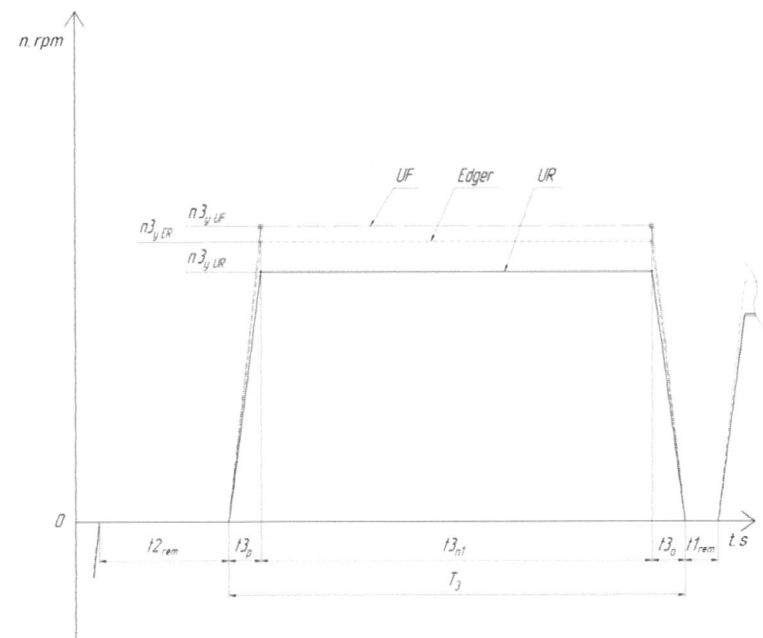

Figure 4. Speed diagram of the third pass.

3. Conclusions

1. In all passes and in all stands, the metal capture is carried out at a constant speed of rolls revolution.

2. When rolling the workpiece in the first and second passes, the speed diagrams have two stages with a constant speed of rolls revolution – reduced and maximum operating speed n_n. At a reduced constant speed (0.42-0.51) n_n, a steady metal capture by the rolls of the first stand occurs, then the workpiece passes interstand distance and the workpiece is captures by the rolls of the next stand. Then there is a simultaneous acceleration of rolls with metal in two stands to the maximum working speed n_n, and metal is rolled at the maximum working speed $n_n = n_{max}$.

3. Duting the third pass the metal capture, rolling and metal release are carried out at a constant working speed equal to n_n=(0.6-0.7)n_{max}, where n_{max} is the maximum number of revolutions of the mill main drive.

4. In connection with the specific features of equipment, the metal release in the first pass is carried out at the maximum constant speed, and the release in the second pass from ER2 gauge at a constant high speed and from UR2 gauge during the rolls braking.

5. The rolls braking in UR and ER stands after the first and second passes without metal is performed in two stages. In the first pass: for UR stand at the first stage of braking there is a reduction in speed from 297.55 to 161.49 rpm, at the second stage of braking the revolution speed of the rolls decreases from 161.49 to 0 rpm; for ER stand in the first stage of braking the number of rolls revolutions is reduced from 338.65 rpm to 181.52 rpm.

6. After that the rolls of ER stand rotate at a constant speed. At the second stage of braking, the rolls revolution in ER decreases from 181.52 to 0 rpm. In the second pass: at the first stage of braking for ER stand the speed is reduced from 269.67 rpm to 144.11 rpm, and for UR stand the revolution speed is reduced from 309.92 rpm to 168.22 rpm.

7. Then the rolls rotate at a constant speed. The second stage of braking of the rolls takes place in ER stand from 144.11 rpm to 0 rpm and in UR stand from 168.22 rpm to 0 rpm.

8. The introduction of these results allowed the productivity of the rolling mill to be increased by 10%.

Acknowledgements

The authors are grateful to administration of EVRAZ ZSMK, namely, to Managing Director, Doctor of Technical Sciences A.B. Yuriev; Director of Rail Production, Ph.D. A.V. Golovatenko, Chief Calibrator, Doctor of Technical Sciences V.V. Dorofeev for their cooperation.

References

[1] Tyagunov V A 1954 *Rolling Modes on Reversible Mills* (Sverdlovsk, Moscow: Metallurgizdat) p 136
[2] Shilov V A et al 2003 *Rolling Production* **7** 30–33
[3] Polyakov V V and Artamonova E A 1989 *Development of Rails Production Abroad* (Moscow: TsNIITEIChM) p 29
[4] Matveev B N 2006 *Ferrous Metallurgy. Bul. Inst. "Chermetinformatsi"* **2** 40–43
[5] Smirnov V K et al 2010 *Calibration of Forming Rolls* (Moscow: Teplotehnik) p 490
[6] Peretyatko V N et al 2008 *Izv. Vuzov. Ferrous Metallurgy* **10** 68–69
[7] Peretyatko V N et al 2015 *Rolling Production* **7** 20–27
[8] Peretyatko V N and Smetanin S V 2015 *Production of Rolled Metal* **12** 29–37
[9] Golovatenko A V et al 2014 *Production of Rolled Products* **2** 25–39
[10] Golovatenko A V et al 2014 *Bulletin "Ferrous Metallurgy"* **6** 32–38

IOP Conf. Series: Materials Science and Engineering **411** (2018) 012072 doi:10.1088/1757-899X/411/1/012072

Information modelling system for diagnostics of different types of blast-furnace smelting deviations from normal conditions

N A Spirin, O P Onorin, A S Istomin, V V Lavrov and I A Gurin

Ural Federal University n.a. the first President of Russia B.N. Yeltsin, 51 Lenina ave, Ekaterinburg, 620075, Russia

E-mail: n.a.spirin@urfu.ru

Abstract. A logic basis for assessment of normal progress of blast-furnace smelting and identification of different types of deviations from normal smelting conditions were developed, including peripheral gas stream, central (axial) gas stream, hot and cold run, tight furnace operation, upper and lower suspension of burden in the blast furnace. Both the complex of controlled characteristics and design parameters of the model-based decision support system (URFU-MMK model of blast-furnace process) are used for assessment. The main complex design parameters for diagnostics of blast-furnace smelting progress characterize heat, blast, gas dynamic and slag conditions, blast-furnace smelting rate in terms of quantity. Identification of Blast-Furnace Smelting Deviation Type software was developed in accordance with modern principles of application program development (functionality, expandability, database integration, user-friendly interface, security, data estimation). It is efficient to use the developed software in computer decision support systems for blast-furnace smelting control in real-time mode.

1. Introduction

High technical and economic performance of the blast furnace can be achieved only in a stable process of blast-furnace smelting without significant deviations from normal conditions of blast furnace operation.

Characteristics of normal operation of each furnace are, as a rule, well known and described in appropriate process instructions of the blast furnace plant. These instructions also describe possible deviations from normal conditions, characteristics of these deviations and measures to be taken to recover the normal conditions of blast-furnace smelting.

In conditions of quick-changing world and overall automation the problem of blast-furnace smelting diagnostics and control system implementation has not been solved yet.

Application of the automated information logic system for assessment of blast-furnace smelting progress and identification of deviations from normal conditions can be a solution to the problem of blast-furnace smelting state diagnostics. The logic basis for identification of deviations from normal conditions of blast-furnace smelting is described in papers [1-8].

2. Probability estimate of normal smelting operation

The calculation method is as follows. A set of characteristics is considered for two periods of furnace operation, i.e. base period and estimated period. The base period represents a set of assigned (reference) values characterizing the optimum conditions of furnace operation. The estimated period is a set of parameters for the last period between iron and slag tappings.

At normal operating conditions the deviation for the module of i-th characteristic ΔX_i describing furnace operation in the base $X_i^{Б}$ (assigned values) and estimated periods $X_i^{П}$ shall not exceed the allowable value $\Delta X_i^{доп}$, which is a model setting.

$$\Delta X_i = \left| X_i^{Б} - X_i^{П} \right| \leq \Delta X_i^{доп}. \quad (1)$$

If the condition (1) is fulfilled ("True"), the i-th identifier of P_i characteristic is assigned the value "1", otherwise ("False") – the value "0". All the characteristics herewith are being ranked. Each of them is assigned the value of its rank R_i varying in the range from 0 to 1, which is determined by the expert assessment method.

Probability of normal blast furnace operation ($B_н$) is calculated by the following proportion:

$$B_н = \sum_{i=1}^{n} \left(P_i \times \frac{R_i}{\sum_{i=1}^{n} R_i} \right) = \sum_{i=1}^{n} \left(P_i \times \alpha_i \right), \quad (2)$$

where α_i is a weight coefficient of i-th characteristic identifier varying in the range from 0 to 1; n is a number of characteristics.

The number of controlled characteristics used in the model for identification of normal conditions of blast furnace operation is 12.

Unlike well-known works of other authors, this study integrates sets of controlled characteristics and design parameters of blast-furnace smelting. The mathematical and program software uses additionally a set of nine main design parameters used in the model of blast-furnace process (URFU-MMK model of blast-furnace process) [9-16] and adapted with respect to conditions of Magnitogorsk Iron and Steel Works (MMK). Thus, the number of characteristics reaches 21.

The main complex design parameters for diagnostics of blast-furnace smelting operation refer to the following conditions:

- heat conditions – generalized parameters describing the thermal state: in the upper (thermal state index in the furnace stack) and lower (thermal state index in the bottom) parts of the blast furnace;
- gas dynamic conditions – degree of gas utilization for burden balancing in the upper and lower parts of the furnace, in particular ring areas, etc. Prediction and calculation of critical gas dynamic parameters of blast-furnace smelting;
- slag conditions – viscosity and viscosity polytherms of final slag: slag viscosity at the designed temperature, slag viscosity gradient within the range from 2.5 to 0.7 Pa*s, Pa*s/ °C, slag viscosity gradient within the range from 1400 to 1500°C, Pa*s/°C;
- smelting rate (volume of melted burden per time unit, m3/min).

3. Diagnostics of different types of blast-furnace smelting deviations from normal conditions

In case of deviations from normal furnace operation, the main types of deviations are analysed.

1. Loss of gas stream stability (peripheral, central and channel gas streams).
2. Disturbance of heat conditions of blast-furnace smelting (hot and cold run, furnace cooling and heating).

3. Failure of smooth burden descent in the furnace (upper and lower suspension of burden, tight operation).

To identify the type of smelting deviation from normal conditions of blast-furnace operation, we found it reasonable to compare parameters of two periods. The first period is the base period whose smelting parameter values $X_i^Б$ are a model setting and describe the normal operating conditions. The second period for identification of deviations from normal conditions of blast furnace operation is the estimated period where data are collected during the last period between tappings.

If there is a deviation from normal conditions of blast-furnace smelting, the difference between the values of *i*-th characteristic ΔX_i describing furnace operation in the base $X_i^Б$ and estimated $X_i^П$ periods exceeds the allowable value $\Delta X_i^{доп}$.

$$\Delta X_i = \left(X_i^Б - X_i^П\right) \geq \Delta X_i^{доп}. \quad (3)$$

Identification probability of deviations from normal conditions, value $B_н^{отк}$ is calculated by the proportion similar to equation (2).

Diagnostics of deviations from normal conditions of blast-furnace smelting, number of controlled and design (by model) parameters for assessment of these conditions are given in table 1.

Table 1. Number of controlled and design (by model) parameters for assessment of deviations from normal conditions of blast-furnace smelting.

Type of deviation from normal conditions	Number of controlled parameters	Number of complex design parameters	Total
Loss of gas stream stability:			
– peripheral gas stream;	9	3	12
– central gas stream;	8	2	10
Disturbance of heat conditions of blast-furnace smelting:			
– hot run of smelting;	8	4	12
– cold run of smelting;	7	5	12
Failure of smooth burden descent in the furnace:			
– upper suspension of burden;	6	3	9
– lower suspension of burden;	3	3	6
– tight furnace operation;	6	6	12
Total:	47	26	73

4. Software implementation and example use of information logic system

Following development of the information logic system, it is logical to implement this system as a computer software enabling technical personnel to get on-line data about blast-furnace smelting operation and react appropriately in case of any deviations.

Software implementation is made with the use of functional modelling, .NET technology in C# language in Microsoft Visual Studio 2015. Due to convenience and huge functionality, this set of tools has recently become a de facto standard for development of desktop applications [17-19].

Table 2 shows a fragment of one variant for assessment of normal operation of Blast Furnace No.10 of MMK.

The probability value of normal progress of blast-furnace smelting equal to 0.63 implies normal conditions of smelting at the blast furnace in question.

The modelling results given in table 3 show that probability of deviations of any type from normal conditions does not exceed threshold limit values taken equal to 0.9.

IOP Conf. Series: Materials Science and Engineering **411** (2018) 012072 doi:10.1088/1757-899X/411/1/012072

Table 2. Probability calculation of normal progress of blast-furnace smelting.

No.	Characteristic, unit of measurement	Base period	Estimated period	Parameter deviation	Allowable deviation	Rank of characteristic, R_i	α_i	P_i
P_1	Blast low rate, m3/min	3422	4019	597	603	0.9	0.06	1
P_2	Blast temperature, °C	1147	1154	7	50	0.5	0.03	1
P_3	Total pressure drop, kPa	131	140	9	5	0.7	0.04	0
P_4	Lower pressure drop, kPa	92	98	6	3.5	0.8	0.05	0
P_5	Upper pressure drop, kPa	39	42	3	1.5	0.7	0.04	0
P_6	Circumferential distortion of gas temperature, °C	193	332	139	100	0.6	0.04	0
P_7	Non-uniform distribution of top gas temperature in gas pipes, °C	95	95	0	100	0.8	0.05	1
P_8	Deviation of averaged top gas temperature, °C	184	161	-23	100	0.8	0.05	1
P_9	Si content in hot metal, %	0.72	0.948	0.228	0.1	0.9	0.06	0
P_{10}	CO_2 content in top gas, %	20.26	19.21	-1.05	1	0.8	0.05	0
P_{11}	Hot metal temperature, °C	1416	1475	59	50	0.8	0.05	0
P_{12}	Slag basicity (CaO+MgO)/SiO_2, ratio	1.221	1.214	-0.007	0.05	0.9	0.06	1
P_{13}	Thermal state index at the bottom, ratio	1	1.1	0.1	0.05	0.9	0.06	0
P_{14}	Thermal state index at the furnace top, ratio	0.7	0.7	0	0.05	0.6	0.04	1
P_{15}	Degree of gas utilization for burden balancing, ratio	0.55	0.58	0.03	0.1	0.9	0.06	1
P_{16}	Degree of gas utilization for burden balancing in the upper part of the furnace, ratio	0.3	0.3	0	0.05	0.9	0.06	1
P_{17}	Degree of gas utilization for burden balancing in the lower part of the furnace, ratio	0.65	0.7	0.05	0.1	0.9	0.06	1
P_{18}	Viscosity of final slag at temperature 1500 °C, Pa*s	0.35	0.3	0.05	(0.05)	0.7	0.04	1
P_{19}	Slag viscosity gradient within the range (2.5-0.7 Pa*s), Pa*s/°C	0.0175	0.0185	0.001	0.005	0.7	0.04	1
P_{20}	Slag viscosity gradient within the range from 1400°C to1500°C, Pa*s/°C	0.005	0.0055	0.0005	0.001	0.8	0.05	1
P_{21}	Volume of melted burden per time unit, m3/min	6.5	6.75	0.25	0.5	0.8	0.05	1
	Probability of normal furnace operation, ratio							0.63

Table 3. Calculation results of blast-furnace smelting deviation probability in the estimated period.

No.	Deviation type	Deviation probability, ratio
1	Peripheral stream	0.169
2	Axial stream	0.208
3	Hot run	0.191
4	Cold run	0.172
5	Tight operation	0.265
6	Upper suspension of burden	0.097
7	Lower suspension of burden	0.000

5. Conclusion

The mathematical model of state diagnostics and blast-furnace smelting prediction was improved due to development of the logic model block using a set of controlled and design parameters.

A software was developed which enables to asses progress of blast-furnace smelting in real-time mode as well as types of deviations from normal conditions when necessary.

References

[1] Spirin N A et al 2011 *Model-Based Decision Support Systems in ICS of Blast-Furnace Smelting* (Ekaterinburg: UrFU) p 426
[2] Spirin N A et al 2014 *Mathematical Modelling of Metallurgical Processes in ICS* (Ekaterinburg: OOO UIPTs) p 558
[3] Lida O et al 1992 *Kawasaki Steel Techn Dept* **26** 30–37
[4] Ueda S et al 2010 *ISIJ International* vol 50 **7** 914–923
[5] Vapaavuori E 1997 *Expert Systems with Applications* vol 12 **3** II
[6] Frenkel M M et al 1992 *Steel* **7** 15–18
[7] Onorin O P et al 2014 *Izv. Vuzov. Ferrous Metallutgy* **8** 42–47
[8] Yasuo Omori 1987 *Blast Furnace Phenomena and Modelling* (London and New York: Elsevier Applied Science) p 631
[9] Spirin N A 2016 *Metallurgist* vol 60 **5-6** 471–477
[10] Rybolovlev V Y et al 2015 *Metallurgist* vol 59 **7** 653–658
[11] Spirin N A 2011 *Metallurgist* vol 54 **9-10** 566–569
[12] Lavrov V V and Spirin N A 2016 *IOP Conf. Series: Materials Sci. and Eng.* **150** 012010
[13] Spirin N A 2016 *IOP Conf. Series: Materials Sci. and Eng.* **150** 012011.
[14] Spirin N A et al 2015 *AISTech 2015 Iron and Steel Technology Conference and 7th International Conference on the Science and Technology of Ironmaking, ICSTI 2015* (USA: Cleveland) vol 1 article No. 113707 pp 1225–32
[15] Onorin O P et al 2005 *Computer Methods of Blast-furnace Process Modelling* (Ekaterinburg: UGTU-UPI) p 301
[16] Onorin O P et al 2017 *AISTech 2017 Iron and Steel Technology Conference Proceedings* (USA: Nashville) vol 1 article No. 128474 771–778
[17] Dubeykovsky V I 2009 *Efficient Modelling with CA ERwin Process Modeler* (BPwin; AllFusion Process Modeler) (M.: Dialog-MIFI) p 384
[18] Troelsen A and Japikse P 2016 *C# 6.0 and the .NET 4.6 Framework* 6th ed in tr (M.: OOO I.D. Williams) p 1440
[19] Richter D 2017 *CLR via C#. Full Coverage of Multicore Programming* 4th ed (SPb.: Piter) p 896

Study of transition processes of blast-furnace smelting by the mathematical model method

N A Spirin, O P Onorin, A S Istomin, V V Lavrov and I A Gurin

Ural Federal University n.a. the first President of Russia B.N. Yeltsin, 51 Lenina ave, Ekaterinburg, 620075, Russia

E-mail: n.a.spirin@urfu.ru

Abstract. In work possibilities of the developed computer model system for forecasting of a thermal condition of a blast furnace (on change of the maintenance of silicon in pig-iron) in a mode of real time are considered. The basis of the model is the fundamental knowledge of the theory and practice of the modern blast-furnace process, the use of regularities in the processes of heat and mass transfer, gas dynamics, slag formation processes in modern blast furnace melting. When solving the dynamic problem, it is envisaged to use the analytical (obtained on the basis of fundamental knowledge), but the linearized model of the domain process, the principle of small deviations and the natural-mathematical approach. In the algorithm for solving problems in transient blast furnace processes, the calculation of the dynamics of the change in the heating of the bottom of a blast furnace (based on the silicon content in cast iron) was performed according to the additivity rule by summing the predicted silicon concentrations in cast iron when the loading parameters (ore load) and the control actions on the thermal load the state of the furnace from the bottom - the consumption of natural gas, the concentration of oxygen in the blast and the humidity of the blast. Examples of calculations of the transient processes of the gas-dynamic resistance of the batch layer, the thermal state, and the productivity of the blast furnace are shown in the context of the changes in the properties and composition of the iron-ore materials being loaded and the parameters of the combined blast are applied to the conditions of PJSC "Magnitogorsk Iron and Steel Works" (MMK).

1. Introduction

The paper investigates transition processes of blast-furnace smelting using analytical and theoretical models describing regularities of heat and mass transfer with consideration of kinetics of processes proceeding in the blast furnace; by methods of passive and active experiments, according to the data of blast furnace operation with the use of statistical methods as well as based on results of specially setup experiments or by selecting significant periods of normal blast furnace operation containing input signals of the required type. Each of these directions has obvious advantages and disadvantages widely described in scientific literature [1–17].

Analytical models based on physics of the processes enables to theoretically evaluate the nature of dynamic characteristics of the processes in any conditions of blast-furnace smelting without interferences [1, 10–13]. However, complexity of blast-furnace smelting processes, problematical character of the complete description of the processes, complexity of the mathematical solution of the differential equation system does not allow the developed models to fully consider characteristics of the processes. Data on actually used dynamic analytical models of this class over the last years are not virtually available.

It should be noted that the common regularity summarizing the results of mathematical modelling of unsteady heat transfer processes in the layer consists in an aperiodic (non-oscillatory) response of transition processes [14, 15]. However, this provision is only valid if the ratio between thermal capacities of material and gas flows along the height of the concerned area of the layer is fixed, which does not reflect development of heat transfer processes along the height of the blast furnace. Due to the above, the results of modelling of unsteady heat transfer processes in the layer considering thermal effects of reduction processes give a quantitative estimation of transition process duration and a general idea about the nature of transition processes only in particular areas of the layer but not for the blast furnace as a whole.

2. Description of dynamic model for blast-furnace process

The dynamic model intended for prediction of blast-furnace smelting progress in a real-time mode is based on fundamental theoretical and practical knowledge of the modern blast-furnace process, use of regularities of heat and mass transfer processes, gas dynamics, slag generation processes in modern blast-furnace smelting. In addition, mathematical, algorithmic and program software developed earlier by the authors of the paper for blast-furnace smelting control is widely used [16-20].

For solution of the dynamic problem, an analytical (based on fundamental knowledge) and linearized model of the blast-furnace process, principle of small deviations and natural mathematical approach shall be used [21]. In this regard it is necessary to assess preliminarily the range of input parameter variation against the underlying value within which the coefficients of the dynamic model can be taken as constant [16].

Dynamic modelling of the blast-furnace process was performed in a specific sequence. Using the averaged data of blast furnace operation in the basic period (10 hours), the following parameters are calculated:

- blast-furnace burden, including the specific rate of coke and iron-ore materials for both basic and predicative options [22]. The specific output of slag and slag composition as well as flux flow rate at a given slag basicity and slag properties are determined;
- set of parameters characterizing gas dynamic conditions of blast-furnace smelting [16, 17];
- arrangement and configuration of the softening and melting zone of iron-ore materials [18];
- coefficients of transfer along different exposure pathways for performance of factor analysis using URFU's algorithm; application areas of linearized models are assessed using methods presented in papers [16, 17];
- time from the start of burden charging to the moment when burden reaches the tuyeres (time of one burden cycle) [17, 22, 23];
- transition processes of gas dynamic resistance in the upper and lower zones of the burden layer when the composition and properties of iron ore burden are changed;
- transition processes of blast-furnace smelting for different exposures: change of ore load, natural gas flow rate, blast humidity and temperature, oxygen content in blast as well as blast-furnace slag basicity;
- in accordance with variation of basic parameters in time in the design period, prediction of gas dynamic and heat conditions of blast-furnace smelting in real time is made for duration of the design period equal to 8 hours.

Considering the hypothesis of linearity of the object in question, the reaction of the system to any sum *n* of input exposures can be calculated as per following equation:

$$\Delta[\mathrm{Si}](t) = \sum_{j=1}^{n} \int_{0}^{t} h_{\mathrm{Si}}^{j}(t-\tau) \frac{dX_{\mathrm{BX}}^{j}(\tau)}{d\tau} d\tau ,$$

where $\Delta[\mathrm{Si}](t)$ is a variation of increment in [Si] with time; $h^{j}_{\mathrm{si}}(\tau)$ is a transition function of the system for *j*-th exposure; $X^{j}_{\mathrm{BX}}(\tau)$ is a variation of the increment time of *j*-th input exposure.

The delay time when properties and flow rate of burden materials are changed is taken equal to the time of one burden cycle. To determine the type of the transition process (aperiodic, oscillatory, readjustment value), the calculation results of transition processes as per mathematical model presented in the papers were used [14, 15].

3. Calculation results of transition processes using dynamic model

Calculation examples of transition processes of gas dynamic resistance in the burden layer, thermal state and blast furnace performance with changes in properties and composition of charged iron ore materials and parameters of combined blast are made in relation to conditions at Magnitogorsk Iron and Steel Works (MMK).

Dynamics of the lower gas pressure drop along the height of the burden layer with a bigger portion of pellets in the iron ore part of the burden is shown in figure 1, a. As the temperature range of melting for nonfluxed pellets is wider than for fluxed agglomerate, an increase in the portion of iron ore pellets in the burden of the blast furnace will be accompanied by growth of gas dynamic resistance in the cohesive zone. Addition of agglomerate with a bigger content of fines will be accompanied by growth in the upper gas pressure drop along the height of the layer (figure 1, b). The growth in the upper gas pressure drop along with the growth in the total pressure drop will be observed immediately after charging burden with a bigger portion of fines in agglomerate and will continue up to arrival of this burden in the slag formation zone, i.e. in a melting time of one furnace volume.

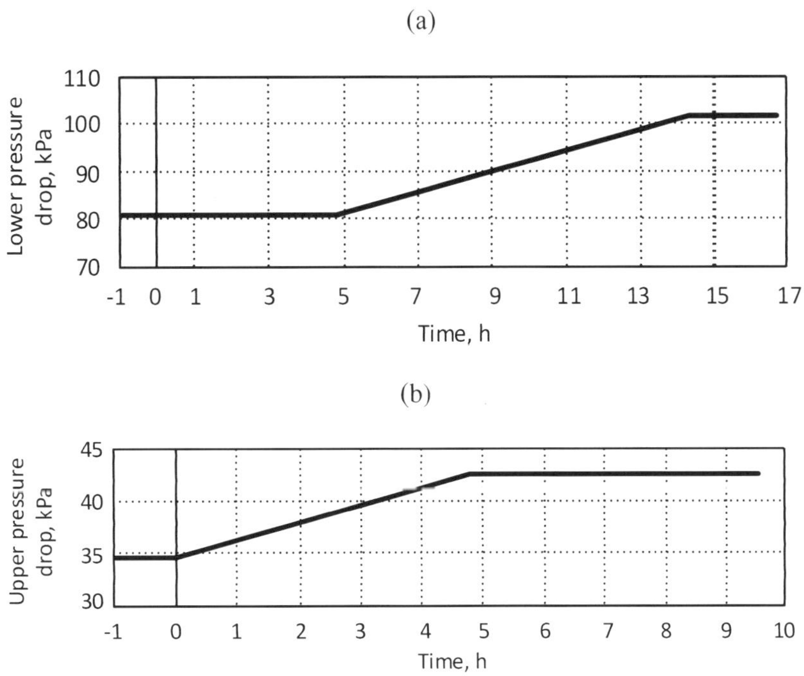

Figure 1. Change in gas pressure drop along the height of the burden layer:
a – when the portion of pellets in the iron ore part of burden is increased from 0.34 to 0.54 t/t; – when the portion of pellets in the iron ore part of burden is increased from 0.34 to 0.54 t/t;
b – when the content of fines in agglomerate (0-5 mm fraction size) is increased from 6.1% to 10.8%.

When ore load is changed, the start of a change in furnace performance over time corresponds to changing of one burden volume in the furnace (figure 2, *a*); the transition period time is approximately equal to the same period. A change in intensity of melting products heating characterized by silicon content in hot metal starts in a period of time which lasts one burden cycle in the furnace and the total time of the transition process up to a new value of silicon content is 3 to 3.2 burden cycles.

A change in the natural gas flow rate and oxygen content in blast causes an alternating nature of Si content change in hot metal (figure 3). This is stipulated by almost instantaneous impact of these parameters on changing the thermal level in the furnace hearth and by inertia influence on reducing and heating processes. At the initial time, when the natural gas flow rate decreases due to an increased temperature in the hearth, which is related to reduction in heat input for natural gas decomposition, Si content in hot metal hot metal increases. After a certain time period equal to the time of one burden cycle in the furnace, we observe a decrease in heating of the lower zone due to increased direct reduction of iron oxides. At the initial time, after the oxygen content in blast increases, due to a temperature growth in the furnace hearth during one burden cycle, Si content in hot metal increases. However, further on we observe a decrease in heating of the lower furnace zone and decrease in Si content due to a temperature drop in the furnace stack and increased direct reduction of iron oxides. Thus, an alternating nature of Si content change in hot metal is formed for these actions.

When slag basicity (CaO/SiO_2) is increased, we observe a decreased Si concentration in hot metal. This is related to the fact that addition of the basic oxide (CaO) to slag is accompanied by binding of silicon oxide in slag into a strong entity, i.e. calcium silicate ($CaO{\cdot}SiO_2$), where silicon reduction requires much more efforts in comparison with silicon reduction from "pure" silica. When calculating a dynamic pattern of Si content change in hot metal with introduction of disturbance due to a change in slag basicity (figure 4), we took into account that Si content change in hot metal starts in a time period equal to the time of one burden cycle in the furnace and the time of the transition period is also equal to the time of one burden cycle.

(a)

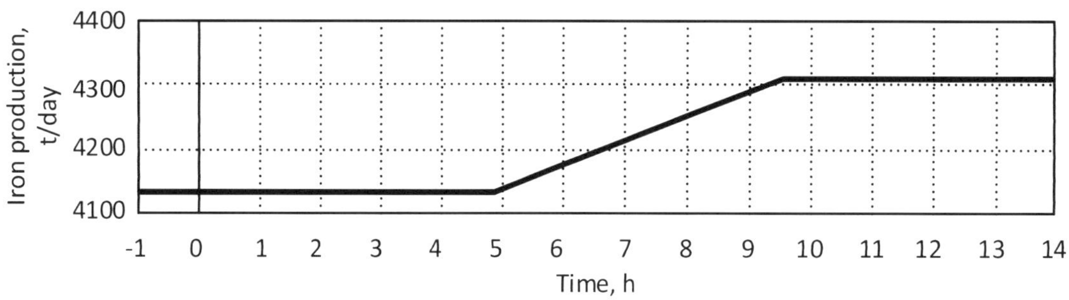

(b)

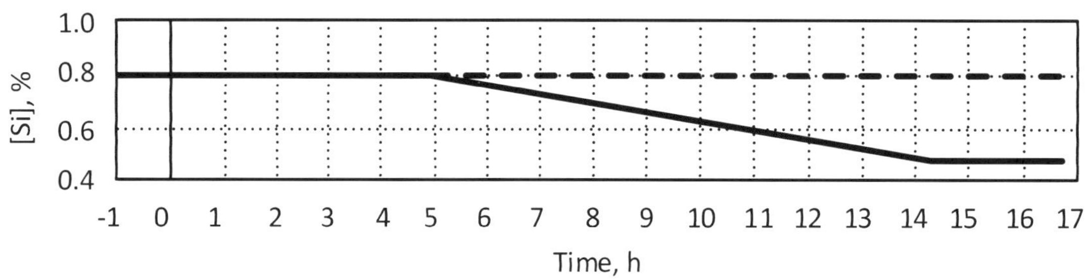

Figure 2. Change in furnace performance *(a)*, silicon content in hot metal *(b)* when ore load is increased from 3.76 to 3.86 t/t.

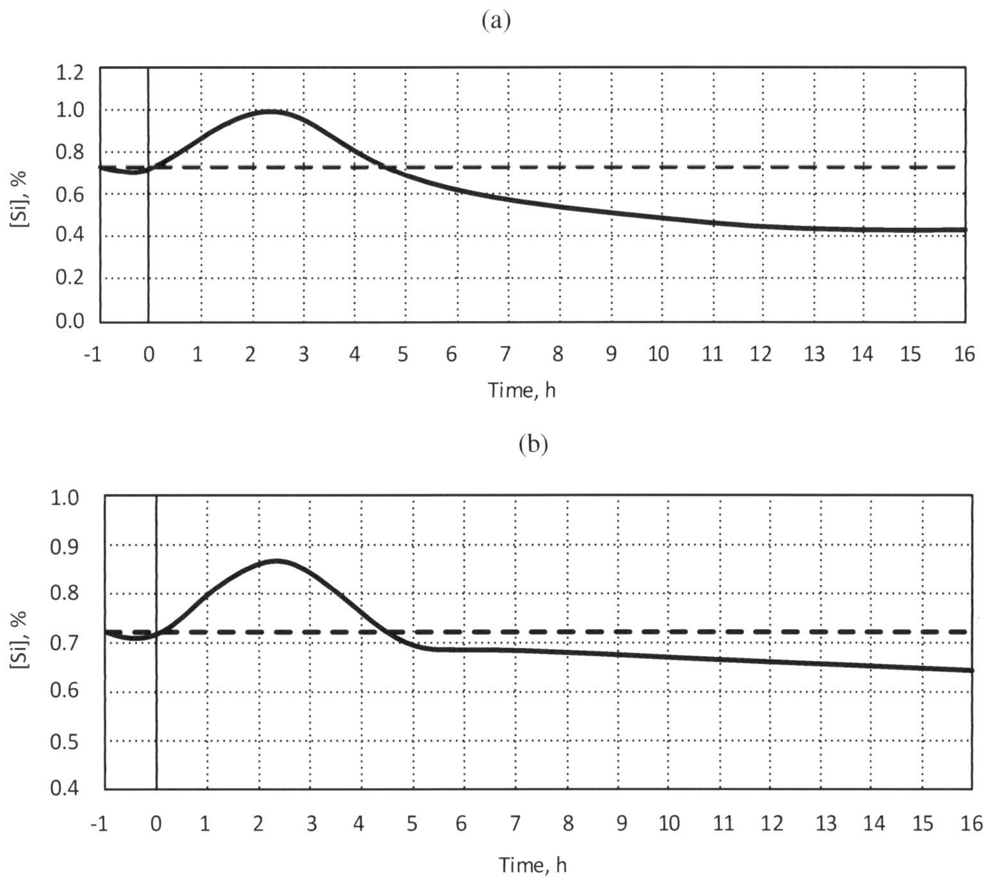

Figure 3. Si content change in hot metal: *a* – when the natural gas flow rate is reduced by 10 m^3/t of hot metal; *b* – when the oxygen concentration in blast is increased from 27.5% to 28.5%.

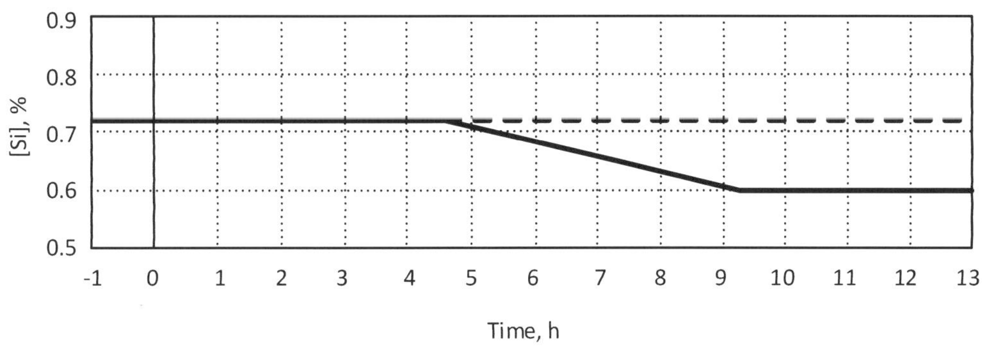

Figure 4. Dynamic pattern of Si content in hot metal when slag basicity is increased from $CaO/SiO_2 = 0.98$ to $CaO/SiO_2 = 1.04$.

With a change in blast temperature, we observe a change in heating of the furnace bottom and, as a result, a change of Si content in hot metal (figure 5). With introduction of disturbance due to a change in the blast temperature, the thermal condition of the lower furnace zone changes immediately after the change in this parameter. The time of the transition period is equal to the time of one burden cycle.

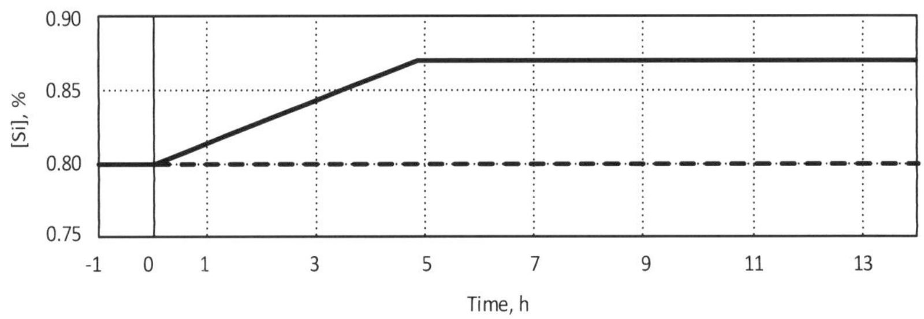

Figure 5. Dynamic pattern of Si content in hot metal when the blast temperature is increased from 1110 to 1120 °C.

To test the model for adequacy, we made calculations with respect to operational conditions of Blast Furnace No.10 of MMK throughout the duration of 17 iron tappings (26 hours of continuous furnace operation). We selected two periods of furnace operation – at the start (first period) Si content in hot metal was continuously growing [Si] = 0.78 % to [Si] = 1.402 %; then, in a result of decisions taken, Si content in hot metal was monotonously going down from [Si] = 1.402 % to [Si] = 0.574 %. The specific feature of these periods is that the chemical composition of iron ore burden components and coke properties virtually remained unchanged during this period. At the same time, corrections were made in ore load, natural gas flow rate, oxygen content in blast and hot blast temperature.

Figure 6 shows dynamic patterns of main input parameters of ore load, natural gas flow rate, oxygen content in blast, hot blast temperature at a simultaneous change of input parameters and figure 7 shows a comparison between the actual and design change of Si content in hot metal for the cases under study. The comparison between the actual and design change of Si content in hot metal demonstrates a quite satisfactory concurrence of predicted and actual values of Si content in hot metal.

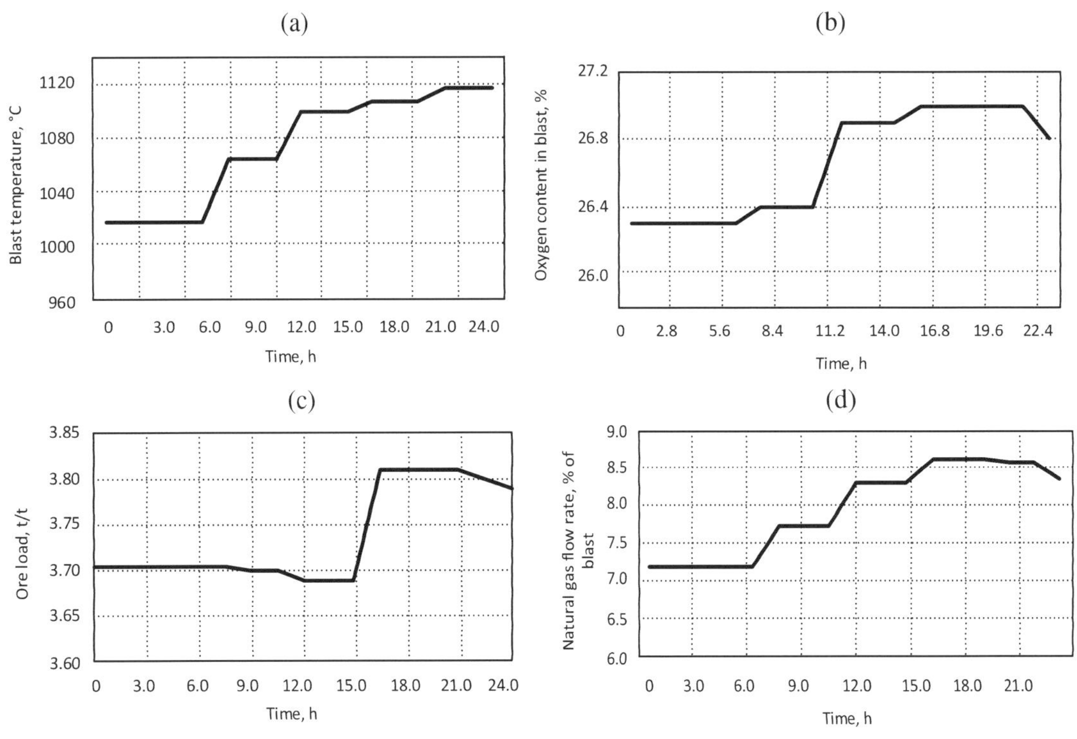

Figure 6. Dynamic patterns of main input parameters during blast-furnace smelting:
a – blast temperature, °C; *b* – oxygen content, %; *c* – ore load, t/t; *d* – natural gas flow rate, % of blast.

Thus, the suggested algorithm for solving a problem concerning study of dynamic patterns of the thermal state during smelting (by Si content in hot metal) describes the dynamics of the process state quite well with actions aimed both at furnace heating increase and furnace heating decrease.

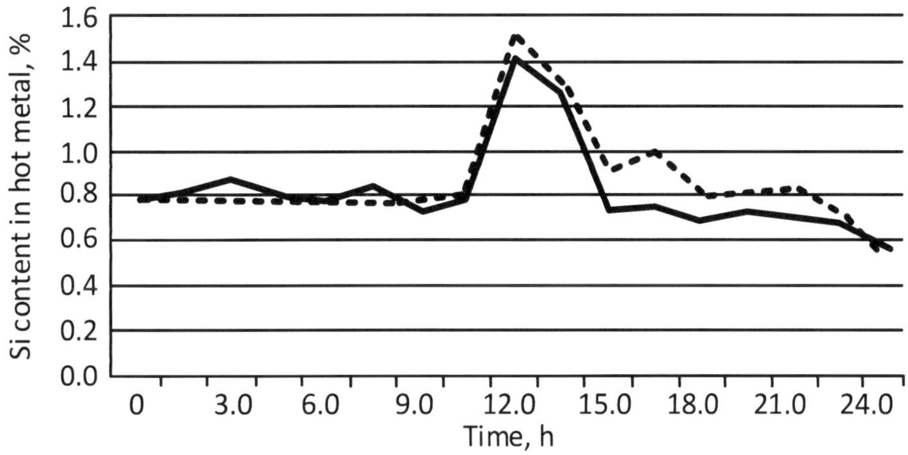

Figure 7. Comparison between the actual change (continuous line) of Si content in hot metal (%) and design (dashed line) values with introduction into the smelting process of parameters causing an increase and decrease of Si content in hot metal.

4. Conclusions

1. The oscillatory transition process is observed in the blast furnace provided that after introduction of a disturbance into the process this disturbance will have an opposite influence on the thermal state of the lower and upper heat-exchange stages. The more the difference is in size and sign, the more the value of readjustment will be in this case.
2. The most predictable parameters influencing Si content variations in hot metal are changes in ore load as well as change in slag basicity. It is certain that Si concentration in hot metal is also determined by other factors, i.e. blast temperature and humidity. However, use of these parameters for immediate influence on heat conditions of blast-furnace smelting is limited as the blast temperature is kept at a maximum level and the change in humidity is used as a parameter to control the heat conditions of blast-furnace smelting when there are deviations from normal heat conditions.
3. It is not desirable to use a change of oxygen concentration in blast and natural gas flow rate as parameters for control of Si content in hot metal. This is related to an alternating influence of these parameters on heat conditions of blast-furnace smelting. An unreasoned decision to use the parameters of natural gas flow rate and oxygen concentration in blast for control of Si content in hot metal can lead to the opposite results.

References

[1] Tovarovsky I G 2009 *Blast-furnace Smelting* (Dnepropetrovsk: Porogi) p 768
[2] Ulakhovich V A et al 1975 *Steel* **1** 9–14
[3] Mkrtchan L S et al 1975 *Steel* **11** 981–986
[4] Bulter D and Mattheis K 1970 *Ferrous Metals* **22** 3–11
[5] Mattheis K et al 1974 *Ferrous Metals* **22** 31–38
[6] Kurunov I F et al 1973 *Issues of Automatic Control of Blast-furnace Production: col. issue 3*

(Kiev: Automation Institute) pp 85–87
[7] Moykin V I et al 1984 *Dynamic Characteristics of Blast Furnace According to the Results of Mathematical Modelling: Subject Collection of Scientific Works MChM USSR (IChM) Issues of Iron Production in Blast Furnaces* (M.: Metallurgy) pp 46–52
[8] Jian Xu et al 2012 *Proc. the 6th Int. Congress on the Science and Technology of Ironmaking* (Rio de Janeiro, RJ, Brazil. October 14–18, 2012) pp 414–424
[9] Ramm A N 1980 *Modern Blast-furnace Process* (M.: Metallurgy) p 304
[10] Yamamoto T et al 1982 *Transactions of the Iron and Steel Institute of Japan* vol 22 issue 10 pp 774–781
[11] Taguchi S et al 1982 *Tetsu-To-Hagane. Journal of the Iron and Steel Institute of Japan* vol 68 issue 15 2303–10
[12] Hatano M et al 1982 *Tetsu-To-Hagane. Journal of the Iron and Steel Institute of Japan* vol 68 issue 15 2369–76
[13] Abe Y et al 1982 *Transactions of the Iron and Steel Institute of Japan* vol 23 issue 3 b–81
[14] Ovchinnikov Yu N et al 1989 *Non-steady Processes and Improvement of Blast-furnace Smelting Efficiency* (Chelyabinsk: Metallurgy) p 120
[15] Kitaev B I et al 1978 *Heating Engineering of Blast-furnace Process* (M.: Metallurgy) p 248
[16] Onorin O P et al 2005 *Computer Methods of Blast-furnace Process Modelling* (Ekaterinburg: UGTU-UPI) p 301
[17] Spirin N A et al 2014 *Mathematical Modelling of Metallurgical Processes in ICS* (Ekaterinburg: OOO UIPTs) p 558
[18] Onorin O P et al 2013 *Steel in Translation* vol 43 issue 6 335–340
[19] Spirin N A et al 2016 *Metallurgist* vol 60 issue 5-6 471–477
[20] Lavrov V V and Spirin N A 2016 *IOP Conf. Series: Material Sci. and Eng.* **150** 012010
[21] Emelyanov S V et al 2008 *Theory and Practice of Prediction in Control Systems* (Kemerovo; M.: Russian Universities Publishing Union: Kuzbassvuzizdat – ASTech) p 487
[22] Spirin N A et al 2016 *IOP Conf. Series: Materials Sci. and Eng.* **150** 012011
[23] Onorin O P et al 2017 *Metallurgist* vol 61 issue 1-2 121–126

Mathematical modelling of ladle furnace thermal regime with bubble melt blowing by gas

N A Spirin, V S Shvidkii, Y G Yaroshenko and V V Lavrov

Ural Federal University n.a. the first President of Russia B.N. Yeltsin, 51 Lenina ave, Ekaterinburg, 620075, Russia

E-mail: n.a.spirin@urfu.ru

Abstract. The differential equations of energy of steel and gas with the 3-d rank of boundary conditions were used to evaluate the regularities of steel heating in a ladle furnace. The developed mathematical model, however, cannot be used as a control tool in real process time. The analogy between heating of the porous media in a fixed bed was applied to describe the heat transfer mechanism during the gas bubbles movement through the melt. This approach provides accurate results and with some modifications could be used as a part of the expert system to control ladle furnace operation.

1. Introduction

Production of a high quality steel is one of the challenging problems of iron and steel industry. Finishing, heating and alloying of steel in a ladle furnace before casting is a common practice at BOF and EAF steelmaking shops. The heat transfer between heating/steering gas and liquid steel plays a major role in metal treatment. Introduction of the parameter "ϕ" the ratio of the volume of gas bubbles to volume of steel (melt) in a ladle furnace (m^3/m^3) allows to apply regularities of heat transfer between moving gas and fixed bed of burden materials at pelletizing machine or sinter strand to describe the heat transfer in a ladle furnace [1, 2]. With such an approach to the physical phenomena the mathematical solution of heat transfer problem in a ladle furnace is based on the definition of volumetric heat transfer coefficient "α_v"[kJ/(m^3 K·sec)] between gas bubbles and surface of pseudo steel particles in 1 m^3 volume of the liquid bath. The volumetric heat transfer coefficient is defined by the following formula: $\alpha_v = \alpha_f \cdot F$, where α_f (kJ/(m^2·K·sec) is a heat transfer coefficient from gas to the surface of pseudo macro particles of liquid steel and F (m^2/m^3) – is a surface of pseudo macro particles of liquid steel in 1 m^3 volume of the liquid bath. Parameter "F" also could be interpreted as a overall surface of gas bubbles in 1 m^3 of steel bath.

2. Mathematical model

The physical formulation of the problem of the heat transfer between gas bubbles and liquid bath is as follows. Gas bubbles are moving through 1 m^2 area of the steel bath with the velocity at empty cross-section of the ladle furnace $W_g = \phi\ \omega_g$, passing the distance dy along the furnace height during the time period $d\tau$. Here ω_g – is actual gas velocity. Assuming constant gas velocity ω_g = (0.44+0.507)/2 = 0.475 [3] and infinite heat conductivity of the liquid bath the change in enthalpy of the elemental volume of the gas could be described by the following differential equation:

$$d^2Q = c_g \rho_g \varphi \left(\frac{dt_g}{d\tau} \right) dy d\tau = \\ = c_g \rho_g \varphi \frac{\partial t_g}{\partial y} \cdot \frac{\partial y}{\partial \tau} dy d\tau + c_g \rho_g \varphi \frac{\partial t_g}{\partial \tau} dy d\tau. \quad (1)$$

Here $\frac{\partial y}{\partial \tau} = \omega_g$ and $\varphi \cdot \omega_g = W_g$, where ω_g is a velocity of bubbles, m/s; c_g – gas specific heat, J/(kg·K); ρ_g – gas density, kg/m^3. Taking into account these definitions equation (1) can be simplified:

$$d^2Q = c_g \rho_g \left(W_g \frac{\partial t_g}{\partial y} + \varphi \frac{\partial t_g}{\partial \tau} \right) dy d\tau. \quad (2)$$

The same amount of enthalpy provided by gas to steel could be expressed by the following equation:

$$-d^2Q = \alpha_V \left(t_g - t_{melt} \right) dy d\tau + d^2Q_{loss}. \quad (3)$$

Equalization of equations (2) and (3) assuming that heat exchange is an adiabatic process and $d^2Q_{loss} = 0$ allows deriving the following equation:

$$-\alpha_V \left(t_g - t_{melt} \right) = c_g \rho_g \left(W_g \frac{\partial t_g}{\partial y} + \varphi \frac{\partial t_g}{\partial \tau} \right). \quad (4)$$

Similar consideration of the heat transfer in a metal bath, assuming that the enthalpy of macro particle of steel changes only in time (d_τ), allows to conclude that

$$-\alpha_V \left(t_g - t_{melt} \right) = c_{melt} \rho_{melt} (1 - \varphi) \frac{\partial t_{melt}}{\partial y}, \quad (5)$$

where: $c_{melt} = c_{melt}^{phys} + c_{melt}^{chem}$. Here c_{melt}^{phys} is a mass heat capacity of fusion, J/(kg·K), and $c_{melt}^{chem} = \sum q_{chem} / \left[m_{melt} \left(t_g - t_{melt} \right) \right]$; $\sum q_{chem}$ is a total thermal effect of oxidation reactions, J.

System of equations (4) and (5) should be supplemented by initial and boundary conditions to complete the mathematical model:

$$y = 0,\ t_g = t_{g0}; \\ \tau = 0,\ t_{melt} = t_{init} = 1100\ ^\circ C. \quad (6)$$

Thus, the complete mathematical model of heat transfer in the ladle furnace with the gas bubble blowing incorporates equations (4), (5) and (6).

3. Mathematical model transformation

The numerical solution of this system of equations could be done with following assumptions:

1. Gas bubbles are evenly distributed in a volume of ladle furnace, maintaining their spherical shape and having average diameter 13.57 mm [4];
2. Steel bath is stagnant and the bubbles velocity is constant along the height and equal to 0.475 m/s;
3. There is no chemical reactions in a melt and because of this $c_{melt}^{chem} = 0$;

4. The "F" value is simply determined by the parameter "ϕ" and the size of the bubble R_{bub}:

$$\varphi = \frac{\sum V_{bub}}{V_{melt}} = \frac{n \cdot \frac{4}{3}\pi R_{bub}^3}{V_{melt}}, \tag{7}$$

$$F = \frac{\sum F_{bub}}{V_{melt}} = \frac{n \cdot 4\pi R_{bub}^2}{V_{melt}}, \tag{8}$$

where "n" is an instantaneous amount of bubbles in a bath; $\sum V_{bub}$ – is a total volume of bubbles, m^3; $\sum F_{bub}$ – is a total surface of bubbles, m^2.

Joint solution of equations (7) and (8) allows estimation of value "F":

$$F = \frac{3\varphi}{R_{bub}} = \frac{3 \cdot 2 \cdot \varphi}{13.57 \cdot 10^{-3}} = 0.442 \cdot 10^3 \varphi. \tag{9}$$

The statistics of bubble's parameters in a ladle furnace during bottom blowing is presented in papers [5, 6]. The amount of bubbles in a bath as a function of the height and time in the course of different technological functions of ladle furnace operation could be found in literature [7, 8]. These data are in good correlation with estimated parameters [9].

Let's assume "ϕ"= 11.3 %, which correspondence with the real conditions. The values of various coefficients in a system of equations (4)–(6) are as follows (air is assumed as a stirring gas):

$\alpha_V = \alpha_F \cdot F = 442 \cdot \varphi \cdot \alpha_F$ and $\alpha_F = Nu \cdot \lambda / (2 \cdot R_{bub})$.

Here "Nu" – is the Nusselt number:

$Nu = 2 + 0.6 \cdot \mathrm{Re}^{0.5} \cdot \mathrm{Pr}^{0.33}$ [10].

After substitution of numerical values of specific heat and conductivity coefficient at the average temperature of the t_{melt} = 1200°C [11], the following values of similarity numbers and heat transfer coefficient are obtained: Re = 28.8; Pr = 0.724; Nu = 4.894; α_F =33.02, W/(m^2*K) and α_V=442·φ·33.016= 14593.3·φ, W/(m^3*K).

As a result instead of system of equation (4) - (6) the following two differential equations represent the heat transfer in a liquid bath of the ladle furnace [12]:

$$\begin{aligned} &\frac{\partial t_g}{\partial \tau} + 0.475\frac{\partial t_g}{\partial y} = \frac{14593.3}{1210 \cdot 1.27}\left(t_{melt} - t_g\right); \\ &513.9 \cdot 7790 \cdot (1-\varphi)\frac{\partial t_{melt}}{\partial \tau} = \\ &= 14593.3 \cdot \varphi \cdot \left(t_g - t_{melt}\right). \end{aligned} \tag{10}$$

System of equations (10) was numerically resolved for the following values of determining parameters: ϕ = 0.1, 0.15, 0.2, 0.5; t_{g0}, °C = 1500, 1600, 1800, 2000; τ = 0, t_{melt}=$t_{initial}$=1100 °C. This system of equations (10) needs to be converted to the standard type to use standard solutions and define dimensionless parameters:

$$\begin{aligned} &\frac{\partial t_g}{\partial \tau} + 0.475\frac{\partial t_g}{\partial y} = \frac{14593.3}{1210 \cdot 1.27}\left(t_{melt} - t_g\right) = \\ &= 9.496518 \cdot \left(t_{melt} - t_g\right); \\ &\frac{513.9 \cdot 7790 \cdot (1-\varphi)}{4593.3 \cdot \varphi} \cdot \frac{\partial t_{melt}}{\partial \tau} = \left(t_g - t_{melt}\right). \end{aligned} \tag{11}$$

or

$$0.105302 \cdot \frac{\partial t_g}{\partial \tau} + 0.0500183 \frac{\partial t_g}{\partial y} = t_{melt} - t_g; \tag{12}$$

$$\frac{(1-\varphi)}{0.00364533 \cdot \varphi} \cdot \frac{\partial t_{melt}}{\partial \tau} = t_g - t_{melt}; \tag{13}$$

$$\frac{\partial t_{melt}}{\partial\left(\frac{0.00364533 \cdot \varphi \cdot \tau}{1-\varphi}\right)} = t_g - t_{melt}. \tag{14}$$

The dimensionless time can be derived from equation (13):

$$Z = \frac{0.00364533 \cdot \varphi \cdot \tau}{1-\varphi}. \tag{15}$$

The following two points [13] need to be addressed with respect to the equation (12):

- The first component of this equation becomes significant only at $\tau < \phi y/W_g$ or until gas will not reach the level "y". At the real conditions of ladle furnace operation (H = 1.812 m; $\tau_{heating}$ = 3.75 h = 13500 s; $W_g = 0.475 \cdot \phi$), this component could be neglected;
- The dimensionless height of steel bath Y = y/0.0500183 unlike the dimensionless time does not depend on the number of bubbles.

Taking into account these comments, the equations (16) were derived to describe the distribution of gas and steel bath temperature along the ladle furnace height.

For the complete height of steel bath H = 1,812 m values of dimensionless height are as follows:

y, m	0	0.302	0.906	1.51	1.812
Y	0	6.038	18.113	30.19	36.27

$$\vartheta = \exp(-Y)\int_0^Z I_0\left(2\sqrt{Y_\zeta}\right)\cdot \exp(-\zeta)d\zeta;$$
$$t_{melt} = 1100 + (t_{g0} - 1100)\cdot \vartheta;$$
$$\theta = 1 - \exp(-Z)\int_0^Y I_0\left(2\sqrt{Z_\eta}\right)\cdot \exp(-\eta)d\eta; \tag{16}$$
$$t_g = 1100 + (t_{g0} - 1100)\cdot \theta.$$

Here $\vartheta=(t_{melt} - t_{init})/(t_{g0} - t_{init})$; $\theta=(t_g - t_{init})/(t_{g0} - t_{init})$. The values of dimensionless time are determined by the set volume of bubbles. At ϕ = 0.1 the value of Z equal to Z= 4.05037·τ:

τ, sec	0	2700	5400	10900	13500
Z	0	1.094	2.19	4.374	5.468

4. Results of mathematical modeling

Mathcad is the most convenient package to solve system of equation (16). The dimensionless gas temperature distribution along the furnace height as a function of dimensionless time Z is presented in figure 1 for the fraction of gas bubbles φ=0.1 and initial gas temperature t_{g0}=1500 °C.

IOP Conf. Series: Materials Science and Engineering **411** (2018) 012074 doi:10.1088/1757-899X/411/1/012074

Figure 1 results allow to conclude that at assumed fraction of the gas bubbles in the bath (φ=0.1) enthalpy of the incomming stirring gas is sufficient only for heating bottom part of the steel bath – approximately 1/3 of the overall height.

The increase in gas bubbles fraction to φ=0.2 does not change significantly this conclusion. Even at initial gas temperature of 2000 °C the temperature of the gas on the top of the bath remains equal to the initial steel temperature.

At φ=0.5 the situation changes dramatically. The enthalpy of the gas in this case is sufficient to heat steel even at initial gas temperature t_{g0}=1500 °C (figure 2).

In this case dimensionless time is equal Z = 0.00364533·τ. It is evident that the limit of the gas volume is a function of the melt temperature at the top of the bath. The change in bath temperature in time along the furnace height at ϕ = 0.5 is presented in figure 3.

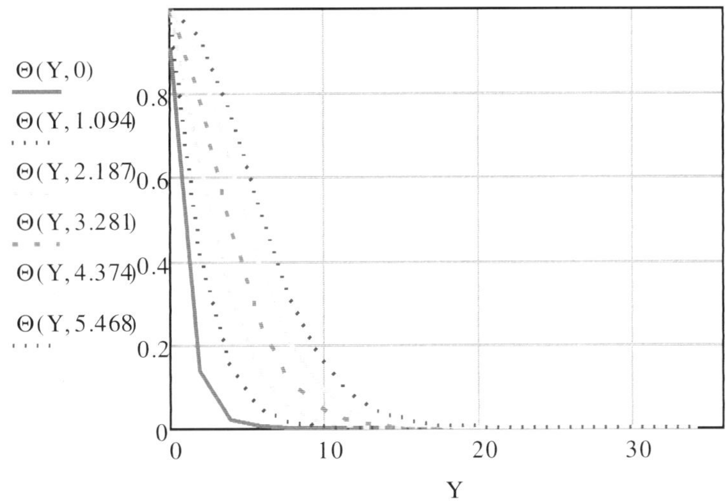

Figure 1. Gas temperature as a function of steel bath height and dimension less time φ=0.1; t_{g0}=1500 °C.

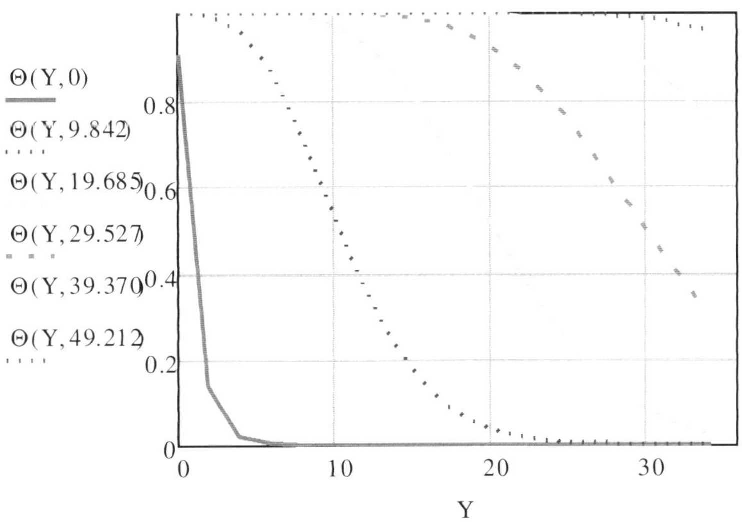

Figure 2. Gas temperature as a function of steel bath height and dimension less time φ=0.5; t_{g0} = 1500 °C.

The influence of parameters of blowing on average melt temperature was studied in time interval of 1.01.5 hours. Since dimensionless height of the steel bath Y = y/0.0500183 does not depend on number of bubbles or their fraction "φ" and gas temperature at the entrance to the bath is t_{g0}, the average bath temperature in all possible variants will be determined by the same formula:

$$t_{melt} = \frac{1}{36.267} \int_0^{36.267} \exp(-\xi) \int I_0 \left(2\sqrt{\xi u}\right) \exp(-\mu) d\mu d\xi. \quad (17)$$

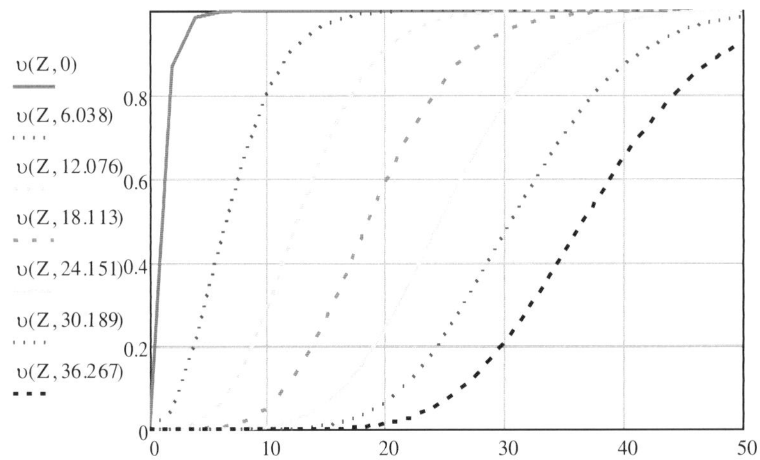

Figure 3. Bath temperature distribution as a function of bath height and time at φ=0.5.

For the time of blowing τ=1.25 hour the integral for average melt temperature could be transformed to simple arithmetical expression

$$t_{melt} = 1231.133 - 1614.243 \cdot \varphi - 0.072 \cdot t_{g0} + \\ +765.789 \cdot \varphi^2 + 1.031 \cdot \varphi \cdot t_{g0} - 4.801 \cdot 10^{-9} \cdot t_{g0}^2, \text{ °C}. \quad (18)$$

Results of calculations are presented in figure 4, where each curve represents the average melt temperature.

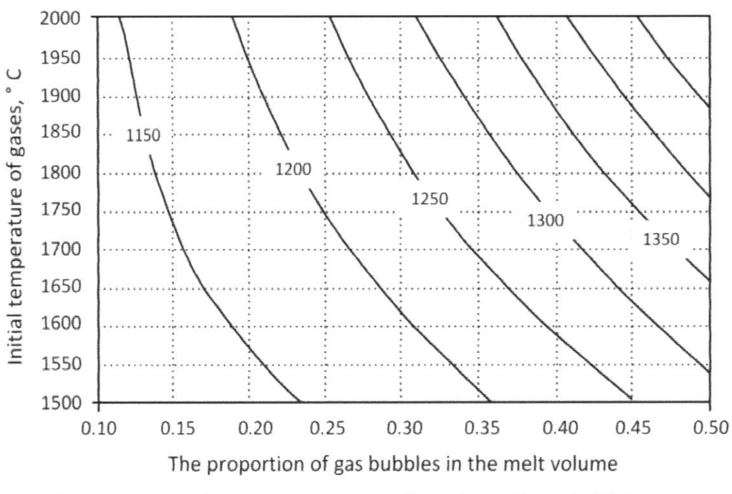

Figure 4. Average bath temperature as a function of gas initial temperature and fraction of gas bubbles at τ = 1.25 hour.

Based on results of figure 4 it is possible to conclude that the desirable average temperature of melt, for example, t_{melt}^{assign}= 1250 °C, could be achieved at φ=0.25 and t_{g0} =2000 °C or φ=0.45 and t_{g0}=1500 °C.

Thus, if time of blowing is set, then it is easy to run calculations. For one hour blowing time the average bath temperature could be estimated by the following expression:

$$\begin{aligned} t_{melt} &= 1206.764 - 1305.325 \cdot \varphi - 0.058 \cdot t_{g0} + \\ &+ 624.18 \cdot \varphi^2 + 0.831 \cdot \varphi \cdot t_{g0} + 3.14 \cdot 10^{-8} \cdot t_{g0}^2, \ °C. \end{aligned} \quad (19)$$

For 1.5 hour blowing time solution of integral (17) gives similar to (18) and (19) expression (20):

$$\begin{aligned} t_{melt} &= 1256.704 - 1952.495 \cdot \varphi - 0.084 \cdot t_{g0} + \\ &+ 932.594 \cdot \varphi^2 + 1.243 \cdot \varphi \cdot t_{g0} - 7.616 \cdot 10^{-7} \cdot t_{g0}^2, \ °C. \end{aligned} \quad (20)$$

5. Conclusions

- The mathematical model of heat transfer between gas bubbles and liquid bath was developed base on approximation of well known Shuman problem of heat transfer between moving gas and the fixed layer of material.
- Influence of bubbling gas initial temperature and blowing time on distribution of the gas and bath temperature along the furnace height at different moments of time was studied and discussed.
- While numerical analysis was applied to ladle furnace semi steel refining, developed mathematical model and approach are absolutely applicable for analysis of gas bubble movement and temperature distribution in steelmaking ladle furnace.

References

[1] Novokreschenov S, Shvidkii V et al 2009 *Proc. of Int. Conf. on Sci. Heritage of B.I. Kitaev* (Ekaterinburg, Russia) pp 324–328
[2] Novokreschenov S, Shvidkii V et al 2013 *Proc. of the Conf. on Modern Metallic Materials and Technology* (S. Petersburg, Russia)
[3] Novokreschenov S et al 2013 *Izvestia Vuzov. Non-ferrous Metallurgy* **5** 58–60
[4] Novokreschenov et al 2011 *Izvestia Vuzov. Non-ferrous Metallurgy* **4** 61–62
[5] Gizatulin R 2017 *Izvestia Vuzov. Ferrous Metallurgy* **8** 26–29
[6] Gizatulin R 2016 *Bulletin of South-Ural's University. Series "Metallurgy"* **10(65)** 63–68
[7] Sokovnin O, Zagoskin N and Zagoskin S 2012 *Theoretical Foundation of Chemical Technology* vol 46 **5** 540
[8] Protopopov E 2002 *Izvestia Vuzov. Ferrous Metallurgy* **4** 9–13
[9] Zgukov V et al 2013 *Izvestia Vuzov. Non-ferrous Metallurgy* **3** 58
[10] Gordon Y et al 1992 *Heat Transfer Calculations for Metallurgical Furnaces* (Moscow: Metallurgia) p 355
[11] Belousov V, Klevtsov A, Pribitkov I and Sborschikov G 1993 *Heat Transfer and Heat and Power Engineering* (Moscow: Metallurgia) p 336
[12] Chirkin V 1967 *Thermo-physical Properties of Nuclear Technology Materials* (Moscow: Atomizdat) p 474
[13] Telegin A, Shvidkii V and Yaroshenko Yu 2002 *Heat- and Mass Transfer* (Moscow: Akademkniga) 263

IOP Conf. Series: Materials Science and Engineering **411** (2018) 012075 doi:10.1088/1757-899X/411/1/012075

A study on the metal loss of spring steel 40s2 heated for rolling and peculiarities of the structure, chemical and phase composition of its scale

M V Temlyantsev[1], K S Konoz[1], O V Kuznetsova[1], V B Deev[2] and E Ya Zhivago[1]

[1]Siberian State Industrial University, 42 Kirova street, Novokuznetsk, 654007, Russia
[2]National University of Science and Technology "MISIS", 4 Leninskiy ave, Moscow, 119049, Russia

E-mail: uchebn_otdel@sibsiu.ru

Abstract. The article presents the results of study on kinetics of high-temperature oxidation of steel 40S2. For the implementation of predictive calculations, an equation is obtained that allows metal loss to be determined depending on the temperature and the heating time. The structural features, chemical and phase composition of the scale, formed during high-temperature heating of 40S2 steel are studied. It is established that the hematite layer bordering on the furnace atmosphere has a minimum thickness of about 40 – 60 μm, the magnetite layer has a thickness of about 250 – 280 μm, the largest thickness 800 – 900 μm has a wuestite layer with secondary magnetite appeared upon cooling. The chemical and phase compositions of hematite and magnetite layers are relatively uniform in thickness and perimeter. The scale layer adjacent to the steel surface is characterized by a much greater macrostructural, chemical, and phase inhomogeneity. In addition to the wustite-magnetite mixture, iron silicates with an increased silicon content are present.

1. Introduction

Medium- and high-carbon siliceous steels are widely used for the manufacture of springs of various purposes and rod griffons – elastic elements of rail fastenings of rail joints with assembled rails and sleepers [1 - 3]. The rolled metal for their production is obtained by hot rolling of continuous castings, which are heated in the continuous furnaces. High-temperature heating is accompanied by scaling, leading to losses in the useful metal mass and the formation of a scale layer on the surface of the workpieces, which can lead to a decrease in the quality of round products surface [4, 5].

The structure and properties of the scale formed play an essential [5, 6]. When developing the heating technology and assigning temperature modes that provide metal-saving heating [7 – 9], information is needed on the kinetics of the steel oxidation. Since precise analytical solutions for finding the dependence of the oxidation intensity of steels of different chemical composition are missing, the only way to obtain such dependences is empirical.

The analysis of special technical literature [4, 5, 9 - 11] shows that the oxidation kinetics of 40S2 silicon steel has been little studied, and there are no quantitative data on the influence of the temperature-time factor on its loss. Thus, obtaining new information on the kinetics of high-temperature oxidation of spring steel 40S2 and the features of its scale structure is of great practical importance.

2. Methods of research

In this paper a study on the metal loss of steel 40S2 (chemical composition,%: 0.377 C, 1.604 Si, 0.65 Mn, 0.204 Cr) and the structural features, phase and chemical composition of its scale is presented. The oxidation kinetics is studied using a gravimetric method on cylindrical samples with a diameter 11 mm, length 20 – 30 mm. The samples were cut from sections of round bars. Before cutting out the samples, the rods were softened and turned on a turning workstation with in order to completely remove the decarburized layer.

In the laboratory experiments the samples heating up to temperatures 850 – 1200 °C (with 50 °C interval) and holding at constant surface temperature for 5, 15 and 25 min was carried out in the electric resistance oven SUOL-0.25.1/12.5-I1 with silicon carbide heaters in the air atmosphere.

Before and after the experiments the geometrical dimensions of the samples were measured with a caliper with 0.1 mm accuracy, the mass was determined on Vibra AF-220CE weighing-machine with 0.1 mg accuracy. To remove scale from the surface, the samples were etched in the warmed up to 40-60 °C 10% sulfuric acid solution with addition of 0.1 g/l thiourea.

When processing the experimental data, it was considered that the dependence of the loss (or the thickness of the oxidized layer) on time obeys the square root law, which reflects the diffusion mechanism of scale formation, the temperature of the metal surface is constant in time, and the dependence of the metal loss on the temperature-time factor obeys the following equation [4]:

$$L = k\sqrt{\tau}, \tag{1}$$

where k – the oxidation rate constant, $g/(cm^2 \cdot min^{0.5})$; τ – metal holding time at constant temperature, min.

For relation (1), it is assumed that the oxidation rate constant obeys the Arrhenius equation, which can be represented in the form [4]

$$k = A\exp\left(-\frac{Q}{RT}\right), \tag{2}$$

where A – the empirical coefficient, which depends on the chemical composition of steel, the atmosphere composition and other factors, $g(cm^2 \cdot min^{0.5})$; Q – energy of activation or loosening of the oxide lattice, J/kmol; R = 8.3143 – universal gas constant, J/(kmol·K); T – the temperature, K.

Assuming $B = Q / R$, relation (2) is written in the form

$$k = A\exp\left(-\frac{B}{T}\right), \tag{3}$$

where the coefficient B has the dimension K.

3. Analysis of the results of experimental studies

Figure 1 shows the dependence of the logarithm of oxidation rate constant on temperature for steel 40S2.

For predictive calculations the dependence was obtained that makes it possible to determine the metal loss L as a function of the temperature and the heating time:

$$L=22371.9 \cdot \exp(-18435/T) \cdot \sqrt{\tau} \tag{4}$$

The analysis of the obtained results shows that the temperature-time dependence of the metal loss of steel 40S2 is similar to the dependences in chemical composition to steels 60S2 and 60C2KhA [4,

10, 11]. In the temperature range 1050 – 1100 °C there is sharp intensification of metal loss. For steel 40S2 the temperature has a stronger effect on the loss compared with steel 60S2KhA. In particular with a holding time 25 min, an increase in temperature from 850 to 1175 °C for steel 40S2 is accompanied by an increase in the loss from 0.016 to 0.367 g/cm^2, i.e. almost in 23 times, and for steel grade 60S2KhA from 0.019 to 0.270 g/cm^2, i.e. in 14 times.

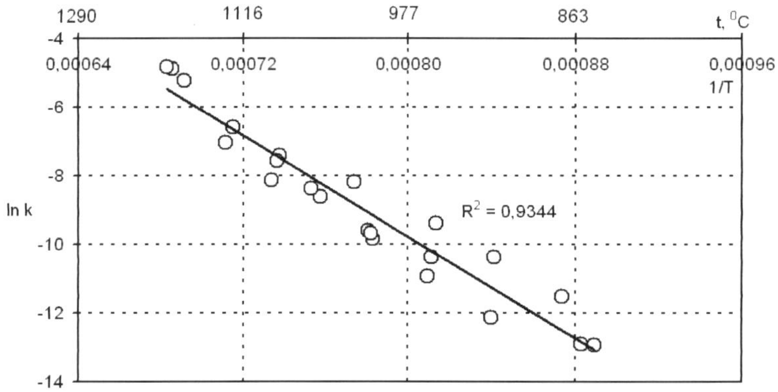

Figure 1. Dependence of the logarithm of oxidation rate constant on temperature for steel 40S2.

For steel grade 40S2 in the temperature range 850 – 950 °C, a less intense scale formation is observed in comparison with the steel grade 60S2KhA, and for temperatures over 1150 °C on the contrary.

A study of the structure, chemical and phase composition of the scale was carried out at the NUST MISIS on the electron scanning microscope TESCAN VEGA 3 with OXFORD AZtec energy dispersive attachment for micro-X-ray analysis. Samples for analysis were taken from scale plates with thickness 1 – 1.5 mm heated to 1175 °C, and the holding time was 25 minutes. After heating the samples with scale were cooled in atmospheric air in the conditions of natural convection.

Figure 2 shows a general view of the scale microstructure. The hematite layer bordering on the furnace atmosphere has a minimum thickness – approximately 40 –60 μm, the magnetite layer has a thickness approximately 250 – 280 μm, the largest thickness 800 – 900 μm has a wuestite layer appearing upon cooling with secondary magnetite bordering on the steel surface.

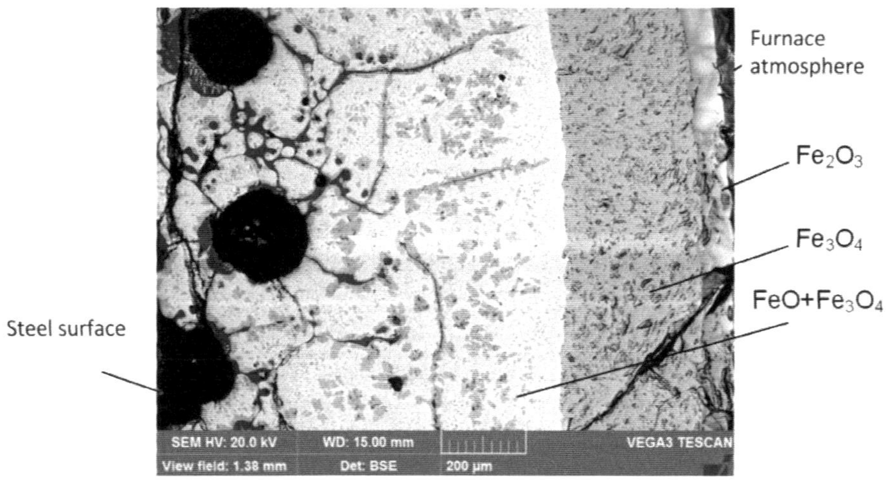

Figure 2. Scale microstructure.

IOP Conf. Series: Materials Science and Engineering 411 (2018) 012075 doi:10.1088/1757-899X/411/1/012075

The chemical and phase compositions of the hematite and magnetite layers are relatively uniform in thickness and perimeter (figure 3). Table 1 shows their compositions at different points in the layers.

The scale layer adjacent to the steel surface (figure 4) is characterized by a much greater macrostructural, chemical and phase inhomogeneity (table 2). In addition to the wustite-magnetite mixture, iron silicates are also present, characterized by an elevated (up to 13 wt.%) silicon content.

As the distance to the steel surface gets smaller, the number of sites enriched with silicon increases.

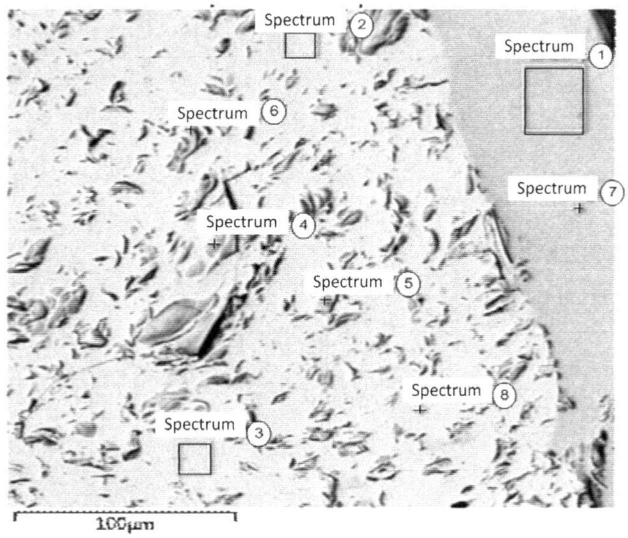

Figure 3. Microstructure of hematite and magnetite layers.

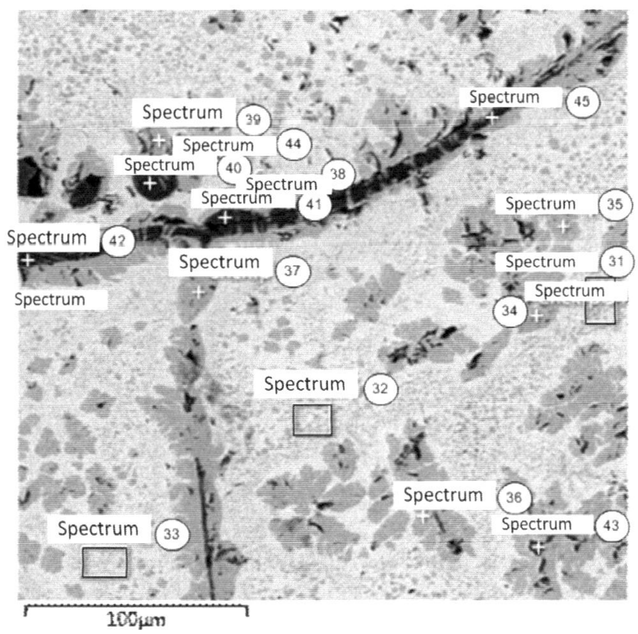

Figure 4. Microstructure of a wuestite layer with secondary magnetite released during cooling.

Table 1. Chemical composition (wt.%) of hematite and magnetite layers.

Spectrum No.	O	Si	P	S	Mn	Fe	Total
1 hematite	33.52	0.09	0	0	0.08	66.31	100
7 hematite	34.11	0.06	0.03	0	0.1	65.7	100
2 magnetite	31.37	0	0.02	0.03	0.44	68.14	100
3 magnetite	31.17	0.06	0	0.02	0.52	68.23	100
4 magnetite	31.52	0	0	0.03	0.42	68.02	100
5 magnetite	31.21	0.02	0	0.01	0.45	68.3	100
6 magnetite	31.86	0.02	0	0.01	0.37	67.74	100
8 magnetite	30.19	0.12	0.03	0.03	0.48	69.16	100

Table 2. Chemical composition (wt.%) of the wuestite layer with secondary magnetite released during cooling.

Spectrum No.	O	Si	P	S	Mn	Fe	Cu	Mo	Total
31 wuestite + magnetite	27.58	0.4	0	0.03	0.55	71.44	–	–	100
32 wuestite + magnetite	27.97	0.06	0.02	0.05	0.42	71.49	–	–	100
33 wuestite + magnetite	27.37	0.04	0.01	0.03	0.43	72.12	–	–	100
34 wuestite + magnetite	31.02	0.02	0	0.05	0.42	68.5	–	–	100
35 wuestite + magnetite	31.77	0.44	0.02	0	0.35	67.43	–	–	100
36 wuestite + magnetite	31.44	0.01	0.01	0.02	0.4	68.12	–	–	100
37 wuestite + magnetite	31.24	0.3	0.02	0	0.31	68.12	–	–	100
38 wuestite + magnetite	31.1	0.33	0	0.04	0.35	68.17	–	–	100
39 wuestite + magnetite	31.67	0.22	0	0	0.36	67.75	–	–	100
40 iron silicates	36.74	13.58	0.08	0.03	1.61	47.96	–	–	100
41 iron silicates	36.26	13.48	0.14	0	1.65	48.47	–	–	100
42 iron silicates	36.04	13.34	0.14	0	1.35	49.14	–	–	100
45 iron silicates	31.87	7.39	0.01	0.02	1.09	59.62	–	–	100
43 iron silicates	26.72	2.12	0	0.05	0.3	70.24	0.22	0.35	100
44 iron silicates	31.77	1.61	0.04	0.04	0.45	66.09	–	–	100

4. Conclusion

The effect of the temperature-time factor on the oxidation kinetics of siliceous spring steel 40S2 is studied. For the implementation of predictive calculations, an equation is obtained that allows the loss to be determined, depending on the temperature and the heating time. It is established that in the temperature range 1050 – 1100 °C a sharp intensification of the metal loss is observed, for steel 40S2 in the temperature range 850 – 950 °C, less intense scale formation is characteristic in comparison with 60S2KhA steel, and for temperatures over 1150 °C, on the contrary.

With the use of electronic scanning microscopy, the structure of scale was investigated. It is established that the hematite layer bordering on the furnace atmosphere has a minimum thickness of about 40 – 60 μm, the magnetite layer has a thickness of about 250 – 280 μm, the largest thickness 800 – 900 μm has a wuestite layer with secondary magnetite precipitated upon cooling. The chemical and phase compositions of hematite and magnetite layers are relatively uniform in thickness and perimeter.

The scale layer adjacent to the steel surface is characterized by a much greater macrostructural, chemical, and phase inhomogeneity. In addition to the wustite-magnetite mixture, iron silicates with an increased silicon content are present.

IOP Conf. Series: Materials Science and Engineering **411** (2018) 012075 doi:10.1088/1757-899X/411/1/012075

References

[1] Rakhshtadt A G 1984 *Spring Steels and Alloys* (Moscow: Metallurgy) p 359
[2] Semikhatsky S A et al 2006 *Forging-stamping Production. Metal Forming* **10** 43–48
[3] Fedin V M 2002 *Volume-surface Hardening of Rolling stock and Track Structure* (Moscow: Intext)
[4] Temlyantsev M V et al 2006 *Oxidation and Decarburization of Steel in the Processes of Heating under Pressure Treatment* (Moscow: Heat Engineer) p 200
[5] Peretyatko V N et al 2008 *Heating of Steel Slabs* (Moscow: Heat Engineer) p 192
[6] Peretyatko V N and Temlyantsev NV 2006 *Bulletin of the Mining and Metallurgical Section of the Russian Academy of Natural Sciences. Department of Metallurgy: Proceedings* (Novokuznetsk: SibSIU) **16** pp 63–70
[7] Temlyantsev M V et al 2005 *Izv. Vuzov. Ferrous Metallurgy* **6** 69–70
[8] Temlyantsev M V et al *2005 Izv. Vuzov. Ferrous Metallurgy* **12** 47–49
[9] Filippova M V et al 2016 *Development and Implementation of Energy and Resource-saving Technologies for Metal Forming* (Novosibirsk: SB RAS) p 269
[10] Temlyantsev M V and Temlyantsev N V 2005 *Procurement Industries in Mechanical Engineering* **3** 50–52
[11] Temlyantsev M V et al 2014 *Bulletin of the Mining and Metallurgical Section of the Russian Academy of Natural Sciences. Department of Metallurgy: Proceedings* (Novokuznetsk: SibSIU) **33** pp 55–63

The problems of mastering the production of special purpose railway products in the conditions of universal rail-and-structural steel mill

A A Umansky[1], A V Golovatenko[2] and V V Dorofeev[2]

[1]Siberian State Industrial University, 42 Kirova street, Novokuznetsk, 654007, Russia
[2]EVRAZ Consolidated West Siberian Metallurgical Plant, 16 Kosmicheskoye shosse, 654043, Russia

E-mail: umanskii@bk.ru

Abstract. The paper contains the analysis results of technological features of asymmetric rail profiles production in the conditions of a universal rail-mill, as well as the stages of the development of intensified rolling mode of switch point rails, the introduction of which allowed the productivity of the rail-and-structural steel mill to be increased and the quality of the finished rails to be improved. On the example of the operating rail-and-structural steel mill of EVRAZ ZSMK JSC the technological difficulties in the production of asymmetric rail profiles with the use of a continuous universal train are illustrated. Ways for optimization of rolling modes for asymmetric rail profiles (switch point rails are taken as an example) are shown using the developed technique for calculating the resistance to deformation of rail steels and modeling the flow of metal using modern software systems. The existing standards regulating the quality parameters of switch point rails is analyzed and the recommendations for its improvement are given.

1. Introduction

Domestic and foreign practice of metallurgical production testifies that the initial period after commissioning of rolling mills is characterized by the most intensive development of new sizes of produced profiles [1-4]. The completeness of such machines with modern equipment opens up wide opportunities for the production of rolling profiles in accordance with international requirements, which in its turn opens new promising markets for finished products.

In 2013 at EVRAZ ZSMK JSC the first in the national metallurgy universal rail-and-structural steel mill was launched. The rolling mill consists of two successive two-roll reversing crimping stands (BD1 and BD2) and a universal tandem mill displaced relative to the rolling line, consisting of two universal and one auxiliary two-roll stand, as well as a separately located finishing universal stand (figure 1).

The main assortment of the universal rail-and-structural steel mill of EVRAZ ZSMK JSC is the railway rails of wide track (including long ones up to 100 m in length) of rails of types R65 and R50 supplied in heat-strengthened and non-heat-strengthened condition. However, the production of rails for industrial railways (RP65, RP50), tramway grooved rails (T62), crane rails (KR80, KR120), railway rail counter rail (PK50, PK65), switch point rails (OR43, OR50, OR65). It should be noted that the greatest difficulties arise when mastering the production of profiles that have a pronounced asymmetric shape, including switch point rails (figure 2).

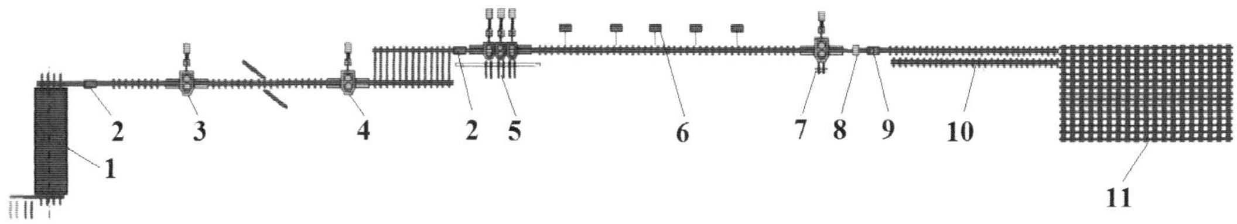

Figure 1. The layout of the equipment of the universal rail-and-structural steel mill at JSC EVRAZ ZSMK: 1 – heating furnace; 2 – hydraulic descaling device; 3, 4 – crimping stands BD1 and BD2; 5 – tandem train; 6 – saws of hot cutting; 7 – finishing stand; 8 – laser profiler; 9 – marker; 10 – section of differentiated hardening; 11 – section of the refrigerator.

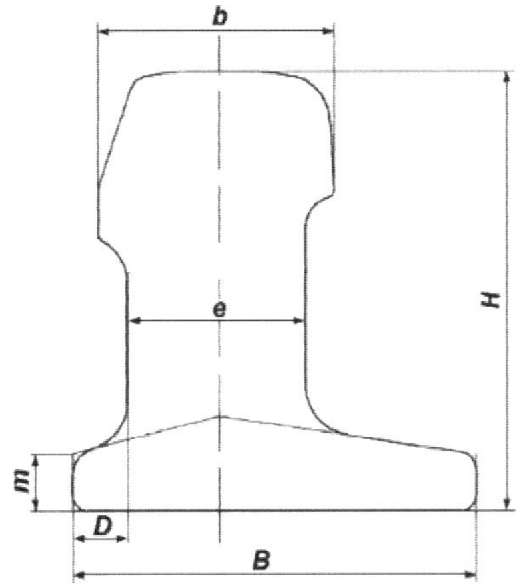

Figure 2. Cross-section profile of switch point rails (H – rail height; b – width of the rail head; e – rail web thickness; B – width of the rail foot; D – width of the short arm of rail foot; m – height of the rail base blade).

The difficulty in developing calibrations of rolling rolls for the production of such profiles is exacerbated by the presence of a continuous train in the universal rail-and-structural steel mill. It also should be noted that the tendency towards the increase in the speed of railway trains [5-7] in recent years has caused a corresponding increase in the requirements for the quality of railway switching, which are the most rapidly wearing part of the track upper structure.

In particular, rails switching requirements for high-speed traffic (speed above 200 km/h) need to meet higher requirements in terms of geometric parameters (unevenness on the rolling surface of rails). Since the initial workpiece for the production of switch points are switch point rails, improvement of the quality of the latter is an urgent task. This thesis is supported by the available data that the main causes of defects in the manufacture of switch point rails are unsatisfactory geometry and the dispersion of the residual stress values in the initial blanks – the switch rails [8].

2. Results and discussion

In accordance with the contract calibration of the equipment supplied for the rail-and-structural steel mill (by SMSMeer), the rolling scheme for the switch point rail of OR65 type included the following main steps:

- seven passes in the box gauges of stand BD1 to obtain a rectangular semifinished rolled products;
- seven passes in BD2 stand, including the first pass in the "lying trapezoid" gauge, the second and third passes – in the "standing trapezoid" gauge, the fourth and fifth passes – in the cut rail gauge of the closed type, the sixth and seventh passes – in the preparatory rail gauge of open type;

- rolling in the tandem mill cages in three passes, including the first pass – in the universal stand UR and the auxiliary (edger) stand E, the second pass – in the UR stand; the third pass – successively in all three stands of the tandem mill (UR, E, UF).

Practical experience in the production of switch point rails according to the scheme described above made it possible to identify a number of its significant shortcomings:

- during rolling in stand BD2 there was a twisting of the rolls at the exit from the gauges;
- in the closed groove of the angular wire gauge, intensive adherence of metal along the perimeter of the walls occurred;
- the impossibility of restoring an angular wire gauge of closed type to the width with a regulated metal removal during the reconditioning of roll.

To overcome the above disadvantages, a new rolling method has been developed (figure 3) for switch point rails with intensified crimping mode in stand BD1, reduced number of passes in stand BD2 and the use of an inclined angular wire gauge with diagonal connectors instead of a angular wire gauge of closed type.

The possibility for intensification of the deformation regime is preliminarily justified by calculations of the rolling force, carried out using the technique developed earlier by the authors for calculating the resistance of deformation of rail steels [9, 10]. According to this technique, the deformation resistance is determined from the expression:

$$\sigma = A \cdot e^{m_1 \cdot t} \cdot \varepsilon^{m_2} \cdot e^{\frac{m_3}{\varepsilon}} \cdot (1+\varepsilon)^{m_4 \cdot t} \cdot e^{m_5 \cdot \varepsilon} \cdot u^{m_6 \cdot t}, \quad (1)$$

where A, m_1-m_6 – the coefficients of the equation, depending on the chemical composition of the steel;
e – the Euler number;
t – the rolling temperature, °C;
ε – logarithmic degree of deformation;
u – the deformation rate, s^{-1}.

$$A = 4365.4 + 69118 \cdot [S], \quad (2)$$

$$m_1 = -0.0033 - 0.0043 \cdot [V], \quad (3)$$

$$m_2 = 0.2607 - 5.7663 \cdot [P], \quad (4)$$

$$m_3 = -0.0025 + 0.00308 \cdot [C] + 0.00025 \cdot [Mn], \quad (5)$$

$$m_4 = -0.0015 + 0.0475 \cdot [P], \quad (6)$$

$$m_5 = -0.407 + 0.655 \cdot [Mn], \quad (7)$$

$$m_6 = 0.0002 - 0.0012 \cdot [V], \quad (8)$$

where [*S*], [*V*], [*P*], [*C*], [*Mn*] – the content in the steel of sulfur, vanadium, phosphorus, carbon and manganese, respectively.

Calculations were carried out for steel grade E76KhSF according to GOST R 55820-2013. The actual chemical composition of this steel grade from the current production melts was used for the calculations (table 1), the rolling temperature for the passages was selected based on the temperature of billet preheating at 1170 °C and the temperature drop by 5 °C per each pass.

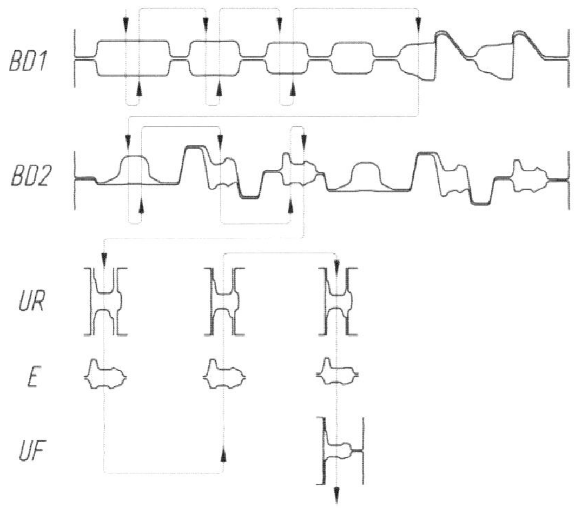

Figure 3. Improved rolling scheme of switch point rails of OR65 type at the universal rail-and-structural steel mill of EVRAZ ZSMK JSC.

Table 1. Chemical composition of steel E76KhSF for the production of rails OR65.

	Element content, %						
	C	Mn	Si	V	Cr	P	S
Actual average composition	0.76	0.90	0.60	0.07	0.60	0.015	0.015
Requirements of GOST R 55820-2013	0.69-0.80	0.70-1.00	0.3-0.7	0.03-0.15	0.35-0.80	not more than 0.020	not more than 0.020

According to the calculated data, confirmed by the actual experience of rail-and-structural steel mill operation, when using the new rolling mill mode of switch point rails, the rolling force does not exceed 80% of the permissible rolling force (figure 4).

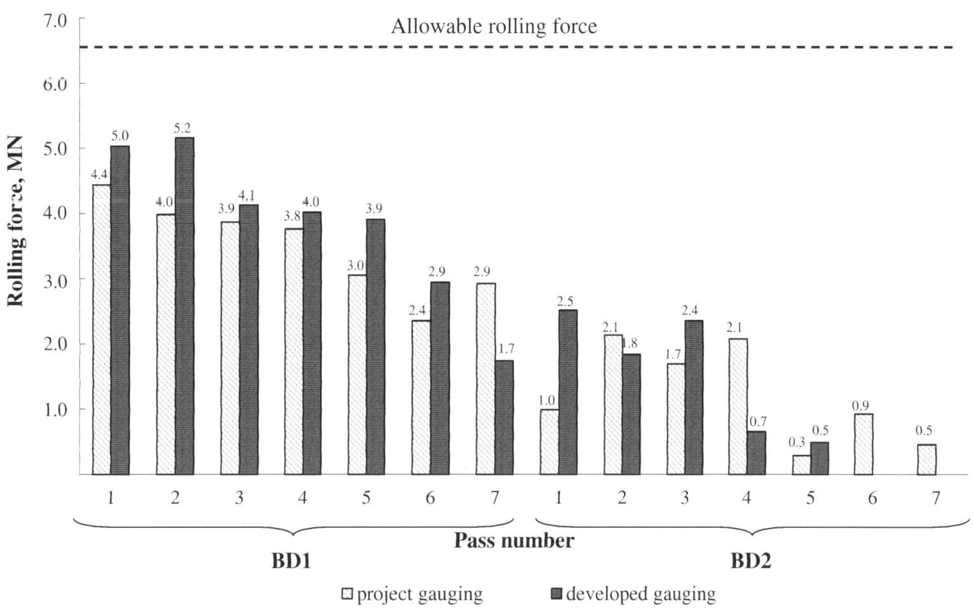

Figure 4. Calculated rolling forces during deformation in the break-down mills of the universal rail-and-structural steel mill at EVRAZ ZSMK JSC.

As a result of the introduction of the new mode of rolling RR65 rails, by reducing the number of passes in BD2 stand, a reduction in the rolling cycle was achieved by 22.5 s, which conditioned the increase in the mill's productivity during production of this product by 39.8 tonne/h. The economic effect from increasing the mill's productivity was 29.7 million rubles per year.

The use of the inclined arrangement of the rail gauge in the new rolling scheme made it possible to achieve the following advantages in comparison with the contract calibration:

- to ensure the rolls resharpening without changing the width of the gauges;
- to increase the intensity of breakdown;
- to eliminate the danger of the "shackle" of rolls by rolling due to the diagonal arrangement of closed and open flanges;
- to facilitate the task of rolling into rolls due to the large output of the gauge;
- to abandon the rolling scheme with different diameters of the lower and upper rolls.

According to the actual data, the simplification of the task conditions of rolling into the rolls with the use of a new rolling regime made it possible to reduce the rejection of the finished rails OR65 by the defect of "captivity" by 0.5%.

Analyzing the processes of metal flow and the formation of a final shape of rails using the developed rolling scheme, it should be noted that the control of the width of the sole flanges and the profile head of the processed workpiece after the first and third four-roll gauges is carried out in the auxiliary open two-roll gauges of the edger stand. The final design of the profile of the switch point rail is carried out in a three-roll finishing universal gauge.

Groovs of this gauge forming the rail head are designed according to the same rules as for two-roll finishing gauges with a roll connector in the middle of the rolling surface of the profile head, and the sole is formed in a three-roll section of the gauge where the width of the sole is made by expansion of the metal in the roll connectors (figure 5 *a*).

Thus, the broadening of the short and long arm of the profile sole (ΔB_f) is made up of natural broadening (ΔB_e), due to the basic breakdown of these elements in the gauge, and the forced broadening that results from additional breakdown of the local metal deposits at the tips of the short and long arm of the sole (ΔB_b), obtained in the result of deformation of these elements in the auxiliary gauges of the horizontal edger stand:

$$\Delta B_f = \Delta B_e + \Delta B_b \tag{9}$$

$$\Delta B_e = \frac{2.54 \cdot \Delta t \cdot (\varepsilon_f - \varepsilon_w)}{\varepsilon_f} \tag{10}$$

$$\Delta B_b = \frac{k \cdot \Delta H \cdot t'}{\lambda \cdot t} \tag{11}$$

where $\Delta t = t' - t$ – average flange breakdown in thickness;

ε_f and ε_w – relative compression of the flanges and web, respectively;

$\varepsilon_f = \Delta t$; $\varepsilon_w = d' - d$

$\Delta H_f = H'_f - H_f$ – breakdown of flanges in height in the auxiliary stand;

λ – drawing coefficient in the universal gauge;

k – coefficient that takes into account the proportion of metal in the flange broadening (it is assumed 0.5 for breakdown mill and 0.7 for finishing stands).

The rail foot formation of the switch rail in the three-roll section of the gauge makes it possible to eliminate the moment typical for rolling in the two-roll gauge (figure 5 *b*), caused by the pinching of metal in the closed upper flange of the gauge that forms the long arm of the rail sole (M_3), which coincides with the direction of the twisting moment (M_C) caused by the difference in speeds along the perimeter of the final pass gauge of the switch point rail (figure 5 *a*), and reduce the torsionality when leaving the rolls.

At the same time, the attention should be paid to the limitations inherent in the developed rolling mode of switch point rails in the conditions of universal rail-and-structural steel mill. As shown above, the final formation of the rails profile in the finishing universal gauge by the vertical roller and lateral surfaces of grooves of horizontal rolls provides the formation of the width of the profile sole due to free broadening of metal upon deformation of this element (figure 5 *a*). As a result, the width of the profile sole in the final pass is achieved not by a tool, but by a free metal broadening, which depends on many technological parameters and can not be predicted with a high degree of accuracy. In this regard, there are natural difficulties in meeting the requirements of regulatory documentation on the accuracy of the size of the profile elements. The available production data of domestic and foreign metallurgical enterprises make it possible to speak about the achievable accuracy of the dimensions of the elements of rolling profiles obtained by hot rolling with the presence in the rolling scheme of free metal broadening ± 1.0 mm.

The value of this parameter is also confirmed by the results of modeling the processes of metal flow in the program complex "DEFORM-3D". At the same time, according to the current Russian GOST R 55820-2013, the maximum tolerance limits for the width of the short arm of rail foot for the accuracy classes of the "high grade" (X) and "first grade" (Y) profiles are set within ±0.3 mm and ±0.5 mm, respectively. In accordance with the foregoing, the fulfillment of this requirement is impossible, which requires the adjustments in the current regulatory and technical documentation for the production of switch point rails.

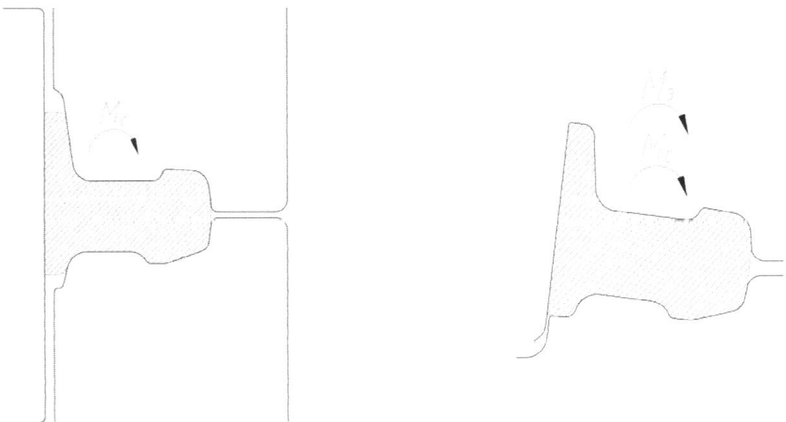

(a) (b)

Figure 5. General view of the universal three-roll (a) and finished two-roll (b) gauges for rolling switch point rails.

3. Conclusion

Based on the analysis of the production technology of switch point rails in the conditions of the universal rail-and-structural steel mill at EVRAZ ZSMK JSC, an intensified rolling mode has been developed that ensures the increase in mill productivity and improves the quality of rail products, as well as recommendations on improving the regulatory documentation governing the quality parameters of switch point rails.

References

[1] Shilov V A , Shvarts D L and Skosar' E O 2016 *Metallurgist* **60(3-4)** 260–266

[2] Stalinskii D V and Rudyuk A S 2011 *Steel in Translation* **41(5)** 73–77
[3] Svejkovsky U 2002 *AISE Steel Technology* **2** 33–39
[4] Former F E 2002 *AISE Steel Technology* **11–12** 27–35
[5] Samoilovich Yu A 2012 *Metallurgist* 55**(11-12)** 903–911
[6] Sinel'nikov V A and Filippov G A 2001 *Metallurgist* **45(9-10)** 403–407
[7] Golovatenko A V, Umansky A A and Dorofeev V V 2016 *IOP Conference Series: Materials Science and Engineering* **150** 012002
[8] Kogan A G and Shur E A 2002 *Proc. of Jubilee Rail Commission of "KMK"* (Novokuznetsk) pp 42–48
[9] Golovatenko A V, Umansky A A and Kadykov V N 2016 *IOP Conference Series: Materials Science and Engineering* **150** 012028
[10] Umansky A A, Golovatenko A V, Kadykov V N and Dumova L V 2016 *IOP Conference Series: Materials Science and Engineering* **150** 012029

A study on the influence of the ratio of pig iron and metal scrap in the of electric melting charge on the technical and economic performances of the rail steel production

A A Umansky and L V Dumova

Siberian State Industrial University, 42 Kirova street, Novokuznetsk, 654007, Russia

E-mail: umanskii@bk.ru

Abstract. The article contains the results of the statistical study on the influence of the ratio of pig iron and metal scrap in the charge for the production of rail steel on the main technical and economic performances of the electric furnace shop. The regression curves obtained and substantiated from a technological point of view indicate a linear decrease in the specific consumption of electricity and an increase in the specific consumption of oxygen according to the parabolic law with an increase in the proportion of both hot and cold iron in the metal scrap. Also, a linear reduction in the manganese content at the tapping was observed with an increase in the proportion of pig iron in the metal charge. The dependence of the duration of melting on the ratio of the charge components in the metal scrap has a pronounced nonlinear character and indicates the presence of a pronounced minimum. The statistical model constructed on the basis of the dependencies obtained is an effective tool for optimizing the share of pig iron in the metal charge in the production of rail steel for a specific period of time, taking into account the level of prices for materials and energy carriers.

1. Introduction

At present, in the world of metallurgy, the bulk of the steel of rail grades is smelted in oxygen converters. In the domestic metallurgy the production of rail steel in the arc steelmaking furnaces is carried out only at "EVRAZ ZSMK" JSC.

The main reason for the rarity of rails production from electric steel is the problem for ensuring a low and stable level of impurities of residual elements (chromium, nickel, copper) in the steel, which can not be solved without significant additional costs or a sharp limitation of the raw material base.

Nevertheless, the production of rails from electric steel seems promising for a number of reasons [1, 2]:

- the electric arc furnace is a universal unit that allows a wide range of technological operations to be carried out with high efficiency, and it is well combined with out-of-furnace processing units and continuous casting plants.
- the broader technological capabilities of electric arc furnaces allow the production of improved metallurgical rails to be organized, including those intended for operation in particularly rough conditions.

In fact, the electric arc furnace is a more flexible unit than the oxygen converter in the following parameters [3-5]:

- maintenance of the given rate of melting, heating, decarburization, homogenization in temperature and chemical composition;
- the possibility of alloying (including direct alloying of steel) of a metallic melt;

- provision of a predetermined melting rate for slag-forming constituents, formation of homogeneous slag with a high refining capacity for carrying out dephosphorization in the furnace, and with the release of a slag melt from the furnace and desulphurization, which facilitates the removal of non-metallic inclusions and the degassing of the steel.

At "NKMK" (which is currently a part of "EVRAZ ZSMK" JSC), the technology of melting rail steel in 100-tonne arc electric furnaces with the use of hot iron was developed and has been implemented since 2000 [6-8]. Beginning in 2010, due to the closure of its own blast furnace production, cold (pig iron) iron is used in the metal charge.

It should be noted that the replacement of part of the scrap with cast iron is dictated, first of all, by the need to meet the requirements for the concentration of impurities of non-ferrous metals. At the same time, the technical and economic effectiveness of such a replacement is not always obvious.

In particular, the technology of using cast iron in electric melting has such technological and economic disadvantages as:

- increase in the slag number due to the high content of silicon in the cast iron, which in turn causes an increase in the cost of slag utilization, increase in ferro-alloys loss;
- increase in the lime consumption to obtain the required basicity of the slag, as well as an increase in heat losses with slag;
- increase in the consumption of manganese-containing ferroalloys due to a lower manganese content in cast iron than scrap metal;
- increase in the costs for metal charge due to higher cost of pig iron as compared to scrap metal;
- increase in the consumption of oxidants due to the need to remove carbon introduced by pig iron from the melt.

Although, at the same time replacing part of the scrap with cast iron in the metal charge makes it possible to achieve a number of significant advantages:

- reduce the fuel consumption (natural gas);
- reduce the consumption of carburizing agents (coke);
- reduce costs for the preparation of scrap metal for smelting;
- reduce the concentration of nitrogen in the steel, due to the intensification of the oxidation period and the overall decrease in the duration of melting;
- reduce the consumption of refractory products due to the decrease in temperature differences and the absence of mechanical destruction of the lining when heavy metal charge is loaded;
- reduce the contamination of steel by exogenous non-metallic inclusions.

In general, the assessment of the technical and economic efficiency of the use of cast iron in the metal scrap of electric arc furnaces and the choice of the optimal consumption of cast iron should be carried out for the conditions of the specific unit and taking into account the current level of prices for materials and energy carriers. Thus, to obtain quantitative dependencies that adequately reflect the real situation, a necessary condition is statistical processing of the data array for a specific aggregate over a period of time with stable production technology.

2. Methods of research

The effect of the composition of the metal charge on the technical and economic performance of the electric furnace shop at "EVRAZ ZSMK" JSC was studied by regression analysis. As a basis for the analysis, an array of 1000 batches of rail steel E76F (2007) and E76HF (2016) was used. The choice of these brands is due to their largest share in the assortment of the shop in question.

Indicators of production of rail steel were analyzed in periods characterized by the following production scheme [9, 10]:

- smelting in the arc steel smelting furnace using 30-60% of hot iron with subsequent out-of-furnace processing of steel in the unit "ladle-furnace" and casting on the continuous casting machine into billets with a cross section of 300×330 mm (2007);

- smelting in the arc steel smelting furnace using 20-50% cold (pig iron) iron, successive out-of-furnace processing on two ladle-furnace units and a vacuum furnace and casting on the reconstructed continuous casting machine into 300x360 mm billets(2016).

The influence of the metal scrap composition on the following performances indicators of electric furnaces during the smelting of rail steel was studied: the specific consumption of electricity, oxygen, lime, manganese and silicon content on the tapping (characterizes the consumption of ferroalloys), irrecoverable metal losses, duration of melting.

3. Results and discussion

As a result of the analysis, a significant influence of the metal charge composition on the following parameters was established: the specific consumption of electricity and oxygen, the manganese content at the tapping, and the duration of melting.

According to the data obtained, with an increase in the share of both hot and cold iron in the metal scrap, the specific energy consumption decreases linearly (figure 1).

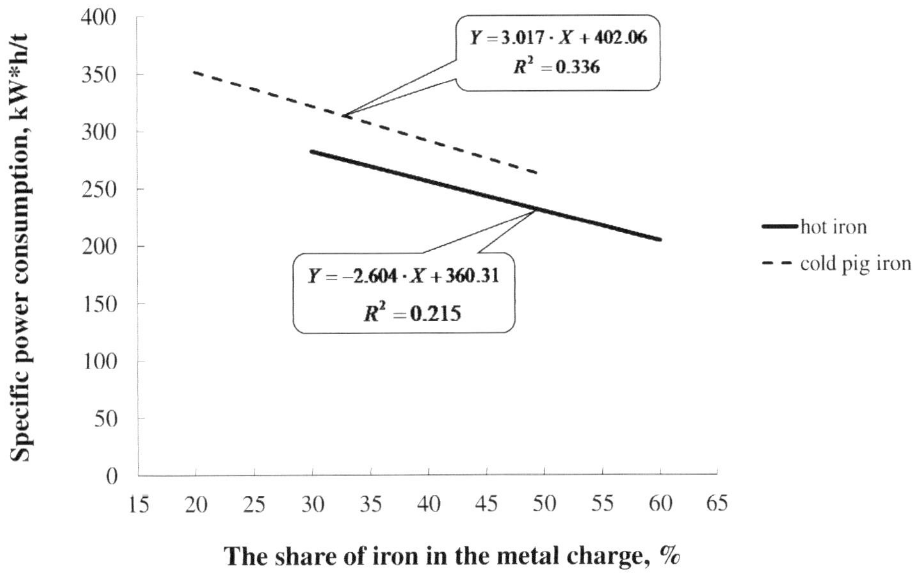

Figure 1. Dependence of the specific consumption of electricity on the share of cast iron in the metal scrap.

At the same time, the change in the consumption of hot iron slightly more significantly affects the power consumption compared to cold iron. This difference is due to the increased heat content of molten iron, which causes a reduction in the melting period. The lower melting point of cold iron as compared to the metal scrap also results in the decrease of melting time, which, however, is not as significant as in the case of hot iron.

The specific consumption of oxygen with an increase in the share of pig iron in the scrap metal increases according to the parabolic law (figure 2), regardless of the aggregate state of the cast iron used. The increased oxygen consumption is due to the need for oxidation of additional carbon added by the cast iron, and the parabolic nature of the obtained dependence indicates the increase in the intensity of oxygen blowing to reduce the duration of the oxidation period and the melting as a whole.

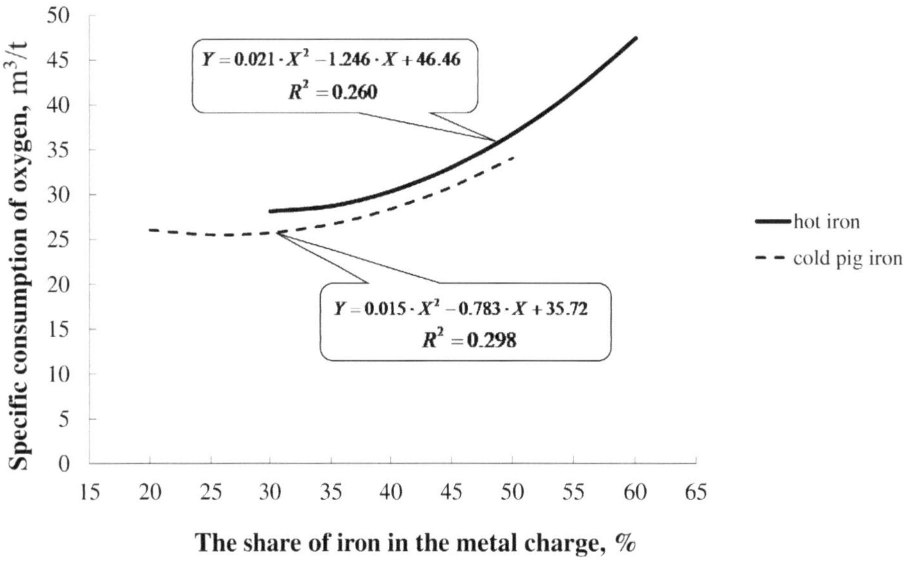

Figure 2. Dependence of the specific consumption of oxygen on the share of cast iron in the metal scrap.

The obtained results indicate a decrease in the manganese content at the tapping with the increase in the flow rate of cast iron to smelting (figure 3), which is explained by the reduced manganese content in pig iron relative to the scrap and increased manganese oxidation due to the increase in the duration and intensity of oxygen blowing.

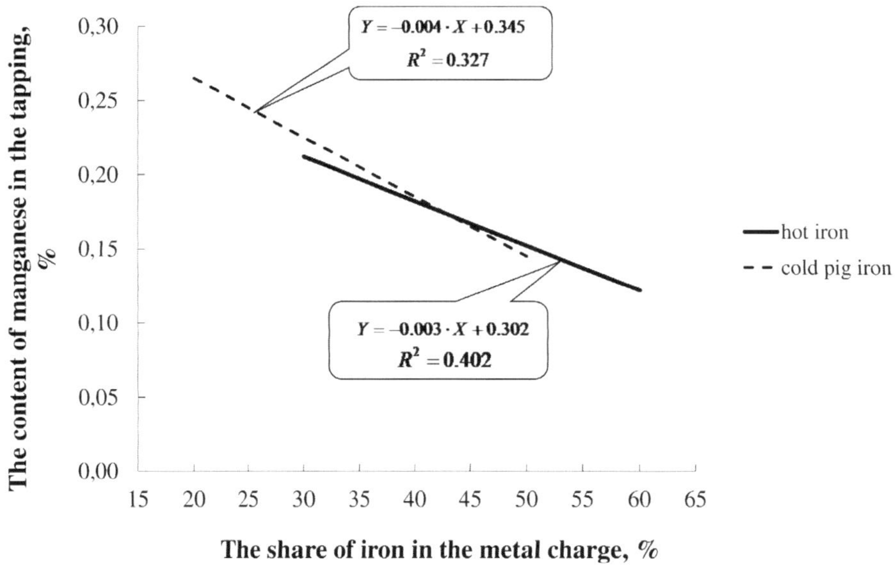

Figure 3. Dependence of manganese content at the tapping from the share of cast iron in the metal scrap.

The dependence of the smelting duration on the ratio of the charge components in the metal scrap shows that there is a pronounced minimum on the considered interval of variation in the consumption of cast iron (figure 4).

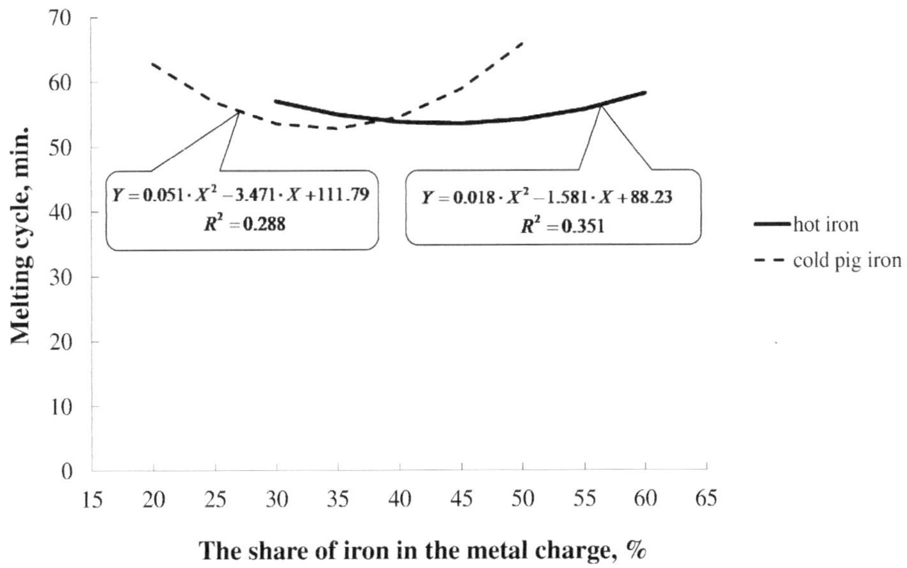

Figure 4. Dependence of the duration of electric melting on the share of cast iron in the metal scrap.

At the same time, when using hot iron this minimum is in the range of 35-40%, and when using cold iron it is 30-35%. This dependence is evidently due to the multidirectional influence of the increase in the consumption of pig iron on the duration of the melting periods (melting period, oxidation period, amount of additional charge of scrap) and nonlinear change in the duration of the periods themselves.

Thus, with the increase in the consumption of cast iron, the duration of the melting period decreases, while the duration of the oxidation period increases. In this case, as shown above, the increase in the consumption of hot iron has a stronger effect on the acceleration of melting of the charge compared to cold iron. Also, an increase in the consumption of cast iron leads to a reduction or total failure of scrap metal scrap, which naturally reduces the duration of melting.

Based on the obtained dependencies, a statistical model is constructed for the effect of the composition of the metal charge on the technical and economic performance of MSP during the smelting of rail steel. The following parameters we choose as optimization ones: total costs for cost items, which depend on the composition of the metal charge and the capacity of the shop in terms of proper continuous cast billets.

As forming costs the total costs for the components depending on the composition of the metal charge were: the cost for charge materials (pig iron and scrap), the cost for electricity, oxygen, ferroalloys:

The forming costs for the components, which depend on the composition of the metal charge, were as forming costs: the cost of the charge materials (pig iron and scrap), the cost of electricity, oxygen, ferroalloys:

$$C_{TOTAL} = (C_{P.IRON} \cdot P_{P.IRON} + C_{SCRAP} \cdot P_{SCRAP} + C_{EL.} \cdot P_{EL.} C_{OX.} \cdot P_{OX.} + C_{FER.} \cdot P_{FER.}), \tag{1}$$

where C_{IRON}, C_{SCRAP}, $C_{EL.}$, $C_{OX.}$, $C_{FER.}$ – specific costs of pig iron, scrap, electricity, oxygen and ferroalloys, respectively; $P_{P.IRON}$, P_{SCRAP}, $P_{EL.}$, $P_{OX.}$, $P_{FER.}$ – the price of pig iron, scrap, electricity, oxygen and ferroalloys, respectively.

The consumption of manganese-containing ferroalloys is inversely proportional to the content of manganese in the tapping. Taking into account the manganese content in the ferroalloy and its loss, the specific consumption of ferroalloys will be determined by the following formula:

$$SC_{FER.} = \frac{10 \cdot (C_{FIN} - C_{TAPPING})}{C_{FER.} \cdot (100 - L)}, \text{ t/t} \tag{2}$$

where C_{FIN} – content of manganese in the finished steel, %;

$C_{TAPPING}$ – manganese content in the tapping, %;

C_{FER} – manganese content in the ferroalloy, %;

L – manganese loss, %.

The productivity of the shop was determined by the formula:

$$P = \frac{1440 \cdot M}{100 \cdot T} \cdot Y \cdot F, \tag{3}$$

where M – is the mass of the metal scrap, t;

Y – metal yield,%;

F – worktime fund of the shop, day/year.

T – melting cycle, min.

The model is implemented in the spreadsheet program "Excel" and works as follows. As changing input data, the following are used: prices for pig iron, metal scrap, electricity, oxygen and manganese-containing ferroalloys, as well as parameters such as the average weight of the metal scrap, the yield, the actual operating time of the shop, the average manganese content in ferroalloys, manganese loss during deoxidation and alloying.

After entering the initial data, the program calculates the forecast values of electricity, oxygen, manganese content in the tapping, and the duration of melting using the regression equations shown in figures 1-4. Then, the forecasted value of total costs is calculated by the line items, depending on the ratio of pig iron and scrap in the metal charge and the predicted productivity of the workshop. The received data are displayed on the screen in the form of graphs.

The use of the developed model allows sound recommendations on the optimal share of pig iron in the metal scrap to be developed for the current price level for materials used for electric smelting and energy carriers and also taking into account the change in the shop productivity.

4. Conclusion

The statistical studies carried out for the conditions of the electric steelmaking shop of "EVRAZ ZSMK" JSC established qualitative and quantitative dependences of the technical and economic indicators of the production of rail steel on the ratio of pig iron and scrap in the metal charge. The resulting regression relations formed the basis of the model, the use of which makes it possible to develop sound recommendations on the optimal proportion of pig iron in the metal charge for a particular period, taking into account the current level of prices for materials and energy carriers, and also taking into account the change in the productivity of the shop.

References

[1] Katunin A I, Kozyrev N A, Obsharov M V, Timmerman N N and Sychev P E 2000 *Metallurgist* **44(11-12)** 560–562

[2] Godik L A, Kozyrev N A, Shabanov P A, Timmerman N N and Zakharova T P 2009 *Steel in Translation* **39(5)** 412–415

[3] Lipuhin Yu V, Molchanov O E and Baldaev B Ya 1997 *Stal'* 7 26–27

[4] Milyuts V G, Bocharnikov A F and Kulikov V V 1997 *Stal'* **8** 30–32
[5] Uymanov V A, Baldaev B Ya, Shurygin A V and Gromov G I 1997 *Proceedings of the Fourth Congress of Steelmakers* (Moscow: Chermetinformatsiya) pp 148–149
[6] Katunin A I, Godik L A, Obsharov M V and Kozyrev N A 2000 *Metallurgist* **44(6)** 299–301
[7] Katunin A I, Godik L A, Kozyrev N A, Timmerman N N and Sychev P E 2001 *Steel in Translation* **31(1)** 19–21
[8] Godik L A, Kozyrev N A, Danilov A P, Zakharova T P and Timmerman N N 2002 *Ehlektrometallurgiya* **1** 9–14
[9] Ayzatulov R S, Umanskiy A A, Chernysheva and N A 2009 *Proceedings of Mining and Metallurgy Section of the RANS* (Moscow – Novokuznetsk) **23** 39–46
[10] Dumova L V and Umanskiy A A 2017 *Proc. of II Int. Conf. on The Role of Technical Sciences in the Development of Society* (Kemerovo: KuzSTU) pp 131–134

Improvement of the extra-furnace rail steel processing on the "ladle-furnace" unit in order to increase the operational stability of railway rails

A A Umansky[1], NA Kozyrev[1], D V Boykov[2] and L V Dumova[1]

[1]Siberian State Industrial University, 42 Kirova street, Novokuznetsk, 654007, Russia
[2]EVRAZ Consolidated West Siberian Metallurgical Plant, 16 Kosmicheskoye shosse, 654043, Russia

E-mail: umanskii@bk.ru

Abstract. The article contains the results of studies of the influence of technological parameters of the extra-furnace rail steel processing in the conditions of the electric furnace shop of "EVRAZ ZSMK" JSC on the oxygen content in the finished rails and on their mechanical properties. The studies carried out using the multiple regression analysis showed that the oxygen concentration in the rail steel was reduced by 2 ppm with an increase in the duration of the inert gas blowing on the ladle-furnace unit to 100 min. At the same time, a decrease in the concentration and length of the rows of oxide brittle fracture nonmetallic inclusions, which have the most negative effect on the operational properties of the railway rails, was also recorded. In addition, the carried out experimental studies showed that the decrease in the concentration of oxygen in the rail steel by the specified amount, due to an increase in the duration of the blowing of the melt with an inert gas, makes it possible to increase the following mechanical properties of the rails: ultimate strength, elongation, relative constriction, hardness. Based on the conducted research, the technology of extra-furnace rail steel processing was developed and introduced at the electric steelmaking shop of "EVRAZ ZSMK" JSC in series on two "ladle-furnace" units, which allowed the recommended values of the duration of blowing with inert gas to be achieved without reducing the shop productivity.

1. Introduction

The results of comprehensive research of VNIIZhT and IMET RAS named after A.A. Baykov [1-4] testify that the main reason for the premature decommissioning of rails is the formation of contact fatigue defects in the working fillet of the rail head, the emergence of which is associated with the accumulation of nonmetallic inclusions, namely, brittle-decomposed oxide inclusions.

The brittle decomposed oxide inclusions in the degree of plasticity during deformation occupy an intermediate position between the so-called "plastic" inclusions, which change their shape and stretch into the rows during rolling (sulfide inclusions, plastic silicates) and "brittle" inclusions, which, during the similar treatment do not stretch, but are crushed (alumina, silica).

The fragile oxide inclusions, which are mainly complex oxides (aluminates, silicates, spinels), can stretch in the rows during deformation, but when a certain limiting value is achieved, the degrees of deformation get destroyed. The bulk of brittle-grained oxide inclusions in rail steel, in particular calcium aluminates (figure 1) are formed when it is deoxidized. And this fact takes place, despite the refusal to use aluminum as a deoxidizer for rail steels – aluminum as a by-product is introduced by ferroalloys (ferrovanadium, silicomanganese).

Figure 1. Calcium aluminates in the rail electric steel.

The formation of these defects occurs according to the following scheme (figure 2): at the initial moment a longitudinal inclined crack appears in the zone of accumulation of brittle fractured oxide inclusions, the development of which leads to the metal delamination – the metal spalling on the side fillet head (figure 2 a); then this defect is converted into transverse cracks (figure 2 b).

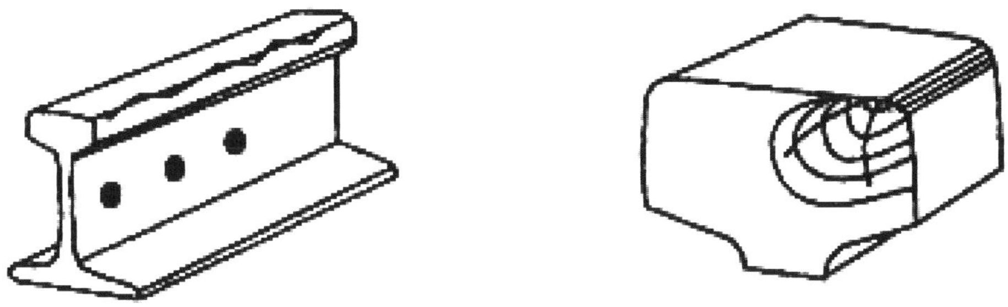

(a) (b)

Figure 2. Stages of formation of contact fatigue defects in the head of rails in the process of operation: a – formation of a longitudinal inclined crack; b – formation of transverse cracks.

Due to the sharply negative influence of these nonmetallic inclusions on the operational stability of rails, the current regulatory documentation, covering the rails production of various categories, provides for a strict regulation of not only the total oxygen content in the finished rails (not more than 20 ppm), but also oxygen bound to brittle fractured oxide inclusions (no more than 10 ppm).

At “EVRAZ ZSMK” JSC, which is currently one of the leading domestic manufacturers of railroad rails [5, 6], in recent years the work has been purposefully carried out to reduce the contamination of rail steel with oxygen and oxide nonmetallic inclusions. So for the period from 2005 to 2013 due to the introduction of a complex of technical and technological measures to improve the technological modes of smelting, deoxidation, extra-furnace rail steel processing and evacuation [7-9], it became possible to reduce the total oxygen content in rails by more than 2 times – to 14 ppm (figure 3).

However, in order to ensure compliance with the requirements of standards for the concentration of oxygen bound to brittle fractured oxide inclusions (not more than 10 ppm), further reduction of the total oxygen content in the rail metal to a level not more than 12-13 ppm is necessary.

2. Methods of research

In order to determine the nature and extent of the effect of technological parameters on the concentration of oxygen in the rail steel, multiple regression analysis was performed. The object of the study was a random sample from 190 melts of rail steel E76KhF of the current production.

The influence of the following technological parameters on the content of total oxygen in rail rolling was analyzed:

- the content of silicon in steel in the first sample on the “ladle-furnace” unit;

- carbon content in the tapping;
- consumption of coke breeze during steel is deoxidation in the tapping;
- consumption of silicomanganese during the steel deoxidation in the tapping;
- duration of blowing with inert gas during processing on the "ladle-furnace" unit;
- duration of steel processing in the vacuum unit.

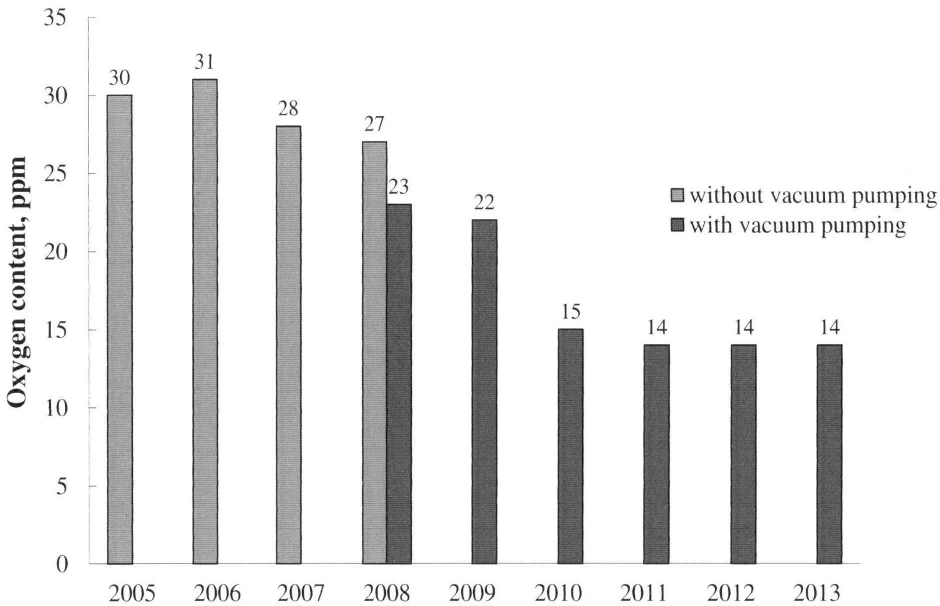

Figure 3. Dynamics of the total oxygen content in the finished rails produced by "EVRAZ ZSMK" JSC.

3. Results and discussions

The analysis established a significant effect on the content of total oxygen in the metal for the duration of metal blowing in the ladle furnace. According to the data obtained, with an increase in the duration of steel blowing on the "ladle-furnace" unit in the range 40-128 minutes, a decrease in the concentration of total oxygen in the metal is observed (figure 4). At the same time, a decrease in the content of total oxygen is achieved due to the decrease in the oxygen content bound to nonmetallic inclusions.

The dependence obtained is explained by the well-known phenomenon of intensification of nonmetallic inclusions transfer processes to the slag with increasing blowing duration. Since the inclusions transfer process by gas bubbles is due to the general capacity of non-wettable solids to adhere to the gas bubbles in the liquid – flotation, the adhesion forces exert a significant influence on the degree of melt refining, which in turn are determined by the dimensions of non-metallic inclusions and their composition.

Thus, the choice of the optimum blowing duration depends on the parameters influencing the composition of the resulting non-metallic inclusions, in particular on the steel grade, the deoxidizers used and the alloying additives. So, in relation to the conditions for the production of E76KhF rail steel be the electric steelmaking shop of "EVRAZ ZSMK" JSC, a significant decrease in the concentration of brittle fractured oxide inclusions is observed with an increase in the blowing duration more than 100 min.

The absence of a relation between the oxygen content in the steel and the parameters characterizing the oxidation of the melt (the carbon content at the furnace outlet and the silicon content in the first ladle sample at the ladle-furnace unit) is due to the order of addition of the deoxidants used. At present the small coke additive is carried out after the addition of silicomanganese, that is, in the deoxidized metal. The absence of the dependence of oxygen content in the steel on the slag basicity during the processing in the ladle-furnace unit is due to a rather narrow interval of values variation of this parameter (2.2-2.8).

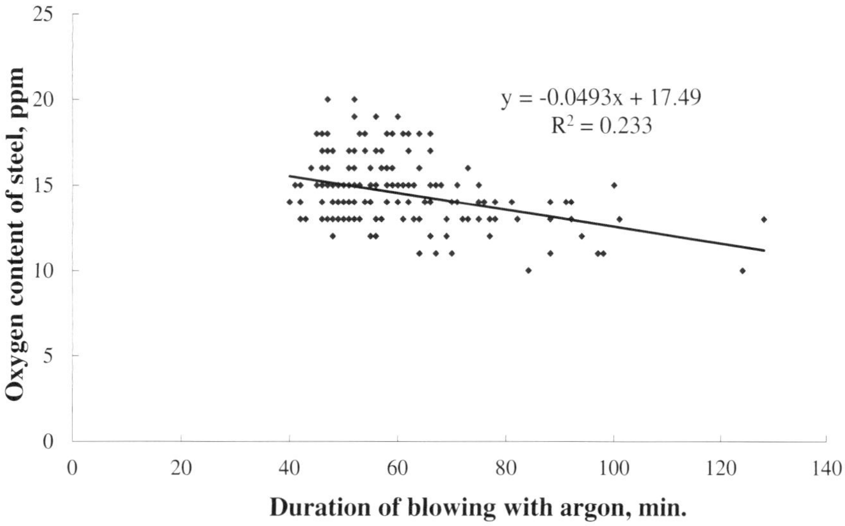

Figure 4. Dependence of oxygen content in the rail steel on the duration of blowing with inert gas during extra-furnace rail steel processing.

To analyze the changes in the technical and economic parameters of steel production, using a technology of secondary treatment with an extended duration of blowing with inert gas, a series of 15 experimental meltings was carried out, for which blowing duration was at least 100 minutes. As a basis for the comparative analysis, 19 smelting of the current production, produced in the same period, were used.

According to the data obtained (table 1), an increase in the average duration of metal blowing on the "ladle-furnace" unit from 63 min up to 122 min led to an increase in the specific electricity consumption by an average of 27.3 kW×h/t (from 64.9 kW×h/t to 92.2 kW×h/t steel); the specific consumption of argon increased by 0.32 m^3/t (from 0.24 m^3/t to 0.46 m^3/t).

At the same time, the content of total oxygen and oxygen bound into brittle fractured nonmetallic inclusions in rails obtained from the steel of experimental melts is lower by 2 ppm in comparison with rails from melts of current production. The length of the lines of nonmetallic inclusions on rails obtained from the steel of experimental meltings is lower by 0.004 mm.

Table 1. Analysis of changes in technical and economic indicators of the technology with increased blowing duration.

Indicator name	Mean values of the indicators: experimental melts	Mean values of the indicators: melts of the current production
Duration of blowing on LFU, min.	126.4	63.0
Argon consumption for blowing, m^3/t	0.46	0.24
Temperature, °C:		
- at the beginning of processing on the "ladle-furnace" unit	1549	1545
- at the end of processing on the "ladle-furnace" unit	1620	1620
Specific power consumption for automatic transmission, kW·h/t	92.2	64.9
Consumption of nitrated ferrovanadium, kg/t	1.44	1.44

Total oxygen content in rolled rail stock, ppm	12	14
The content of oxygen bound in brittle fractured nonmetallic inclusions, in rolled rail stock, ppm	8	10
Length of a line of brittle fractured oxide inclusions, mm	0.008	0.012

An analysis was also performed concerning the effect of the new technology on the mechanical properties of the rails after rolling (before hardening) with an extended duration of blowing with inert gas. As a result, it was established (table 2) that the rails obtained from the steel of experimental smelting have slightly better results compared to the rails obtained from the smelting of current production (an increase in such indicators as the ultimate strength, yield strength, relative contraction, elongation, hardness).

Table 2. Analysis of the influence of new technology of extra-furnace steel processing on the mechanical properties of the rails after rolling (before hardening).

Indicator name	Average values of the indicator		Requirements of GOST (not less than)
	experiment al melts	melts of the current production	
1. Yield strength (σ_t), N/mm^2	700	620	–
2. Strength limit (σ_v), Н/мм^2	1150	1130	1111
3. Specific elongation (δ), %	11.5	10.5	8.1
4. Contraction ratio, %	18.3	14.5	–
5. Hardness, HB	339	333	–

Increase in the duration of blowing with an inert gas up to 100 min and more, as was shown above, contributes to the decrease in the concentration of brittle fractured oxides in steel, which leads to an increase in the strength and plastic characteristics of rail products.

4. Conclusion

The experiments conducted in the conditions of the electric steelmaking shop of "EVRAZ ZSMK" JSC shows that when the blowing duration of the melt with an inert gas is increased during the processing on the "ladle-furnace" unit to 100 min, there is a decrease in the oxygen concentration in the rail steel grade E76KhF by 2 ppm, including oxygen, bound in the oxide brittle fractured nonmetallic inclusions, by the same amount. In this case, an improvement in the mechanical properties of the rails is achieved. Based on the obtained results, the technology of out-of-furnace processing of rail steel was developed and implemented in series on two "ladle-furnace" units, which allows the recommended values of the duration of blowing with inert gas to be achieved without reducing the productivity of the shop.

References

[1] Deryabin A A and Dobuzhskaya A V 2000 *Stal'* **11** 38–43
[2] Deryabin A A, Rabovskiy V A and Shur E A 2000 *Stal'* **11** 82–85
[3] Deryabin A A and Mogil'nyi V V 2010 *Steel in Translation* **40**(5) 464–471
[4] Deryabin A A, Semenkov V E, Matveev V V, Dobuzhskaya A B and Karimov Kh I 2004 *Stal'* **1** 58–61
[5] Golovatenko A V, Umansky A A and Dorofeev V V 2016 *IOP Conference Series: Materials Science and Engineering* **150** 012002
[6] Golovatenko A V, Umansky A A and Kadykov V N 2016 *IOP Conference Series: Materials Science and Engineering* **150** 012028

[7] Yurev A B, Kozyrev N A, Boikov D V, Feiler S V and Zakharova TP 2013 *Steel in Translation* **43(2)** 42–46
[8] Kozyrev N A, Protopopov E V, Aizatulov R S and Boikov D V2012 *Steel in Translation* **42(2)** 110–113
[9] Korneva L V, Yunin G N, Kozyrev N A, Atkonova O P and Polevoi E V 2010 *Steel in Translation* **40(12)** 1047–50
[10] Kozyrev N A, Protopopov E V, Umanskii A A and Boikov D V 2015 *Steel in Translation* **45(10)** 717–722

IOP Conf. Series: Materials Science and Engineering **411** (2018) 012079 doi:10.1088/1757-899X/411/1/012079

Experience in the use of technogenic materials for the production of fluxes used for forming rolls surfacing

A A Umansky, N A Kozyrev, R E Kryukov, P D Sokolov and L V Dumova

Siberian State Industrial University, 42 Kirova street, Novokuznetsk, 654007, Russia

E-mail: umanskii@bk.ru

Abstract. The article contains the results of studies on the feasibility of using slag from silicium manganese production and flux additives based on aluminum purification waste products in the manufacture of welding fluxes for surfacing forming rolls. Based on the conducted experimental studies, the regularities of the influence of fractional and component composition of welding fluxes made using the above-mentioned technogenic wastes on the mechanical properties and quality of the microstructure of the deposited metal layer were established. In particular, it is shown that the maximum impact strength is achieved at a specific fraction content less than 0.45 mm in the charge for welding flux production at a level of 15 to 30%. It was established that with the additional use of liquid glass as a binder component in the charge for the production of welding fluxes, the highest mechanical properties of the weld metal layer are achieved at a flow rate of liquid glass within 20-30% of the mass of the charge. It is shown that the increase in the proportion of the flux additive from 2% to 8% in the charge for the production of welding fluxes causes the increase in the impact strength of the weld metal layer by more than 2 times and leads to the decrease in contamination with nonmetallic inclusions. The technology of rolls surfacing using the developed new fluxes has undergone industrial testing and has been introduced into production.

1. Introduction

In recent years, technologies have been actively introduced to restore the working surface of forming rolls by arc surfacing under a flux. The use of deposition can significantly increase the durability of the forming rolls gauges, which causes a reduction in the specific consumption of rolls for the production of rolled products, a reduction in the duration of downtime due to the increase in the duration of the overhaul period, and, ultimately, reduces the cost of finished rolled products while increasing the productivity of rolling mills. In addition, increasing the durability of forming rolls positively affects the quality of finished products.

According to the available literature and production data [1-4], the efficiency and quality of the surfacing work is largely determined by the welding-technological properties of the flux cored wires and welding fluxes used. On the basis of earlier studies with the participation of the authors [5, 6], new compositions of powder wires for surfacing forming rolls have been developed, the distinctive feature of which is the introduction of technogenic waste into the initial charge, namely carbon-fluorine-containing dust. As part of the development of the field for improving the composition of surfacing materials using technogenic waste, new welding fluxes have been proposed [7, 8], in the manufacture of which slag of silicomanganese of the following composition was used,%: 6.91-9.62 Al_2O_3; 22.85-31.70 CaO; 46.46-48.16 SiO_2; 0.27-0.81 FeO; 6.48-7.92 MgO; 8.01-8.43 MnO; 0.28-0.76 F; 0.26-0.36

Na_2O; to 0.62 K_2O; 0.15-0.17 S; 0.01 P. It should be noted that single attempts to use slag, which forms manganese-containing ferroalloys, for the production of welding materials have also been undertaken by a number of foreign researchers [9].

2. Methods of research

In order to substantiate the optimal component and fractional composition of the fluxes, a series of experimental studies was carried out. The deposition under flux of various compositions was produced on samples of 500×75 mm in size with a thickness of 16 mm made of steel grade 09G2S. At the same time, Sv-08GA wire was used using the ASAW-1250 welding tractor in the following conditions: Iw = 700 A; Ua= 30 V; Vw = 35 m/h. Metallographic studies of microsections were carried out using an optical microscope OLYMPUS GX-51 in a bright field at various magnifications after etching in an alcoholic nitric acid solution as well as in a solution of hydrofluoric acid. The chemical composition of the weld metal was determined using an XRF-1800 X-ray fluorescence spectrometer and atomic emission spectrometer DFS-71.

3. Results and discussion

At the first stage, the following flux compositions were used in the study (table 1).

Table 1. The fractional and component compositions of the investigated fluxes.

Sample number	Ratio, %, of fractions, mm
1	100 % fractions 0,45 – 2,5
2	95 % of fractions 0.45 – 2.5 + 5 % of fractions <0.45
3	90 % of fractions 0.45 – 2.5 + 10 % of fractions < 0.45
4	85 % of fractions 0.45 – 2.5 + 15 % of fractions < 0.45
5	80 % of fractions 0.45 – 2.5 + 20 % of fractions < 0.45
6	70 % of fractions 0.45 – 2.5 + 30 % of fractions < 0.45
7	60 % of fractions 0.45 – 2.5 + 40 % of fractions < 0.45
8	60 % of silicomanganese slag + 40 % of liquid glass
9	70 % of silicomanganese slag + 30 % of liquid glass
10	80 % of silicomanganese slag + 20 % of liquid glass
11	85 % of silicomanganese slag + 15 % of liquid glass

The results analysis of the samples mechanical tests made it possible to establish that the maximum impact strength at the test temperature 20°C is reached when the flux contains 15-30% of the fraction less than 0.45 mm (figure 1).

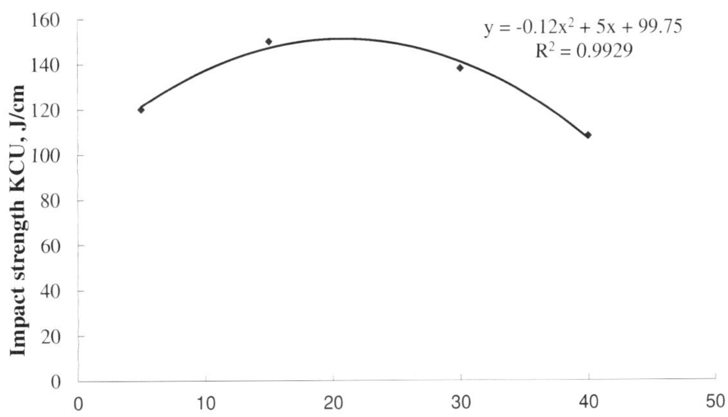

Figure 1. The dependence of the impact strength of the welded metal layer on the proportion of fraction <0.45 mm in the flux.

The carried out metallographic analysis made it possible to establish that in the metal structure of all samples ferrite is present in the form of non-equal grains stretched in the direction of heat removal (figure 2).

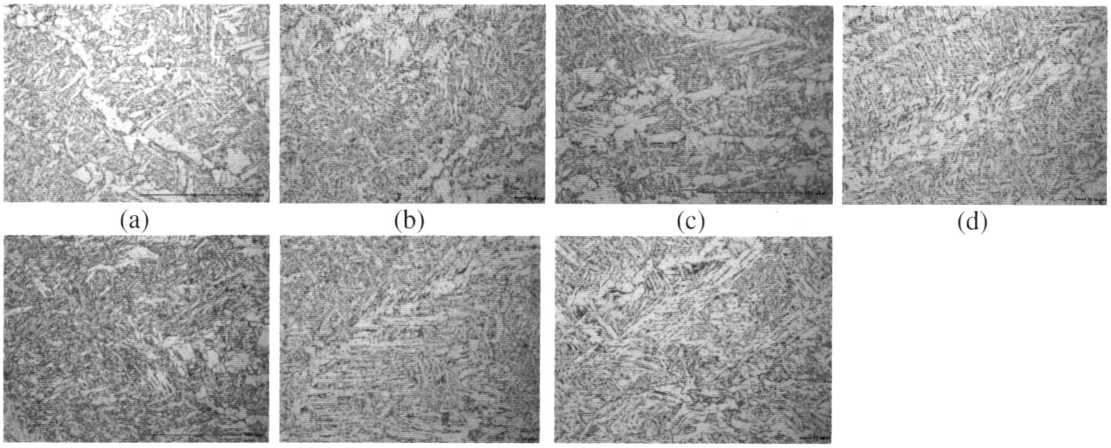

Figure 2. The microstructure of the deposited layer of samples: a – No. 1; b – No. 2; c – No. 3; d – No. 4; e – No. 5; f – No. 6; g – No. 7.

In this case, the transition from a uniform ferrite-pearlite structure to the structure of perlite and ferrite of the vismanthtite orientation is noticeable. It should also be said that in the samples there was no significant change in grain size (table 2).

Table 2. The grain size of the welded metal layer according to GOST 5639-82.

Sample number	The grain size on the grain scale
1	No. 4, No.5
2	No.5, No.4
3	No.4, No.5, No.6
4	No.4
5	No.5, No.4
6	No.4
7	No.4
8	No.5, No.4
9	No.4, No.5
10	No.4
11	No.4, No.5

According to the data obtained, the optimal content of liquid glass in the flux from the point of view of ensuring high mechanical properties of the deposited layer is 20-30% (figures 3-5).

It should be noted that a significant drawback of the developed fluxes is an increase in the level of contamination of the welded layer by nonmetallic inclusions, since the studied fluxes are oxidizing and are built on the principles of silica-manganese-oxidation-reduction processes.

In order to eliminate this shortcoming, the effectiveness of introducing a FD-UFS additive made on the basis of gas cleaning waste from aluminum production in the amount of 2, 4.6 and 8% in the new flux was conducted.

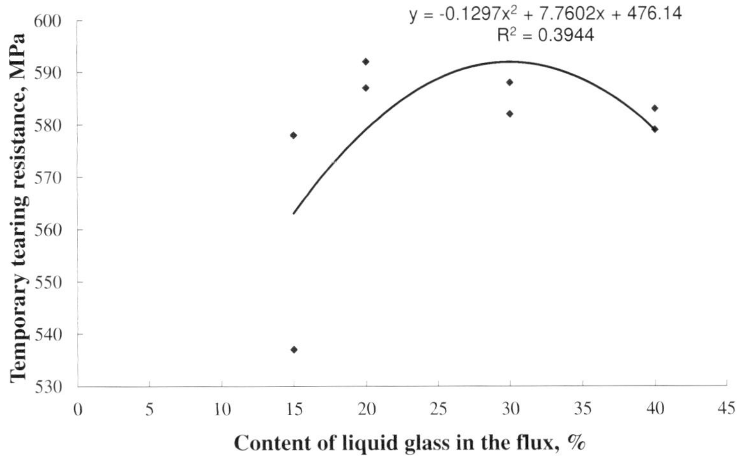

Figure 3. Dependence of the temporary tearing resistance of the welded metal layer on the content of liquid glass in the flux.

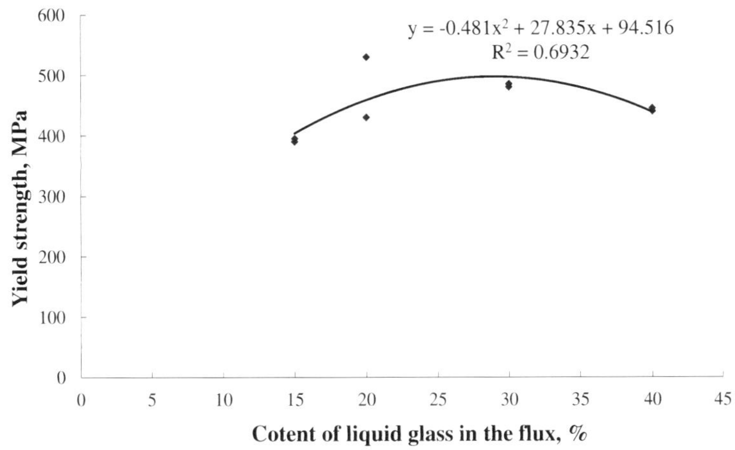

Figure 4. Dependence of the yield strength of the welded metal layer on the content of liquid glass in the flux.

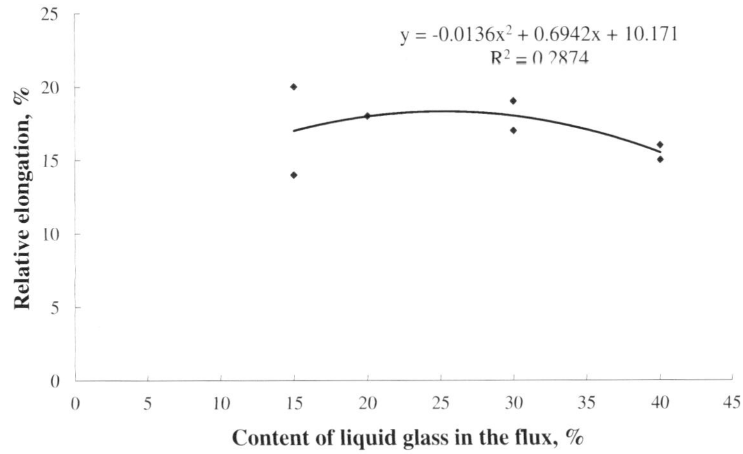

Figure 5. Dependence of the relative elongation of the welded metal layer on the content of liquid glass in the flux.

Chemical composition of FD-UFS additive,%: 21-46 Al_2O_3; 18-27 F; 8-15 Na_2O; 0,4-6 K_2O; 0.7-2.3 CaO; 0.5-2.5 SiO_2; 2.1-3.3 Fe_2O_3; 12.5-30.2 C_{tot}; 0.07-0.9 MnO; 0.06-0.9 MgO; 0.09-0.19 S; 0.10-0.18 P. The chemical composition of fluxes using this additive is given in table 3.

Table 3. Chemical composition of the studied mixtures of fluxes, %.

The content of FD-UFS additive in the flux, %	FeO	MnO	Ca	SiO_2	Al_2O_3	MgO	Na_2O	K_2O	S	P	ZnO	F
2	0.40	8.01	15.80	50.08	11.55	7.39	0.77	0.63	0.22	0.008	0.002	1.30
4	0.91	7.90	17.72	46.63	10.32	6.63	1.10	0.68	0.24	0.01	–	1.95
6	0.81	7.68	16.79	43.64	11.27	5.71	2.25	0.65	0.34	0.01	0.003	4.04
8	0.46	7.46	16.00	43.64	11.86	5.56	2.30	0.60	0.33	0.01	0.002	3.96

The results of metallographic studies of the deposited layer of samples using a flux with the addition of FD-UFS (table 4, figure 6) indicate a decrease in contamination with nonmetallic inclusions with an increase in the proportion of this flux additive in an amount up to 8%. It should be noted that there are no brittle silicates in the structure of the metal.

Table 4. Non-metallic inclusions in the deposited layer.

Content of FD-UFS additive in the flux,%	Content of nonmetallic inclusions, point	
	non-deforming silicates	point oxides
2	2b, 4b, 5a	1a, 2a
4	2b, 4b	1a, 2a
6	2b, 4b, 1b	1a, 2a
8	2b	1a, 2a

(a) (b) (c) (d)

Figure 6. Non-metallic inclusions in the weld zone of samples with a flux additive in the amount of: 2% (a); 4% (b); 6% (c); 8% (d).

The analysis of the microstructure of the deposited layer showed that the introduction of a flux additive of FD-UFS grade in the amount of up to 8% does not affect the size and morphology of the structural components (figure 7).

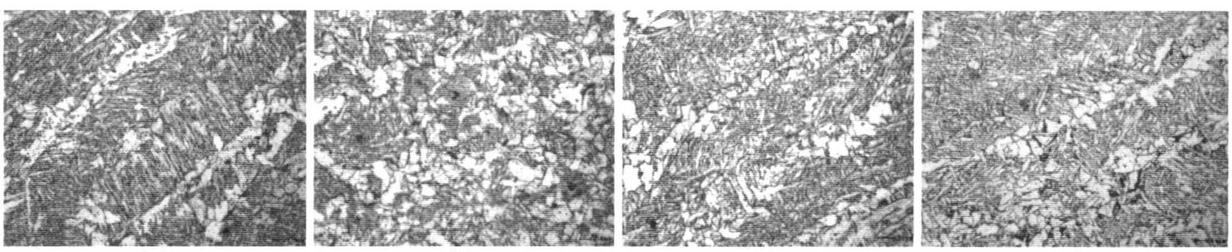

(a) (b) (c) (d)

Figure 7. Microstructure of welded joints of samples with a flux-additive in the amount: 2% (a); 4% (b); 6% (c); 8% (d).

A study of the mechanical properties of the samples showed that when the amount of the FD-UFS additive is increased from 2% to 8%, the impact strength increases by more than 2 times. The technology of surfacing rolls using the developed new fluxes has undergone trial-industrial testing and has been introduced into production.

4. Conclusion

The experimental studies have shown the efficiency of simultaneous use of slag from the production of silicomanganese and flux additives based on gas purification waste from aluminum production during manufacture of welding fluxes for surfacing forming rolls. The optimum fractional and component composition of the charge for the production of welding fluxes based on these technogenic wastes is determined, which ensures the improvement of operational and structural characteristics of the welded metal layer.

References

[1] Kondratiev I A and Ryabtsev I A 2014 *The Paton Welding Journal* **6** 95–96
[2] Golyakevich A A, Orlov L N, Malinov L S and Titarenko V I 2016 *The Paton Welding Journal* **9** 33–37
[3] Crespo A C, Puchol R Q, González L P, Gómez Pérez C R, Cas Tellanos G, Díaz Cedré E and Ortíz T 2009 *Welding International* **23(2)** 130–131
[4] Shebanits E N, Omelyanenko N I, Kurakin Yu N, Matvienko V N, Leshchinskii L K, Dubinskii B Eand Stepnov K K 2012 *Metallurgist* **56(7-8)** 613–617
[5] Kozyrev N A, Kibko N V, Umanskii A A, Titov D A and Nikitin A G 2016 *Steel in Translation* **46(10)** 711–717
[6] Kozyrev N A, Kibko N V, Umanskii A A, Titov D Aand Bashchenko L P 2016 *Steel in Translation* **46(11)** 781–787
[7] Kozyrev N A, Kryukov R E, Kozyreva O E, Lipatova U I and Filonov A V 2016 *IOP Conference Series: Materials Science and Engineering* **125** 012034
[8] Kozyrev N A, Kryukov R E, Lipatova U I and Kozyreva O E 2016 *IOP Conference Series: Materials Science and Engineering* **150** 012032
[9] Cruz-Crespo A, Quintana-Puchol R, Perdomo González L, Gómez-Pérez CR, García-Sánchez L L, Ejiménez-Vielsa G and Cores-Sánchez A 2005 *Welding International* **19(7)** 544–551

Evaluation of technological risks in the work of rolling equipment

A R Fastikovskiy[1], V V Dorofeev[2] and O Yu Efimov[3]

[1]Siberian State Industrial University, 42 Kirova street, Novokuznetsk, 654007, Russia
[2]EVRAZ Consolidated West Siberian Metallurgical Plant, 16 Kosmicheskoye shosse, 654043, Russia
[3]LLC "PROMMEST", 4A per. Ulyany Gromovoy, St. Petersburg, 191036, Russia

E-mail: omd@sibsiu.ru

Abstract. The approach to serviceability evaluation of the system rolling stand – rolling accessories is considered. The criteria defining the upper and lower levels of the considered system operability are proposed. Practical use of the developed provisions can be illustrated by rolling a circle with diameter 12, 14 mm and accessories No. 12, No. 14 in the conditions of a continuous small-section mill 250-1 of JSC "EVRAZ ZSMK" section-rolling shop. The received graphic data allow problem gaps to be revealed and the differentiated monitoring of the considered system to be organized.

1. Introduction

Forming rolls and their accessories form a single interconnected complex in the rolling mill, the performance and quality of the products being largely dependent on the precise operation of its work. In connection with this, when adjusting and operating the rolling mill-accessory system, it is necessary to take into account the forming conditions in the deformation center of the operating rolling stand. Ignoring the potential of the deformation center often leads to incidents in the system under consideration [1]. Guided by the conception under consideration about the inextricable link between the deformation center and the roll accessories, a technique for assessing the degree of technological risks in the system forming rolls – accessories was developed.

2. Methods of research

In practical implementation of the developed procedure, the first step determines the magnitude of the longitudinal force provided by the friction force reserve of the operating stand, taking into account the deformation conditions and design features of the gauge, and is compared with values corresponding to the system operability criteria [2 - 4]. To determine the performance criteria of the system forming rolls – accessories, we consider two limiting cases: the first one – the longitudinal force resulting from the incidents is so great that it can lead to breakage of the details of the roller accessories, the second one – the roller accessories must perform a large set of operations (holding the bar in the given position, correction before and after the center of deformation, tapering with twisting, etc.), which requires a considerable amount of longitudinal force, which can not provide a reserve of frictional forces in the deformation center of the stand, which leads to bar sticking. The considered situations characterize the upper and lower limits of the system operability.

IOP Conf. Series: Materials Science and Engineering **411** (2018) 012080 doi:10.1088/1757-899X/411/1/012080

The upper limit is limited by the strength of the parts of the roller accessories, the most important of which is the reinforcing bar, the lower limit is the magnitude of the longitudinal force required for the accessories to perform the functions assigned to it. Comparing the values of the longitudinal forces corresponding to the upper and lower level of the system's operability with the amount of effort that the friction force reserve can provide in the deformation area of the serviced stand, the degree of technological risk of the system is estimated.

Dependencies were obtained to determine the upper limit of operability, based on the strength of the reinforcing bar.

For the case of placing the tested gauge in the strictly defined place on the body of roll:

$$Q_{ad} = \frac{[\sigma]}{\sqrt{\left(\frac{6(l_m - a_{\max})a_{\max}}{l_m b_m h_m^2}\right)^2 + 3\left(\frac{d_m a_{\max}}{l_m \alpha_m \eta_m b_m h_m^2}\right)^2}}, \tag{1}$$

where l_m – the length of the reinforcing bar, mm; b_m, h_m – respectively, the width and height of the bar ($b_m > h_m$), mm; d_m – distance from the rolling line to the reinforcing bar, mm; a_{max} – the maximum distance from the force application to one of the supports, mm; $[\sigma]$ – allowable tension, MPa; α_m, η_m – correction factors, depending on the ratio of the sides b_m/h_m.

When rolling one workpiece for the case of cutting on the rolls of the same gauges:

$$Q_{ad} = \frac{[\sigma]}{\sqrt{\left(\frac{1{,}5 l_m}{b_m h_m^2}\right)^2 + 3\left(\frac{d_m}{2\alpha_m \eta_m b_m h_m^2}\right)^2}}. \tag{2}$$

When rolling into two blanks with the same gauges on the body of roll:

$$Q_{ad} = \frac{[\sigma]}{\sqrt{\left(\frac{6\left(l_m - \frac{B_K}{2} - b_{clamp}\right)\left(\frac{B_K}{2} + b_{clamp}\right) + 1{,}5 l_m^2}{l_m b_m h_m^2}\right)^2 + 3\left(\frac{2d_m\left(l_m - \frac{B_K}{2} - b_{clamp}\right) + d_m l_m}{2 l_m \alpha_m \eta_m b_m h_m^2}\right)^2}}, \tag{3}$$

where b_{clamp} – the clamps width at the edge of the body of roll, mm.

The lower level of the operability evaluation of the system rolling stand – roller accessories can be determined on the basis of the tasks solved by the roller accessories during operation.

Roller accessories solves the following main tasks [5]:

1) workpiece supply in a strictly fixed position to a certain gauge and metal removal from the gauge;

2) retention of the workpiece in the required position, in the roll throat during rolling, preventing stalling, providing the given direction of motion after rolling;

3) correction of the roll;

4) workpiece tapering;

5) small plastic deformation of individual thickened parts of the rolling.

IOP Conf. Series: Materials Science and Engineering **411** (2018) 012080 doi:10.1088/1757-899X/411/1/012080

All the operations listed above require a certain amount of longitudinal force, which determines the lower level of performance in the system rolling stand – roller accessories. Let us consider how to determine the values of the longitudinal force necessary to perform the tasks of the roll accessorie.

The component of longitudinal force ensuring the prevention of stalling, twisting (Q_t), can be determined by the formula:

$$Q_t = M_{pk}\Theta_s, \tag{4}$$

where M_{pk} – the torque necessary for twisting, N·m; Θ_c – the relative angle of twisting while keeping from stalling, rad.

The component of the longitudinal force needed to straighten the front or rear ends (Q_f), can be found:

$$Q_b = \frac{W_{pl.b.}\sigma_s}{x_e\left(1-\frac{x_e}{l_b}\right)}\mu_b, \tag{5}$$

where $W_{pl.b.}$ – the moment of resistance under plastic bending, mm^3; x_e – the distance from the front end of the wiring to the point where the force is applied when straightening, mm; l_e – length of the wiring, mm; μ_n – the coefficient of friction between the bar and the wiring.

The component of the longitudinal force required for plastic deformation (Q_d) of individual thickened parts of the workpiece in the roller accessories:

$$Q_d = \sigma_s b_0 h_0 \ln\lambda, \tag{6}$$

where σ_s – the deformation resistance, MPa; b_0, h_0 – respectively, the width and thickness of the bar before entering the rolling stand, mm; λ – the stretching ratio of the thickened parts of the workpiece.

The longitudinal force necessary for the tapering (Q_k) can be determined by the formula:

$$Q_t = \frac{M_{p.k}\varphi_{tap}l_b}{l_c l_0}, \tag{7}$$

where $M_{p.k}$ – the twisting moment, n·m; φ_{tap} – the full angle of tapering, rad; l_b – distance between the axes of work rolls and tapering rolls, mm; l_0 – the distance between adjacent stands, mm; $l_з$ – the value of the capture arc during tapering, mm.

The value of the longitudinal force necessary for the operation of the roller accessories (Q_n) can be determined as the sum of all components by the following general formula:

$$Q_n = Q_s + Q_f + Q_d + Q_t. \tag{8}$$

The above formula is general and for each specific pass it should be specified taking into account the features of rolling and the problems solved by the roll accessories.

The condition of trouble-free operation of the system rolling stand – roller accessories can be formulated as an inequality:

$$Q_{ad} \geq Q_1 \geq Q_n, \tag{9}$$

where Q_l is the longitudinal force provided by the friction reserve, MPa.

Carrying out the above inequality ensures the trouble-free operation of the system. Exceeding the indicated limits creates a situation of technological risk, in which the failure of the system is very probable. The possible technological risk for non-fulfillment of inequality (9) is controlled, so if $Q_n > Q_l$, then this situation can be corrected by increasing the degree of deformation in the serviced stand by redistributing the drawing through the passes or reducing the load on the roller accessories when the gauge is changed. In the case where $Q_l > Q_{ad}$, the situation can be corrected by redistributing the drawing coefficients in the stands, reinforcing the elements structure of the roller accessories in multistrand rolling with some time overlap.

3. Results and discussion

Method for assessing the system performance rolling stand – rolling accessories is developed using MS Excel application with results interpretation in a graphical form, which is very convenient for analysis. The developed methodology and this software application allow the range of products of the mill to be visually analyzed and problem gaps to be identified from the point of view of the system operation reliability rolling stand – rolling accessories.

The possibilities of the developed technique will be considered using the example of analyzing the working gauges of circles with a diameter of 12, 14 mm and accessories No. 12, No. 14 rolled on a continuous small-section mill 250-1 of a long product rolling mill at JSC "EVRAZ ZSMK". Calculations were carried out using the above dependencies, the magnitude of the longitudinal force provided by the friction force reserve according to the data of [6] is shown in figure 1. The graph shows the upper and lower limits of the system operability under consideration.

All stands in which the magnitude of the longitudinal force is comparable with the allowable value or exceeds the indicated limits are high risk stands in the system rolling stand – rolling accessories, and require additional attention during the installation of the accessories and during operation. As it follows from figure 1, when rolling circles with a diameter of 12, 14 mm, accessories No. 12, No. 14 on a continuous small-section mill 250-1 of a long product rolling mill at JSC "EVRAZ ZSMK" at an incident on one line in the first stand, the magnitude of the longitudinal force exceeds the permissible value, obtained based on the strength of the reinforcing bar, with the simultaneous incident on two lines in stands from A to 5, the same pattern is observed that must be taken into account when setting and operating the accessories.

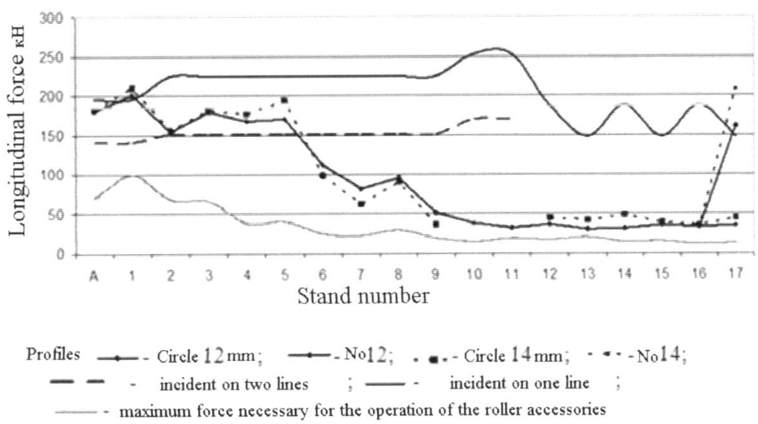

Figure 1. Evaluation of the system performance rolling stand – roller accessories, for rolling circles 12, 14 mm and accessories No. 12, No. 14 on the continuous small-section mill 250-1 of a sort-rolling shop at JSC "EVRAZ ZSMK".

In all passes of the considered example the magnitude of the longitudinal force, necessary for the accessories to perform their functions, is less than the longitudinal force that can provide the reserve forces of friction in the deformation center of the serviced cage, which guarantees the normal operation of the accessory.

4. Conclusion

A new technique for assessing technological risks in the operation of rolling equipment is proposed, which makes it possible to identify potentially hazardous passes and to organize a differentiated monitoring of the system operation rolling stand – roller accessories, to perform sound calculations of details of the roller accessories, to evaluate and correct the gauges in terms of reducing the number of possible incidents.

References

[1] Fastikovskiy A R and Saveliev A N 2015 *Features of the Design and Trouble-free Operation of the Roller Accessories of Mills* (Moscow: Teplotekhnik) p 170
[2] Fastikovskiy A R and Peretyatko V N 2001 *Izv. Vuzov. Ferrous Metallurgy* **12** 5–8
[3] Fastikovskiy A R and Peretyatko V N 2001 *Metallurg* **12** 43–44
[4] Fastikovskiy A R and Peretyatko V N 2002 *Izv. Vuzov. Ferrous Metallurgy* **4** 22–24
[5] Fastikovskiy A R and Savelyev A N 2008 *Designs and Calculations of the Equipment of Rolling Stands of High-quality and Sheet Mills* (Novokuznetsk: SibSIU) p 316
[6] Fastikovskiy A R, Efimov O Yu, Chinokalov V Ya and Kopylov I V 2008 *Steel* **2** 63–64

IOP Conf. Series: Materials Science and Engineering **411** (2018) 012081 doi:10.1088/1757-899X/411/1/012081

Performance analysis of the rolling equipment using computer modeling

A R Fastykovskiy[1], V N Peretyatko[1], V V Evstifeev[2] and O Yu Efimov[3]

[1]Siberian State Industrial University, 42 Kirova street, Novokuznetsk, 654007, Russia
[2]Siberian State Automobile and Highway University, 5 Prospekt Mira, Omsk, 64080, Russia
[3]LLC "PROMMEST", 4A per. Ulyany Gromovoy, St. Petersburg, 191036, Russia

E-mail:omd@sibsiu.ru

Abstract. The possibilities of modern computer technologies for the strength analysis of rolling equipment are shown. The example of calculation of a frameless stand 630 medium-section mill 450 at JSC "EVRAZ ZSMK" using the strength analysis module APMFEM Compass-3D is considered.

1. Introduction

Rolling equipment is widely used at metallurgical plants to produce finished products. The operating conditions of this equipment are associated with large static and dynamic loads, high temperature. Such operating conditions create high risks of incidents, significantly affect the quality of products. The cost of the main rolling equipment and the cost of its operation are very high and therefore all unforeseen stops and breakdowns significantly affect the cost and competitiveness of rolling products.

2. Methods of research

One of the possible ways to extend the service life of rolling equipment reducing the costs for its operation is a thorough calculation of strength with maximum consideration of all factors of influence. The issue of strength calculations of rolling equipment in the technical literature has been given great attention [1 - 3]. Available calculation techniques allow a preliminary assessment of the performance of rolling equipment to be performed, but their accuracy is not enough, calculations are cumbersome, many factors are left without attention. With the development of information technologies it became possible to solve the problems of equipment strength analysis at a new modern level, taking into account the influence of a great variety of factors.

The latest versions of Compass-3D, SolidWorks software contain modules for strength analysis of equipment and structures. Modules of strength analysis are based on the principles of three-dimensional finite element modeling and allow the distribution of stresses, safety factors for the yield strength and stability of a part, temperature at various points by volume to be found.

The use of the above mentioned software packages allows the designer to make correct and valid constructive decisions already at the initial stages of design using 3D models built. This, of course, improves the quality and saves time spent on designing.

The main task of finite element modeling in the software packages under consideration is the analysis of the strength of parts and assemblies, for which it is important to evaluate quickly the reliability of elements with possible optimization of the design, using the associative relationship of geometric and computational models.

3. Results and discussion

Let us illustrate it with the example of the strength calculation of a frameless stand 630 medium-section mill 450 JSC "EVRAZ ZSMK" using APMFEM application of the strength analysis system Compass-3DV16. At the initial stage it is necessary to create a three-dimensional assembly of the upper roll chock and anchors and set the boundary conditions for fastening. Apply the calculated loads and generate a finite element grid with a given condensation coefficient on the surface and a dilution factor in the volume (figure 1). The resulting grid can be viewed both on the surface and in the depth of the body of the part by changing the position of the "visit depth" slider.

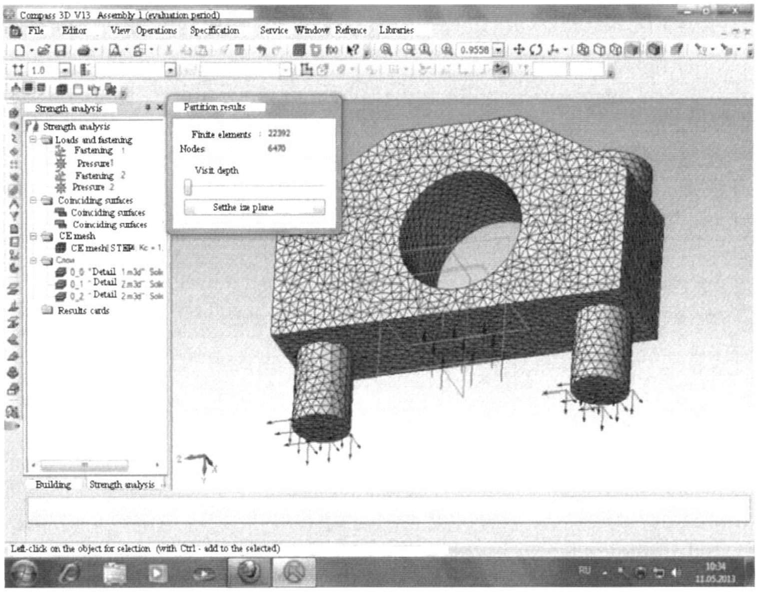

Figure 1. The generated finite element grid.

After carrying out preparatory operations, calculations are made. The information obtained in the calculations is displayed in the form of color maps, which shows the distribution of the studied value at any point in the volume of the structure under study. As an example, figure 2 shows the distribution of safe coefficients by the yield strength in the assembly body.

To evaluate the reliability of the results obtained, the safe coefficients were calculated by the yield strength using traditional methods [4] and compared with machine calculations. When carrying out verification calculations for real structures using APMFEM application, there was a problem with a limited library of steel grades. As the default one the application uses 10KP steel. This inconvenience can be eliminated by introducing a correction factor (K_σ).

$$K_\sigma = \frac{\sigma_{base}}{\sigma_{calc}} \tag{1}$$

,

where σ_{base} – the yield strength of the material from which the corresponding part is made; σ_{calc} – the yield point of the material used in the calculations.

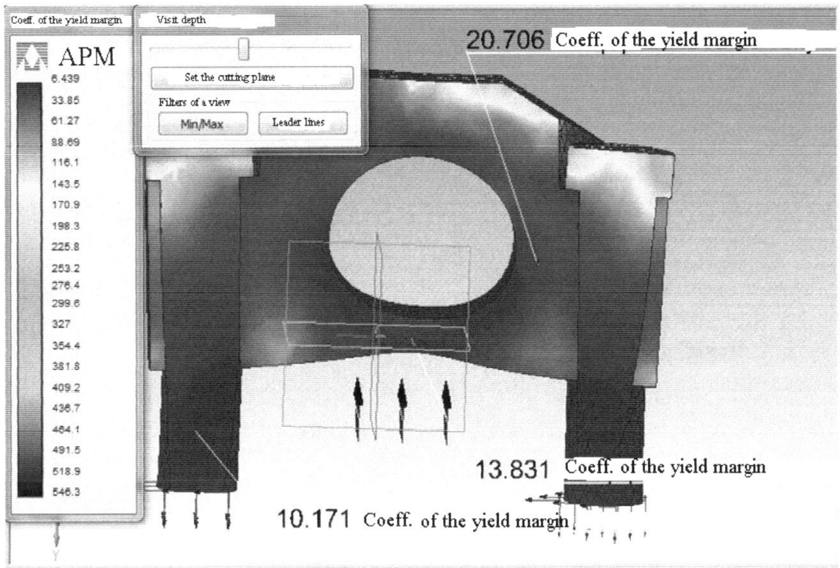

Figure 2. Distribution of safe coefficients by yield stress in the body of the assembly.

Using the value of the correction factor, the true value of the safety coefficient by the yield strength (K_t) for any material can be determined using the formula:

$$K_t = K_\sigma K_{calc}, \tag{2}$$

where K_{calc} – the calculated value of the safety coefficient by the yield strength when the material of the workpiece is used as default (steel 10KP).

Taking into account the correction coefficient, the minimum value of the margin factor for the yield strength in the computer simulation for the anchor was 46 for the roll chock 31. When calculating by the conventional method, the safety coefficient values for the yield strength were 48 and 34, respectively. A comparison of the obtained data shows that the results are close, divergence is not more than 10 %. However, the possibilities of computer modeling allow us to identify sections of the structure with a great safety coefficient by yield strength and reduce their metal consumption, and in areas with a low safety coefficient to focus on the planned inspections.

4. Conclusion

The results presented in the paper show the possibilities and advantages of modern computer technologies in assessing the performance of the main rolling equipment.

References

[1] Korolev A A 1987 *Mechanical Equipment of Rolling Tube Shops* (Moscow: Metallurgy) p 480
[2] Fatykovskiy A R, Saveliev A N 2015 *Features of Designing and Trouble-free Operation of Roller Accessories of Section Mills* (Moscow: Teplotehnik) p 170
[3] Fatykovskiy A R, Efimov O Yu, Chinokalov V Ya and Kopylov I V 2008 *Steel* **2** 63–64
[4] Fatykovskiy A R and Savelyev A N 2008 *Constructions and Calculations of the Equipment of Rolling Stands of Section and Rolling Mills* (Novokuznetsk: SibSIU) p 316

IOP Conf. Series: Materials Science and Engineering **411** (2018) 012082 doi:10.1088/1757-899X/411/1/012082

Improvement of continuous forming the "Conform" method

A R Fastikovskiy[1], E V Selivanova[1], A A Fedorov[1], V N Peretyatko[1], V V Evstifeev[2] and O Yu Efimov[3]

[1]Siberian State Industrial University, 42 Kirova street, Novokuznetsk, 654007, Russia
[2]Siberian State Automobile and Highway University, 5 Prospekt Mira, Omsk, 64080, Russia
[3]LLC "PROMMEST", 4A per. Ulyany Gromovoy, St. Petersburg, 191036, Russia

E-mail:omd@sibsiu.ru

Abstract. Improvements of continuous forming method "Conform" are offered, consisting in the possibility of changing the slope angle of the bearing plate relative to the wheel, allowing a number of restrictions to be removed, the setup and operation of the equipment to be substantially simplified. Conditions are determined according to which the workpiece capture and the mounting of the bearing plate at various angles are possible. The reliability of the dependences obtained is confirmed by the studies carried out on the laboratory installation.

1. Introduction

Modern development of technology initiates the need for the development and production of complex shapes of increased accuracy, the use of which allows competitive processes and equipment to be created. Usually such shapes are obtained by pressing. However, the classical pressing process has a low productivity due to the large time spent on preparatory operations, often exceeding the processing time, high energy intensity, since all metal volume is subjected to deformation at the same time and as a result it is possible to realize only on expensive metal-intensive equipment.

The disadvantages of the classical pressing process make it difficult and in some cases make it impossible to obtain complex shapes of great length, a high price of the products limits its use, installation and operation of expensive and complex equipment increases the payback period and production cost, which is not desirable in the system of market relations.

The cardinal solution to the above problems is the development of a continuous pressing process based on the use of active frictional forces in the deformation zone [1 - 4]. Among the known methods of continuous pressing, the most interesting from a practical point of view is the method called "Conform" [5].

The advantages of this method include the possibility of welding the metal in the deformation zone, the simplicity of the kinematic scheme, high technological flexibility due to the rapid tool change, deformation does not require heating to high temperatures, since during the shaping process heat is generated ensuring high plasticity of metal. However, the practical implementation of the "Conform" method is restrained by insufficient technical and technological development of this method, little theoretical study and, in this connection, the lack of well-grounded recommendations for the selection of energy-efficient modes of shaping.

2. Method of research

To implement the "Conform" method, a rotating roll and a stationary bearing plate are used, covering up to a quarter of the roll circumference. The roll and the bearing plate form a closed channel with a constant cross-section along the length. At the end, the channel is blocked by the stop, and to form the finished shape a matrix is used, usually located in the upper part of the bearing plate in the zone of the stop.

Under the action of active frictional forces the metal is drawn into the channel on the roll surface, where the pressure necessary for pressing is created. Analysis of the known device designs for the implementation of the "Conform" method made it possible to identify their shortcomings, imposing limitations and creating difficulties in practical implementation:

1. The section of the workpiece must correspond exactly to the section of the channel. Otherwise, the active frictional forces will not be sufficient for pressing;
2. At the entrance to the channel, it is necessary to place the master device providing the initial stage of the process;
3. The amount of active frictional forces can not be adjusted and, as a rule, considerably exceeds the required value, which in turn leads to unjustifiably high energy costs;
4. The absence of deformation of the metal between the roller and the bearing plate makes it impossible to use this method to produce simple shapes, flattened tape, transfers the entire load of form the finished shape on the matrix.

The identified shortcomings and limitations can be eliminated by changing the structure of the bearing plate fastening. It is necessary to equip additionally the installation that implements the "Conform" method with a device that changes the slope angle of the bearing plate. Such a constructive solution will allow the shape of the channel formed by the roller and the bearing plate to be changed. The height of the channel will decrease from the entrance to the matrix, which will provide additional deformation of the workpiece in the channel reducing the load on the matrix. Metal deformation in the channel will allow restrictions on the size and shape of the section of the workpiece to be removed.

Changing the slope angle of the bearing plate will allow the area of the deformation center to be changed where active frictional forces act, thereby creating optimal process conditions for energy costs. The possibility of deformation in the narrowing channel formed by the roll and the bearing plate will make it possible to form simple shapes without the use of a matrix, which will significantly reduce energy consumption and make this method of production of simple shapes economically efficient, thereby expanding the range of products.

3. Results and discussion

For the practical implementation of the proposed "Conform" method, it is necessary to determine the conditions under which it is possible to implement it. One of the most important is the condition of capture. The workpiece capture will occur when the amount of retracting active frictional forces on the roll surface is equal to or greater than the reactive friction forces acting on the surface of the bearing plate. Implementation of these recommendations is possible with different friction conditions on the working surfaces of the roll and the bearing plate:

$$f_c \geq f_{b.p.}, \tag{1}$$

where f_c – the coefficient of friction on the surface of the roll; $f_{b.p.}$ – the coefficient of friction on the surface of the baring plate

$$f_{b.p.} = nf_c, \tag{2}$$

where n – a coefficient of proportionality that is less than one.

IOP Conf. Series: Materials Science and Engineering **411** (2018) 012082 doi:10.1088/1757-899X/411/1/012082

Knowing the value of the coefficient n, it is possible to determine at what values of the friction coefficients $f_{b.p.}$ and f_c the realization of the studied process is possible. Let us consider the conditions of workpiece capture during deformation by the improved "Conform" method with the changing position of the bearing plate. We will design the forces acting at the point of contact of the sample with the bearing plate and the roll on the axis that coinciding with the axis of the sample. The equation of equilibrium of forces will have the form:

$$\tau_{av1} - \tau_{av2}\cos\omega - p_{av}\sin\omega = 0, \qquad (3)$$

where τ_{av1} – contact frictional forces acting at the point of contact between the roll and the sample; τ_{av2} – contact forces of friction acting in the place of contact between the sample and the bearing plate, p_{av} – average normal stress; ω – the slope angle of the bearing plate.

Assuming that the frictional law of the Amonton-Coulomb acts on the contact surfaces, taking into account (2), we obtain:

$$n \le \frac{f_c - \sin\omega}{f_c\cos\omega}. \qquad (4)$$

If condition (4) is not fulfilled, natural capture is not possible. It should be noted that the obtained condition of capture is valid for both a roll with a smooth surface and with a pass. As follows from the dependence (4) the value of the coefficient n depends on the slope angle of the bearing plate. Figure 1 shows a graph of the change in the coefficient n from the slope angle of the bearing plate ω.

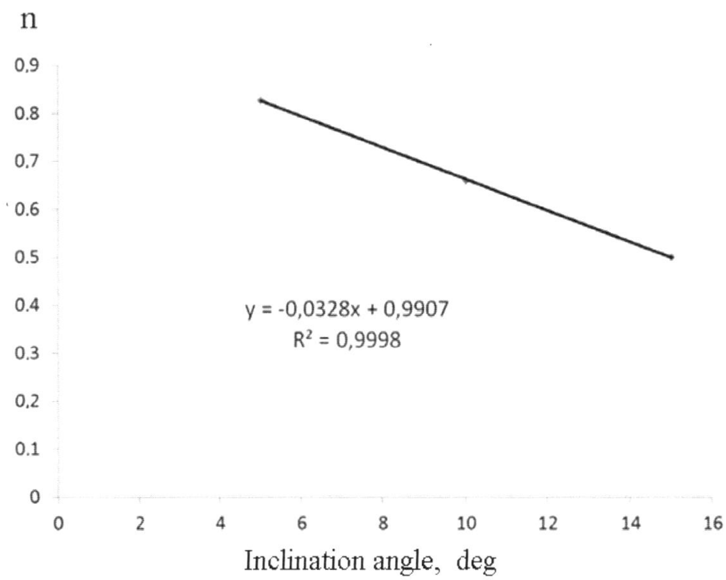

Figure 1. Graph of the dependence of the coefficient n on the slope angle of the bearing plate ω.

To study the possibilities of the improved "Conform" method, an experimental unit with an upper position of the bearing plate was designed. The mechanism of adjusting the bearing plate allows the distance between the roll and the bearing plate to be changed in the horizontal plane, as well as the angle of inclination. The friction conditions on the surfaces of the bearing plate and the roll were adjusted due to the different roughness and different material of the bearing plate (cast iron) and the roll (steel). The conducted experiments confirmed the reliability of the obtained condition for the improvement of "Conform" method.

When carrying out experiments using a bearing plate and a roll with smooth surfaces and the deformation of a round billet 12 mm in diameter between them, a great value of the broadening value took place (figure 2). The results are given in figure 2, where the sample obtained by the conventional method of rolling is shown at the top, at the bottom – by the "Conform" method using a roll and a bearing plate with smooth surfaces. This phenomenon can be explained by the large reactive frictional forces acting on the surface of the bearing plate and creating an obstacle to the metal movement.

Processes that provide large values of broadening are of interest when implementing the technology of flattening. The main task of flattening is to obtain strips from a round billet with the maximum value of the broadening. Products obtained by flattening are in demand, however, the possibilities of the rolling process used for this purpose are limited and the search for new solutions in particular the application of the "Conform" method is promising and relevant.

Figure 2. The broadening value after deformation in the rolls (the upper sample) and by the "Conform" method (the lower sample).

4. Conclusion

Technical solutions have been developed that make it possible to expand significantly the possibilities of continuous pressing by the "Conform" method. The conditions for the implementation of the improved "Conform" method are determined and confirmed experimentally, as well as the perspective directions of its use are discussed. It is established that deformation between a smooth roll and a bearing plate leads to large values of broadening. This circumstance is of interest for obtaining a strip from a round billet.

References

[1] Danilin V N 2009 *Forging and Stamping Production* **9** 30– 37
[2] Fastikovskiy A R 2004 *Forging and Stamping Production. Processing of Metals by Pressure* **2** 3–6
[3] Fastikovskiy A R 2004 *Izv. Vuzov. Non-ferrous Metallurgy* **2** 67–70
[4] Fastikovskiy A R 2003 *Izv. Vuzov. Non-ferrous Metallurgy* **5** 47–51
[5] Gorokhov Yu V 2004 *Metals* **4** 113–116

IOP Conf. Series: Materials Science and Engineering **411** (2018) 012083 doi:10.1088/1757-899X/411/1/012083

Complex technology of stemping details "hub" from ball blank

M V Filippova, V N Peretyatko, E E Prudkiy, M V Temlyantsev and A G Nikitin

Siberian State Industrial University, 42 Kirova street, Novokuznetsk, 654007, Russia

E-mail: kafkshp@sibsiu.ru

Abstract: In order to reduce the amount of metal wastes and improve the quality of press forgings, the new energy-and resource-saving technological complex of warm volumetric forging was developed, which has the following features: the metal is divided into billets on the rolling mill of cross-screw rolling; the induction heating of the billet to the temperature of warm forging (850-900°C); pressing of metal is performed, as a rule, in one pass in closed dies.

1. Introduction

Currently, the most common method of manufacturing the main parts of machines and units subjected to increased dynamic loads is forging and stamping. The products obtained in this way have high strength, reliability and durability. However, due to the low accuracy of the initial blanks, the finished forging is characterized by relatively large allowances and overlaps. This is due to the presence of large tolerances in the diameter of hot-rolled metal and the difficulties of accurately measuring the length and evenness of the ends of blanks cut from the bar on presses [1].

Since in the cost of engineering products the cost of metal consumed for the manufacture of parts reaches 80%, the minimum consumption of metal should be taken as the main criterion for reducing the cost of products in the production of forgings. To reduce metal losses it is necessary to use complex non-waste production technologies in which resource saving is taken into account at all stages: cutting the initial billet into measured lengths, heating the metal for punching, stamping.

Modern technologies for obtaining axisymmetric forgings of round-shaped (stamped into the butt) hot stamping by serial and mass production include the following operations [2]:

- Separation of metal into billet. The most common way of dividing a bar into billets is cutting metal in dies or using a shear press. The advantage of this separation method lies in the non-waste process of cutting, rapid rearrangement of the die for different cutting lengths, low cost of cutting, simple design of the cutting tool. Significant disadvantages include errors in the form of deviation of the dimensions of the workpiece from the correct cylinder and the deviation of the mass of the workpiece. Moreover, the deviation of the mass of the billet from the nominal value when cutting metal is 7-10%. Thus, blanks obtained by cutting metal with shears, can be used only for stamping in open dies [3, 4].
- Heating of the workpiece for stamping. Heating of the workpiece for stamping is carried out to temperatures 1150-1200°C. The advantage of heating to high temperatures is a decrease in the resistance of the deformation metal and an increase in the plastic properties of the metal. The disadvantages of high-temperature heating include the formation of scale on the surface of the

workpiece, which reduces the quality of the forging surface and requires additional scale removal operations and a decarburized surface layer of the metal, which requires additional machining. Losses of metal due to scale make up 1-3% of the mass of the billet [3, 5, 6].
- Form change of the workpiece in the stamp. In connection with the non-precise dimensions of the billet, stamping in the open dies occurs with the burr formation. Losses of metal due to burr cutting make up to 25% of the forging weight [3].

Thus, the existing technology of hot stamping has big drawbacks and does not meet modern requirements for energy and resource saving.

2. Methods of research

In order to reduce the amount of metal waste and improve the quality of stamped forgings, a new energy-saving and resource-saving integrated technology for semi-hot volume forging [7] of a "hub" forging (figure 1) was developed, which includes:

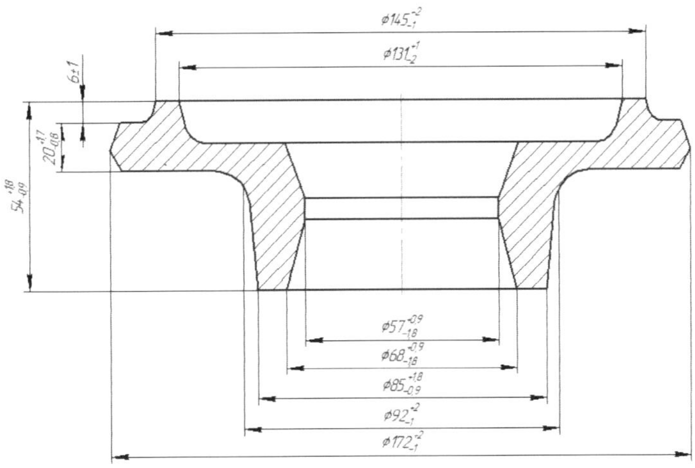

Figure 1. Draft of the forging "hub".

- Separation of metal into billets on a ball rolling mill. It is carried out on a rolling mill of cross-screw rolling. The resulting blank has the shape of a ball or a cylinder, the mass fluctuation does not exceed 3% [1]. The ball blank has a good surface quality, is evenly heated in the inductor, stably centered in the die, does not require additional pre-settling operation and fills the die bed contour well. Statistical processing of the experimental data on ball weighing showed that the average mass of a ball with a diameter of 93 mm is: for rolling on a new rolls of a rolling mill (3307±19.63) g, for rolling on worn rolls of a rolling mill (3336±21.30) g and the maximum relative error is equal to ±0.63% [8].
- Induction heating of the workpiece to the temperature of half-hot forging (850-900°C). This heating technology ensures the production of forgings of good quality (by reducing the amount of scale) and reduces energy consumption up to 20% (due to a lower heating temperature). The preliminary calculation of the optimal temperature of the ball blank heating from 40Kh steel showed that the optimum temperature of the blank heating, taking into account the resistance of the deformation metal, the plastic properties of the metal and the formation of scale, is 860 ± 5°C. In the production environment, the temperature of the ball blank heating was increased to 900°C. The additional studies carried out on the effect of decreasing the final heating temperature showed that with a decrease in the temperature of the workpiece heating from 1200°C to 900°C, the metal deformation resistance of steel 40Kh increased by 52%, plastic properties decreased by 34%, metal burnt decreased by 92%, decarburization is absent.

- Stamping of a heated blank without burr in the closed die, as a rule, in one transition. Such a resource-saving, non-burr stamping makes it possible to save metal due to the burr absence. Investigation of metal flow during stamping, filling of a stamp, as well as stress-strain state of a metal was carried out by computer simulation. For the numerical realization of mathematical models of stamping, the finite element method was used [8-11]. Simulations were performed using QForm-3D and Deform 3D software products. These software products make it possible to simulate the filling of a stamp with metal, the formation and development of defects, the change in the time and volume of the components of the stress tensor and deformations, the invariant characteristics of stress and strain, the change in the time of the deforming force, the temperature conditions of metal deformation, the wear of the die and other values in the course of metal shaping.

Modeling conditions: the initial ball blank with a diameter of 90 mm from steel grade 40Kh; deformation of the workpiece is performed on a crank press in the closed die at a temperature of 900°C with a coefficient of friction 0.4.

Industrial researches were carried out at the the rolling-mill and the crank hot-stamping press AK KB8544.

3. Results of computer simulation

The simulation results of stamping the hub from a ball blank showed that the metal flow during the filling of the die occurs without the formation of clamps and cracks. The stamp is completely filled with metal, there are no unfilled corners. Hub stamping in the closed die can be carried out in one pass.

4. Industrial research

According to the existing technology of stamping "hub" forgings, the rolled bars with a diameter of 75 mm and a length of 3000 mm after preliminary heating in the furnace to a temperature of 400-600°C on a crank press are cut into blanks of 106±2 mm in length. The cylindrical part of the workpiece is 24 mm, the oblique cut is 8 mm and the crumpling of a cylindrical part is up to 6.5 mm (figure 2). According to the technological map, the mass of the cylindrical billet must be at least 3675 g.

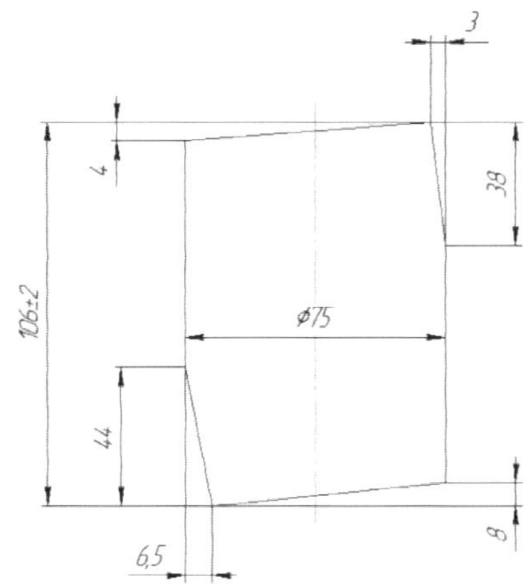

Figure 2. Cylindrical blank for hub stamping.

Further, the cylindrical blank is heated in the induction heater to a forging temperature of 1150-1200 °C. The heated billet is stamped in three passes in open dies: preliminary settling of the blank for descaling, stamping in the blocker, and then in the finishing impression. Further, the forging is transferred on the conveyor to the edged crank press, where the burr cutting and wad removal are performed.

The mass of the forging is 2800 g, the mass of the burr is 675 g, the mass of the wad is 145 g, and the mass of the loss is 55 g.

The hot stamping of the pilot batch of hub forgings from a ball billet with a diameter of 93 mm and a mass of 3300 g (according to the factory technology 3675 g) was performed to test the possibility of stamping the hub from a ball blank in the industrial conditions. Heating and stamping were carried out by the factory technology. Heating: in the induction heater to the temperature of 1200°C with a discharge rate of 16-18 s. Stamping on a crank hot-stamping press was carried out in three steps: pre-settling of the billet, stamping in the blocker, and then in the finishing stamp. For accurate centering of the ball, the design of the blocker and finishing impression was changed – the configuration of the marking for the punching. After stamping the cutting of bur and wad removal were performed. Inspection and measurement of forgings showed that all dimensions in height are fully satisfied, the quality of the surface meets the requirements of the technical control department. The mass of the burr decreased almost twofold.

Several forgings of the hub were stamped from the ball blank without preliminary settling. The measurements showed that, despite the absence of preliminary settling and a smaller mass of the ball in comparison with the cylinder, all dimensions of the finished forging are satisfied.

Thus, based on the results of computer simulation and experimental studies, the technology of the non-burr hub stamping from a pre-rolled ball blank was recommended for implementation.

After carrying out a number of preliminary studies of individual operations, an industrial approbation of the technology of warm forging of a hub from the ball blank was carried out:

A spherical blank with a nominal diameter of 90 mm was heated in the induction heater KINZ 750/2.4, installed with a slope of 15°, so the balls were easily moved by rolling along the guides of the inductor. The ball rolling promoted a good warm-up of the workpiece along the section, and also eliminated the possibility of jamming in the inductor. In addition, this way of moving the metal protects the guide tubes of the inductor from premature wear. The heating temperature was 900 °C. The temperature of the heating was monitored by means of an optical pyrometer. The measurements of the surface temperature of the sphere with an optical pyrometer and the central layer of the ball (thermocouple) showed that the ball blank was heated up faster than the cylindrical blank, therefore, the heating time of the ball billet could be reduced to 12-14 s.

The heated spherical blanks were fed to the crank hot-stamping press AK KB8544 and stamped in one non-burr finishing gauge. The draft of the new stamp is shown in figure 3.

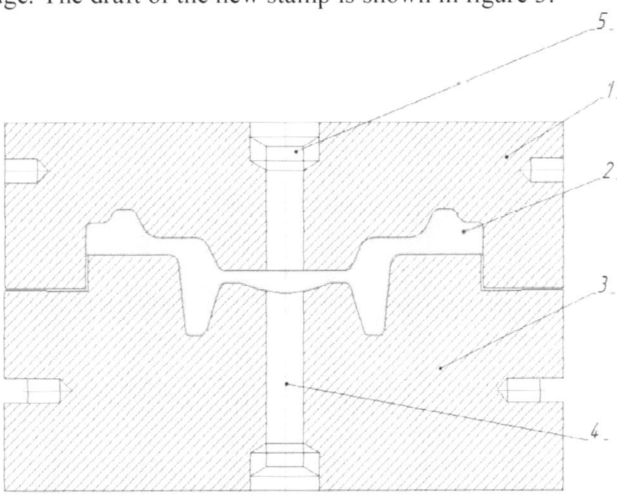

Figure 3. The design of the stamp for the non-burr stamping of the hub: 1 – punch; 2 – forging; 3 – matrix; 4, 5 – ejector.

The process of unbroken hub stamping showed that the die is filled with metal completely, the metal is not displaced into the burr. After removing the wad on the edging press, inspecting and measuring the forging of the hub, it was established that the forging dimensions correspond to the dimensions of the forging drawing, the clamping of destruction and other defects are absent.

The possibility of forgings without preliminary settling is due to the design of the finishing gauge of the die, in which the accuracy of the ball blank centering is effected by the cavity in which the ball blank is formed. The expediency of excluding the operation of settling from the technological process is that this stream is used to bring down the scale and ensure the forging a good fixation in the subsequent gauge. But on the ball blank, which is heated to lower temperatures and having a smaller surface area than the cylindrical blank less scale is formed.

Figure 4 shows the operational drafts of the hot hub stamping according to the existing factory technology with the burr from the cylindrical blank (4a), and the new non-burr hub stamping from the ball blank (4b). As can it be seen from the figure, according to the existing technology the cut cylindrical billet is heated in the inductor, its settling takes place, then stamped in the blocker and finishing impression, the burr is cut.

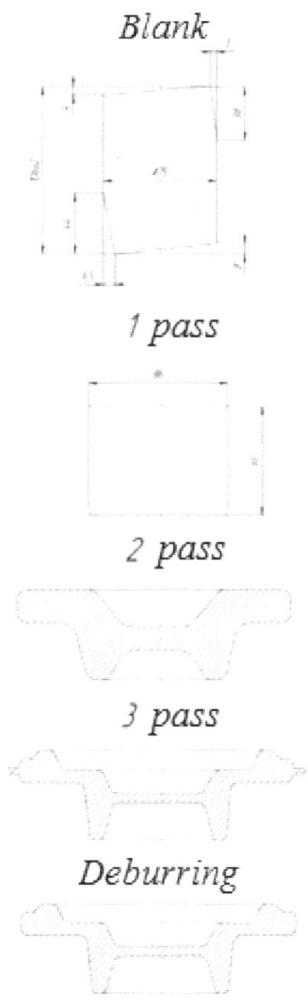

(a)

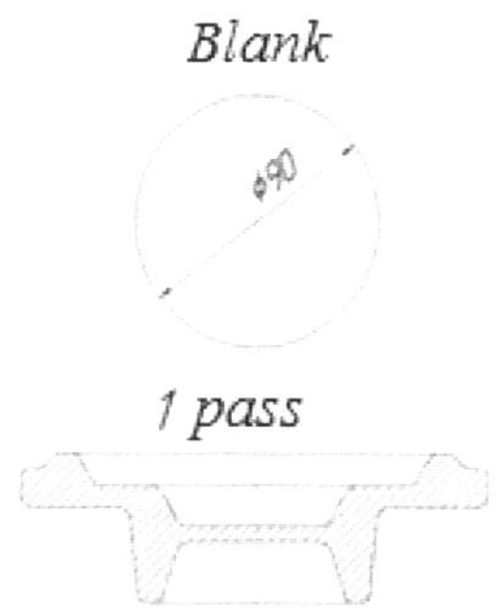

(b)

Figure 4. Operating drafts of a hub stamping: a – cylindrical blank, b – ball blank.

It can be noted that when stamping from a ball blanket, the number of operations is reduced by two passes. The press for cutting with a force of 2500 kN is not used in full production in the technological process, and is only needed to make a the hub hole.

5. Conclusions

1. A new energy and resource-saving technology for forging a hub has been developed, which can be characterized as a complex technology of warm volume forging from a ball blank.

2. The ball blank obtained by rolling on a cross-screw rolling mill has a stable mass and it can be used for the non-burr stamping of axisymmetric forgings.

3. Induction heating of the blank from steel 40Kh to the temperature of warm forging (850-900 °C) ensures the production of forgings of good quality (by reducing the amount of scale) and reduces energy consumption by 20-25%.

4. Resource-saving non-burr die stamping allows 26% of the metal to be saved due to the absence of burr.

6. References

[1] Tselikov A I et al 1971 *Special Rolling Mills* (Moscow: Metallurgy) p 336
[2] Sosenushkin E N 1991 *Improvement of Technological Preparation of Production of Cold and Warm Forging* (Moscow: Mechanical Engineering) p 108
[3] Edwardov M S 1971 *Stamping in Closed Stamps* (L: Mechanical Engineering) p 240
[4] Solovtsev S S 1985 *Waste-free Cutting of Long Products in Stamps* (Moscow: Mechanical Engineering) p 176
[5] Temlyantsev M V et al 2006 *Oxidation and Decarburization of Steel in Processes of Heating under Metal Treatment with Pressure* (Moscow: Teplotehnik) p 200
[6] Peretyatko V N et al 2010 *Development of Theory and Practice of Metallurgical Technologies* (Moscow: Teplotehnik) p 352
[7] Filippova M V et al 2016 *Conf. Proc. in Metallurgy: New Technologies, Management, Innovation and Quality* pp 227–229
[8] Filippova M V et al 2016 *Development and Implementation of Energy and Resource-saving Technologies in the Processing of Metals by Pressure* (Novosibirsk: SB RAS) p 269
Peretyatko V N 2012 *Blank Production in Mechanical Engineering* **3** 17–19
[9] Rybin Yu I et al 2004 *Mathematical Modeling and Design of Technological Processes of Metal Forming* (SPb.: Nauka) p 644
[10] Volodin I M 2006 *Modeling of Processes of Hot Volumetric Forging* (Moscow: Mashinostroenie-1) p 253

IOP Conf. Series: Materials Science and Engineering **411** (2018) 012084 doi:10.1088/1757-899X/411/1/012084

Effect of phase transition temperature and particle size on residual stresses and properties of laser cladding layer

Shuwen Xu[1], Xizhang Chen[1] and Sergey Konovalov[2]

[1]School of Mechanical and Electrical Engineering, Wenzhou University, Chashan University Town, Wenzhou, 325035, China
[2]Samara National Research University, 34, Moskovskoye sh., Samara, 443086, Russia

E-mail: chenxizhang@wzu.edu.cn.

Abstract. Laser cladding is one of the main technologies for material surface strengthening technologies. Residual tensile stress on the surface of the cladding layer reducing the fatigue and wear performances, is difficult to control. The phase transition dilation during the cooling process with the laser cladding technique is proposed to counteract the volume shrink and then reduce the residual tensile stress and partially replace the post cladding treatment. The influence of low temperature phase change alloy on the residual stress of the cladding layer was investigated by comparing with that of the 17-4PH stainless steel powder cladding layer. The micro hardness and friction and wear properties of the cladding layer with different particle size were also analyzed. The result shows that the volume expansion caused by phase change counter-acts the heat shrinkage produced during the cooling of the cladding layer, and the hardness and abrasion resistance of the cladding layer are improved.

1. Introduction

As a new surface repairing and manufacturing technology, laser cladding with high energy density, metallurgical bonding cladding and substrate, high melting efficiency, has been widely used in molds, tools and mechanical parts repair, antiwear and corrosion-resistant coating processing and so on [1,2]. Laser cladding process is a rapid state transition process. The certain difference in thermal properties between the substrate and cladding power results in its solidification process of crystallization imbalance which tends to produce uneven tensile stress in the cladding surface [3,4]. The presence of residual tensile stress increases the cracking sensitivity of the cladding layer and reduces its fatigue, wear resistance and other properties [5,6]. The post-treatment method such as local heating method, impact method and shot peening method can reduce the residual stress to improve fatigue and wear resistance, but these methods increase the production cost and reduce the production efficiency.

Iron and steel materials in the cooling process will be organized into the transformation, accompanied by different volume expansion, when the volume expansion in the constraints of the conditions, can produce compressive stress [7]. The low temperature phase material reduce the phase transition temperature of the martensite by controlling the alloying element content so that the volume expansion produced by martensitic transformation can counteract some or all of the tensile stress caused by the heat shrinkage and achieve the goal of improving or eliminating harmful residual tensile stress[8]. In this paper, aiming at the problem that the mechanical properties of the parts are reduced due to the residual stress on the surface of the laser cladding layer, the method of laser cladding low temperature phase transformation alloy powder is proposed to improve the residual stress on the cladding surface in order to improve the mechanical strength of the cladding layer.

2. Experimental materials and methods

2.1. Experimental materials
In order to investigate the effect of martensitic transformation on the residual stress and mechanical properties of the cladding layer after the fusion of the low transformation temperature alloy material, the cladding layer formed by conventional 17-4PH stainless steel powder was used as the comparative sample. The effect of particle size on the residual stress and mechanical properties of the cladding layer was analyzed by the different particle sizes of the low transformation temperature alloy powder. The low temperature phase change alloy powders with the meshes of 200 (LTT1) and 500 (LTT2) and the 17-4PH stainless steel powders with the mesh of 200 were adopted. The basic chemical compositions of the three kinds of deposited alloy powders are shown in table 1.

Table 1. Chemical compositions of laser cladding alloy powders (wt%).

Powder	C	Si	Mn	Cr	Ni	Mo	V	Cu	Ti	B	Nb	Fe
17-4PH	0.01	0.72	0.097	16.95	4.06	--	--	4.05	--	--	0.31	Bal.
LTT1	0.1	0.2	0.5	5	3	0.3	0.2	0.5	0.2	0.002	0.01	Bal.
LTT2	0.1	0.2	0.5	5	3	0.3	0.2	0.5	0.2	0.002	0.01	Bal.

2.2. Text methods and equipment
Germany ROFIN FL-060 type 6kW fiber laser with small laser spot (5mm × 5mm), combined with ABB IRE4600 robot processing system, was used in the experiment. The laser cladding experiment was carried out on Q235 steel with the sizes of 100mm × 100mm × 10mm and the preparation parameters are shown as follows: the laser power is 2800kW, the laser scanning speed is 28mm/min, the focus is 3mm and the Ar flux is 5L/min. The cladding process is protected by high-purity argon. The GTV PF2 / 2 laser cladding powder feeder ensures the uniformity and continuity of the powder in the cladding process. Powder delivery method is coaxial argon powder, as shown in figure 1.

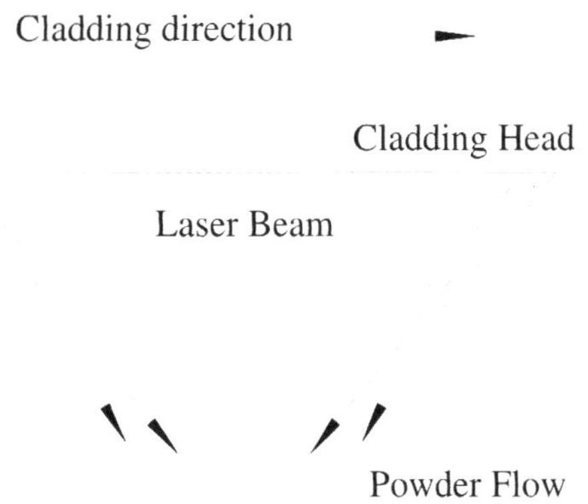

Figure 1. The schematic diagram of laser cladding.

Secondly, the three kinds of cladding layers were tested on the MFT-5000 friction and wear tester. The wear specimen was 20mm × 10mm × 10mm. Before the experiment, the surface of the test was polished smooth and flat with sandpaper. The friction pair is a 440C stainless steel ball with the diameter of 9.524mm. The load is 100N and the friction slip speed is 8mm/s, with the accuracy of 0.1mg electronic balance to measure the wear and tear weight of the situation.

The residual stress of the different kinds of cladding layers was measured by the Canadian Proto LXRD X-ray diffraction stress analyzer. Before the test, the instrument was calibrated with stress-free iron powder. To the end of the cladding 5mm as a starting point, the residual stress was measured by the points in increments of 0.5 mm in the 0 ° direction, a total of 11 points. The test was shown in figure 2.

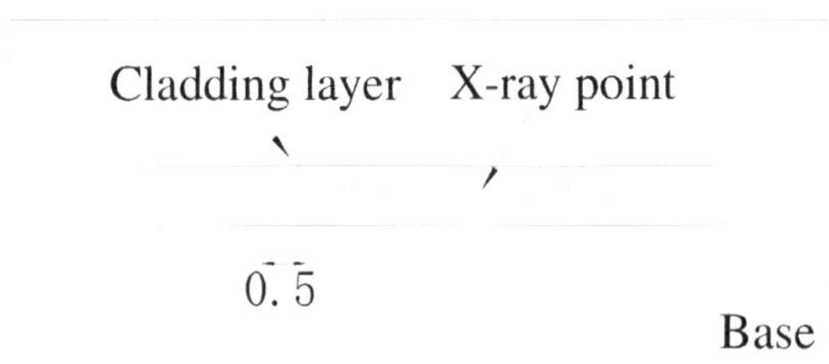

Figure 2. Sketch map of residual stress test.

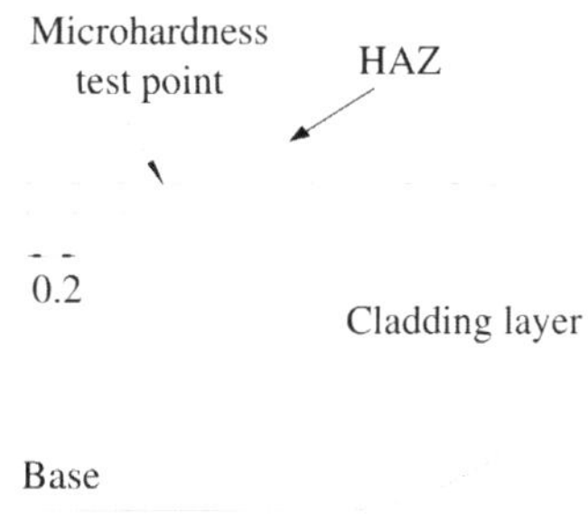

Figure 3. The schematic diagram of micro-hardness test.

Finally, after cutting, grinding, and polishing, the cladding layer produced in the experiment was corroded using picric acid solution (1g picric acid + 4% nitric acid + 96% alcohol). The microstructure of clad-ding layer is observed under an optical microscope. The phase of the cladding layer was analyzed by X-ray diffractometer, and the micro-hardness of the cladding layer, the heat affected zone and the base metal was measured using an HV-1000 micro-hardness tester, as shown in figure 3.

3. Results and discussion

3.1. Macroscopic morphology and microstructure analysis of cladding layer

Figure 4 shows the cladding of low temperature phase change alloy powder and 17-4PH ordinary stain-less steel powder under the same laser welding process conditions. It is clearly seen from Figure 4 that the cladding of the three powders is well formed and the droplets on both sides are less, and the surface of the 17-4PH cladding layer is brightest. In the case of flatness, compared with figure 4 (a), (b) and (c), it was found that the surface of the LTT2 cladding layer with finer powder size was more wrinkled.

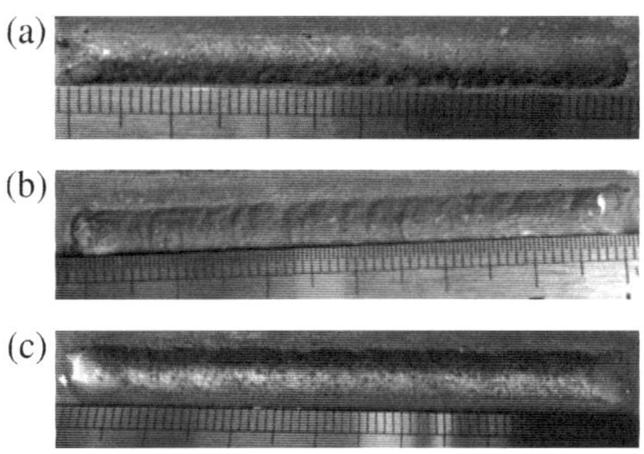

Figure 4. The macroscopic morphology of cladding layer: a) Cladding layer of LTT1; b) Cladding layer of LTT2; c) Cladding layer of 17-4PH.

The analysis shows that the total surface area is large with the fine powder. The contact between the fine particles is more, producing a larger internal friction, resulting in poor fluidity of deposited metal which affecting the ability to produce powder during the cladding process [9]. When the particle size of the powder is fine, the uniformity of the delivery of the powder alloy is poor and the deposited metal is intermittently sprayed into the laser irradiation area. The formation of the bridging effect makes the cladding layer easy to form wrinkles affecting the final shape. It is further explained that when the particle size of the LTT alloy powder is small, it is not conducive to obtain the cladding layer with better forming effect.

Figure 5 (a) and (b) show the microstructures of LTT1 and LTT2 cladding layers, respectively. It can be seen from figure 5 that the two powder cladding layers are composed of lath martensite (black) and the retained austenite (white) between them. After reaching room temperature, a small amount of austenite that cannot completely transform to the martensite was remained and formed as the residual austenite. Small amount of retained austenite structure are formed to enhance the toughness of the cladding layer. Comparing figure 5 (a) and (b), it can be seen that the density in LTT2 cladding layer structure was higher. The analysis considers that the powder with small particle size has fine structure with more grain boundaries distributed with a lot dislocation defects. The powder is rapidly crystallized after melting and the finer grains are formed into dense structures.

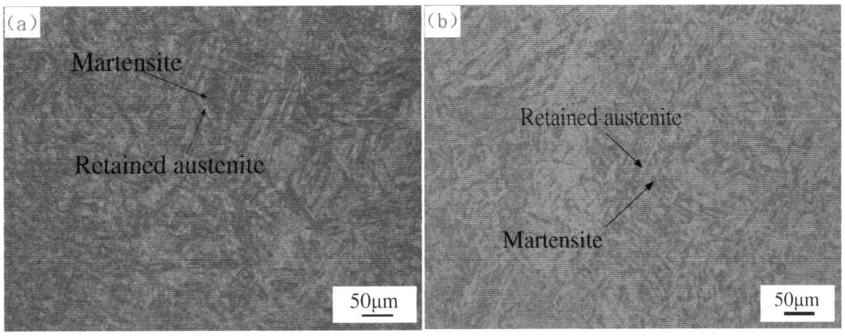

Figure 5. The microstructure of cladding layer: a) LTT1 cladding microstructure; b) LTT2 cladding microstructure.

The XRD diffraction pattern of LTT2 powder cladding is shown in figure 6. Microstructure shows that the cladding layer is mainly martensite and retained austenite. According to the analysis of jade software, the diffraction peak mainly contains Fe element and Cr element. It is judged that the C and Cr elements are sol-id-dissolved into the Fe element to form a solid solution structure of martensite and austenite. Through com-paring the PDF card, indices relationship of crystal face can be inferred that the (110) peak and (211) peak rep-resent the martensite, and the (200) peak represents the austenite. It is found that the main structure of the clad-ding layer is martensite and a small amount of retained austenite after laser cladding the LTT alloy.

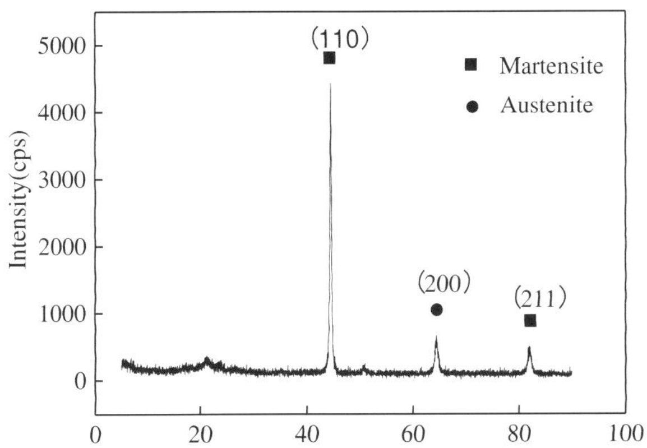

Figure 6. The XRD diffraction pattern of LTT2 cladding layer.

3.2. Micro-hardness analysis of section of cladding

The micro-hardness distribution of the three cladding layers is shown in figure 7. It is seen from figure 7 that the micro-hardness of the three kinds of powder cladding layers is a gradient distribution. Along the normal direction of the cladding layer, the micro-hardness value increases first and then decreases gently. The analysis considers that in the laser cladding coaxial argon power feeding was used (shown in figure 1). During the cladding process, the powder leaves the fritted nozzle into the air and some of the powder is not able to move into the laser beam due to air resistance, but is free fall. This led to that part of the powder can only accept the west heat melting of the cladding layer and , cladding surface density decreased, hardness decreases. This led to the part of the powder can only melt by accepting the formation of cladding layer of waste heat, followed by a decrease in the surface density of the cladding layer and a decrease in hardness. The sub-surface powder melts fully and the state transforms rapidly, which result in density cladding layer and increasing hardness. Due to the rapid melting and cooling, the cladding layer has a small grain size, a dense structure, and thus a higher hardness than the substrate. With the increase of the depth of the cladding layer, reaching the fusing area, the hardness value decreases due to the dilution of the matrix, and the slow drop zone corresponding to the heat affected zone. The hardness value drops to the lowest after reaching the substrate.

Compared with the micro-hardness distribution of the two kinds of composition powders, it was found that the low temperature phase change powder cladding layer had better hardness enhancement effect than stainless steel powder. The micro-hardness of LTT2 powder cladding layer is as high as 431.7HV, which is much higher than that of substrate about 160HV, and the micro-hardness is increased by about 2.7 times. The maximum hardness of the stainless steel powder cladding layer is 367.3HV and the micro-hardness is improved by about 2.3 times. The residual compressive stress in the cladding layer of the low temperature phase change powder reduces the shear stress of the indentation pressed by the micro-hardness gauge and improves the micro-hardness of the cladding layer. The maximum hardness of LTT1 powder cladding is 419HV. This is mainly because the LTT2 powder cladding layer of the structure is more com-pact, making its micro-hardness value slightly higher. The high micro-hardness value is beneficial for the improvement of the wear resistance of the cladding layer.

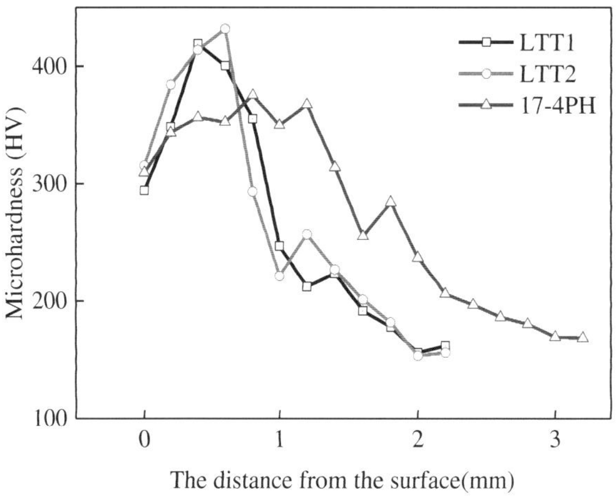

Figure 7. The micro-hardness of the cladding layer.

3.3. Analysis of residual stress on cladding surface

Figure 8 shows the residual stress distribution on the surface of the three powder laser cladding layers. It can be seen from figure 8, compared with the low temperature phase change alloy powder cladding layer, the surface residual stress of 17-4PH cladding layer is higher and the maximum value is about 236MPa. The surface residual stress of the low temperature phase change alloy powder laser cladding layer is negative, that is, the surface stress state is compressive stress. It is concluded that the martensitic transformation occurs during the solidification of the three alloy powders, and the martensitic transformation will produce volume expansion [10,11]. When the martensitic transformation occurs at a lower temperature, the deposited metal is constrained, and the volume expansion resulting from the martensitic transformation counteracts the heat shrinkage produced during the cooling of the cladding layer. If the phase transition pressure produced by the phase change expansion is greater than the tensile stress generated during the heat shrinkage, the cladding layer would exhibit a residual compressive stress. The martensitic transformation of the 17-4PH cladding layer occurs at a higher temperature, where the deposited metal is in a plastic state and the volumetric expansion of the martensitic transformation is released without being restricted. No compressive stress is produced, and finally the residual tensile stress is expressed.

Comparing LTT1 and LTT2 powder cladding surface residual stress, it can be seen that LTT1 exhibits a larger residual compressive stress and the maximum compressive stress is 288.4MPa. The residual stress distribution of LTT2 cladding layer is gentle and the maximum compressive stress is 135.1MPa. There are two reasons for the low compressive stress of LTT2 cladding. On the one hand, the grain size of the austenite is small when the original size is small, so the grain boundary migration is small in the process of martensitic transformation. The formation of the grain size is small and the volume expansion is reduced accordingly. On the other hand, due to small particle size of LTT2 powder, the burning of the LTT2 powder is more serious. Some of the martensitic transformation point (Ms) reduction element content reduces resulting in that the final martensitic transformation expansion do not achieve the best effect.

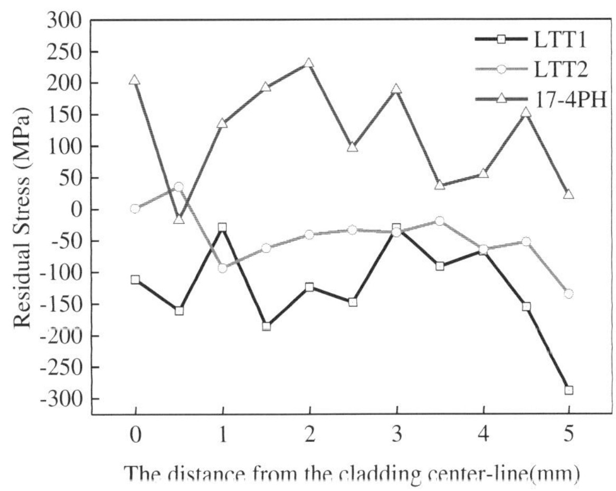

Figure 8. The residual stress distribution of laser cladding layer.

3.4. Cladding layer friction and wear properties

According to the change of friction coefficient and wear weight, the wear resistance of the three cladding layers was analyzed. Figure 9(a) shows the weight loss of the three powders at different wear periods. As can be seen from the figure, with time increases, the wear amount of the cladding layer of the three powders increase gradually, but the wear per unit time is gradually reduced. In 15 minutes of the wear time, the wear rate is large, in which the wear rate of LTT2 powder is minimum, 4.5mg. Analysis shows that at the beginning of the wear time, there are many particulate on the surface of the layer, which are contact with the friction side first, and produce large stress [12,13]. The presence of these stresses leads to the destruction of the microburst giving a rise to the initial mass wear.

Figure 9(b) shows the variation trend of friction coefficient of cladding layer with time in three kinds of powders under the same cladding process. The average coefficient of friction of 17-4PH is about 0.67. The friction coefficient of LTT1 is close to that of 17-4PH powder, and the average friction coefficient is about 0.6. The smallest friction coefficient appears in the LTT2 powder coating with an average value of about 0.4. The friction coefficient fluctuates greatly at the initial stage of friction, and the friction coefficient tends to be stable with the increase of time.

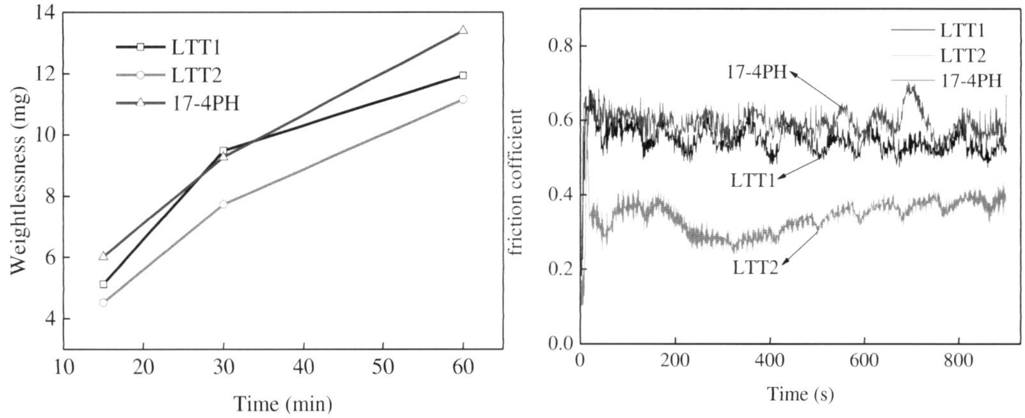

Figure 9. (a)The weightlessness of the cladding layer. (b) The Variation of friction coefficient of different powder cladding layer with time.

Compared with the wear and friction coefficient of the three kinds of powders, the wear amount of LTT2 powder cladding layer is the lowest and the friction coefficient is the lowest, which shows the best wear resistance. As the LTT2 alloy powder particle size is smaller, after the absorption of laser energy, the powder particles quickly melt and curd resulting in smaller grains. In the case where the cladding layer is the same as the martensite structure, the wear resistance of the cladding layer is higher than that of the powder having a larger particle size. Compared with the weight loss of 17-4PH power cladding, the weight loss of LTT1 powder cladding showed a tendency to increase first and then decrease. The friction coefficients of the two coating also present a small difference, showing a similar wear resistance. The wear resistance of the LTT1 cladding layer is slightly higher because the residual compressive stress in the cladding layer relaxes the stress concentration and the shear stress caused by the friction pair on the surface is reduced, thereby improving the wear resistance. Due to the higher hardness of the LTT1 powder cladding layer, the abrasion resistance is slightly higher than that of the stainless steel powder cladding layer. Moreover, the residual compressive stress in the cladding layer relaxes the stress concentration and then the shear stress caused by the friction pair on the surface was reduced. This improves its wear resistance to a certain extent.

4. Conclusions

In this paper, 17-4PH stainless steel alloy powder and two kinds of low temperature phase change alloy powders with different particle sizes were adopted for comparison. The residual stress and microstructure of cladding layer were analyzed:

(1) During the laser cladding of the low temperature phase change alloy powder, the volumetric expansion of martensitic transformation counteracts the heat shrinkage stress of the cladding layer, and the surface of the cladding layer shows residual compressive stress of -228.4MPa improving the wear resistance of the cladding layer.

(2) The microstructure of the low temperature phase change cladding layer is mainly martensite and a small amount of retained austenite. The martensitic transformation starts at a lower temperature. A small amount of retained austenites distributed between the lath martensites improve the toughness of the cladding layer.

(3) Due to its dense structure, the micro-hardness of the LTT alloy power cladding layer was improved significantly. The cladding layer with high density has better wear resistance, and the increase of micro-hardness also has a beneficial effect on the wear resistance of the cladding layer.

Acknowledgments
This work is sponsored by the National Natural Science Foundation of China under Grant no. 51575401, this work is also partly sponsored by Zhejiang Provincial Natural Science Foundation (Grant no. LY16E050007).

References
[1] Zhang J, Wu W and Zhao L 2013 *Hot Working Technology* **42** 131–4
[2] Li Y L, Jin H X, Bai X B, et al 2009 *Heat Treatment Technology and Equipment* **30** 1–5
[3] Fang Y Y 2016 *Residual Stress and Corrosion Performance of Pipeline Steel with Dissimilar Welding Wire Jiangsu University*
[4] Alam M K, Edrisy A, Urbanic J, et al 2017 *J. Mater. Eng. Perform.* **26** 1–9
[5] Wang D Z, Hu Q W and Zeng X Y 2015 *Suef. Coat. Technol* **274** 51–9
[6] Shiozaki T, Tamai Y and Urabe T 2015 *Int. J. Fatigue* **80** 324–31
[7] Wang W X, Huo L X and Zhang Y S 2003 *China Mechanical Engineering* **756** 5–20
[8] Kromm A, Dixneit J and Kannengiesser T 2014 *Welding in the World* **58** 729–41
[9] Li H S 2004 *Theoretical and Experimental Study on the Interaction of Laser and Metal Powder Flow in the Laser Remanufacturing Tianjing Polytechnic University*
[10] Chen X Z, Fang Y Y, Li P et al 2015 *Mater. Des.* **65** 1214–21
[11] Kannengiesser T and Kromm A 2009 *Soldagem & Inspeção* **14** 74–81
[12] Chen X Z, Hu K and Yuan Q B 2016 *China Surface Engineering* **29** 118–24
[13] Song J, Zhang Q M, Lin X C and Liao J H 2008 *Chinese Journal of Lasers* **35** 776–81

IOP Conf. Series: Materials Science and Engineering **411** (2018) 012085 doi:10.1088/1757-899X/411/1/012085

The formation and control of porosity during laser DP780 dual-phase galvanized steels

Lei Huang[1], Xizhang Chen[1,2], Sergey Konovalov[2] and Hongbo Ma[3]

[1]School of Mechanical and Electrical Engineering, Wenzhou University, Chashan University Town, Wenzhou, 325035, China
[2]Samara National Research University, 34, Moskovskoye sh., Samara 443086, Russia
[3]Jiangsu Beiren Robot System CO., LTD, Suzhou, 215123, China

E-mail: kernel.chen@gmail.com

Abstract. DP780 dual-phase galvanized steels with 0.8mm thickness were lap welded by 4kw continuous fiber lasers to investigate the formation of porosity. By observing the welding parameters and recording the welding process with high speed cameras, the formation mechanism of internal porosity and external porosity was proposed. Based on the formation mechanism of porosity, we established the escape channel model of zinc vapor leaving the molten pool during the welding process, by controlling the dynamic balance of the pores of each channel to reduce the welding internal pores and external porosity defects, improve the welding quality.

1. Introduction

Car body lightweight is one of the most important development tendency for vehicle manufacturing, advanced high strength steels (AHSS) is one of the most widely used material in the automotive industry for lightweight and higher qualities. These steel grades include dual phase (DP) steels, transformation-induced plasticity (TRIP) steels, high hole expansion (HHE) steels, complex-phase (CP) steels, martensitic steels (MS), and twining induced plasticity (TWIP) steels [1, 2], Additionally, these steels are galvanized in order to improve the surface corrosion resistance for automotive parts. However, it is still a great challenge to laser weld galvanized steels in a zero-gap lap joint configuration, since the zinc coatings at the faying interface will vaporize; due to the lower boiling point (907 ℃) of zinc as compared to the melting temperature of steel (above 1500 ℃), causing the highly pressurized zinc vapor to expel the liquid metal out of the weld pool, resulting in spatter on adjacent surfaces in addition to blowholes and pores which dramatically decrease the mechanical properties of the weld [3, 4].

In order to study the influence and kinetics of the molten pool on the hole escape at the keyhole, a large number of scholars have introduced the on-line monitoring means during the welding process. For example, Norman used high speed cameras to observe the zinc vapor on the laser welding process [5]. Panwisa was C established a physics-based model for keyhole welding to simulate keyhole and porosity formation during laser welding of Ti-6Al-4V titanium alloy [6]. Kong applied spectroscopy to study the correlations between the optical emission of the plasma and zinc vapor induced welding defects in the laser welding of galvanized steel for a lap joint configuration [7]. However, there is relatively little research on the laser trapping channel of galvanized sheet laser welding, and the new steam escape has important influence on the quality of the welded joint. Therefore, it is necessary to study the formation mechanism of porosity of laser welding of galvanized sheet.

In this paper, DP780 galvanized dual phase steels were laser lap welded by 4 kW continuous fiber lasers. The effect of pores on the molten pool was studied. The mechanism of formation of internal porosity and external porosity was explained and analyzed. Based on the analysis of the formation mechanism of porosity, we established the escape channel model of zinc vapor leaving the molten pool during the welding. We used the high speed camera to verify the correctness of the model. We were controlling the dynamic balance of the zinc vapor of each channel to reduce the welding internal porosity and external porosity defects, improving the welding quality.

2. Test materials, equipment and research methods

2.1. Experimental materials

In this experiment, laser welding of galvanized high-strength steel (DP780) with the size of 100×30×0.8mm was conducted. The chemical composition and mechanical properties of DP 780 steel are shown in table 1 and table 2, respectively.

Table1. Mechanical properties of DP780 galvanized high-strength steel.

Yield strength Re /(MPa)	Tensile strength Rm /(MPa)	Elongation ψ/(%)
500~650	≥780	≥10

Table2. Chemical composition of DP780 galvanized high-strength steel (wt%).

C	Si	Mn	P	S	Al
0.094	0.140	1.990	0.0056	0.0023	0.036

2.2. Experimental equipment

The equipment used in this experiment includes KUKA robot laser welding system, high speed camera system, material stretching machine, microscope, laser microscope, high-speed camera and x-ray instrument (figure 1). Fiber laser welding fixed on KUKA robot was applied to weld the high-strength galvanized steel, the welding process was observed by the CR600X2 high-speed camera. The KUKA robot for welding system was composed of 4KW fiber laser, water cooling system, optical path system, robot system, gas protection system and welding head system. The fiber laser with 0.68mm spot diameter was produced by IPG company.

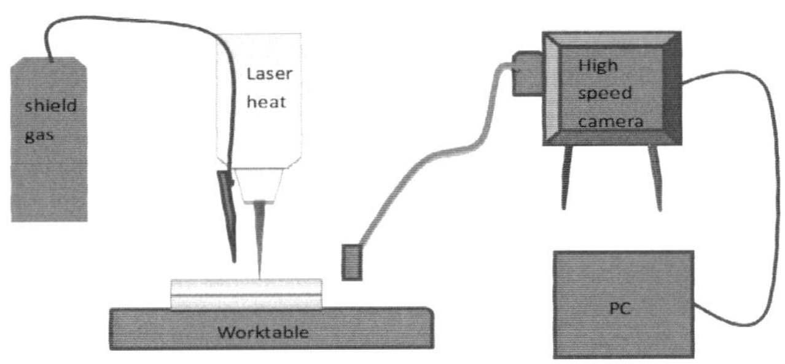

Figure 1. High speed camera recording during laser welding.

3. Observation and analysis of porosity of laser welding of DP780 galvanized steel

The formation of porosity in the weld seam of the galvanized steel sheet was mainly affected by the solubility of the gas in the molten metal, gas partial pressure, temperature, viscosity of the molten pool and the settling time of the molten pool. The boiling point of zinc is 906 °C, much lower than that of steel. Some zinc vapor escapes outside while the other part keeping the gaseous state enters into the molten pool [7].

It was often accompanied with splash and internal porosity during laser welding of galvanized steel. In order to explore its relationship with the zinc escape and the welding process, a high-speed camera is used to observe the phenomenon. The relevant picture was concluded as follows. A typical pores escape and the formation were recorded by high-speed camera. As shown in figure 2, the process of pores escape was diverse, and it had impact on the welding forming the appearance and internal structure.

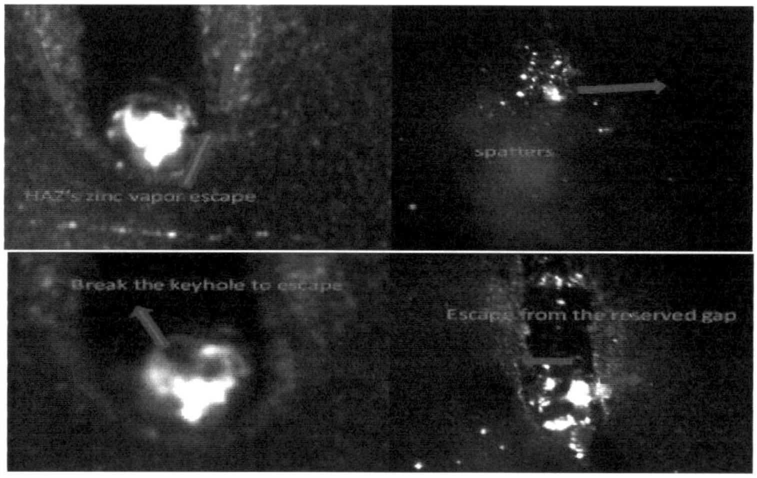

Figure 2. Zinc vapor escape form.

The shape of the zinc vapor escaped the pool in a variety of forms, some of it pooled at the keyhole was like a boiled water, from the keyhole in a pointed mouth, or in a large group of clouds [8]. Laser radiation of zinc vapor completely escaped. Some zinc vapor escaped from the gap between the two plates. When the gap was small, the zinc vapor couldn't gently escape from the reserved gap, the remaining zinc vapor accumulated at the keyhole were to splashing out and breaking the keyhole to escape. As shown in the following figure 3, (a) (b) (c) shown the external porosity and internal porosity of the welded joint.

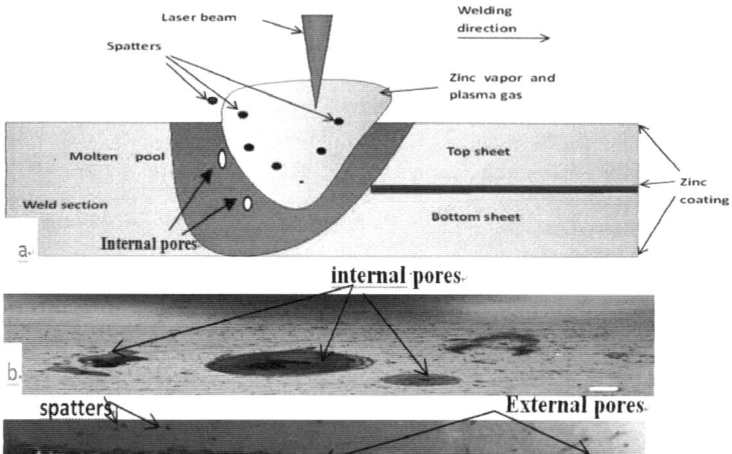

Figure 3. The escape of vapor and formation of pores. (a) A schematic cross-Sectional View of the laser welding of galvanized steel in a zero gap lap-joint configuration. (b) representative photo of internal pores Micrograph. (c) representative photo of laser welding exhibition spatter and external pores within the laser weld.

IOP Conf. Series: Materials Science and Engineering **411** (2018) 012085 doi:10.1088/1757-899X/411/1/012085

4. Mechanism of porosity formation and dynamic equilibrium of zinc steam escape channel

4.1. Mechanism of porosity formation

When observed the forming the internal porosity, we proposed a new porosity formation mechanism. As shown in the figure 4 and figure 5, zinc vapor led to two parts, the internal porosity and the external porosity. The moving of zinc vapor caused the formation of internal porosity. During the welding process, zinc vapor escaped when the zinc-coating temperature beyond the zinc melting point, before the melting of two-phase steels. Due to the low solubility in the solid, the zinc vapor would move in the direction of the metal liquid behind the melting pool and remained in the welded joint to form the internal porosity after curing.

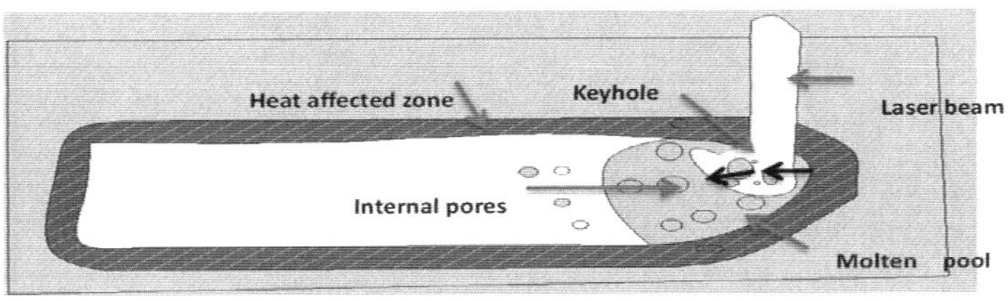

Figure 4. Internal porosity of the formation of the process.

As shown in figure 6, when the buoyancy of the zinc vapor at the keyhole was greater than the sum of viscosity of the liquid metal and the gravity of the zinc vapor itself, the zinc vapor escaped with the molten metal solution, causing splashing. when the buoyancy of the zinc vapor at the keyhole does not reach the sum of viscosity of the liquid metal and the gravity of the zinc vapor itself, the zinc vapor stayed in the melting pool. The corresponding escape form and process have the following effects on the shape of the welded joint. When the zinc vapor escapes from the molten pool with the molten metal melt, the welded joint would have an external pore defect, and when the zinc vapor fails to escape in time, it would form the internal porosity inside the pool, affecting the strength of welded joints.

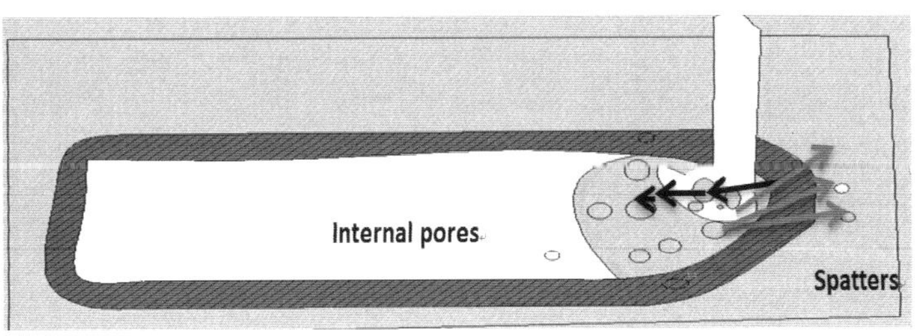

Figure 5. External porosity of the formation of the process.

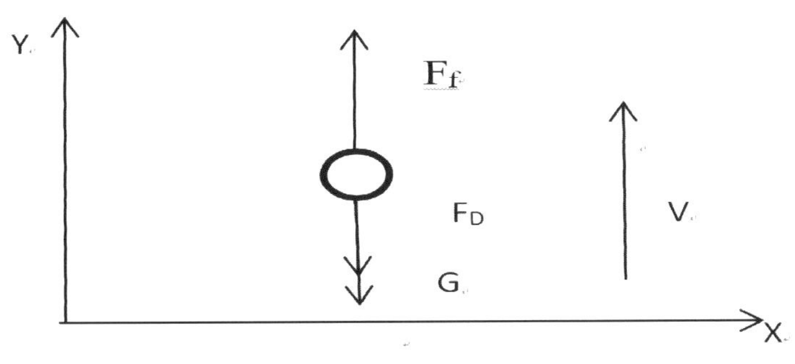

Figure 6. Analysis of the force of zinc vapor in the molten pool.

IOP Conf. Series: Materials Science and Engineering **411** (2018) 012085 doi:10.1088/1757-899X/411/1/012085

4.2. Dynamic equilibrium of zinc steam escape channel

When the reserved gap was small, it couldn't rule out the zinc vapor through it, causing a large amount of zinc vapor accumulated at the keyhole without being eliminated in time. This led to the result that the buoyancy of the zinc vapor at the keyhole was greater than the sum of viscosity of the liquid metal and the gravity of the zinc vapor itself, splashing at the same time.

From the welding process of High-speed camera, it could be found that zinc vapor escaped from three-channel, as shown below (figure 7).

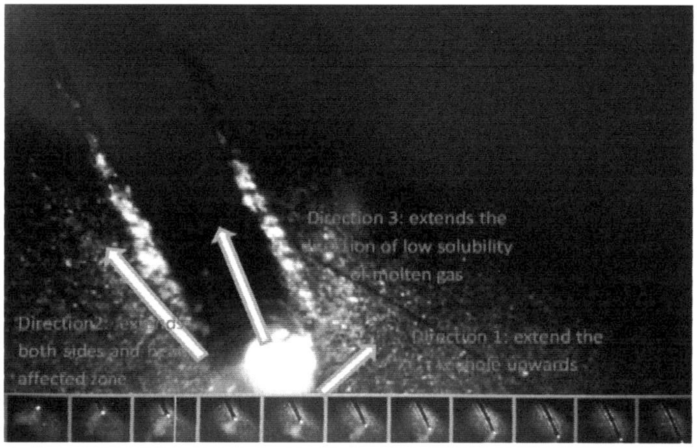

Figure 7. Zinc vapor escape three - channel equilibrium diagram

As shown in the figure 7, the zinc vapor accumulated in the keyhole, and the orifice escape direction 1 escaped upward, which was caused by a large number of zinc vapor flow. When the buoyancy of the zinc vapor was greater than the sum of viscosity of the liquid metal and the gravity of the zinc vapor itself, the zinc vapor escaped outside. In this point, when the zinc vapor was more, the zinc vapor to the molten metal out of the pool will form external pores, which has a great impact on the strength and aesthetics of the welded joints.

As to escape direction 1, when the zinc vapor was more, it escaped from the keyhole with a pointed mouth, bringing out the molten metal. At this time, many melt was out of the keyhole, which caused obvious hole on the surface of galvanized sheet, causing welding of external porosity. When the pores moved from the channel 3, there was no escape before the melt was solidified, the zinc vapor remain in the welding joint was formed the internal porosity.

Reserved gap can greatly ease the zinc vapor in the keyhole at the gas pressure, so it was helpful for the entire process of zinc vapor escape dynamic equilibrium that we can control zinc vapor to escape as much as possible from it [9]. When the channel 2 was blocked, the keyhole was prone to welding splash. When the direction of 2 sets a larger gap, Reserved gap enhance the channel 2 of the exhaust capacity, but in this time it was prone to welding incompatibility. Tests have shown that the channel 2 can be controlled by adjusting the reserved gap of the weld to change the formation of the pores. When the Reserved gap was 0 or 0.1mm, the channel 2 was smaller at this time. The zinc vapor was accumulated through the keyhole. It was the limited exhaust capacity of the channel 2. It was occurring serious splash that zinc vapor escape channel failed to dynamic balance. When the galvanized sheet was laser welded at a reserved gap of 0.2-0.25mm, the zinc vapor formed at the keyhole and quickly discharged through the channel 2. At this time, the welding joint was beautiful. zinc vapor was achieve dynamic balance equilibrium. when the reserved gap was greater than 0.3mm, the welding molding was poor, there was not penetration of joint in the situation. At this time welding metal did not quickly fill the welding channel, ausing welding joints subsidence.

When the dynamic equilibrium control of zinc vapor escape channel was better, the process of zinc steam escape was more stable and the porosity was less [10]. As shown in the figures 8, 9, it was the typical diagram of the welded joint what used the zinc vapor escape control.

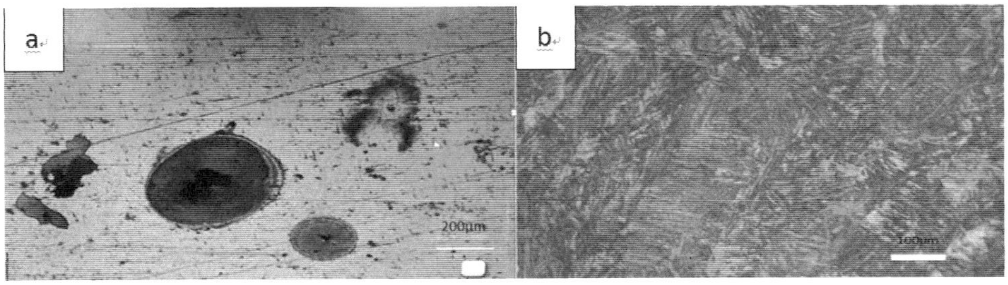

Figure 8. Microstructures of welded joints under dynamic equilibrium and non-dynamic equilibrium conditions. (a) Microstructures at non-dynamic equilibrium conditions. (b) Microstructures at welded joints under dynamic equilibrium.

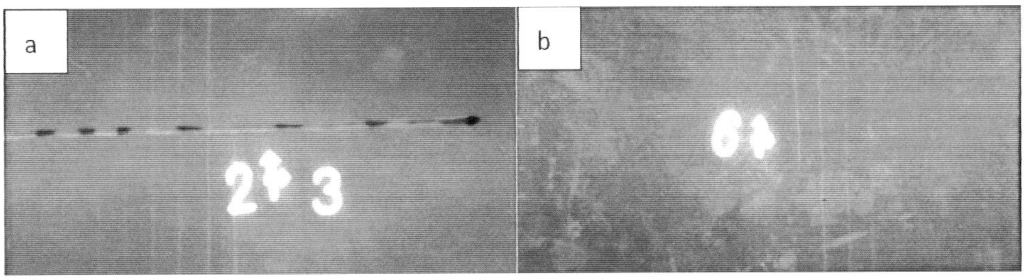

Figure 9. Welded joints under dynamic equilibrium and non-dynamic equilibrium conditions X-ray. (a) Welded joint under dynamic equilibrium conditions x-ray equilibrium. (b) Welded joint under dynamic equilibrium conditions x-ray equilibrium.

During the test, the welding process parameters of control zinc vapor in the three-channel escape to achieve dynamic balance was conducive to the effective escape of zinc vapor. Experiments showed that when the size of the reserved gap value effectively controlled the escape of zinc vapor with the reserved gap of 0-0.1mm, the welding channel 2 was seriously blocked resulting a large number of zinc vapor gathered in the keyhole, the increasing of pressure, zinc vapor splashed out of molten metal. When the reserved gap was 0.2mm-0.25mm, the three-channel escape process to achieve dynamic balance, the zinc vapor was better to escape, so the welding joint shape was better and no splash occurred. When the reserved gap was greater than 0.3mm, it appeared welding incompatibility, because the molten pool of metal was not enough to set aside the gap.

5. Conclusions

Through the above experimental study and analysis, we draw the following conclusions:

(1) This paper presented the formation mechanism of internal porosity and external porosity. The formation mechanism of the internal pores was that the zinc vapor always moves toward the metal liquid with greater solubility. The formation mechanism of the external porosity was that the zinc vapor dynamic equilibrium of the molten pool was broken, then the buoyancy of the zinc vapor at the keyhole reach the sum of viscosity of the liquid metal and the gravity of the zinc vapor itself, showing outward splash or escape.

(2) summarized the three channels of welding zinc vapor escape, through reserve gap to adjust the controlled affect the zinc vapor channel to achieve dynamic balance, adjusted to the zinc vapor can

effectively escape the molten pool, so that the welding joints of the external porosity and internal porosity reduced, thus enhance the welding quality, improve the welding appearance.

Acknowledgments
Financial support of the research was provided under State Agreement #14.578.21.0228 of Ministry of Education and Science of the Russian Federation (unique identifier of the project RFMEFI57817X0228).

References

[1] Pouranvari M and Marashi 2013 *Science and Technology of Welding and Joining* **18** 361–403
[2] Ma J 2013 *Experimental and numerical studies on the issues in laser welding of galvanized high-strength dual-phase steels in a zero-gap lap joint configuration* (Dallas: Southern Methodist University) p 3
[3] Volpp J 2017 *Production Engineering* **11** 9–18
[4] Atabaki M M, Ma J, Liu W and Kovacevic R 2015 *Materials & Design* **67** 509–21
[5] Norman P, Eriksson I and Kaplan A 2009 Proc. on *FORCE Technology* (Copenhague) pp 24–6
[6] Panwisawas C, Perumal B, Ward R M, Turner N, Turner R P, Brooks J W and Basoalto H C 2016 *Acta Materialia* **126** 251–63
[7] Kong F, Ma J, Carlson B and Kovacevic R 2012 *Optics & Laser Technology* **44** 2186–96
[8] Huang L, Chen X Z and Zheng H Z 2016 *Journal of Shanghai Jiaotong University* **50** 121–4
[9] Chen Z, Yang S, Wang C, Hu X, Shao H and Wang J 2014 *Journal of Materials Processing Technology* **214** 1456–65
[10] Stützer J, Zinke M and Jüttner S 2017 *Welding in the World* **61** 351–9

IOP Conf. Series: Materials Science and Engineering **411** (2018) 012086 doi:10.1088/1757-899X/411/1/012086

Modeling of complex metallurgical systems using dissipative structures and system dynamics

V P Tsymbal and P A Sechenov

Siberian State Industrial University, 42 Kirova street, Novokuznetsk, 654007, Russia

E-mail: tsymbal33@mail.ru

Abstract. The principles of physical and chemical processes modeling in a column spray-emulsion reactor in the form of dissipative structures are considered, which simplifies greatly the modeling. Comparison of two fundamentally different numerical methods of flow modeling is presented: Smooth Particle Hydrodynamics (SPH) method and method of Adaptive Mesh Refinement (AMR). The advantage of SPH method over the AMR grid methods is the high resolution at a constant number of particles. To solve the problem of simulation, a method similar to the method of smoothed Lagrangian particles was chosen. However, in the problem the mechanisms of interaction between particles, slag, metal, and also particle propagation are simulated. The isolation of dissipative structures made it possible to sharply reduce the dimension of the problem, since at the next higher level in this case, information is used only about the input and output parameters of this structure, complex interactions are closed within the dissipative structure. For the modeling the object-oriented programming language Action Scipt 3.0 was used, which allows the motion, the interaction of particles in a gravity separator to be visualized and animated in real time.

1. Introduction

The object of the study is gas-dynamic, physical and chemical processes in a column spray-emulsion reactor based on the system model of a complex of dissipative structures.

The creation of a continuous steelmaking process was an old dream of metallurgists. A.M. Bigeyev [1] and a number of other researchers considered the existence of separate chambers (zones) for the realization of necessary technological operations as an important and necessary advantage of such a process, along with the absence of intermediate losses of energy and raw materials. However, in connection with the process close to the equilibrium state (and, consequently, low rates of chemical reactions), these chambers turned out to be rather bulky, and the unit as a whole was capital intensive.

In the spray-emulsion reactor due to the transition to the gas suspension and emulsion zone, it was possible to create certain zones (in contrast to the physical chambers discussed above) in the form of dynamic dissipative structures that exist only at the moment of the process in a certain mode [2]. The zone model of a spray-emulsion reactor, includes the following zones (figure 1): 1 – core of the seal, 2 – reactor-oscillator, 3 – connecting channel, 4 – dynamic cushion, 5 – relatively dense gas-slag emulsion, 6 – dissipative gravity separator, 7 – wall layer with a negative (reverse) velocity, 8 – layer of liquid metal.

The most deviated from equilibrium is the zone 6 - the dissipative gravitational separator. Let's consider this zone in more detail.

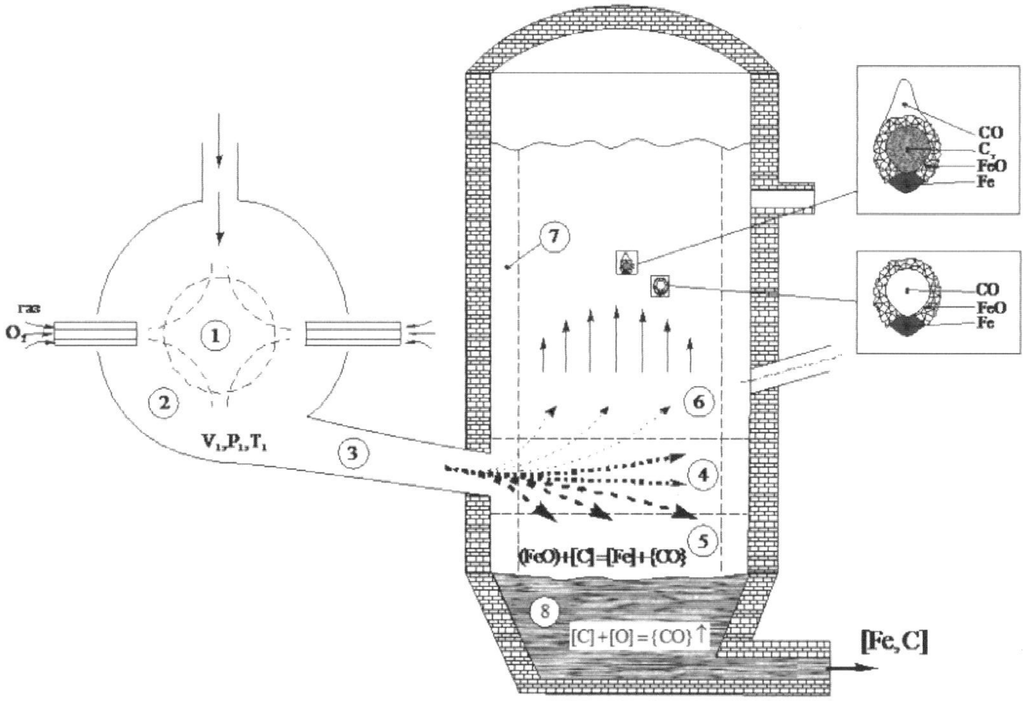

Figure 1. Zone model of the spray-emulsion reactor.

The dissipative gravitational separator (zone 6) is a sufficiently high layer of froth gas-slag-metallic emulsion suspended above zone 4. This zone occupies the overwhelming part of the column reactor. Due to the influence of the gravitational component and the emulsion velocity in the cross section of the column reactor, separation of the particles takes place depending on their density, size and collisions with other particles. At the same time, the particles of reduced iron, having a density two or three times greater than that of the iron oxide particles, "roll down" to the periphery of the flow and form a wall layer (zone 7). This phenomenon was observed by us on the physical model.

2. Choice of numerical methods

The problem of physical and chemical interaction of the dispersed particles of the charge and the reactions products in the flow of the carrying high-temperature gas has no analytical solution. One of the possible ways to solve it is simulation modeling using the approach similar to the "first principles" method in combination with stochastic modeling. Below we will consider the numerical methods used to simulate two-phase flows.

There are two fundamentally different approaches to numerical methods of flow modeling [3]: Smooth Particle Hydrodynamics (SPH) methods and methods of Adaptive Mesh Refinement (AMR).

The smoothed particle hydrodynamics method is an effective meshless Lagrangian method used to calculate the structure of flows with an unknown free boundary, in particular, high-speed processes in media with a significantly changing topology of simulated objects [4].

In this method, N particles of spherical shape are placed in the simulated physical space, each particle has mass, internal energy, velocity and moves in accordance with the laws of mechanics [5]. Hydrodynamic quantities, such as density and pressure, are defined as the result of the contribution of particles in a certain region [4].

In connection with the fact that the task of describing physical and chemical processes does not have an analytical solution, the method on adaptive meshes, using differential equations in the basis of its solution, can not be applied.

The advantage of SPH method in comparison with the mesh methods of AMR is high resolution at a constant number of particles, while for grid methods it is required to increase the number of calculated points in space in order to refine the results (for example, with a non-uniform flow density).

The disadvantage of SPH method is the finding of an effective method for searching neighbors in the space.

To solve the problem of simulation, a method similar to the method of smoothed Lagrangian particles was chosen. However, in the problem being developed, collisions between particles are taken into account, mechanisms of interaction between particles, slag, metal, and also the particle propagation are simulated. The Monte Carlo method was applied to model the mechanisms of interactions between particles, generation of radius and composition of particles.

3. Isolation of dissipative structures for folding information about the object of research

Isolation of dissipative structures allows the dimension of the problem to be sharply reduced, because at the next higher level, in this case, only the input and output parameters of this structure are used. Complex interactions are closed within a dissipative structure.

In the simulation model under consideration, at a low level (of the floating and interacting particles), the dissipative gravitational separator model (zones 6 and 7) is considered. Zone 5 – relatively dense slag emulsion and zone 8 – a layer of liquid metal, are considered as dissipative structures with input and output parameters.

4. How, and at what level is column gravitational separator simulated?

The program is written in the object-oriented programming language Action Script 3.0, which allows the motion, the interaction of particles in a gravity separator to be visualized and animated in real time.

What objects are allocated and implemented in the program (the appropriate classes are developed)?

The space of the column spray-emulsion reactor is divided into three dissipative structures that interact with each other: a gravitational separator, a slag, and a metal. In this case, for the gravitational separator, as mentioned above, the particle method is used. In this connection, the following objects (classes) are displayed that reflect the physicochemical phenomena in the gravitational separator: condensed particles and gas particles. At the level of dissipative structures, the objects are: the slag below and the metal below.

In turn, the object (condensed particles) is divided into three types: ore, metal and slag.

Let us consider the input-output parameters of each of the selected structures (figure 2).

As it can be seen from figure 2, the input and output objects for each of the allocated structures are particles. At the same time the gravitational separator leads to the formation, collision, melting and formation of new particles. The gas particles formed in the gravity separator together with the initial gas composition are discharged into the gas outlet. And the composition of the gas in it, at this point in time, is determined by the composition of the outgoing gas particles. As a result of the melting of iron ore particles in the gravitational separator, slag particles are formed that can fly to the slag collector or fall into a dense layer of slag. The composition of the dense layer of slag is formed from the slag particles trapped in it, and the composition of the metal is formed from the composition of the metallic particles that have fallen into it.

In dissipative structures, such as slag and metal, internal processes are not considered at the particle level. Only processes occurring at the slag-metal interface (diffusion transition) are described, as a result of which gas, metal and slag particles are formed. The composition of the metal depends on the composition of the metal particles trapped in the collector and the decarburization rate at the slag-metal interface.

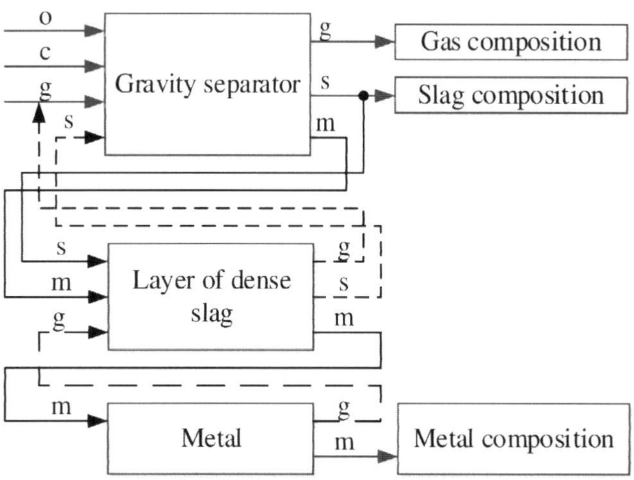

Figure 2. Input-output parameters of the simulation objects of the column reactor (particles): o – ore; c – coal; g – gas; m – metal; s – slag.

Thus, we obtain a system of interconnected objects. In this case, the most deviated from the equilibrium object – the gravitational separator – is given special attention. It is considered and modeled on the basis of particles. This in turn allows all the required characteristics from the change in the space and time of the composition of the heterogeneous two-phase flow to be studied, and in the composition of the main output product – metal, depending on the constructive and mode parameters.

5. Conclusion

In the paper the object of research is described. The choice of numerical methods for modeling heterogeneous processes in a column reactor is substantiated. It is shown that the isolation of dissipative structures allows the number of connections and calculations of the modeled object to be reduced. The object-oriented programming language Action Script is chosen as a modeling tool. In this connection, the main modeling subobjects are selected and their input and output parameters are shown.

References

Bigeyev A M 1986 *Continuous Steelmaking Processes* (Moscow: Metallurgy) p 136
Tsymbal V P, Mochalov S P, Rybenko I A et al 2014 SER *Process – Metallurgical Spray-emulsion Reactor* (Moscow: Metallurgizdat) p 488
Lazareva G G 2010 *Bulletin of NSU. Series Mathematics, Mechanics, Information* **10(1)** 40–64
Afanasiev K E 2010 *Proc. of the Sci. School "Information and Computing Technologies in Numerical Calculations and Management of the University"* (Kemerovo: INT) 245–57
Davydov M N and Kedrinsky V K 2013 *Applied Mechanics and Technical Physics* (Novosibirsk: SB RAS) **54(6)** 17–26

IOP Conf. Series: Materials Science and Engineering **411** (2018) 012087 doi:10.1088/1757-899X/411/1/012087

Nonequilibrium dissipative structures and control over the carbon content in the spray-emulsion unit

V P Tsymbal[1], P A Sechenov[1], I A Rybenko[1] and A A Olennikov[2]

[1]Siberian State Industrial University, 42 Kirova street, Novokuznetsk, 654007, Russia
[2]Tyumen State University, 15a Perekopskaya street, Tyumen, 625003, Russia

E-mail: tsymbal33@mail.ru

Abstract. The principles of creating a self-organizing spray-emulsion reactor (SER) are considered, the basis of which is the organization of forced motion of a two-phase working (reaction) mixture in the closed system under pressure, which allows a significant deviation from thermodynamic equilibrium to be created and nonequilibrium dissipative structures to be obtained. The role of dissipative structures in solving problems of controlling the chemical composition of the metal, as well as in creating a process and an unit having a small specific volume and energy capacity is shown.

1. Introduction

Studies devoted to the creation of continuous metallurgical processes especially intensively were conducted in the 70-80's of the last century. In the collection of reports [1], entirely devoted to the development of continuous processes, units with sequentially connected baths, converter-type processes, etc., are presented. Of particular interest are studies on spray-emulsion processes, the results of which have not lost their value even today. A.M. Bigeyev, who made a great contribution to the development of the continuous steelmaking process [2, 3], considered the existence of separate chambers (zones) for the realization of the necessary technological operations (decarburization, desulfurization, dephosphorization, etc.) important and necessary along with the absence of intermediate losses of energy and raw materials. However, at that time, due to the process occuring close to the equilibrium state and, as a consequence, low rates of chemical reactions, these chambers turned out to be rather cumbersome, and the unit as a whole was capital-intensive.

In this paper it is shown that the simultaneous solution of the task of controlling the metal chemical composition and reducing the specific volume of the unit is achieved through the constructive and mode organization of dynamic dissipative structures [4], which play the role of peculiar "chambers" that were planned to be created in the initial designs of units.

2. Basic principles and solutions

The following principles and decisions were laid down in the basis of the constructive implementation of the process and unit below [4, 5]:

- organization of the forced movement of the working (reaction) mixture in the closed system under pressure, which made it possible to create a significant deviation from thermodynamic equilibrium and simultaneously solve the problem of internal transport of reaction products through all the devices connected in series;

- creation of a large reaction surface and a two-phase working mixture (gas suspension or emulsion), that is, the conversion of the process to the gas dynamics zone, which made it possible to obtain high rates of physical and chemical processes;
- use of non-linear dependence of the flow velocity of a two-phase medium on the gas content for creating a pressure potential due to the gas-dynamic locking effect of the connecting channel;
- creation of dissipative structures that are substantially deviated from thermodynamic equilibrium, which made it possible to obtain greater control over the chemical composition of metal and slag.

Below we briefly consider these dissipative structures [4], with reference to the band model shown in figure 1: 1 – core of the compaction; 2 – reactor-oscillator; 3 – connecting channel; 4 – dynamic cushion; 5 – relatively dense gas-slag emulsion; 8 – layer of liquid metal; 6,7 – dissipative gravitational separator.

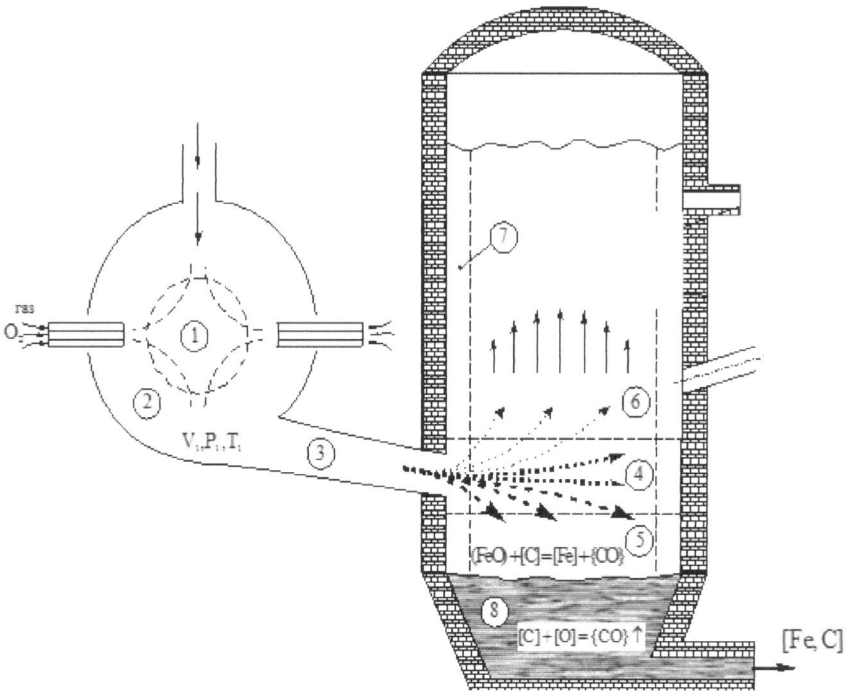

Figure 1. Zone model of the spray-emulsion process.

1. *The core of compaction* on countermoving gas jets (zone 1) makes it possible to organize an intensive dispersion of the burden falling from above.

2. *Reactor-oscillator* (figure 2), including the said structure 1, and also zone 2 (gaseous suspension created as a result of dispersion by counter-streams) and a connecting channel 3 playing the role of gas-dynamic gate. As a result of using the effect of the nonlinear dependence of the speed of sound on gas content [6], an oscillator-reactor was created in which internal feedback is realized due to the critical outflow of a two-phase medium [4] and the influence of pressure (according to the Le Chatelier-Brown principle) on the course of chemical reactions with gas evolution, which leads to a change in the volumetric gas content in the reactor, and then to a change in the flow velocity of the two-phase medium in the connecting channel 3 by the formula:

$$\alpha_g = \frac{v_g}{v_g + v_c}, \tag{1}$$

where v_g – volume of gas, m^3; v_c – volume of the condensed phase, m^3.

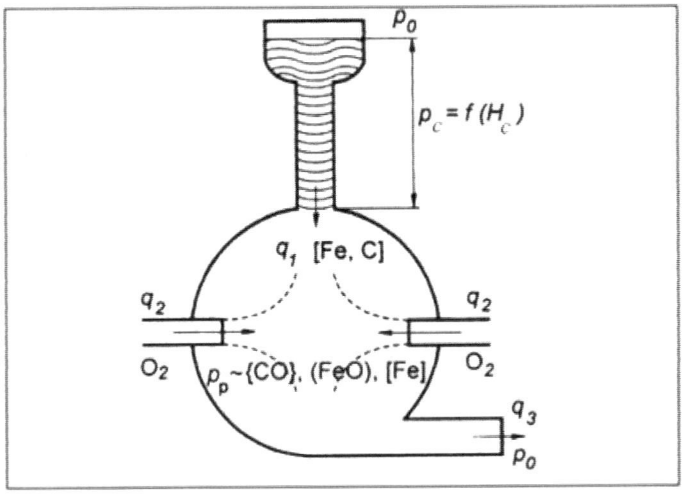

Figure 2. Reactor-oscillator: p_0, p_p, p_c – respectively atmospheric pressure, in the reactor and charge column; q_1, q_2, q_3 – streams of charge, oxygen and the two-phase medium flowing out of the reactor.

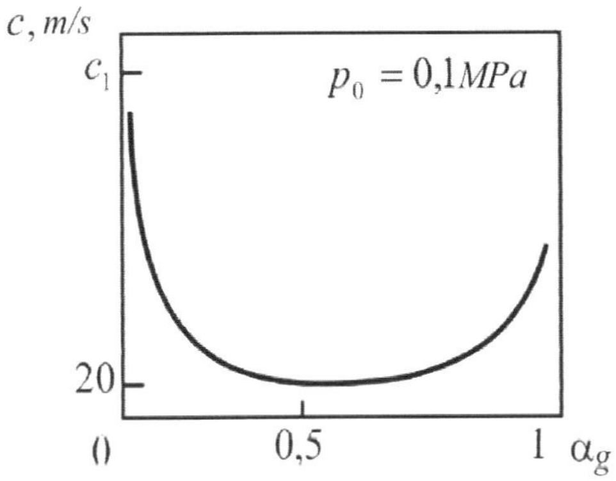

Figure 3. Dependence of the sound velocity in the water-air mixture on the gas content.

The process as it were "runs" along the right branch of the nonlinear dependence of the sound velocity c on the gas content α_g, presented in figure 3. As a result it was possible to create a stationary oscillatory mode at any given pressure level. This is the most important link in the task of creating a self-organizing process.

3. Thus, on the basis of the oscillator-reactor, a flow stimulator (a kind of carburetor-compressor) was created, by means of which (in combination with the transfer of the process to the gas suspension and emulsion region) it was possible to create an internal transport of the working mixture through all the elements of the unit, heat recovery devices (waste heat boiler, gas turbine, fluidized bed, reformer of smoke into synthesis gas). This allows conditions for the deep use of the energy of the initial fuel to be created.

3. *Dynamic "cushion"*, playing the role of a failure grating (zone 4), formed as a result of the transformation of the kinetic energy of the spray emitted from the reactor-oscillator into the potential energy of pressure. This cushion separates and maintains zones 6 and 7, where mainly regenerative reactions occur, from zones 5 and 8, where oxidative reactions can occur.

4. *Relatively dense gas-slag emulsion* of highly foamed converter or open-hearth slag in the layer adjacent to the metal (zone 5). This zone can play a decisive role in the processes of self-organization in the refining settler, since here is the most nonequilibrium component of the process, associated with a large content of iron oxides in the slag.

5. *The layer of liquid metal*, zone 8 – is intended for the accumulation of metal, which is recovered in zone 6 and descending through zone 5. It should also be emphasized that oxidation reactions can occur at the boundary of zones 5 and 8, especially if relatively large pieces come to this boundary, for example, ores.

6. *Dissipated gravitational separator* (zone 6). This is a sufficiently high layer of foamy gas-slag-metallic emulsion suspended above zone 4, in which particles of solid oxides and carbon can also be

present. This zone occupies the overwhelming part of the column reactor (refining settler). Due to the influence of the gravitational component and the velocity of the emulsion moving close to the parabolic distribution in the cross section of the column reactor, separation of the particles takes place depending on their density. Due to this phenomenon, the particles of reduced iron, having a density two to three times greater than that of the iron oxide particles, "roll" to the periphery of the flow and form a wall layer with a negative (reverse) velocity (zone 7). This phenomenon was observed visually by us on the physical model.

It is not possible to consider the system model of the complex of dissipative structures presented above [4, 7] within the framework of this article. But it is important to note that the decomposition of the process and its model in the form of a set of dissipative structures made it possible to substantially simplify the mathematical description of this complex process. Due to the fact that self-organization takes place within each dissipative structure, or at least self-stabilization, information, in fact, folds [8], and structures are exchanged between themselves and the external environment only by the flows of matter and energy. This allows models of each dissipative cell us to be built separately and the number of necessary connections between the elements (subsystems) to be significantly reduced.

Below we briefly touch on the model of only one very important structure, dissipative gravitational separator of metal, slag and gas.

Taking into account the important role of the marked dissipative structure, which connects all other zones and causes the circulation of substances in the unit, an imitation model was created for its reproduction and study [9], in which disperse particles of charge with the real specified granulometric distribution are the "first principles" – elements of the lower level of the hierarchy determined by the grain size distribution. The graphic interface of the program is shown in figures 4 a,b.

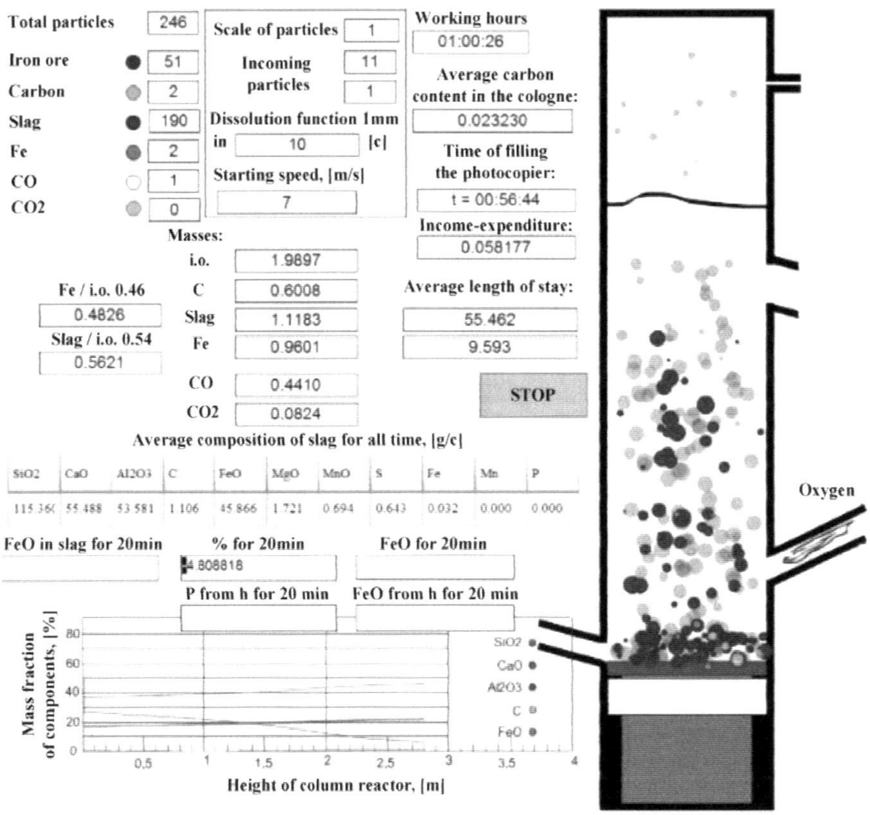

(a)

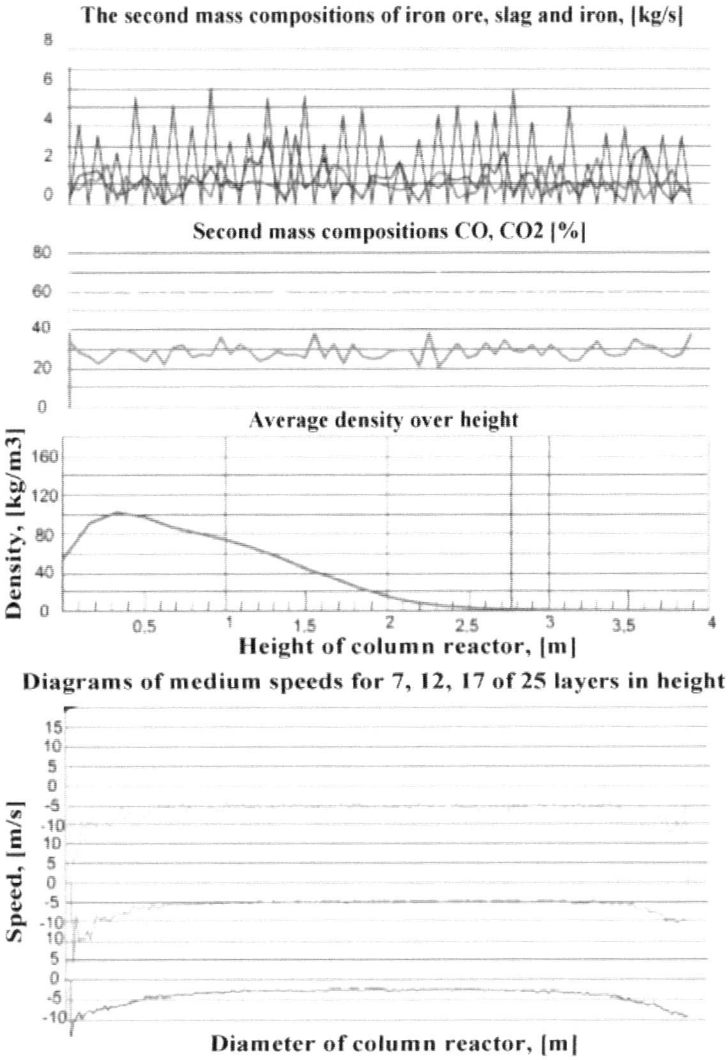

(b)

Figure 4. Fragment of implementation of the simulation model gravity separator.

The core of this model is the problem of the flow of a single particle by a vertical flow of a carrier reaction gas. With the help of a specially designed algorithm and a computer program using the Monte Carlo method, a large number of elastic and inelastic collisions of particles of charge materials and reaction products are lost, taking into account the processes taking place in them. The developed model, tested on the basis of the results of a large number of computer statistical tests coincident with the conservation laws, proved to be quite adequate "virtual reality" of the process under study). It allows the internal mechanism of complex interactions to be revealed that occur in the column reactor: to evaluate the residence time of the initial substances and reaction products, including metal and slag, in the unit; distribution of density of suspended dispersed materials along the height of the unit; influence on the course of the process of various control actions, including the granulometric composition of the metal, the ratio of ore materials to the fuel-reducing agent and other. Being a substitute object, this model allows a numerical method to solve the problems of system dynamics, lose and develop the projected technologies. In more detail this model deserves a separate consideration [9].

3. Control of the chemical composition of metal

After getting acquainted with the basic principles of the functioning of the process and the role of dissipative structures, let us dwell on the control of the chemical composition of the metal, primarily the carbon content [10]. To do this, we again turn to the band model (figure 1). Stressing that chemical reactions occur under nonequilibrium conditions with the removal of reaction products, we note that the reactions of reducing iron oxides with solid carbon:

$$(Fe_2O_3) + C_s = 2(FeO) + \{CO\}, \quad (2)$$

$$(FeO) + C_s = [Fe] + \{CO\}, \quad (3)$$

$$C_s = [C], \quad (4)$$

have a preferential development in zone 6 (the top of the column reactor), and reactions with the carbon dissolved in the metal:

$$(Fe_2O_3) + [C] = 2(FeO) + \{CO\}, \quad (5)$$

$$(FeO) + [C] = [Fe] + \{CO\}, \quad (6)$$

occur at the boundary of zones 5 and 8 (the turbulent layer at the metal-slag boundary) and the velocity is an order of magnitude larger than the reactions (2-4). In this case, zones 5 and 8 are separated by a dynamic "*cushion*" (zone 4) from zone 6.

The proportion of iron oxides processed at the slag-metal interface (zones 5 and 8) can be increased by increasing the share of larger ore materials and additional induction heating of the copier, and in zone 6, increase the recovery potential by feeding coal to the column reactor to the upper level of the gas-slag emulsion.

Thus, the possibility of creating the dynamical dissipative structures described above under certain gas dynamic modes allows us to control separately the ratio of the rates of the reducing and oxidizing processes, and, consequently, the carbon content in the resulting metal.

4. Structural and technological features of the SER unit

The use of the above described approaches and principles made it possible to create a unit with a very small specific volume and energy capacity [4, 5, 10]. The technological scheme and brief description of the SER (self-organizing spray-emulsion reactor) is presented below (figure 5). The basis of the technological scheme of the mini-module is: the charging system 1-5, the reactor-oscillator 6, the connecting channel with gas-dynamic self-locking 7, the refining settler 8, simultaneously playing the role of the first stage of wet gas scrubbing, the collector 9, and the garrison cooling system 10, the channel 11 for the delivery of a gas slag emulsion and a channel 12 for the transfer of a portion of the gas, a slag receiver 13 with a granulator 14, the system of heat recovery in the fluidized bed or the transition of flue gases into synthesis gas 17, and a gas cleaning system 18.

The high-pressure potential created in the reactor-oscillator 6, as well as the complete isolation of the process from the atmosphere, allows the products of reactions to be pushed through all the energy reclamation devices located behind the main technological unit without the use of high-temperature flow stimulators.

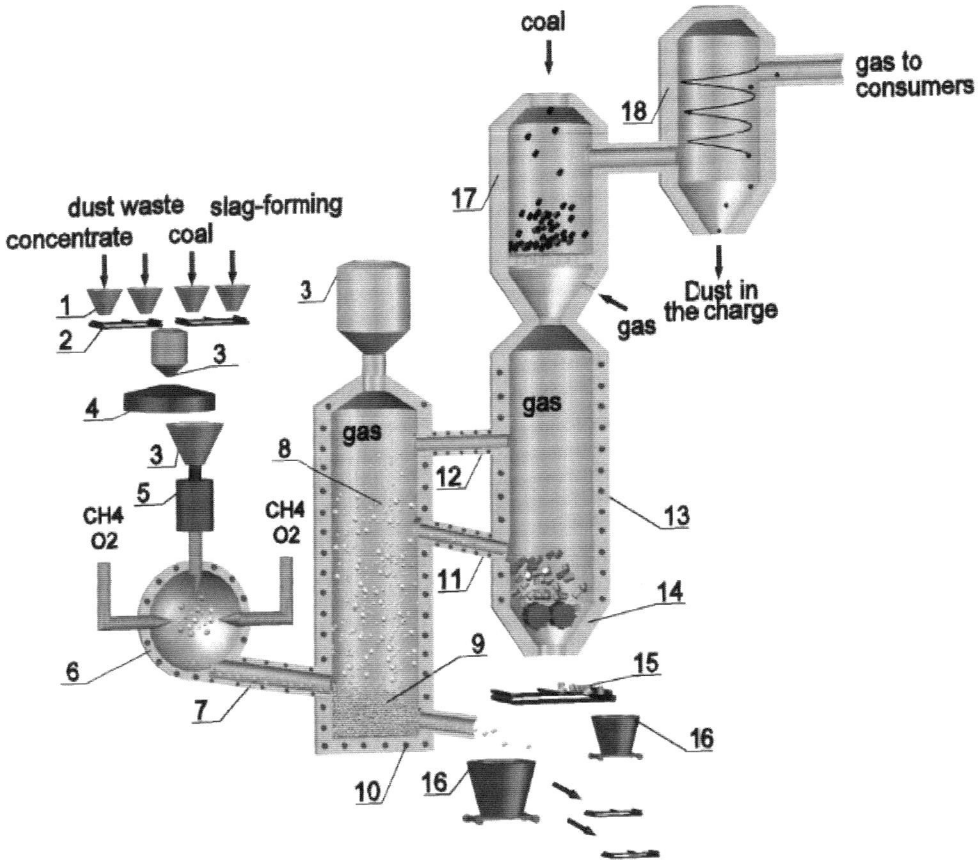

Figure 5. Technological scheme of the mini-module based on the SER process.

The dusty charge, consisting of a mixture of metal oxides and solid reducing agents, is efficiently dispersed in the reaction chamber 6. A gas suspension with a gas volume fraction of about 0.99 is produced here. Due to the dependence of the flow velocity of the two-phase medium on the gas content (Figure 3), an aerodynamically locked gate (jet braking) is formed in the connecting channel 7, and in combination with feedback on the gas content (due to changes in the conditions of chemical reactions, and oxygen) it is possible to create a stationary oscillatory mode (self-organizing reactor-oscillator).

The use of a vertical column reactor 8 as a refiner with a lower feed through the channel 7 of the reaction gas suspension combined with a significant deviation of the processes from the thermodynamic equilibrium is the most important factor that conditions the gravity separation of the iron-carbon metal stream flowing along the periphery of the column reactor and settling in the collector 9, and a stream of iron-depleted slag discharged along the inclined channel 11 to the slag receiver 13.

5. Results

When creating this unit, for the first time in world metallurgy [10] an attempt was made to use some ideas of the theory of self-organization (synergetics), such as the principles of subordination and the least coercion, a large deviation from thermodynamic equilibrium and other [11]. To realize these principles a number of physical effects was applied, such as the dispersion of the charge by counter gas jets, the creation of a self-organizing reactor-oscillator by using the critical flow of the two-phase flow and feedback on the change in the gas content of the reaction products, the lower supply of the working mixture from the oscillator-reactor to the column reactor, the organization of forced movement and internal pneumatic transport of a two-phase working mixture.

6. Conclusion

The combination of these factors and measures made it possible to create dynamic dissipative structures in the unit, which led to a sharp decrease in the specific volume of the unit (by 10-15 times) and ensured considerable control flexibility, including the ratio of the rates of reduction and oxidation reactions, with a fairly wide range of carbon content. Currently, several dozens of new direct reduction units are being developed in the world, but in none of them, including COREX, it is possible to produce an iron-carbon alloy with a carbon content below 2%, due to the fact that the processes in them are close to the state of equilibrium and it is impossible to control separately the composition of metal and slag.

Below a comparison of technical and economic indicators of the SER process with the closest and most advanced analogue in the world – COREX unit is provided (table 1).

Table 1. Comparison of technical and economic indicators of the processes of SER and COREX.

Position	Indicators	Process COREX	Process SER	Advantage
1	Energy intensity, GJ/t	29	15-17	in 1.7 times
2	Specificvolume, t/m^{3}day.	1.1	11	in 10 times
3	Capital expenditures, $/t.year	350	120-150	in 2.5 times

References

[1] 1975 *Proceedings of the meeting "Continuous processes of smelting ferrous and non-ferrous metals"*, November 26 - 28, 1973, USSR Academy of Sciences, Scientific Council "Physical and Chemical Foundations of Metallurgical Processes" (Moscow: Nauka) p 204
[2] Bigeev A M 1986 *Continuous Steelmaking Processes* (Moscow: Metallurgy) p 136
[3] Bigeev A M 2000 *Theory and Technology of Steel Melting* (Magnitogorsk: MSTU) p 344
[4] Tsymbal V P, Mochalov S P, Rybenko I A et al 2014 *SER Process – Metallurgical Jet-emulsion Reactor* (Moscow: Metallurgizdat) p 488
[5] Tsymbal V P, Kozhemyachenko V I, Rybenko I A, Padalko A G and Olennikov A A 2014 *Proc. XIII Int. Cong. of Steel Makers in The Use of the Principles of Self-organization and Dissipative Structures in the Creation of a New Spray-emulsion Metallurgical Process* (Moscow: Polevskoy) pp 472–477
[6] Nakoryakov V E, Pokusaev B G and Shreiber I R 1990 *Wave Dynamics of Gas and Liquid – Vapour Environments* (Moscow: Energoatomisdat) p 248
[7] Tsymbal V P, Sechenov P A and Olennikov A A 2016 *Modeling and Knowledge-intensive Information Technologies* (Novokuznetsk: SibSIU) pp 52–59
[8] Khaken G 1985 *Synergetics. Hierarchy of Instability in Self-organizing Systems and Devices* (Moscow: Mir) p 419
[9] Tsymbal V P, Sechenov P A, Olennikov A A and Padalko A G 2015 *Proc. of the XIX Int. Sci. and Prac. Conf. in Metallurgy: Technologies, Innovations, Quality* (Novokuznetsk: SibSIU) vol 1 pp 73–80
[10] Tsymbal V P, Mochalov S P and Shakirov K M 2015 *Metallurgist* **59** 119–125
[11] Nikolis G and Prigozhin I 1979 *Self-organization in Nonequilibrium Systems* (Moscow: Mir) p 512

IOP Conf. Series: Materials Science and Engineering **411** (2018) 012088 doi:10.1088/1757-899X/411/1/012088

Multi-factor regression analysis of the process of rails contact welding on K1000 machine

R A Shevchenko, N A Kozyrev, R E Kryukov, A O Patrushev and A A Usoltsev

Siberian State Industrial University, 42 Kirova street, Novokuznetsk, 654007, Russia

E-mail: kozyrev_na@mtsp.sibsiu.ru

Abstract. The article describes the technology of welding railroad rails with an electrocontact method by a pulsating reflow on a K1000 machine and production tests for a static three-point transverse bending. The analysis of the technology of welding rail rails on K1000 machine made it possible to determine the parameters of the welding mode and to reveal the essential (determining) characteristics of the technology for welding rails. The evaluation of the influence of the welding parameters on the value of the flexural force P_{bend} (kN) and the value of the deflection f_{def} (mm) was carried out by means of multiple regression analysis, which makes it possible to study the regularities of the change in the resultant index depending on the behavior of various factors. Based on the available data, a mathematical model of the process of contact welding of rails was created. The functional dependence of the influence of technological parameters on the force P_{bend}. Using the regression model, it is proposed to predict the quality of the welded seam and control the technological parameters of the process of contact welding of rails.

1. Introduction

Various methods of welding rails have been developed for the construction and repair of a seamless rail track. The most applicable technologies are: electrocontact and aluminothermic welding of rails. In the Russian Federation the most widely used electrocontact method of welding rails. Most of the rails are welded at stationary rail welding enterprises and a small part of joints by rail self-propelled machines [1]. In stationary conditions at the enterprise LLC "RSP-M" (RSP-29) welding is performed with the help of MCP 63.01, K1000, K1100 machines.

2. Research methods

The process of welding of full-profile rails on the K1000 machine was monitored with the recording of the main parameters in the computer memory. These parameters were fixed during the welding process and stored in the database in a tabular form, and, if necessary, they can be represented graphically.

The technology of welding rails on RSP includes a mandatory welding of control samples at the beginning of the work shift. Two samples for each welding machine are welded, after which the non-heat-treatable joints are tested for three-point static bending according to the industry standard of RZD 1.08.002-2009 "Railway rails welded by the electrocontact method". Static bending tests were carried out on a press of PMS-320 type.

Control samples are tested after welding and deburring in the welding machine without additional processing of the joints. In tests for static transverse bending, the test piece should have a length of at least 1200 mm with a welded joint in the middle. The load was applied in the middle of the span of the control sample at the place of the welded joint with the distance between the supports of 1 m (figure 1).

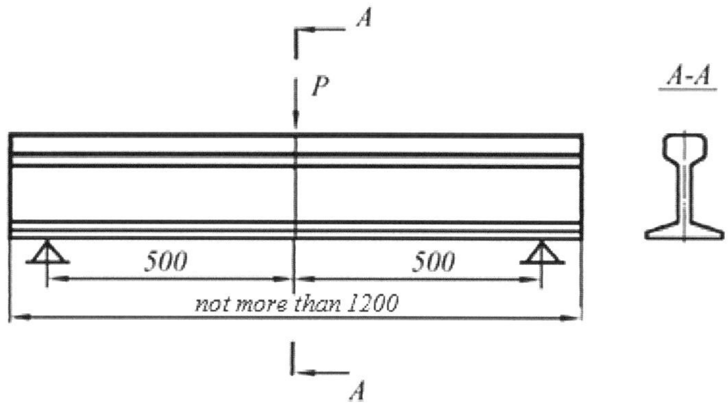

Figure 1. Scheme of the test for static transverse bending.

3. Results and discussion

The process of welding rails by the electrocontact method on K1000 machine is divided into several stages: 1st stage of reflow, 2nd stage of reflow, forcing, upset, flash removal. Each stage is divided into several sections: the first stage of reflow consists of two sections (in table 1 they are numbered 1 and 2), the second stage of reflow consists of 4 sections (in table 1 they are numbered 3, 4, 5, 6), the forcing is divided into 2 sections (in table 1 they are numbered 7 and 8).

Each stage is characterized by welding modes: current strength I, A; voltage U, B; the amount of movement of the movable bed S, mm; pressure in the system P, atm; moving speed of the movable bed V, mm/s; duration of stage T, s [2].

Table 1. Defined parameters of the welding mode on K1000 machine.

Section No.	1	2	3	4	5	6	7	8
U, B	U_1	U_2	U_3	U_4	U_5	U_6	U_7	U_8
S, mm	S_1	S_2	S_3	S_4	S_5	S_6	S_7	S_8
I_1, A	I_{11}	I_{12}	I_{13}	I_{14}	I_{15}	I_{16}	I_{17}	I_{18}
I_2, A	I_{21}	I_{22}	I_{23}	I_{24}	I_{25}	I_{26}	I_{27}	I_{28}
I_3, A	I_{31}	I_{32}	I_{33}	I_{34}	I_{35}	I_{36}	I_{37}	I_{38}
V_m, B·10	V_{m1}	V_{m2}	V_{m3}	V_{m4}	V_{m5}	V_{m6}	V_{m7}	V_{m8}
V_n, B·10	V_{n1}	V_{n2}	V_{n3}	V_{n4}	V_{n5}	V_{n6}	V_{n7}	V_{n8}
Upset under current, s				O_I				
Upset, mm·10				O				
Forging, s·10				Pr				

The test of one control sample is performed with the load applied to the head (elongation in the base), the second control sample is loaded on the base (elongation in the head). The results of the test are the values of force during bending P_{bend}, kN and the deflection force f_{def}, mm at which the control sample crashes, or the maximum values of these indicators if the sample does not fail during the tests. After the actual indicators are compared with the normative ones. If the sample has passed the regulatory requirements, welding of the rails into the line begins. If the sample does not withstand the regulatory requirements, then it is re-welded. If the repeated tests also have an unsatisfactory outcome, actions are undertaken to find the causes of low indicators and adjust the technological modes [3, 4].

The method of statistical modeling was used to describe the process of rails contact welding in this paper. To implement it, it is necessary to establish independent (input) variables, as well as to identify the dependent (output) parameters. To create a mathematical model of the process of contact welding of rails, the existing database of parameters of welding modes and test results of welded joints was used.

As input variables were chosen: the average value of the current in each section I_{av}, A; the average value of the voltage in each section is U_{av} B; duration of setup under current O_1, s; amount of setup O, mm; duration of forging Pr, s; resistance of the joint R, Ohm. The results of tests for static transverse bending were selected As output parameters: the force arising from bending P_{bend}, kN and the values of the deflection f_{def}, mm.

For each section the corresponding path S (mm) is set, with which the welding machine sets the values U, I_1, I_2, I_3, V_m, V_n. From the array of data the parameter values for each section were selected (U_{ji}, I_{ji}) and the mean of these parameters was determined ($\overline{U_j}, \overline{I_j}$):

$$\overline{U_j} = \frac{\sum U_{ji}}{p}, \tag{1}$$

where p – the number of values U_{ji} fixed by the sensor in this section; U_{ji} – value of voltage, fixed by the sensor at the moment of fusion on the j-th section; $\overline{U_j}$ – the average value of the current in the j-th section.

$$\overline{I_j} = \frac{\sum I_{ji}}{p}, \tag{2}$$

where p – the number of values I_{ji} fixed by the sensor in this section; I_{ji} – value of voltage, fixed by the sensor at the moment of fusion on the j-th section; $\overline{I_j}$ – the average value of the current in the j-th section.

After processing the production control data and parameters of the welding modes, data were obtained that reflect the main actual values of the welding modes on the machine of contact butt welding K1000 and the test results of weld joint consisting of 60 observations.

Based on the received production data they were statistically processed using the STATISTICA 10.0 package. The evaluation of the effect of the welding parameters on the amount of force P_{bend} (kN) and the deflection value f_{pr} (mm) was carried out by means of multiple regression analysis, which makes it possible to study the regularities of the change in the resultant index depending on the behavior of various factors [5-7].

The coefficients of the regression equations were calculated by the matrix method using the STATISTICA 10.0 package. The regression equation is presented in the following form:

$$Y = b_0 + b_1 \cdot X_1 + b_2 \cdot X_2 + \cdots + b_n \cdot X_n, \tag{3}$$

where Y – the dependent (output) parameter;
$X_1, X_2, \ldots X_n$ – independent (input) variables;
$b_0, b_1, b_2, \ldots b_n$ – coefficients of regression.

The obtained model describing the process of rails welding R65 DT350 on K1000 machine No. 01 in the conditions of RSP-M (RSP-29) has the form:

$$P_{bend} = 18441.49 - 2.69\ I_{1av} - 73.72\ U_{1av} - 1.06\ I_{2av} + 9.98\ U_{2av} + 3.98\ I_{3av} - 11.51\ U_{3av} + 1.64\ I_{4av} + 29.45\ U_{4av} + 1.49\ I_{5av} - 27.48\ U_{5av} + 19.8\ U_{6av} - 0.54\ I_{7av} + 5.19\ U_{7av} + 593.77\ O_1 + 8.03\ R;\ R^2 = 0.72.$$

The adequacy of these regression equations was verified by the average error of approximation:

$$\bar{\varepsilon} = \frac{1}{m}\sum_{i=1}^{m}\left|\frac{Y_i - \tilde{Y}_i}{Y_i}\right| \cdot 100, \quad (4)$$

where m – the number of observations; $\tilde{Y}_i$ – the calculated value of the resulting indicator;

Y_i – the actual value of the resulting indicator.

For P_{bend}, the average approximation error is equal to: 4.45%.

4. Conclusion

On the basis of experimental data, a mathematical model of the process of contact welding of rails on the K1000 machine has been developed, which makes it possible to evaluate the the influence of the parameters of the process of rails contact welding on the quality of the welded seam.

It is proposed to predict the quality of the welded seam using the regression model and control the technological parameters of the process of rails contact welding.

References

[1] Kozyrev N A, Usoltsev A A, Kryukov R E, Shevchenko R A and Shishkin P E 2017 *Proc. of the Int. Sci. and Pract. Conf. in Innovations in the Fuel and Energy Complex and Machine Building* (Kemerovo: KuzSTU) pp 123-127

[2] Shevchenko R A, Kozyrev N A, Usoltsev A A, Patrushev A O and Shishkin P E 2017 *Bulletin of SibSIU* **1(19)** 4–8

[3] Shevchenko R A, Kozyrev N A, Usoltsev A A, Bashchenko L P and Knyazev S V 2017 *Bulletin of SibSIU* **1(19)** 12–15

[4] Shevchenko R A, Shishkin P E and Patrushev A O 2017 *Proc. of the All-Russian Sci. Conf. "Students, Postgraduates and Young Scientists Science and youth: problems, searches, solutions"* (Novokuznetsk: SibSIU) **21-1** pp 229–232

[5] Shevchenko R A, Kozyrev N A, Shishkin P E, Kryukov R E and Usoltsev A A 2017 *Proc. of the Int. Sci. and Pract. Conf. in Innovations in the Fuel and Energy Complex and Mechanical Engineering (TEK-2017)* (Kemerovo: KuzSTU) pp 205–211

[6] Kozyrev N A, Usoltsev A A, Shevchenko R A, Kryukov R E and Shishkin P E 2017 *Izv. Vuzov. Ferrous Metallurgy* **60(10)** 785–791

[7] Kozyrev N A, Usoltsev A A, Kryukov R E, Shevchenko R A, Gizatulin R A and Valueva A V 2017 *Key Engineering Materials* **736** 116–121

Structure-phase state evolution of 100-m differentially hardened rails in long-term usage

A A Yuriev[1], B E Gromov[2], K V Morozov[1], Yu F Ivanov[3, 4], S V Konovalov[5] and A P Semin[2]

[1]EVRAZ Consolidated West Siberian Metallurgical Plant, 16 Kosmicheskoye shosse, 654043, Russia
[2]Siberian State Industrial University, 42 Kirova street, Novokuznetsk, 654007, Russia
[3]Institute of High-Current Electronics of Siberian Branch of Russian Academy of Sciences, 2/3, Academicheskii ave, Tomsk, 654052 Russia
[4]National Research Tomsk State University, 36 Lenina ave, Tomsk, 634050, Russia
[5]Academican S.P. Korolev Samara National Research University, 34 Moskovskoye shosse, Samara, 443086, Russia

E-mail: gromov@physics.sibsiu.ru

Abstract. By methods of optic and transmission electron diffraction microscopy the structure-phase states evolution of surface layers' head of differentially hardened rails to the depth up to 10 mm along the fillet after the passed tonnage of 691.8 mln. t. brutto was examined.

1. Introduction

The long-term usage of rails is accompanied by the considerable change in structure and properties of the surface layer [1 3]. The structure-phase states with anomalous high values of microhardness are formed in the surface layers even at a comparatively small operating load of 100-500 mln.t. of passed tonnage.

Under this surface intensive plastic deformation the cementite plates are either arched or fractured at interfaces. The dissolution of cementite and austenite formation is noted at the expense of the reverse $\gamma \rightarrow \alpha$ transformation [4-6]. The understanding of the processes running in steel in this case is one of the essential conditions of material state control making possible to predict the operating of rails.

The purpose of the research is a comparative layer-by-layer analysts of surface structure and phase composition of differentially hardened rails in long-term usage.

2. Materials and methods

The test materials were the samples of differentially hardened rails of E76CrV steel of DT 350 category fabricated at open joint-stock company "Evraz ZSMK" after operating load tonnage of 691.8 mln.t. brutto in the processes of field-tests at experimental ring "VNIIZhT".

The chemical composition of rail material meet the State Standard requirements P51685-2013 for E76CrV steel. The microstructure of rails' metal was tested according to the requirements the of State Standard P51685-2013 at templet cut from rails in transverse direction, after etching in 50% water solution of hydrochloric acid. The microstructure analysis was done after electrolytic polishing of micro metallographic section in 5% acetic solution of perchloric acid followed by etching in 4%

alcoholic solution of nitric acid. The investigation of phase composition and defect substructure of rails was carried out by methods of diffraction electron microscopy. The foils for testing were manufactured by method of electrolyte thinning of plates cut by electro-spark method from the fillet region at 2 mm and 10 mm distance and near the fillet surface according to figure 1.

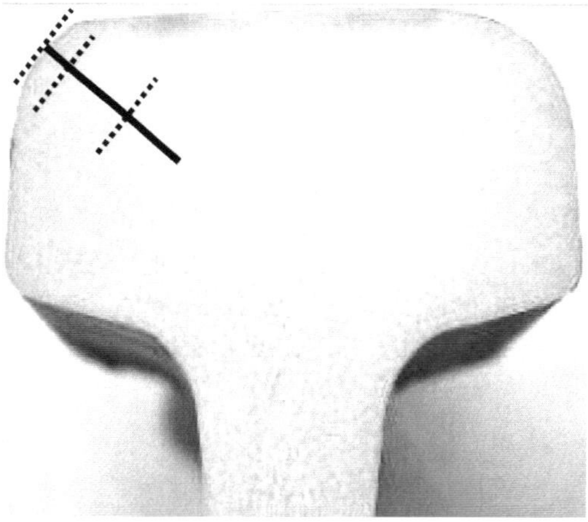

Fig. 1. Diagram of rail sample preparing on testing its structure by methods of electron diffraction microscopy. The direction along the fillet is designated with solid line; the regions of metal layers' location used for foil preparation are designated with dotted lines.

3. Results and discussion

By methods of metallography it is established that a considerably deformed structure to depth up to 200 μm is observed on the etched metallographic sections from the surface of the working fillet. The decarbonized layer value from the surface along the solid ferrite network does not excud 0.25 μm. In the initial state (before the usage) the following structural constituents were marked out by the morphologic feature by methods of electron diffraction microscopy: the lamellar pearlite, the grains of ferrite-carbide mixture and the grains of structurally free ferrite. The main structural types of the tested steel are the grains of lamellar pearlite with their relative content in the material being 0.7; the relative content of grains of ferrite-carbide mixture is 0.25; the balance is the grains of structurally free ferrite.

After the passed tonnage of 691.8 mln.t. this state is conserved only at 10 mm depth from the fillet surface. The distinctive feature of the structure at this distance is a large number of bend extinction contours being indicative of the elastic-plastic distortions of the material's crystal lattice caused by the intensive mechanical effect on rails' material in the process of usage.

The stress concentrators of the tested steel are the intraphase and interphase interfaces (1) of ferrite and pearlite grains (figure 2, a); (2) plates of cementite and ferrite of pearlite colonies (figure 2, b, c); (3) particles of globular cementite and ferrite (figure 2, d).

The multiple changes in rail metal structure at the microlevel are manifested in the formation of small crack network of the contact-fatigue origin being revealed on the tread surface of rail's head after etching. At the region of defects of contact-fatigue origin the continuity violations filled with corrosion products passing at an acute angle to the surface to the depth up to 140 μm are observed. The distance between the defects measures 700-110 μm.

Figure 3 shows the image of cementite plates of pearlite colony located in the surface layer of rail head fillet after usage. The studies performed by the methods of dark field analysis show that the usage is accompanied by the fragmentation of cementite plates followed by their destruction. The carbide phase reflections in microelectron diffraction pattern obtained from the plates have both the radial and azimuthal spreading indicative of the high level of cementite crystal lattice imperfection and the change in crystal lattice parameter due to the escape of carbon atom as well.

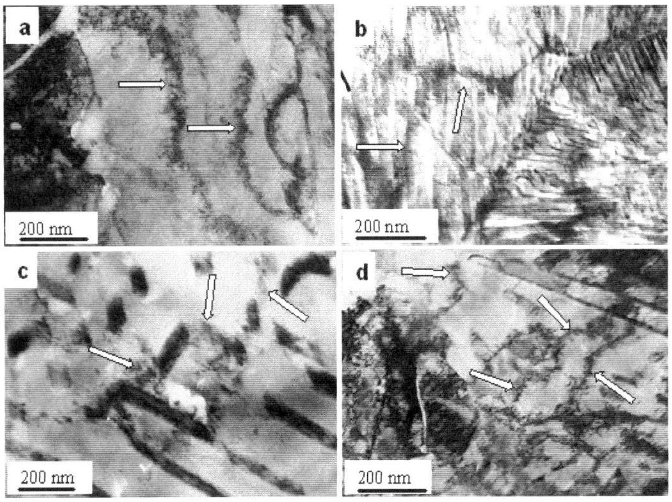

Figure 2. Bend extinction contours (designated by the arrows) in grains of ferrite (a); plastic (b, c) and ferrite-carbide mixture (d).

Rail operation is accompanied by significant changes in the defect substructure of the structure-free ferrite grains, within which band substructure is observed. The distance between the bands is 30-40 nm. At the band structure, we see carbide particles in the range of 5-10 nm. These findings may indicate that two competing processes occur in rail operation: (1) fragmentation of the cementite particles, with their subsequent entrainment in the ferrite grains or plates (in the pearlite structure); (2) fragmentation and subsequent solution of the cementite particles, with transfer of the carbon particles to dislocations (Cottrell atmospheres) and transportation of carbon atoms by dislocations within the ferrite grains (or plates), culminating in the formation of cementite nanoparticles.

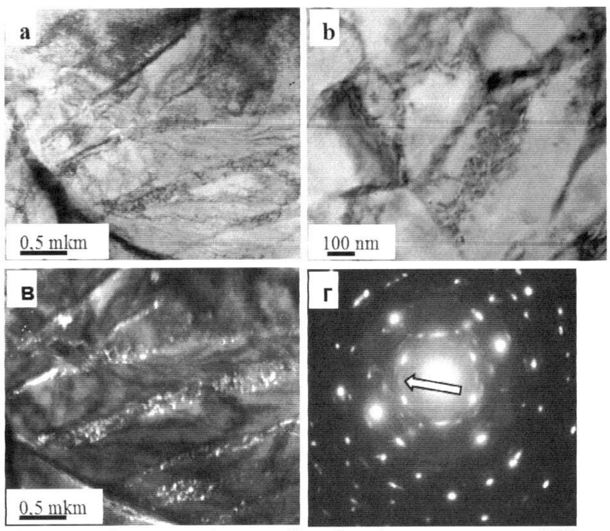

Figure 3. Electron microscopy image of surface layer structure of fillet metal of rail head after usage: a, b – light field images; c- dark field obtained in the reflection [130] Fe_3Cr; d- microelectron diffraction pattern, the arrow designates the reflection in which the dark field was obtained.

4. Conclusion

By methods of optic and electron diffraction microscopy the studies of phase composition, macro- and microdefect structure of fillet metal of differentially hardened rails of category DT 350 from E76CrV steel after the passed tonnage of 691.8 mln.t. brutto in the process of field tests at experimental ring of open joint- stock company "VNIIZhT" were performed. It is shown that rails usage is accompanied by the multiple transformation of steel structure. At the macrolevel it is manifected in the formation of microcraks passing at an acute angle to the surface to the depth up to 140 μm, and the formation of decarbonized layer; at the microlevel – in the formation of elastic- plastic stress fields and the cementite plate destruction of pearlite colonies. In the grains of structurally free ferrite the fragmentation of cementite particles and their dissolution is observed.

Acknowledgements
The research is financially supported by the Russian research fund, project № 15-12-00010.

References
[1] Sheinman E and Friction J 2012 *Wear* **33(4)** 308–314
[2] Gromov V E, Peregudov O A, Ivanov Y F, Morozov K V, Alsaraeva K V and Semina O A 2016 *Surf J Investig X-ray Synchrotron Neutron Techniq* **10(1)** 76–82
[3] Gromov V E, Yur'ev A B, Morozov K V and Ivanov Yu F 2016 *Microstructure of Quenched Rails* (Cambridge Int. Sci. Publ., Cambridge)
[4] Ivanisenko Yu and Fecht H J 2008 *Steel Tech.* 19–23
[5] Ivanisenko Yu, Maclaren I, Souvage X, Valiev R Z and Fecht H J 2006 Acta Mater. **54** 1659–69
[6] Gavriljuk V G 2003 *Mater. Sci. and Eng. A* **345** 81–89